# Geometric Mechanics and Symmetry

OXFORD TEXTS IN APPLIED AND ENGINEERING MATHEMATICS

*Books in the series*

* G. D. Smith: *Numerical Solution of Partial Differential Equations, third edition*
* R. Hill: *A First Course in Coding Theory*
* I. Anderson: *A First Course in Combinatorial Mathematics, second edition*
* D. J. Ancheson: *Elementary Fluid Dynamics*
* S. Barnett: *Matrices: Methods and applications*
* L. M. Hocking: *Optimal Control: An introduction to the theory with applications*
* D. E. Ince: *An Introduction to Discrete Mathematics, Formal System Specification, and Z, second edition*
* O. Pretzel: *Error-Correcting Codes and Finite Fields*
* P. Grindrod: *The Theory and Applications of Reaction-Diffusion Equations: Patterns and waves, second edition*

1. Alwyn Scott: *Nonlinear Science: Emergence and dynamics of coherent structures*
2. D.W. Jordan and P. Smith: *Nonlinear Ordinary Differential Equations: An introduction to dynamical systems, third edition*
3. I.J. Sobey: *Introduction to Interactive Boundary Layer Theory*
4. A.B. Tayler: *Mathematical Models in Applied Mechanics, reissue*
5. L. Ramdas Ram-Mohan: *Finite element and Boundary Element Applications in Quantum Mechanics*
6. Bernard Lapeyre, Étienne Pardoux, and Rémi Sentis: *Introduction to Monte-Carlo Methods for Transport and Diffusion Equations*
7. Isaac Elishakoff and Yongjian Ren: *Finite Element Methods for Structures with Large Stochastic Variations*
8. Alwyn Scott: Nonlinear Science: *Emergence and dynamics of coherent structures, second edition*
9. W.P. Petersen and P. Arbenz: *Introduction to Parallel Computing: A practical guide with examples in C*
10. D.W. Jordan and P. Smith: *Nonlinear Ordinary Differential Equations, fourth edition*
11. D.W. Jordan and P. Smith: *Nonlinear Ordinary Differential Equations: Problems and Solutions*
12. Darryl D. Holm, Tanya Schmah, and Cristina Stoica: *Geometric Mechanics and Symmetry: From Finite to Infinite Dimensions*

Titles marked with (*) appeared in the 'Oxford Applied Mathematics and Computing Science Series', which has been folded into, and is continued by, the current series.

# Geometric Mechanics and Symmetry

## From Finite to Infinite Dimensions

Darryl D. Holm

*Imperial College London*

Tanya Schmah

*Macquarie University and The University of Toronto*

Cristina Stoica

*Wilfrid Laurier University*

With solutions to selected exercises by

David C. P. Ellis

*Imperial College London*

OXFORD
UNIVERSITY PRESS

**OXFORD**
UNIVERSITY PRESS

Great Clarendon Street, Oxford OX2 6DP

Oxford University Press is a department of the University of Oxford.
It furthers the University's objective of excellence in research, scholarship,
and education by publishing worldwide in

Oxford New York

Auckland Cape Town Dar es Salaam Hong Kong Karachi
Kuala Lumpur Madrid Melbourne Mexico City Nairobi
New Delhi Shanghai Taipei Toronto

With offices in

Argentina Austria Brazil Chile Czech Republic France Greece
Guatemala Hungary Italy Japan Poland Portugal Singapore
South Korea Switzerland Thailand Turkey Ukraine Vietnam

Oxford is a registered trade mark of Oxford University Press
in the UK and in certain other countries

Published in the United States
by Oxford University Press Inc., New York



First published 2009



British Library Cataloguing in Publication Data

Data available

Library of Congress Cataloging in Publication Data

Data available

Typeset by Newgen Imaging Systems (P) Ltd., Chennai, India
Printed in Great Britain
on acid-free paper by
CPI Antony Rowe, Chippenham, Wiltshire

ISBN 978–0–19–921290–3 (Hbk)
ISBN 978–0–19–921291–0 (Pbk)

10 9 8 7 6 5 4 3 2 1

# Preface

This is a textbook for graduate students that introduces geometric mechanics in finite and infinite dimensions, using a series of archetypal examples.

Classical mechanics, one of the oldest branches of science, has undergone a long evolution, developing hand in hand with many areas of mathematics, including calculus, differential geometry and the theory of Lie groups and Lie algebras. The modern formulations of Lagrangian and Hamiltonian mechanics, in the coordinate-free language of differential geometry, are elegant and general. They provide a unifying framework for many seemingly disparate physical systems, such as $N$-particle systems, rigid bodies, fluids and other continua, and electromagnetic and quantum systems.

The first part of this book concerns finite-dimensional conservative mechanical systems. The modern formulations of Lagrangian and Hamiltonian mechanics use the language of differential geometry. Some advantages of this approach are: (i) it applies to systems on general manifolds, including configuration spaces defined by constraints; (ii) it is coordinate-free, or at least independent of a particular choice of coordinates; (iii) the geometrical structures have analogues in infinite-dimensional systems. Just as importantly, the geometric approach provides an elegant and suggestive viewpoint. For example, rigid body motion can be seen as geodesic motion on the rotation group. Symmetries of mechanical systems are represented mathematically by Lie group actions. The presence of symmetry allows a reduction in the number of dimensions of a mechanical system, in two basic ways: by grouping together equivalent states; and by exploiting conserved quantities (momentum maps) associated with the symmetry. The book discusses Lie group symmetries, Poisson reduction and momentum maps in a general context before specializing to systems where the configuration space is itself a Lie group, or possibly the product of a Lie group and a vector space. For systems, such as the free rigid body, whose symmetry group is also its configuration space, an especially powerful reduction theorem exists, called Euler–Poincaré reduction. An extension of this theorem covers systems where the Lie group configuration space is augmented by a vector space describing certain 'advected quantities', such as the gravity covector in the heavy top example.

The second part of the book treats what might be considered the infinite-dimensional versions of the rigid body and the heavy top by replacing the action of the rotation group by the action of a group of diffeomorphisms. Roughly speaking, passing from finite to infinite dimensions in geometric mechanics means replacing matrix multiplication by composition of smooth invertible functions. The book develops these ideas in the setting of Euler–Poincaré theory, based on reduction by symmetry of Hamilton's variational principle. The infinite-dimensional results corresponding to rigid-body and heavy-top dynamics are exemplified, respectively, in geodesic motion on the diffeomorphisms governed by the EPDiff equation and in the action of the diffeomorphisms on vector spaces of 'advected quantities' governed by the equations of continuum dynamics.

EPDiff arises in one spatial dimension as the zero-dispersion limit of the Camassa–Holm (CH) equation for shallow water waves. The CH equation is an approximate model of shallow water waves obtained at one order in the asymptotic expansion beyond the famous Korteweg–de Vries (KdV) equation. KdV and CH are both nonlinear partial differential equations. They each support remarkable solutions called 'solitons' that interact in fully nonlinear wave collisions and whose exact solution may be obtained by the linear method of the inverse scattering transform. In the zero-dispersion limit of shallow water wave theory in which the EPDiff equation arises, these solitons become singular particle-like solutions carrying momentum supported on Dirac delta measures. The EPDiff equation for the Euler–Poincaré dynamics of geodesic motion on the diffeomorphisms also applies in image analysis. In particular, EPDiff applies in the comparison of shapes in morphology and computational anatomy.

The Euler–Poincaré approach that generalizes the heavy-top problem from the action of the rotations on vectors in $\mathbb{R}^3$ to the action of the diffeomorphisms on vector spaces produces yet another rich array of results. In particular, it produces the extensive family of equations for ideal continuum dynamics, whose applications range from nanofluids to galaxy dynamics. Among the many available variants of ideal continuum dynamics, we select a single class for a unified treament. Namely, we treat a class of approximate models of global ocean circulation that are used in climate prediction. Thus, the theoretical development of these parallel ideas in finite and infinite dimensions is capped by the explicit application in the last chapter to derive a unified formulation of the family of approximate equations for ocean dynamics and climate modelling familiar to modern geoscientists.

One may think of moving from the first part of the book to the second part as moving from finite-dimensional to infinite-dimensional geometric mechanics. The analogies between the two types of problems are very close. The first part of the book deals with systems of nonlinear ordinary differential equations (ODE), whose questions of existence, uniqueness and regularity of solutions may generally be answered by using standard methods. The second part of the book deals with nonlinear partial differential equations (PDE) where the answers to such questions are often quite challenging and even surprising. For example, these particular PDE possess coherent excitations and even singular solutions that emerge from smooth initial data and whose nonlinear interactions exhibit particle-like scattering behaviour reminiscent of solitons. Unlike many other PDE investigations, geometric mechanics treats the emergence of these measure-valued, particle-like solutions in the initial-value problem for some of the models as a challenge to be celebrated, rather than a cause for regret.

## Prerequisites and intended audience

The reader should be familiar with linear algebra, multivariable calculus, and the standard methods for solving ordinary and partial differential equations. Some familiarity with variational principles and canonical Poisson brackets in classical mechanics is desirable but not necessary. Readers with an undergraduate background in physics or engineering will have the advantage that many of the examples treated here, such as the motion of rigid bodies and the dynamics of fluids, will be familiar. In summary, the prerequisites are standard for an advanced undergraduate student or first-year postgraduate student in mathematics or physics.

## How to read this book and what is not in it

Part I is meant to be used as a textbook in an upper-level course on geometric mechanics. It contains many detailed explanations and exercises. Although a wide range of topics is treated, the introduction to each of them is meant to be gentle. Part II addresses a more advanced reader and focuses on recent applications of geometric mechanics in soliton theory, image analysis and fluid mechanics. However, the mathematical prerequisites for rigorous treatments of these applications are not provided here. Readers interested in a more technical mathematical approach are invited to consult some of the many citations in the bibliography that treat the subject in that style.

The book focuses on Euler–Poincaré reduction by symmetry, which is a broadly applicable theory, but it excludes many other important topics, such as Lagrangian reduction of general tangent bundles and symplectic reduction. Likewise, it omits many other standard mechanics topics, including integrability, Hamilton–Jacobi theory, and more generally, any questions in the modern theory of dynamics, bifurcation and control.

Most of the notation is consistent with that of the Marsden–Ratiu school. This was a deliberate choice, since one of the intentions in writing the first part was to bridge the gap between the standard classical mechanics books such as *Classical Mechanics* by Goldstein [Gol59] and *Mechanics* by Landau and Lifshitz [LL76] and the more advanced books such as *Foundations of Mechanics* by Abraham and Marsden [AM78] and *Introduction to Mechanics and Symmetry* by Marsden and Ratiu [MR02].

## Description of contents

### Part I

The opening chapter briefly presents Newtonian, Lagrangian and Hamiltonian mechanics in the familiar setting of $N$-particle systems in Euclidean space. A short classical description of rigid body motion is also given. Chapters 2 and 3 build up the prerequisites for the extension of Lagrangian and Hamiltonian mechanics to systems on manifolds. Chapter 2 introduces manifolds, with emphasis on submanifolds of Euclidean space, which appear in mechanics as configuration spaces defined by constraints. This chapter also introduces matrix Lie groups (covered in more detail in Chapter 5). Chapter 3 gives a minimal introduction to differential geometry, including a taste of Riemannian and symplectic geometry. Chapter 4 presents Lagrangian and Hamiltonian mechanics on manifolds, and ends with a brief look at symmetry, reduction and conserved quantities.

In order to study symmetry in more depth, one needs Lie groups and their actions, which are the subject of Chapters 5 and 6. Lie groups and algebras are introduced in both matrix and abstract frameworks. Lie group actions on manifolds, and the resulting quotient spaces, provide sufficient tools to introduce Poisson reduction.

The remaining chapters focus on mechanical systems on Lie groups, that is, mechanical systems where the configuration space is a Lie group. Chapter 7 covers Euler–Poincaré reduction, emphasizing the examples of the free rigid body and the heavy top. Chapter 8 introduces momentum maps. Chapter 9 covers Lie–Poisson reduction, which is the Hamiltonian counterpart of Euler–Poincaré reduction. Chapter 10 applies the results of the preceding three chapters to the example of a pseudo-rigid body. Pseudo-rigid motions provide a link between the rigid motions studied in Part I and the fluid motions that are the subject of Part II.

### Part II

In a famous paper, Arnold [Arn66] observed that Euler ideal fluid motion may be identified with geodesic flow on the volume-preserving diffeomorphisms, with a metric determined by the fluid's kinetic energy. This observation was further developed in a rigorous analytical setting by Ebin and Marsden [EM70]. The methods of geometric mechanics systematically develop this result from the Euler–Poincaré (EP) variational principle for the Euler fluid equations. These methods also generate their Lie–Poisson Hamiltonian structure, Noether theorem, momentum maps, etc. As Arnold observed, the configuration space for the incompressible motion of an ideal fluid is the group $G = \mathrm{Diff}_{Vol}(\mathcal{D})$ of volume-preserving diffeomorphisms (smooth invertible maps with smooth inverses) of the region $\mathcal{D}$ occupied by the fluid. The tangent vectors in $TG$ for the maps in $G = \mathrm{Diff}_{Vol}(\mathcal{D})$ represent the space of fluid velocities, which must satisfy appropriate physical conditions at the boundary of the region $\mathcal{D}$. Group multiplication in $G = \mathrm{Diff}_{Vol}(\mathcal{D})$ is composition of the smooth invertible volume-preserving maps. One of the purposes of this text is to explain how the Euler equations of fluid motion may be recognized as the Euler–Poincaré equations $\mathrm{EPDiff}_{Vol}$ defined on the dual of the tangent space at the identity $T_eG = T_e\,\mathrm{Diff}_{Vol}(\mathcal{D})$ of the right-invariant vector fields over the domain $\mathcal{D}$.

## Applications of EPDiff in Part II

- In the motivating example, Euler's fluid equations emerge as $\mathrm{EPDiff}_{Vol}$ when the diffeomorphisms are constrained to preserve volume so that $G = \mathrm{Diff}_{Vol}(\mathcal{D})$, and the kinetic energy norm is taken to be the $L^2$ norm $\|\mathbf{u}\|^2_{L^2}$ of the spatial fluid velocity. From the viewpoint of constrained dynamics, the fluid pressure $p$ may be regarded as the Lagrange multiplier that imposes preservation of volume.
- Other choices of the kinetic energy norm besides the $L^2$ norm of velocity also produce interesting continuum equations as geodesic flows on $\mathrm{Diff}_{Vol}(\mathcal{D})$. For example, the choice of the $H^1$ norm ($L^2$ norm of the gradient) of the spatial fluid velocity yields the ***Lagrangian-averaged Euler-alpha*** (LAE-alpha) equations when incompressibility is imposed. For more discussion of the LAE-alpha equations, see, e.g., [HMR98a, Shk00].
- The $H^1$ norm also yields one of a family of interesting EPDiff equations when incompressibility is *not* imposed, so that the motion takes place on the *full* diffeomorphism group. In one spatial dimension on the real line (and also in a periodic domain), the EPDiff equation for the $H^1$ norm of the spatial fluid velocity is completely integrable in the Hamiltonian sense and possesses ***soliton solutions***. This equation is the limit of the ***Camassa–Holm*** (CH) equation for shallow-water wave motion when its linear dispersion coefficients tend to zero [CH93]. The CH equation and its peaked-soliton solutions – called peakons – that exist

in its zero-dispersion limit are discussed in Chapter 11. EPDiff solution behaviour in one dimension and its wave-train solutions for various other norms are discussed in Chapter 12. The CH equation and its zero-dispersion limit are integrable Hamiltonian systems that possess soliton solutions. The geometric ingredients underlying integrability and soliton solutions are reviewed in Chapter 13.

- Another application of the family of EPDiff equations that involves various choices for the kinetic energy norm in two and three spatial dimensions arises in the field of morphology, in the modern endeavour of ***computational anatomy*** (CA) [MTY02, HRTY04]. After generalizing the peakon solutions of EPDiff for the $H^1$ norm to higher dimensions in Chapter 14, methods for applying higher-dimensional singular EPDiff solutions for matching image contours in computational anatomy are sketched in Chapter 15.
- After the discussion in Chapter 15 about matching image contours, Chapter 16 discusses matching also what is *inside* the contours. The conceptual parallel between the two chapters reflects the step from pure EP (rigid body, incompressible fluids) to EP with advected quantities (heavy top, compressible fluids). In explaining the geometrical theory of image matching using the ***metamorphosis approach***, Chapter 16 sets up the remaining chapters about the EP formulation of fluid dynamics. Chapter 16 also reinforces the theme of soliton equations in the 1D reductions of this approach in image matching. Finally, Chapter 16 discusses the geometric-mechanical analogy of image matching with the problem of how a falling cat reorients its body in mid-air so that it lands on its paws.
- Chapter 17 discusses the EP theorem for ideal continuum flows carrying advected quantities. The application of the more general EP theorem in the derivation and analysis of models of ***geophysical fluid dynamics*** (GFD) is discussed in Chapter 18. These GFD models are crucial in numerical computations for global climate modelling.

  This family of GFD equations is derived by applying the method of asymptotic expansions to Hamilton's principle for a rapidly rotating, stably stratified incompressible flow in the Eulerian description. Such flows are seen in the ocean, whose slow time scales dictate the dynamics of the global climate. The rapid rate of rotation and thin domain of flow in global ocean circulation are characterized by two small non-dimensional parameters: the Rossby number (ratio of nonlinearity to Coriolis force) and the aspect ratio of the domain. Expansion of Hamilton's principle in these two small parameters provides a unified approach in the derivation of the rich and multifaceted family of approximate equations for geophysical fluid dynamics (GFD). Each of the equations in this family of GFD approximations is both nonlinear and non-local in character. The Euler–Poincaré approach explains their shared properties of energy conservation and circulation dynamics.

# Acknowledgements

We are enormously grateful to our friends, students and teachers, without whom this endeavour would never have been completed. We thank our friends and collaborators, Tony Bloch, Dorje Brody, Roberto Camassa, Colin Cotter, John Gibbon, John Gibbons, Daniel Hook, Boris Kupershmidt, Jeroen Lamb, Peter Lynch, Jerry Marsden, Peter Olver, Vakhtang Putkaradze, George Patrick, Tudor Ratiu, Mark Roberts, Martin Staley, Alain Trouvé, Laurent Younes, Alan Weinstein and Beth Wingate for their help and camaraderie in our research together in geometric mechanics. We especially thank our students Cesare Tronci and David Ellis at Imperial College London for making the teaching of these topics so rewarding. Thanks to Robert Jones for comments and careful proof-reading.

DDH: To Justine, for all our expeditions of discovery and fun.
TS: To Hillary Sanctuary, whose questions, enthusiasm and friendship first encouraged me to write a book on this subject.
TS and CS: Thanks to Macquarie University and Wilfrid Laurier University, respectively, for supporting the writing of this book with teaching releases. Many thanks to Ica for the Turkish coffees she made for us during all the summer days we spent writing.

# Contents

# Part I

The theoretical development of the laws of motion of bodies is a problem of such interest and importance, that it has engaged the attention of all the most eminent mathematicians, since the invention of dynamics as a mathematical science by Galileo, and especially since the wonderful extension which was given to that science by Newton. Among the successors of those illustrious men, Lagrange has perhaps done more than any other analyst, to give extent and harmony to such deductive researches, by showing that the most varied consequences respecting the motions of systems of bodies may be derived from one radical formula; the beauty of the method so suiting the dignity of the results, as to make of his great work a kind of scientific poem.

– W. R. Hamilton

'On a General Method in Dynamics', *Philosophical Transactions of the Royal Society of London*, Vol. 124 (1834)

# 1 Lagrangian and Hamiltonian mechanics

This chapter introduces some of the main themes of geometric mechanics in the setting of systems of $N$ point masses. In addition, the final section introduces rigid body dynamics as motion on the group of $3 \times 3$ orthogonal matrices. For more details, see [Arn78] or [Gol59].

## 1.1 Newtonian mechanics

Once a frame of reference is fixed, space is represented as $\mathbb{R}^d$ where $d$ is the spatial dimension ($d = 1, 2$ or $3$).

**Definition 1.1** *A* ***point mass*** *is an idealized zero-dimensional object that is completely described by its mass and spatial position. Its mass is assumed to be constant and its position varies as a function of time.*

At any given time, the position of a point mass, also called its ***configuration***, is denoted $\mathbf{q} \in \mathbb{R}^d$. For systems formed by $N$ point masses, the ***configuration*** is a multi–vector $\mathbf{q} = (\mathbf{q}_1, \mathbf{q}_2, \ldots, \mathbf{q}_N) \in \mathbb{R}^{dN}$ given by the position vectors of each of the point masses. The set of all possible configurations of a system of $N$ point masses is called its ***configuration space***. In the absence of constraints (which will be studied in Section 1.3), the configuration space is either $\mathbb{R}^{dN}$ or some open subset of $\mathbb{R}^{dN}$ (if collisions of point masses are excluded).

Consider a system of $N$ point masses, each with mass $m_i$. The position of each point mass is described by a time-varying vector $\mathbf{q}_i(t)$. The velocity $\mathrm{d}\mathbf{q}_i/\mathrm{d}t$ and acceleration $\mathrm{d}^2\mathbf{q}_i/\mathrm{d}t^2$ of a point mass are denoted $\dot{\mathbf{q}}_i(t)$ and $\ddot{\mathbf{q}}_i(t)$, respectively.

**Definition 1.2** ***Newton's second law*** *for the motion of a system of $N$ point masses is*

$$\mathbf{F}_i = m_i \ddot{\mathbf{q}}_i \quad \textit{for} \quad i = 1, 2, \ldots, N, \tag{1.1}$$

*where $\mathbf{F}_i$ is the total force on the $i$th point mass.*

**Definition 1.3** *A frame of reference in which Newton's second law applies is called an* ***inertial frame****.*

In this section, we assume an inertial frame of reference.

The following ***dynamical quantities*** play central roles in the mechanics of systems of $N$ point masses. In many situations, these quantities are conserved, meaning that they are constant along any ***trajectory***

$$\mathbf{q}(t) = (\mathbf{q}_1(t), \mathbf{q}_2(t), \ldots \mathbf{q}_N(t))$$

of the system $N$ point masses that satisfies eqn (1.1).

**Definition 1.4 (Dynamical quantities)** *The* ***linear momentum*** *(or* ***impulse****) of a point mass with position* $\mathbf{q}_i$ *is*

$$\mathbf{p}_i := m_i \dot{\mathbf{q}}_i. \tag{1.2}$$

*The* ***(total) linear momentum*** *of a system of N point masses is the sum of the linear momenta of all of the point masses, that is*

$$\mathbf{p} := \sum m_i \dot{\mathbf{q}}_i. \tag{1.3}$$

*The* ***angular momentum****, about the origin of coordinates* $\mathbf{q} = 0$*, of a point mass with position* $\mathbf{q}_i \in \mathbb{R}^3$ *is*

$$\boldsymbol{\pi}_i := \mathbf{q}_i \times m_i \dot{\mathbf{q}}_i = \mathbf{q}_i \times \mathbf{p}_i, \tag{1.4}$$

*where* $\times$ *denotes the three-dimensional vector cross-product. The* ***(total) angular momentum*** *of an spatial system* ($d = 3$) *of N point masses, taken about the origin of coordinates* $\mathbf{q} = 0$*, is the sum of the angular momenta of all of the point masses, that is*

$$\boldsymbol{\pi} := \sum \mathbf{q}_i \times m_i \dot{\mathbf{q}}_i = \sum \mathbf{q}_i \times \mathbf{p}_i. \tag{1.5}$$

*Geometrically, the angular momentum is the sum of the N oriented areas given by the cross-products of pairs vectors* $\mathbf{q}_i$ *and* $\mathbf{p}_i$*. For a planar system (with* $d = 2$*), the total angular momentum is the (pseudo)scalar defined by*

$$\pi := \sum m_i \left( q_i^1 \dot{q}_i^2 - q_i^2 \dot{q}_i^1 \right) = \sum q_i^1 p_i^2 - q_i^2 p_i^1, \tag{1.6}$$

*where* $\mathbf{q}_i = (q_i^1, q_i^2)$*, etc. Note that if the vectors* $\mathbf{q}_i$ *are embedded in* $\mathbb{R}^3$ *as* $(q_i^1, q_i^2, 0)$*, then* $\pi$ *as defined above is the third component of the vector* $\boldsymbol{\pi}$

*as defined in eqn (1.5). Angular momentum is undefined for systems defined on a line ($d = 1$).*

*The (**total**) **kinetic energy** of the system of $N$ point masses is*

$$K := \frac{1}{2} \sum m_i \|\dot{\mathbf{q}}_i\|^2 . \tag{1.7}$$

*Here* $\|\dot{\mathbf{q}}_i\|^2 = \dot{\mathbf{q}}_i \cdot \dot{\mathbf{q}}_i = \sum_j (\dot{q}_i^j)^2$ *denotes the squared **Euclidean norm** of* $\dot{\mathbf{q}}_i$.

**Remark 1.5** *For a planar system, the position in polar coordinates of the ith point mass is given by*

$$\mathbf{q}_i = (r_i \cos\theta_i, r_i \sin\theta_i) \quad \textit{for} \quad i = 1, 2, \ldots, N.$$

*Hence, in polar coordinates, the total angular momentum and kinetic energy are given by*

$$\begin{aligned}
\pi &= \sum m_i \,(r_i \cos\theta_i, r_i \sin\theta_i) \times \left(\dot{r}_i \cos\theta_i - r_i\dot{\theta}_i \sin\theta_i, \dot{r}_i \sin\theta_i + r_i\dot{\theta}_i \cos\theta_i\right) \\
&= \sum m_i r_i^2 \dot{\theta}_i , \\
K &= \frac{1}{2} \sum m_i \left\| \left(\dot{r}_i \cos\theta_i - r_i\dot{\theta}_i \sin\theta_i, \dot{r}_i \sin\theta_i + r_i\dot{\theta}_i \cos\theta_i\right) \right\|^2 \\
&= \frac{1}{2} \sum m_i \left(\dot{r}_i^2 + r_i^2 \dot{\theta}_i^2\right).
\end{aligned}$$

**Theorem 1.6 (Conservation of linear momentum)** *If the total force on a system vanishes, that is, if* $\sum \mathbf{F}_i = 0$, *then the total linear momentum is conserved.*

*Proof* The proof of total linear momentum conservation is verified by a straightforward computation,

$$\frac{d\mathbf{p}}{dt} = \sum m_i \ddot{\mathbf{q}}_i = \sum \mathbf{F}_i = 0 .$$

■

A corresponding conservation law exists for total angular momentum, involving ***torque***:

**Definition 1.7** *The **torque** on a point mass with position* $\mathbf{q} \in \mathbb{R}^3$ *and force* $\mathbf{F}$ *is* $\mathbf{q} \times \mathbf{F}$. *The **total torque** of a system of $N$ point mass is* $\sum \mathbf{q}_i \times \mathbf{F}_i$.

**Theorem 1.8 (Conservation of angular momentum)** *If the total torque on a system vanishes, then its total angular momentum is conserved.*

*Proof*

$$\begin{aligned}\frac{d\boldsymbol{\pi}}{dt} &= \frac{d}{dt}\sum m_i \mathbf{q}_i \times \dot{\mathbf{q}}_i = \sum m_i \left(\dot{\mathbf{q}}_i \times \dot{\mathbf{q}}_i + \mathbf{q}_i \times \ddot{\mathbf{q}}_i\right) \\ &= \sum \mathbf{q}_i \times m_i \ddot{\mathbf{q}}_i = \sum \mathbf{q}_i \times \mathbf{F}_i = 0\,.\end{aligned}$$

■

**Definition 1.9** *A **central force problem** is a one-point system whose force is of the form* $\mathbf{F} = F\left(\|\mathbf{q}\|\right)\widehat{\mathbf{q}}$, *where* $\widehat{\mathbf{q}} = \mathbf{q}/\|\mathbf{q}\|$ *and* $F$ *is some real-valued function. Thus, the motion of the point mass is given by:*

$$m\ddot{\mathbf{q}} = F\left(\|\mathbf{q}\|\right)\widehat{\mathbf{q}}\,.$$

**Example 1.10 (The Kepler problem)** The motion of a planet of mass $m$ in the gravitational field of a sun of mass $M >> m$ is modelled by

$$m\ddot{\mathbf{q}} = -\frac{GmM}{\|\mathbf{q}\|^3}\mathbf{q} = -\frac{GmM}{\|\mathbf{q}\|^2}\widehat{\mathbf{q}},$$

where $G$ is the gravitational constant.

**Proposition 1.11** *In any central force problem, angular momentum is conserved.*

*Proof* Since $\mathbf{q}\times\mathbf{F} = \mathbf{q}\times F\left(\|\mathbf{q}\|\right)\widehat{\mathbf{q}} = 0$, this follows from Proposition 1.8. ■

**Corollary 1.12** *In any central force problem, every trajectory remains in a fixed plane.*

*Proof* The total angular momentum $\boldsymbol{\pi}$ is conserved, and

$$\boldsymbol{\pi}\cdot\mathbf{q} = m\left(\mathbf{q}\times\dot{\mathbf{q}}\right)\cdot\mathbf{q} = 0$$

for all time. Hence, $\mathbf{q}$ must remain in the plane normal to $\boldsymbol{\pi}$. ■

**Remark 1.13** *As a consequence of the previous corollary, any central force problem can be considered to be planar. Recall (from Remark 1.5) that the angular momentum in planar systems, in polar coordinates, takes the form* $\pi = mr^2\dot{\theta}$. *Since* $r\dot{\theta}$ *is the velocity in the direction perpendicular to* $\mathbf{q}$, *it follows that the angular momentum per unit mass,* $r^2\dot{\theta}$, *is twice the rate* *at which area is swept out by the vector* $\mathbf{q}$. *Expressed in these terms, the conservation of* $r^2\dot{\theta}$ *is known as* ***Kepler's Second Law****, which, as we have seen, applies not just to planetary motion but to all central force problems. (See [Gol59] for details.)*

**Definition 1.14 (Forces of inter-particle interaction)** *In many systems, the only forces on the point masses are* ***forces of inter-particle interaction*** $\mathbf{F}_{ij}$, *each parallel to the inter-particle position vector* $\mathbf{d}_{ij} = \mathbf{q}_i - \mathbf{q}_j$, *such that* $\mathbf{F}_{ij} = -\mathbf{F}_{ji}$ *always, and the total force on each point mass i is* $\mathbf{F}_i = \sum_j \mathbf{F}_{ij}$. *Such a system is called* ***closed****. Alternatively, such forces are called* ***internal****, and any other forces are* ***external****.*

**Proposition 1.15** *In any closed system, the total force and the total torque are both zero.*

*Proof* Exercise. ■

**Corollary 1.16** *The total angular momentum of a closed system is conserved.*

Unlike linear and angular momenta, the kinetic energy of a closed systems if not necessarily conserved. However, the kinetic energy is conserved if no forces are present (see Exercise 1.1).

Many important examples of systems of $N$ point mass are of the following type.

**Definition 1.17** *A* ***Newtonian potential system*** *is a system of equations*

$$m_i \ddot{\mathbf{q}}_i = -\frac{\partial V}{\partial \mathbf{q}_i}$$

*for* $i = 1, \ldots, N$, *where* $V(\{\mathbf{q}_i\})$ *is a real-valued function, called the* ***potential energy****. The notation* $\partial V/\partial \mathbf{q}_i$ *means the vector with components* $\partial V/\partial q_i^j$, *for* $j = 1, \ldots, d$, *where d is the spatial dimension of the system. Note that if* $i = 1$, *the equation becomes* $m\ddot{\mathbf{q}} = -\nabla V$ *for potential* $V(\mathbf{q})$.

**Example 1.18** Every central force problem is a Newtonian potential system. Indeed, for any central force $F(\|\mathbf{q}\|)\widehat{\mathbf{q}}$, let $U$ be an antiderivative of $F$, and define

$$V(\mathbf{q}) := -U(\|\mathbf{q}\|).$$

Then, $-\nabla V(\mathbf{q}) = F(\|\mathbf{q}\|)\widehat{\mathbf{q}}$.

**Example 1.19** (*N*-**body problem)** Consider the motion of $N$ point masses under their mutual Newtonian gravitational forces (i.e. inverse-square law). This system is a Newtonian potential system, with

$$V(\mathbf{q}) = \sum_{i,j=1}^{N} \frac{-Gm_i m_j}{\|\mathbf{q}_i - \mathbf{q}_j\|}. \tag{1.8}$$

The general problem of solving this system or determining its characteristics is called the ***Newtonian N-body problem.***

**Definition 1.20** *The **total energy** of a Newtonian potential system with potential energy $V(\mathbf{q})$ is $E := K + V$, where $K$ is kinetic energy.*

**Theorem 1.21 (Conservation of energy)** *In any Newtonian potential system, total energy is conserved.*

*Proof*

$$\begin{aligned}\frac{dE}{dt} &= \frac{d}{dt}\left(\frac{1}{2}\sum m_i \|\dot{\mathbf{q}}_i\|^2 + V(\mathbf{q})\right)\\ &= \sum m_i \dot{\mathbf{q}}_i \cdot \ddot{\mathbf{q}}_i + \sum \frac{\partial V}{\partial \mathbf{q}_i} \cdot \dot{\mathbf{q}}_i = \sum \dot{\mathbf{q}}_i \cdot \left(m_i \ddot{\mathbf{q}}_i + \frac{\partial V}{\partial \mathbf{q}_i}\right) = 0.\end{aligned}$$

■

**Remark 1.22** *For this reason, Newtonian potential systems are also called **conservative.***

**Example 1.23** Consider a central force problem, with $m = 1$ for simplicity, so $\ddot{\mathbf{q}} = -\nabla V(\mathbf{q}) = F(\|\mathbf{q}\|)\widehat{\mathbf{q}}$, where $V(\mathbf{q}) = -U(\|\mathbf{q}\|)$ and $F(t) = dU/dt$, as in Remark 1.18. By Corollary 1.12 and Remark 1.13, the motion is planar and the angular momentum $r^2\dot{\theta}$ is constant. So,

$$r^2(t)\dot{\theta}(t) = C := r^2(0)\dot{\theta}(0). \tag{1.9}$$

By Theorem 1.21, the total energy $K + V$ is conserved. In polar coordinates, we have $V(r, \theta) = -U(r)$ and, from Remark 1.5, we have $K(r, \theta) = \frac{1}{2}\left(\dot{r}^2 + r^2\dot{\theta}^2\right)$. So

$$\frac{1}{2}\left(\dot{r}(t)^2 + r^2(t)\dot{\theta}^2(t)\right) - U(r(t)) = E := \frac{1}{2}\left(\dot{r}(0)^2 + r^2(0)\dot{\theta}^2(0)\right) - U(r(0)).$$

Using (1.9) to eliminate $\dot{\theta}$ gives

$$\frac{1}{2}\left(\dot{r}^2 + \frac{C^2}{r^2}\right) - U(r) = E \quad \text{(a constant).} \tag{1.10}$$

This is a first-order ODE in a single variable, which can be integrated to find $r(t)$.

**Remark 1.24 (The importance of conserved quantities)** *We presume that many readers have seen the preceding example before, but we mention it in order to underline the following: conserved quantities are extremely useful in mechanics. For instance, in the previous example, knowledge of the conservation of angular momentum and energy is sufficient to reduce the problem to a first-order ODE in a single variable. Another example is the motion of a rigid body, which can be understood qualitatively by considering angular momentum and energy, as explained at the end of Section 1.5.*

We have seen that total energy is conserved in all Newtonian potential systems. But what about momentum? It turns out that momentum and angular momentum are conserved only for systems with ***symmetry***, i.e. those that are invariant under certain transformations.

**Definition 1.25** *A function $V : \mathbb{R}^{dN} \to \mathbb{R}$ is **translationally invariant** if*

$$V(\mathbf{q}_1 + \mathbf{t}, \ldots, \mathbf{q}_N + \mathbf{t}) = V(\mathbf{q}_1, \ldots, \mathbf{q}_N),$$

*for any $\mathbf{q} = (\mathbf{q}_1, \ldots, \mathbf{q}_N)$ and any $\mathbf{t} \in \mathbb{R}^d$.*

*A function $V : \mathbb{R}^{dN} \to \mathbb{R}$ is **rotationally invariant** if*

$$V(R\mathbf{q}_1, \ldots, R\mathbf{q}_N) = V(\mathbf{q}_1, \ldots, \mathbf{q}_N),$$

*for any $\mathbf{q} = (\mathbf{q}_1, \ldots, \mathbf{q}_N)$ and any $d \times d$ rotation matrix $R$ (we assume $d = 2$ or $3$).*

**Example 1.26** It is easily checked that the gravitational potential (eqn (1.8)) is both translationally and rotationally invariant.

**Proposition 1.27** *In any Newtonian potential system with a translationally invariant $V$, linear momentum is conserved.*

*Proof* Taking $\mathbf{t} = -\mathbf{q}_1$ gives

$$V(\mathbf{q}_1, \ldots, \mathbf{q}_N) = V(0, \mathbf{q}_2 - \mathbf{q}_1, \mathbf{q}_3 - \mathbf{q}_1, \ldots, \mathbf{q}_N - \mathbf{q}_1) = V\left(\mathbf{d}_2, \mathbf{d}_3, \ldots, \mathbf{d}_N\right),$$

where $\mathbf{d}_i := \mathbf{q}_i - \mathbf{q}_1$, for $i = 2, \ldots, N$. Then the total force on the system is

$$\sum_{i=1}^{N} \mathbf{F}_i = \sum_{i=1}^{N} m_i \ddot{\mathbf{q}}_i = -\sum_{i=1}^{N} \frac{\partial V}{\partial \mathbf{q}_i} = -\sum_{i=1}^{N}\sum_{j=2}^{N} \frac{\partial V}{\partial \mathbf{d}_j}\frac{\partial \mathbf{d}_j}{\partial \mathbf{q}_i}$$

$$= -\sum_{j=2}^{N} \frac{\partial V}{\partial \mathbf{d}_j}\frac{\partial \mathbf{d}_j}{\partial \mathbf{q}_1} - \sum_{i=2}^{N}\sum_{j=2}^{N} \frac{\partial V}{\partial \mathbf{d}_j}\frac{\partial \mathbf{d}_j}{\partial \mathbf{q}_i}$$

$$= \sum_{j=2}^{N} \frac{\partial V}{\partial \mathbf{d}_j} - \sum_{j=2}^{N} \frac{\partial V}{\partial \mathbf{d}_j} = 0.$$

Consequently (by Proposition 1.6), linear momentum is conserved for Newtonian potential systems whose potential $V$ is translationally invariant. ■

**Alternative Proof of Proposition 1.27.** Here is a different proof that introduces a general method. Let $\mathbf{q}(t)$ be an arbitrary trajectory. That is, let $\mathbf{q}(t)$ be a solution of the system

$$m_i \ddot{\mathbf{q}}_i = -\frac{\partial V}{\partial \mathbf{q}_i},$$

with a translation invariant potential. The total momentum $\mathbf{p}$ will be shown to be conserved along this trajectory, by showing that $\mathbf{p} \cdot \boldsymbol{\xi}$ is constant, for every $\boldsymbol{\xi} \in \mathbb{R}^d$. For any $t_0$, let $\mathbf{r}(s) = \mathbf{q}(t_0) + s\mathbf{c}$, where $\mathbf{c} = (\boldsymbol{\xi}, \ldots, \boldsymbol{\xi})$ ($N$ copies of $\boldsymbol{\xi}$) and $s \in \mathbb{R}$. Thus, $\mathbf{r}(s)$ is a steady translation of the positions of all the point masses at a given moment $t_0$ by the same velocity $\mathbf{r}'(s) = \mathbf{c}$. Translational invariance of $V$ implies that

$$\nabla V(\mathbf{q}(t_0)) \cdot \mathbf{c} = \nabla V(\mathbf{r}(0)) \cdot \mathbf{r}'(0) = \left.\frac{\mathrm{d}}{\mathrm{d}s}\right|_{s=0} V(\mathbf{r}(s)) = 0.$$

Hence, at $t = t_0$,

$$\frac{\mathrm{d}}{\mathrm{d}t}(\mathbf{p} \cdot \boldsymbol{\xi}) = \frac{\mathrm{d}\mathbf{p}}{\mathrm{d}t} \cdot \boldsymbol{\xi} = \left(\sum m_i \ddot{\mathbf{q}}_i\right) \cdot \boldsymbol{\xi} = -\sum \frac{\partial V}{\partial \mathbf{q}_i} \cdot \boldsymbol{\xi} = -\nabla V(\mathbf{q}(t)) \cdot \mathbf{c} = 0.$$

Since this holds at any time $t_0$ for any vector $\boldsymbol{\xi}$ and for any trajectory $\mathbf{q}(t)$, it follows that $\mathbf{p}$ is conserved. ■

**Proposition 1.28** *In any Newtonian potential system with a rotationally invariant $V$, angular momentum is conserved.*

The proof is in the same spirit as the "alternative proof" of the previous proposition.

*Proof* Let $\mathbf{q}(t)$ be an arbitrary trajectory of the system. One may show that $\boldsymbol{\pi}$ is conserved along this trajectory, by showing that $\boldsymbol{\pi} \cdot \boldsymbol{\xi}$ is constant, for every $\boldsymbol{\xi} \in \mathbb{R}^d$. For every $t_0$, let $\mathbf{r}(s) = (\mathbf{r}_1(s), \ldots, \mathbf{r}_N(s))$ where, for every $i$, $\mathbf{r}_i(s)$ is the unique path in $\mathbb{R}^d$ with $\mathbf{r}_i(0) = \mathbf{q}_i(t_0)$ and constant angular velocity $\boldsymbol{\xi}$. Consequently, $\mathbf{r}(s)$ is a steady rotation of all of the point masses at the same angular velocity $\boldsymbol{\xi}$, and $\mathbf{r}(0) = \mathbf{q}(t_0)$. Rotational invariance of $V$ implies that $V(\mathbf{r}(s))$ is constant. Therefore, writing $\mathbf{r}'$ for differentiation with respect to $s$, the relation

$$\nabla V(\mathbf{r}(0)) \cdot \mathbf{r}'(0) = \left.\frac{\mathrm{d}}{\mathrm{d}s}\right|_{s=0} V(\mathbf{r}(s)) = 0$$

follows by definition of angular velocity,

$$\mathbf{r}'(0) = \left(\mathbf{r}_1'(0), \ldots, \mathbf{r}_N'(0)\right) = \left(\boldsymbol{\xi} \times \mathbf{r}_1(0), \ldots, \boldsymbol{\xi} \times \mathbf{r}_N(0)\right).$$

Hence, at $t = t_0$,

$$\begin{aligned}
\frac{\mathrm{d}}{\mathrm{d}t}(\boldsymbol{\pi} \cdot \boldsymbol{\xi}) &= \frac{\mathrm{d}\boldsymbol{\pi}}{\mathrm{d}t} \cdot \boldsymbol{\xi} = \sum m_i \left(\mathbf{q}_i \times \ddot{\mathbf{q}}_i\right) \cdot \boldsymbol{\xi} = -\sum \left(\mathbf{q}_i \times \frac{\partial V}{\partial \mathbf{q}_i}\right) \cdot \boldsymbol{\xi} \\
&= -\sum \frac{\partial V}{\partial \mathbf{q}_i} \cdot \left(\boldsymbol{\xi} \times \mathbf{q}_i\right) = -\sum \frac{\partial V}{\partial \mathbf{q}_i} \cdot \left(\boldsymbol{\xi} \times \mathbf{r}_i(0)\right) \\
&= -\sum \frac{\partial V}{\partial \mathbf{q}_i} \cdot \mathbf{r}_i' = \nabla V(\mathbf{r}(0)) \cdot \mathbf{r}'(0) = 0.
\end{aligned}$$

Since this holds for any time $t_0$, and any vector $\boldsymbol{\xi}$, and any trajectory $\mathbf{q}(t)$, the total angular momentum $\boldsymbol{\pi}$ is conserved. ∎

**Exercise 1.1**
Prove that if, in a system of $N$ point masses, the total force on each point mass is zero, then the kinetic energy of the system is conserved.

**Exercise 1.2**
Integrate eqn (1.10) in the case of the Kepler problem, that is for $U(r) = 1/r$.

**Exercise 1.3**

In Example 1.10, on the motion of a single planet around the Sun, assume for simplicity that $G = M = m = 1$, so that $\ddot{\mathrm{q}} = -\widehat{\mathrm{q}}/\|\mathrm{q}\|^2$.

a) Write the equations of motion in polar coordinates, and deduce directly that $r^2(t)\dot{\theta}(t)$ is constant. This constant will be denoted by $C$ in the next parts.

b) Change independent variables, so that $\mathrm{d}/\mathrm{d}t = \dot{\theta}\,(\mathrm{d}/\mathrm{d}\theta)$. (This is legitimate, since $\theta(t)$ is monotone – why?) Set $u = 1/r$, and obtain

$$u'' + u = \frac{1}{C^2}.$$

c) Integrate the above to obtain

$$u(\theta) = b\cos(\theta - \theta_0) + \frac{1}{C^2},$$

where $b$ and $\theta_0$ are constants of integration. Deduce that

$$r(\theta) = \frac{C^2}{1 + e\cos(\theta - \theta_0)},$$

where $e := (bC^2)$. For $0 < e < 1$ this is the polar equation for an ellipse (for more on the shape of the trajectories as function of $e$, see for instance [JS98]). This proves Kepler's First Law.

**Exercise 1.4**

Prove the following:

a) For 1-point systems, the only translationally invariant potentials are the constant functions, corresponding to everywhere zero acceleration.

b) For 1-point systems, $V$ is rotationally invariant, if and only if $V$ depends on $\mathrm{q}$ only through $\|\mathrm{q}\|$.

c) For 2-point systems, $V$ is both translationally and rotationally invariant, if and only if $V$ depends on $\mathrm{q}$ only through $\|\mathrm{q}_1 - \mathrm{q}_2\|$.

**Exercise 1.5**
A point mass in a magnetic field experiences a force

$$\mathbf{F} = e\dot{\mathbf{q}} \times \mathbf{B},$$

where $e$ is its charge (a constant) and $\mathbf{B}$ is the magnetic field. This equation is called the ***Lorentz force law***. Since the force depends on the velocity, it is clear that the force is not the gradient of any function $V(\mathbf{q})$, so this is not a Newtonian potential system. Since the force is perpendicular to the velocity, kinetic energy is conserved:

$$\frac{\mathrm{d}K}{\mathrm{d}t} = m\dot{\mathbf{q}} \cdot \ddot{\mathbf{q}} = \dot{\mathbf{q}} \cdot (e\dot{\mathbf{q}} \times \mathbf{B}) = 0\,.$$

Show that point masses in a constant magnetic field move in helices.

## 1.2 Lagrangian mechanics

The equations of motion of a Newtonian potential system, when expressed in different (non-inertial) coordinates, need not have the form given in eqn (1.17). This section and, later, Section 1.4 introduce two coordinate-independent formulations of conservative mechanics: Lagrangian and Hamiltonian.

From now on, we will describe the positions of all of the point masses with one concatenated position vector $\mathbf{q} = (\mathbf{q}_1, \ldots, \mathbf{q}_N) \in \mathbb{R}^{dN}$.

**Theorem 1.29** *Every Newtonian potential system,*

$$m_i\ddot{\mathbf{q}}_i = -\frac{\partial V}{\partial \mathbf{q}_i}, \quad i = 1, \ldots, N, \tag{1.11}$$

*is equivalent to the* ***Euler–Lagrange equations****,*

$$\frac{\mathrm{d}}{\mathrm{d}t}\left(\frac{\partial L}{\partial \dot{\mathbf{q}}}\right) - \frac{\partial L}{\partial \mathbf{q}} = 0, \tag{1.12}$$

*for the* **Lagrangian** *$L : \mathbb{R}^{2dN} \to \mathbb{R}$ defined by*

$$L(\mathbf{q}, \dot{\mathbf{q}}) := \underbrace{\sum_{i=1}^{N} \frac{1}{2} m_i \|\dot{\mathbf{q}}_i\|^2}_{\textit{Kinetic energy}} - \underbrace{V(\mathbf{q})}_{\textit{Potential}} .$$

*The domain of L is called the* **(velocity) phase space***.*

*Proof* For $L$ as defined in the theorem,

$$\frac{\mathrm{d}}{\mathrm{d}t}\left(\frac{\partial L}{\partial \dot{\mathbf{q}}_i}\right) - \frac{\partial L}{\partial \mathbf{q}_i} = \frac{\mathrm{d}}{\mathrm{d}t}(m_i \dot{\mathbf{q}}_i) + \frac{\partial V}{\partial \mathbf{q}_i} = m_i \ddot{\mathbf{q}}_i + \frac{\partial V}{\partial \mathbf{q}_i} .$$

■

Note that in the expression $L(\mathbf{q}, \dot{\mathbf{q}})$, the $\dot{\mathbf{q}}$ is considered to be an independent variable, so that $L : \mathbb{R}^{2dN} \to \mathbb{R}$, and $L$ has partial derivatives with respect to both $\mathbf{q}$ and $\dot{\mathbf{q}}$. Yet, when evaluating $L$ and its derivatives, we substitute $\dot{\mathbf{q}}(t) = \frac{\mathrm{d}}{\mathrm{d}t}\mathbf{q}(t)$.

**Remark 1.30** *A common procedure in the study of ordinary differential equations is to introduce a new variable to reduce a second-order differential equation to a first-order one. For the system in eqn (1.11), a new variable* $\mathbf{v}$ *can be introduced, together with the equation* $\mathbf{v} = \frac{\mathrm{d}}{\mathrm{d}t}\mathbf{q}$*. This is exactly what has been done in the previous theorem, except for the notation* $\dot{\mathbf{q}}$ *in place of* $\mathbf{v}$*.*

**Proposition 1.31** *The Euler–Lagrange equations are coordinate-independent.*

*Proof* Consider a change of coordinates $\mathbf{q} = \mathbf{q}(\mathbf{r})$, where $\mathbf{q}(\mathbf{r})$ is a smooth map with a smooth inverse. Then

$$\dot{\mathbf{q}} = \frac{\mathrm{d}\mathbf{q}}{\mathrm{d}t} = \frac{\partial \mathbf{q}}{\partial \mathbf{r}} \cdot \frac{\mathrm{d}\mathbf{r}}{\mathrm{d}t} = \frac{\partial \mathbf{q}}{\partial \mathbf{r}} \cdot \dot{\mathbf{r}}, \quad \text{so} \quad \frac{\partial \dot{\mathbf{q}}}{\partial \mathbf{r}} = \frac{\partial^2 \mathbf{q}}{\partial \mathbf{r}^2} \cdot \dot{\mathbf{r}} \quad \text{and} \quad \frac{\partial \dot{\mathbf{q}}}{\partial \dot{\mathbf{r}}} = \frac{\partial \mathbf{q}}{\partial \mathbf{r}}.$$

If we also assume $\mathbf{q}(t)$ is smooth, then the equality of mixed partials gives

$$\frac{\mathrm{d}}{\mathrm{d}t}\left(\frac{\partial \mathbf{q}}{\partial \mathbf{r}}\right) = \frac{\partial \dot{\mathbf{q}}}{\partial \mathbf{r}}, \quad \text{so} \quad \frac{\mathrm{d}}{\mathrm{d}t}\left(\frac{\partial \dot{\mathbf{q}}}{\partial \dot{\mathbf{r}}}\right) = \frac{\mathrm{d}}{\mathrm{d}t}\left(\frac{\partial \mathbf{q}}{\partial \mathbf{r}}\right) = \frac{\partial \dot{\mathbf{q}}}{\partial \mathbf{r}} .$$

If the Euler–Lagrange equations hold for $(\mathbf{q}, \dot{\mathbf{q}})$ then

$$\begin{aligned}
\frac{\mathrm{d}}{\mathrm{d}t}\left(\frac{\partial L}{\partial \dot{\mathbf{r}}}\right) - \frac{\partial L}{\partial \mathbf{r}} &= \frac{\mathrm{d}}{\mathrm{d}t}\left(\frac{\partial L}{\partial \dot{\mathbf{q}}} \cdot \frac{\partial \dot{\mathbf{q}}}{\partial \dot{\mathbf{r}}}\right) - \left(\frac{\partial L}{\partial \mathbf{q}} \cdot \frac{\partial \mathbf{q}}{\partial \mathbf{r}} + \frac{\partial L}{\partial \dot{\mathbf{q}}} \cdot \frac{\partial \dot{\mathbf{q}}}{\partial \mathbf{r}}\right) \\
&= \frac{\mathrm{d}}{\mathrm{d}t}\left(\frac{\partial L}{\partial \dot{\mathbf{q}}}\right) \cdot \frac{\partial \dot{\mathbf{q}}}{\partial \dot{\mathbf{r}}} + \frac{\partial L}{\partial \dot{\mathbf{q}}} \cdot \frac{\mathrm{d}}{\mathrm{d}t}\left(\frac{\partial \dot{\mathbf{q}}}{\partial \dot{\mathbf{r}}}\right) - \frac{\partial L}{\partial \mathbf{q}} \cdot \frac{\partial \mathbf{q}}{\partial \mathbf{r}} - \frac{\partial L}{\partial \dot{\mathbf{q}}} \cdot \frac{\partial \dot{\mathbf{q}}}{\partial \mathbf{r}} \\
&= \frac{\mathrm{d}}{\mathrm{d}t}\left(\frac{\partial L}{\partial \dot{\mathbf{q}}}\right) \cdot \frac{\partial \mathbf{q}}{\partial \mathbf{r}} + \frac{\partial L}{\partial \dot{\mathbf{q}}} \cdot \frac{\partial \dot{\mathbf{q}}}{\partial \mathbf{r}} - \frac{\partial L}{\partial \mathbf{q}} \cdot \frac{\partial \mathbf{q}}{\partial \mathbf{r}} - \frac{\partial L}{\partial \dot{\mathbf{q}}} \cdot \frac{\partial \dot{\mathbf{q}}}{\partial \mathbf{r}} \\
&= \left(\frac{\mathrm{d}}{\mathrm{d}t}\left(\frac{\partial L}{\partial \dot{\mathbf{q}}}\right) - \frac{\partial L}{\partial \mathbf{q}}\right) \cdot \frac{\partial \mathbf{q}}{\partial \mathbf{r}} + \frac{\partial L}{\partial \dot{\mathbf{q}}} \cdot \frac{\partial \dot{\mathbf{q}}}{\partial \mathbf{r}} - \frac{\partial L}{\partial \dot{\mathbf{q}}} \cdot \frac{\partial \dot{\mathbf{q}}}{\partial \mathbf{r}} = 0.
\end{aligned}$$

■

**Remark 1.32** *Many books do not prove this directly, because it is a consequence of Theorem 1.38.*

This coordinate–independence allows us to transform directly into convenient coordinates, by first expressing $L = K - V$ in the chosen coordinates and then computing the Euler–Lagrange equations.

**Definition 1.33** *A* ***Lagrangian system*** *on a configuration space* $\mathbb{R}^{dN}$ *is the system of ODEs in eqn 1.12, i.e. the Euler–Lagrange equations, for some function* $L : \mathbb{R}^{2dN} \to \mathbb{R}$ *called the* ***Lagrangian.***

Not all Lagrangian systems are Newtonian potential systems, as the next example shows.

**Example 1.34** Recall from Exercise 1.5 that the motion of a point mass in a magnetic field is not a Newtonian potential system. Nonetheless, this system is Lagrangian, although the Lagrangian $L$ is not of the form $K - V$.

The ***Lorentz force law*** for a point mass in a magnetic field is

$$\mathbf{F} = e\,\dot{\mathbf{q}} \times \mathbf{B},$$

where $e$ is its charge and $\mathbf{B}$ is the magnetic field. Now assume that $\nabla \cdot \mathbf{B} = 0$, so $\mathbf{B} = \nabla \times \mathbf{A}$ for some vector field $\mathbf{A}(\mathbf{q})$ called the vector potential of $\mathbf{B}(\mathbf{q})$. The following identity, in which $\mathbf{v}$ is assumed to be independent of $\mathbf{q}$, may be verified using coordinate indices,

$$\mathbf{v} \times \big(\nabla \times \mathbf{A}\big) = \nabla(\mathbf{v} \cdot \mathbf{A}) - \mathrm{D}\mathbf{A} \cdot \mathbf{v},$$

$$\text{i.e.} \qquad (\mathbf{v} \times (\nabla \times \mathbf{A}))^i = \sum_j v^j \left(\frac{\partial A^j}{\partial q^i} - \frac{\partial A^i}{\partial q^j}\right).$$

Now let

$$L = \frac{1}{2} m \left\| \dot{\mathbf{q}} \right\|^2 + e\, \mathbf{A}\left(\mathbf{q}\right) \cdot \dot{\mathbf{q}}.$$

Since

$$\frac{\partial L}{\partial \mathbf{q}} = e\, \nabla \left(\dot{\mathbf{q}} \cdot \mathbf{A}\right) \quad \text{and} \quad \frac{\partial L}{\partial \dot{\mathbf{q}}} = m\dot{\mathbf{q}} + e\, \mathbf{A}(\mathbf{q}),$$

the corresponding Euler–Lagrange equation is

$$\frac{\mathrm{d}}{\mathrm{d}t}\left(m\dot{\mathbf{q}} + e\, \mathbf{A}(\mathbf{q})\right) = e\, \nabla \left(\dot{\mathbf{q}} \cdot \mathbf{A}\right)$$
$$\Longleftrightarrow m\ddot{\mathbf{q}} = e\, \nabla \left(\dot{\mathbf{q}} \cdot \mathbf{A}\right) - e\, \left(\mathrm{D}\mathbf{A} \cdot \dot{\mathbf{q}}\right) = e\, \dot{\mathbf{q}} \times \left(\nabla \times \mathbf{A}\right) = e\dot{\mathbf{q}} \times \mathbf{B}.$$

So this Lagrangian system is equivalent to the Lorentz force law for magnetism.

**Definition 1.35** *The **energy function** for a Lagrangian $L(\mathbf{q}, \dot{\mathbf{q}})$ is*

$$E := \frac{\partial L}{\partial \dot{\mathbf{q}}} \cdot \dot{\mathbf{q}} - L\,.$$

**Remark 1.36** *If $L(\mathbf{q}, \dot{\mathbf{q}}) = \sum \frac{1}{2} m_i \|\dot{\mathbf{q}}_i\|^2 - V(\mathbf{q})$, then*

$$E = \sum m_i \|\dot{\mathbf{q}}_i\|^2 - L = \frac{1}{2} \sum m_i \|\dot{\mathbf{q}}_i\|^2 + V = K + V.$$

**Theorem 1.37** *In any Lagrangian system, the energy function is conserved.*

*Proof*

$$\frac{\mathrm{d}L}{\mathrm{d}t} = \frac{\partial L}{\partial \mathbf{q}} \cdot \frac{\mathrm{d}\mathbf{q}}{\mathrm{d}t} + \frac{\partial L}{\partial \dot{\mathbf{q}}} \cdot \frac{\mathrm{d}\dot{\mathbf{q}}}{\mathrm{d}t} = \left(\frac{\mathrm{d}}{\mathrm{d}t}\left(\frac{\partial L}{\partial \dot{\mathbf{q}}}\right)\right) \cdot \dot{\mathbf{q}} + \frac{\partial L}{\partial \dot{\mathbf{q}}} \cdot \ddot{\mathbf{q}} = \frac{\mathrm{d}}{\mathrm{d}t}\left(\frac{\partial L}{\partial \dot{\mathbf{q}}} \cdot \dot{\mathbf{q}}\right).$$

Hence

$$\frac{\mathrm{d}E}{\mathrm{d}t} = \frac{\mathrm{d}}{\mathrm{d}t}\left(\frac{\partial L}{\partial \dot{\mathbf{q}}} \cdot \dot{\mathbf{q}} - L\right) = 0.$$

■

The Euler–Lagrange equations correspond to a *variational principle* on a space of smooth paths, i.e. smooth parameterized curves, in $\mathbb{R}^{dN}$, with fixed endpoints. The main idea is that a path $\mathbf{q} : [a, b] \to \mathbb{R}^{dN}$ is a solution of the Euler–Lagrange equations if and only if it is a stationary point of an *action functional* $\mathcal{S} : C^\infty([a, b], \mathbb{R}^{dN}) \to \mathbb{R}$ (a precise statement will follow shortly).

A good analogy is to imagine a heavy chain of a fixed length $l$ hanging from fixed endpoints, say $A$ and $B$. There are infinitely many ways this chain could hang, taking the shape of a curve $\mathbf{c}(t)$, with $t$ being the arc-length parameter, such that $\mathbf{c}(0) = A$ and $\mathbf{c}(l) = B$. But in the real world, the chain hangs in one way only, taking the shape of a so-called catenary curve. This shape is determined as the minimum of the action functional for this problem (see, for instance, [Gol59]).

Let $\mathbf{q}_0 : [a, b] \to \mathbb{R}^{dN}$ be a smooth path with endpoints $\mathbf{q}_0(a) = \mathbf{q}_a$ and $\mathbf{q}_0(b) = \mathbf{q}_b$. A ***deformation*** of $\mathbf{q}_0$ is a smooth map $\mathbf{q}(t, s)$, $s \in (-\epsilon, \epsilon)$, $\epsilon > 0$, such that $\mathbf{q}(t, 0) = \mathbf{q}_0(t)$ for all $t \in [a, b]$. The ***variation*** of the curve $\mathbf{q}_0(\cdot)$ corresponding to a given deformation $\mathbf{q}(t, s)$ is

$$\delta\mathbf{q}(\cdot) := \left.\frac{\mathrm{d}}{\mathrm{d}s}\right|_{s=0} \mathbf{q}(\cdot, s).$$

The corresponding ***first variation*** of a functional $\mathcal{S} : C^\infty([a, b], \mathbb{R}^{dN})$ at $\mathbf{q}_0(t)$ is:

$$\delta\mathcal{S} := \mathbf{D}\mathcal{S}[\mathbf{q}_0(\cdot)](\delta\mathbf{q}(\cdot)) := \left.\frac{\mathrm{d}}{\mathrm{d}s}\right|_{s=0} \mathcal{S}[\mathbf{q}(\cdot, s)]. \tag{1.13}$$

The path $\mathbf{q}_0$ is called a ***stationary point*** of $\mathcal{S}$ if $\delta\mathcal{S} = 0$ for all deformations of $\mathbf{q}_0$ in a designated class.

If the deformation $\mathbf{q}(s, t)$ has fixed endpoints, meaning that $\mathbf{q}(a, s) = \mathbf{q}_a$ and $\mathbf{q}(b, s) = \mathbf{q}_b$, for all $s \in (-\epsilon, \epsilon)$, then $\delta\mathbf{q}(a) = \delta\mathbf{q}(b) = \mathbf{0}$. We will refer to such variations as 'variations among paths with fixed endpoints'.

**Theorem 1.38** *For any differentiable $L : \mathbb{R}^{2dN} \to \mathbb{R}$, the Euler–Lagrange equations,*

$$\frac{\mathrm{d}}{\mathrm{d}t}\left(\frac{\partial L}{\partial \dot{\mathbf{q}}}\right) - \frac{\partial L}{\partial \mathbf{q}} = 0 \tag{1.14}$$

*are equivalent to* ***Hamilton's principle of stationary action,***[1] *namely, that $\delta\mathcal{S} = 0$ holds, for the* ***action functional***

$$\mathcal{S}[\mathbf{q}(\cdot)] := \int_a^b L(\mathbf{q}(t), \dot{\mathbf{q}}(t))\, \mathrm{d}t, \tag{1.15}$$

*with respect to variations among paths with fixed endpoints.*

[1] Just as in elementary calculus, points where a derivative vanishes are called either stationary or critical points. Hamilton himself used "stationary" [Rou60]. This principle is also called Hamilton's principle of least action or Hamilton's variational principle.

*Proof* By the equality of mixed partials, $\frac{\mathrm{d}}{\mathrm{d}t}\delta\mathbf{q} = \delta\dot{\mathbf{q}}$. Thus, integrating by parts, and taking into account that $\delta\mathbf{q}(a) = \delta\mathbf{q}(b) = 0$, one finds

$$\begin{aligned}
\delta\mathcal{S} := \mathbf{D}\mathcal{S}[\mathbf{q}_0(\cdot)](\delta\mathbf{q}(\cdot)) &= \left.\frac{\mathrm{d}}{\mathrm{d}s}\right|_{s=0} \mathcal{S}[\mathbf{q}(\cdot, s)] \\
&= \left.\frac{\mathrm{d}}{\mathrm{d}s}\right|_{s=0} \int_a^b L(\mathbf{q}(t,s), \dot{\mathbf{q}}(t,s))\,\mathrm{d}t \\
&= \int_a^b \left[\frac{\partial L}{\partial \mathbf{q}} \cdot \delta\mathbf{q} + \frac{\partial L}{\partial \dot{\mathbf{q}}} \cdot \delta\dot{\mathbf{q}}\right] \mathrm{d}t \\
&= \int_a^b \left[\frac{\partial L}{\partial \mathbf{q}} - \frac{\mathrm{d}}{\mathrm{d}t}\left(\frac{\partial L}{\partial \dot{\mathbf{q}}}\right)\right] \cdot \delta\mathbf{q}\,\mathrm{d}t + \left.\frac{\partial L}{\partial \dot{\mathbf{q}}} \cdot \delta\mathbf{q}\right|_a^b \\
&= -\int_a^b \left[\frac{\mathrm{d}}{\mathrm{d}t}\left(\frac{\partial L}{\partial \dot{\mathbf{q}}}\right) - \frac{\partial L}{\partial \mathbf{q}}\right] \cdot \delta\mathbf{q}\,\mathrm{d}t\,,
\end{aligned}$$

for all smooth $\delta\mathbf{q}(t)$ satisfying $\delta\mathbf{q}(a) = \delta\mathbf{q}(b) = 0$. Therefore, $\delta\mathcal{S} = 0$ is equivalent to the Euler–Lagrange equations. ■

**Exercise 1.6**
Write the Lagrangian and the corresponding Euler–Lagrange equations in polar coordinates for the central force problem in $\mathbb{R}^2$, which was introduced in Definition 1.9. Deduce that, since $L$ is independent of $\theta$, the quantity $\partial L/\partial\dot{\theta} = mr^2\dot{\theta}$ is conserved.

## 1.3 Constraints

In many examples of systems of $N$ point masses, the point masses are constrained to move on given surfaces or curves in $\mathbb{R}^3$. The constraints may be the same or different for different point masses. More generally, there may be constraints that describe relationships between the point masses. Here are some examples:

- The bob of a spherical pendulum, attached to the origin by a rigid arm of length $l_1$, is constrained to move on the sphere $x^2 + y^2 + z^2 = l_1^2$.
- If a second bob is suspended from the first by a second rigid arm of length $l_2$, resulting in a double spherical pendulum, the two bobs are constrained to move on the subset of $\mathbb{R}^6$ described by the two equations $x_1^2 + y_1^2 + z_1^2 = l_1^2$ and $(x_2 - x_1)^2 + (y_2 - y_1)^2 + (z_2 - z_1)^2 = l_2^2$.

- If all of the mutual distances between $N$ point masses are constrained to fixed values, then the point masses move together as a *rigid body* that can only rotate and translate, keeping a fixed shape.

Of course, any planar system is a spatial system with the added constraint that the point masses move in a given plane. So without loss of generality, we consider spatial systems. We will assume that the constraints can be described by $k$ scalar equations, $f_j(\mathbf{q}) = c_j$, for $j = 1, \ldots, k$. The subset of $\mathbb{R}^{3N}$ determined by these constraints is called the ***configuration space***, usually denoted by $Q$. The number of ***degrees of freedom*** of the system is the dimension of $Q$. We will assume that the gradient vectors $\nabla f_j(\mathbf{q})$ are all non-zero and linearly independent, for all $\mathbf{q}$ in the configuration space.[2] Constraints of this form, i.e. functions of position only, are called ***holonomic***. These are the only kinds of constraints that we will deal with here. In contrast, ***non-holonomic*** constraints involve velocities $\dot{\mathbf{q}}$, a classic example being the no-slip condition of an object rolling on a surface.

We begin by analysing holonomic constraints in the context of Newtonian mechanics. In a constrained system there exist additional forces that keep $\mathbf{q}$ in the configuration space. Let $\mathbf{C}_i$ be the constraint force on the $i$th point mass, and let $\mathbf{F}_i$ be the total of all other forces on it. Then, Newton's second law says that (in an inertial coordinate system)

$$m_i \ddot{\mathbf{q}}_i = \mathbf{F}_i + \mathbf{C}_i.$$

We cannot find the magnitude of the ***constraint forces*** $\mathbf{C}_i$ directly, but we can make a physically–reasonable assumption about the directions in which they act. Indeed, consider first a single point mass moving on a fixed surface. If we assume that the surface is frictionless, then it seems intuitively reasonable that the force exerted by the surface on the point mass must be normal to the surface (i.e. orthogonal to it). Similar considerations apply to multiple point masses, each with its own independent constraint. The situation is less clear when the constraints describe relationships between point masses. However, there is a good physical reason to assume that the vector of constraint forces $(\mathbf{C}_1, \ldots, \mathbf{C}_N)$ is normal to the configuration space. Indeed, since the velocity $\dot{\mathbf{q}}(t) = \big(\dot{\mathbf{q}}_1(t), \ldots, \dot{\mathbf{q}}_N(t)\big)$ is tangent to the configuration space at the point $\mathbf{q}(t)$, the assumption implies that

$$\sum_{i=1}^{N} \mathbf{C}_i \cdot \dot{\mathbf{q}}_i = 0 \tag{1.16}$$

[2] This ensures that the configuration space is a smooth manifold, as shown in Chapter 2.

for all $t$. But $\mathbf{C}_i \cdot \dot{\mathbf{q}}_i$ is the rate of work done by the force $\mathbf{C}_i$ on a point mass with velocity $\dot{\mathbf{q}}_i$. Thus, the assumption that $(\mathbf{C}_1, \ldots, \mathbf{C}_N)$ is always normal to the configuration space ensures that the total rate of work done by the constraint forces, summed over all the point masses, must be zero. Therefore, the constraints don't affect conservation of energy. The best justification of this assumption is that, in practice, it leads to accurate models of physical systems. For further discussion of this widely–accepted model of constraint forces, see [Blo03, BL05].

We have assumed that configuration space is defined by $k$ scalar equations, $f_j(\mathbf{q}) = c_j$, for $j = 1, \ldots, k$, and that the gradient vectors $\nabla f_j(\mathbf{q})$ are linearly independent. The gradient vectors $\nabla f_j(\mathbf{q})$ span the normal space to the configuration space. Thus, an equivalent assumption on the constraint forces is:

$$(\mathbf{C}_1, \ldots, \mathbf{C}_N) = \sum_{j=1}^{k} \lambda^j \nabla f_j, \quad \text{for some real numbers } \lambda^j. \tag{1.17}$$

We shall assume this relation in what follows. The next theorem corresponds to Theorem 1.29, with the addition of constraints.

**Theorem 1.39** *Every constrained Newtonian potential system,*

$$m_i \ddot{\mathbf{q}}_i = -\frac{\partial V}{\partial \mathbf{q}_i} + \mathbf{C}_i, \quad i = 1, \ldots, N, \tag{1.18}$$

*with constraints $f_j(\mathbf{q}) = c_j$, for $j = 1, \ldots, k$, and constraint forces satisfying eqn (1.17), is equivalent to the following version of the Euler–Lagrange equations,*

$$\frac{\mathrm{d}}{\mathrm{d}t}\left(\frac{\partial L}{\partial \dot{\mathbf{q}}}\right) - \frac{\partial L}{\partial \mathbf{q}} = \sum_{j=1}^{k} \lambda^j \nabla f_j, \tag{1.19}$$

*for the Lagrangian*

$$L(\mathbf{q}, \dot{\mathbf{q}}) := \underbrace{\sum_{i=1}^{N} \frac{1}{2} m_i \|\dot{\mathbf{q}}_i\|^2}_{\textit{Kinetic energy}} - \underbrace{V(\mathbf{q})}_{\textit{Potential}} .$$

*Proof* For $L$ as defined in the theorem,

$$\frac{\mathrm{d}}{\mathrm{d}t}\left(\frac{\partial L}{\partial \dot{\mathbf{q}}_i}\right) - \frac{\partial L}{\partial \mathbf{q}_i} = \frac{\mathrm{d}}{\mathrm{d}t}(m_i \dot{\mathbf{q}}_i) + \frac{\partial V}{\partial \mathbf{q}_i} = m_i \ddot{\mathbf{q}}_i + \frac{\partial V}{\partial \mathbf{q}_i} . \tag{1.20}$$

Assuming eqn (1.17), eqn (1.18) is equivalent to

$$m\ddot{\mathbf{q}} + \frac{\partial V}{\partial \mathbf{q}} = (\mathbf{C}_1, \ldots, \mathbf{C}_N) = \sum_{j=1}^{k} \lambda^j \nabla f_j,$$

which from eqn (1.20) is equivalent to eqn (1.19). ■

The numbers $\lambda^j$ are called ***Lagrange multipliers***. In order to find explicit equations of motion in the variables $q^i$ and $\dot{q}^i$ only, the Lagrange multipliers must be eliminated, which can be difficult (see Exercise 7.7). The problem is much easier if the constraints have simple relationships with the coordinates, which suggests that we consider changes of coordinates. However, the previous theorem applies only to Newtonian potential systems, and these do not have the same form when expressed in different (non-inertial) coordinate systems. To study constrained dynamics in general coordinate systems, we turn to Lagrangian mechanics. We will see that in fact eqn (1.19) applies in general coordinate systems.

Recall from the previous section the ***action functional***

$$\mathcal{S}[\mathbf{q}(\cdot)] := \int_a^b L(\mathbf{q}(t), \dot{\mathbf{q}}(t))\, dt, \tag{1.21}$$

and ***Hamilton's principle of stationary action***, which is: $\delta\mathcal{S} = 0$, with respect to all variations among paths with fixed endpoints. As we saw in the proof of Theorem 1.38, Hamilton's principle is also equivalent to:

$$\int_a^b \left[ \frac{\mathrm{d}}{\mathrm{d}t}\left(\frac{\partial L}{\partial \dot{\mathbf{q}}}\right) - \frac{\partial L}{\partial \mathbf{q}} \right] \cdot \delta\mathbf{q}\, \mathrm{d}t = 0, \tag{1.22}$$

for all variations $\delta\mathbf{q}(\cdot)$ satisfying $\delta\mathbf{q}(a) = \delta\mathbf{q}(b) = 0$. Since the variations are arbitrary, this shows that Hamilton's principle is equivalent to the Euler–Lagrange equations. Now instead of considering *all* deformations, suppose we consider only those that remain within the configuration space $Q$ defined by the holonomic constraints

$$f_j(\mathbf{q}(t,s)) = c_j, \quad \text{for } j = 1, \ldots, k, \quad \text{for all } t, s.$$

The variations corresponding to these restricted deformations are always tangent[3] to $Q$. Let $\mathcal{S}_C$ be the restriction of $\mathcal{S}$ to paths that lie in $Q$. Then $\mathbf{q}$ is a stationary point of $\mathcal{S}_C$ if and only if $\mathbf{D}\mathcal{S}[\mathbf{q}_0](\delta\mathbf{q}) = 0$ for all $\delta\mathbf{q}$ such that $\delta\mathbf{q}(t)$ is tangent to $Q$ for all $t$.

[3] A tangent vector to a surface or general manifold at a given point is a velocity vector at that point of a curve in the manifold. Tangent vectors are discussed in detail in Chapter 2.

If we consider only variations $\delta\mathbf{q}$ of this form, then eqn (1.22) holds if and only if $\frac{\mathrm{d}}{\mathrm{d}t}\left(\frac{\partial L}{\partial \dot{\mathbf{q}}}\right) - \frac{\partial L}{\partial \mathbf{q}}$ is normal to the configuration space, for all $t$. Since the gradient vectors $\nabla f_j(\mathbf{q})$ span the normal space, we conclude that

$$\frac{\mathrm{d}}{\mathrm{d}t}\left(\frac{\partial L}{\partial \dot{\mathbf{q}}}\right) - \frac{\partial L}{\partial \mathbf{q}} = \sum_{j=1}^{k} \lambda^j \frac{\partial f_j}{\partial \mathbf{q}}. \tag{1.23}$$

If we define $\mathcal{L} = L + \sum_j \lambda^j f_j$, then

$$\frac{\partial \mathcal{L}}{\partial \mathbf{q}} = \frac{\partial L}{\partial \mathbf{q}} + \sum_j \lambda^j \frac{\partial f_j}{\partial \mathbf{q}},$$

so eqn (1.23) is equivalent to the Euler–Lagrange equations for $\mathcal{L}$.

In summary, we have shown the following.

**Theorem 1.40** *Let $Q$ be defined by $f_j(\mathbf{q}) = c_j$, for $j = 1,\dots,k$, and suppose that, for every $\mathbf{q} \in Q$, the gradient vectors $\nabla f_j(\mathbf{q})$ are linearly independent. Let $\mathcal{S}_C$ be the action functional as defined in eqn (1.21) but restricted to paths $\mathbf{q}(\cdot)$ that remain in $Q$. Then a path in $Q$ is a stationary point of the action functional $\mathcal{S}_C$ if and only if it is a solution to the Euler–Lagrange equations for $\mathcal{L}$, where $\mathcal{L} = L + \sum_j \lambda^j f_j$.*

**Remark 1.41** *If the coordinates $(q^1,\dots,q^n)$ are such that $Q$ is defined by setting the first $k$ coordinates equal to constants, then the constraint function $f$ does not depend on $q^{n-k+1},\dots,q^n$, so the last $n-k$ Euler–Lagrange equations for $\mathcal{L}$ do not contain any Lagrange multipliers. These equations can (in principle) be solved to determine the motion.*

**Example 1.42** Consider a ***spherical pendulum***: a point mass (the 'bob') of mass $m$, suspended from the origin by a massless rigid rod of unit length, under the influence of a constant gravitational force $-mg\hat{\mathbf{k}}$, where $\hat{\mathbf{k}} = (0,0,1)$. There is one constraint, $f(\mathbf{x}) = 1$, where $f(x,y,z) = x^2 + y^2 + z^2$. The configuration space is $Q := \{\mathbf{x} \in \mathbb{R}^3 : \|\mathbf{x}\| = 1\}$, which is the unit sphere, $S^2$. The unconstrained Lagrangian is

$$L = \frac{1}{2}m(\dot{x}^2 + \dot{y}^2 + \dot{z}^2) - mgz.$$

Define

$$\mathcal{L} = L + \lambda f = \frac{1}{2}m(\dot{x}^2 + \dot{y}^2 + \dot{z}^2) - mgz + \lambda(x^2 + y^2 + z^2).$$

Then, the motion is determined by the Euler–Lagrange equations for $\mathcal{L}$, which can be calculated to be

$$\begin{aligned} m\ddot{x} &= -2\lambda x \\ m\ddot{y} &= -2\lambda y \\ m\ddot{z} &= -2\lambda z + mg. \end{aligned}$$

This system, together with the constraint equation $x^2 + y^2 + z^2 = 1$, can be solved for the accelerations. However, the system is easier to solve in spherical coordinates $(r, \theta, \phi)$ where

$$\begin{pmatrix} x \\ y \\ z \end{pmatrix} = \begin{pmatrix} r \sin\phi \cos\theta \\ r \sin\phi \sin\theta \\ -r \cos\phi \end{pmatrix}.$$

In these coordinates, the constraint is $f(r, \theta, \phi) = r^2 = 1$, which is equivalent to $r = 1$ since $r$ is always positive, and the constrained Lagrangian is

$$\mathcal{L} = L + \lambda f = \frac{1}{2} m \left( \dot{r}^2 + r^2 \dot{\phi}^2 + r^2 \sin^2\phi\, \dot{\theta}^2 \right) + mgr \cos\phi + \lambda r^2.$$

The Euler–Lagrange equations for $\mathcal{L}$ are:

$$\begin{aligned} &\frac{\mathrm{d}}{\mathrm{d}t}(m\dot{r}) = mr\dot{\phi}^2 + mr \sin^2\phi\, \dot{\theta}^2 + mg \cos\phi + 2\lambda r, \\ &\frac{\mathrm{d}}{\mathrm{d}t}(mr^2\dot{\phi}) = mr^2 \sin\phi \cos\phi\, \dot{\theta}^2 - mgr \sin\phi, \\ &\frac{\mathrm{d}}{\mathrm{d}t}(mr^2 \sin^2\phi\, \dot{\theta}) = 0. \end{aligned} \tag{1.24}$$

Because of the constraint, $r$ is eliminated from the last two equations, which determine the motion $(\theta(t), \phi(t))$. Note that the first equation is irrelevant.

**Exercise 1.7**
Show that for the spherical pendulum, the $z$-component of the angular momentum is $m \sin^2\phi\, \dot{\theta}$, which, by eqn (1.24) is a conserved quantity.

## 1.4 The Legendre transform and Hamiltonian mechanics

**Theorem 1.43** *Every Newtonian potential system,*

$$m_i \ddot{\mathbf{q}}_i = -\frac{\partial V}{\partial \mathbf{q}_i}, \quad i = 1, \ldots, N, \tag{1.25}$$

*is equivalent to* ***Hamilton's canonical equations,***

$$\dot{\mathbf{q}} = \frac{\partial H}{\partial \mathbf{p}}, \quad \dot{\mathbf{p}} = -\frac{\partial H}{\partial \mathbf{q}}, \tag{1.26}$$

*for the* ***Hamiltonian***

$$H(\mathbf{q}, \mathbf{p}) := \underbrace{\sum_{i=1}^{N} \frac{1}{2m_i} \|\mathbf{p}_i\|^2}_{\text{Kinetic energy}} + \underbrace{V(\mathbf{q})}_{\text{Potential}} .$$

*The precise equivalence is:* $(\mathbf{q}(t), \mathbf{p}(t))$ *is a solution to eqn (1.26) if and only if* $\mathbf{q}(t)$ *is a solution to eqn (1.25) and*

$$\mathbf{p}(t) = (\mathbf{p}_1(t), \ldots, \mathbf{p}_N(t)) = (m_1 \dot{\mathbf{q}}_1(t), \ldots, m_N \dot{\mathbf{q}}_N(t)),$$

*i.e.* $\mathbf{p}(t)$ *is linear momentum. Note that for such a* $\mathbf{p}(t)$*, we have* $H(\mathbf{q}, \mathbf{p}) = \sum_{i=1}^{N} \frac{1}{2} m_i \|\dot{\mathbf{q}}_i\|^2 + V(\mathbf{q}) = K + V$.

*Proof* The second-order system in eqn (1.25) is equivalent to the following first-order system in variables $(\mathbf{q}, \dot{\mathbf{q}})$,

$$\frac{\mathrm{d}}{\mathrm{d}t} \mathbf{q}_i = \dot{\mathbf{q}}_i, \quad m_i \frac{\mathrm{d}}{\mathrm{d}t} \dot{\mathbf{q}}_i = -\frac{\partial V}{\partial \mathbf{q}_i}, \quad i = 1, \ldots, N. \tag{1.27}$$

Changing variables from $(\mathbf{q}, \dot{\mathbf{q}})$ to $(\mathbf{q}, \mathbf{p})$, with $\mathbf{p}_i = m_i \dot{\mathbf{q}}_i$ gives the system

$$\dot{\mathbf{q}}_i = \frac{1}{m_i} \mathbf{p}_i = \frac{\partial H}{\partial \mathbf{p}_i}, \quad \dot{\mathbf{p}}_i = -\frac{\partial V}{\partial \mathbf{q}_i} = -\frac{\partial H}{\partial \mathbf{q}_i}, \quad i = 1, \ldots, N.$$

■

Since we have seen that Newtonian potential systems are equivalent to the Euler–Lagrange equations with $L = K - V$ (Theorem 1.43), we have now shown that the Euler–Lagrange equations for such Lagrangians are equivalent to Hamilton's equations with $H = K + V$. In fact, this equivalence generalizes to a much larger class of Lagrangians. To show this, we first generalize the definition of linear momentum, $\mathbf{p}_i := m_i \mathbf{q}_i$, as follows.

**Definition 1.44** *The **Legendre transform** for a Lagrangian $L(\mathbf{q}, \dot{\mathbf{q}})$ is the change of variables $(\mathbf{q}, \dot{\mathbf{q}}) \mapsto (\mathbf{q}, \mathbf{p})$ given by*

$$\mathbf{p} := \frac{\partial L}{\partial \dot{\mathbf{q}}}.$$

*The new variables $\mathbf{p}$ are called the **conjugate momenta** (conjugate to the position variables $\mathbf{q}$).*

**Remark 1.45** *If $L(\mathbf{q}, \dot{\mathbf{q}}) = \sum_i \frac{1}{2} m_i \|\dot{\mathbf{q}}_i\|^2 - V(\mathbf{q})$, then $\mathbf{p}_i = \frac{\partial L}{\partial \dot{\mathbf{q}}_i} = m_i \dot{\mathbf{q}}_i$.*

For the moment, it suffices to think of the Legendre transform simply as a change of variables on $\mathbb{R}^{2dN}$; but see Section 4.2 for further discussion. In order to define the appropriate Hamiltonian function, this change of variables must be invertible.

**Definition 1.46** *A Lagrangian $L$ is **regular** if*

$$\det \frac{\partial^2 L}{\partial \dot{\mathbf{q}}^2} \neq 0.$$

*$L$ is **hyperregular** if the Legendre transform for $L$ is a diffeomorphism, that is, a differentiable map with a differentiable inverse.*

**Remark 1.47** *The derivative of the Legendre transform has matrix*

$$\begin{bmatrix} I & 0 \\ \dfrac{\partial^2 L}{\partial \mathbf{q} \partial \dot{\mathbf{q}}} & \dfrac{\partial^2 L}{\partial \dot{\mathbf{q}}^2} \end{bmatrix}. \tag{1.28}$$

*This matrix is invertible if and only if $\frac{\partial^2 L}{\partial \dot{\mathbf{q}}^2}$ is invertible. It follows that if $L$ is hyperregular then $L$ is regular.*

**Remark 1.48** *If $L(\mathbf{q}, \dot{\mathbf{q}}) = \sum_i \frac{1}{2} m_i \|\dot{\mathbf{q}}_i\|^2 - V(\mathbf{q})$, then the Legendre transform, defined by $\mathbf{p}_i := \frac{\partial L}{\partial \dot{\mathbf{q}}_i} = m_i \dot{\mathbf{q}}_i$, is hyperregular, since it is differentiable and has a differentiable inverse given by $\dot{\mathbf{q}}_i = \frac{1}{m_i} \mathbf{p}_i$. It follows that $L$ is regular as well, but we can also check this directly:*

$$\det \frac{\partial^2 L}{\partial \dot{\mathbf{q}}^2} = \det \begin{bmatrix} m_1 & & 0 \\ & \ddots & \\ 0 & & m_N \end{bmatrix} \neq 0$$

*(assuming non-zero masses).*

**Theorem 1.49** *If $L : \mathbb{R}^{2dN} \to \mathbb{R}$ is any hyperregular Lagrangian, then the Euler–Lagrange equations,*

$$\frac{\mathrm{d}}{\mathrm{d}t}\left(\frac{\partial L}{\partial \dot{\mathbf{q}}_i}\right) - \frac{\partial L}{\partial \mathbf{q}_i} = 0, \quad i = 1, \dots, N, \tag{1.29}$$

*are equivalent to* ***Hamilton's equations of motion,***

$$\dot{\mathbf{q}} = \frac{\partial H}{\partial \mathbf{p}}, \quad \dot{\mathbf{p}} = -\frac{\partial H}{\partial \mathbf{q}}, \tag{1.30}$$

*for the* ***Hamiltonian***

$$H(\mathbf{q}, \mathbf{p}) := \mathbf{p} \cdot \dot{\mathbf{q}}(\mathbf{q}, \mathbf{p}) - L(\mathbf{q}, \dot{\mathbf{q}}(\mathbf{q}, \mathbf{p})), \tag{1.31}$$

*where $\dot{\mathbf{q}}(\mathbf{q}, \mathbf{p})$ is the second component of the inverse Legendre transform.*

*Proof* The Euler–Lagrange equations are second-order ODEs in the variables $\mathbf{q} = (\mathbf{q}_1, \dots, \mathbf{q}_N)$, but they can also be interpreted as part of an equivalent first-order system of ODEs in the variables $(\mathbf{q}, \dot{\mathbf{q}})$, with the extra equations $\frac{\mathrm{d}}{\mathrm{d}t}\mathbf{q} = \dot{\mathbf{q}}$. Applying the Legendre transform gives the equivalent system

$$\frac{\mathrm{d}}{\mathrm{d}t}\mathbf{q}_i = \dot{\mathbf{q}}_i(\mathbf{q}, \mathbf{p}), \quad \frac{\mathrm{d}}{\mathrm{d}t}\mathbf{p}_i - \frac{\partial L}{\partial \mathbf{q}_i} = 0. \tag{1.32}$$

We now calculate the partial derivatives of $H$:

$$\frac{\partial H}{\partial \mathbf{q}_i} = \mathbf{p}_i \cdot \frac{\partial \dot{\mathbf{q}}(\mathbf{q}, \mathbf{p})}{\partial \mathbf{q}_i} - \frac{\partial L}{\partial \mathbf{q}_i} - \frac{\partial L}{\partial \dot{\mathbf{q}}_i} \cdot \frac{\partial \dot{\mathbf{q}}(\mathbf{q}, \mathbf{p})}{\partial \mathbf{q}_i} = -\frac{\partial L}{\partial \mathbf{q}_i}$$

$$\frac{\partial H}{\partial \mathbf{p}_i} = \dot{\mathbf{q}}_i(\mathbf{q}, \mathbf{p}) + \mathbf{p}_i \cdot \frac{\partial \dot{\mathbf{q}}(\mathbf{q}, \mathbf{p})}{\partial \mathbf{q}_i} - \frac{\partial L}{\partial \dot{\mathbf{q}}_i} \cdot \frac{\partial \dot{\mathbf{q}}(\mathbf{q}, \mathbf{p})}{\partial \mathbf{p}_i} = \dot{\mathbf{q}}_i(\mathbf{q}, \mathbf{p}).$$

Consequently, the Euler–Lagrange equations are equivalent to

$$\frac{\mathrm{d}}{\mathrm{d}t}\mathbf{q}_i = \frac{\partial H}{\partial \mathbf{p}_i}, \quad \frac{\mathrm{d}}{\mathrm{d}t}\mathbf{p}_i = -\frac{\partial H}{\partial \mathbf{q}_i}. \tag{1.33}$$

These are Hamilton's equations of motion. Since $\dot{\mathbf{q}}$ is no longer being used as a variable, it is safe to use a dot to denote differentiation with respect to time, as is done in the statement of the theorem. ■

**Remark 1.50** *The Hamiltonian in the above theorem is the energy function $E$ expressed as a function of $\mathbf{q}$ and $\mathbf{p}$. Note that this change of variables is valid because the Legendre transform has a differentiable inverse, i.e. $L$ is hyperregular.*

**Remark 1.51** *If* $L(\mathbf{q}, \dot{\mathbf{q}}) = \sum \frac{1}{2} m_i \|\dot{\mathbf{q}}_i\|^2 - V(\mathbf{q})$ *then* $\mathbf{p}_i = m_i \dot{\mathbf{q}}_i$ *and*

$$\begin{aligned} H(\mathbf{q}, \mathbf{p}) &= \mathbf{p} \cdot \dot{\mathbf{q}}(\mathbf{q}, \mathbf{p}) - L(\mathbf{q}, \dot{\mathbf{q}}(\mathbf{q}, \mathbf{p})) \\ &= \sum \frac{m_i}{2} \|\dot{\mathbf{q}}_i\|^2 + V = \sum \frac{1}{2m_i} \|\mathbf{p}_i\|^2 + V \,. \end{aligned}$$

*In summary, if* $L = K - V$ *then* $H = K + V$.

**Example 1.52** A system of two point masses with Lagrangian

$$L = \frac{1}{2} \left( m_1 \dot{q}_1^2 + m_2 \dot{q}_2^2 \right) - V(q_1, q_2)$$

has Hamiltonian

$$H = \frac{1}{2m_1} p_1^2 + \frac{1}{2m_2} p_2^2 + V(q_1, q_2).$$

**Remark 1.53** *Since the Euler–Lagrange equations are coordinate independent (see Proposition 1.31) it follows that, for any Hamiltonian defined as in the theorem above, Hamilton's equations are coordinate independent.*

One may also define Hamilton's equations for any $H$, not necessarily derived from a Lagrangian.

**Definition 1.54** *A* ***Hamiltonian system*** *on configuration space* $\mathbb{R}^{dN}$ *is the system of ODEs in eqn (1.30), i.e. Hamilton's equations, for some function* $H : \mathbb{R}^{2dN} \to \mathbb{R}$ *called the* ***Hamiltonian***.

**Remark 1.55** *More generally, a Hamiltonian may be time dependent as well. However, in this book we are interested only in time-independent, that is* ***autonomous****, systems.*

For many applications in physics, one begins with a Lagrangian model and then optionally converts this to a Hamiltonian system, provided the Lagrangian is hyperregular. A major reason for doing this conversion is that the Hamiltonian approach is particularly suitable for studying conserved quantities. Some physical systems, notably in quantum mechanics, are directly modelled as Hamiltonian systems (in the generalized sense of Chapter 4), but quantum mechanics will not appear in this book. The rigid body, treated in Chapter 6, can be modelled as either a Lagrangian or Hamiltonian system. So it is important to be familiar with both the Lagrangian and Hamiltonian approaches.

**Theorem 1.56** *In any Hamiltonian system, the Hamiltonian is conserved.*

*Proof*

$$\dot{H} := \sum_{i=1}^{N} \left( \frac{\partial H}{\partial \mathbf{q}_i} \cdot \dot{\mathbf{q}}_i + \frac{\partial H}{\partial \mathbf{p}_i} \cdot \dot{\mathbf{p}}_i \right) = \sum_{i=1}^{N} \left( \frac{\partial H}{\partial \mathbf{q}_i} \cdot \frac{\partial H}{\partial \mathbf{p}_i} - \frac{\partial H}{\partial \mathbf{p}_i} \cdot \frac{\partial H}{\partial \mathbf{q}_i} \right) = 0.$$

■

**Definition 1.57** *A function $F(\mathbf{q}, \mathbf{p})$ is a* ***conserved quantity*** *for a given Hamiltonian system (synonyms: a* ***constant of the motion*** *or a* ***first integral****), if $F$ is constant along any solution of the system, i.e. $F(\mathbf{q}(t), \mathbf{p}(t))$ is constant for any solution $(\mathbf{q}(t), \mathbf{p}(t))$.*

**Definition 1.58** *The* ***canonical Poisson bracket*** *of two functions $F(\mathbf{q}, \mathbf{p})$ and $G(\mathbf{q}, \mathbf{p})$ is*

$$\{F, G\} := \sum_{i=1}^{N} \left( \frac{\partial F}{\partial \mathbf{q}_i} \cdot \frac{\partial G}{\partial \mathbf{p}_i} - \frac{\partial F}{\partial \mathbf{p}_i} \cdot \frac{\partial G}{\partial \mathbf{q}_i} \right). \tag{1.34}$$

For any $F$, and along any solution of a Hamiltonian system,

$$\dot{F} = \{F, H\},$$

by a calculation similar to the proof of Theorem 1.56. Therefore, $F$ is a conserved quantity if and only if $\{F, H\} = 0$.

**Remark 1.59** *The canonical Poisson bracket is clearly anti-symmetric. Thus, if we have two Hamiltonians, say $G$ and $H$, then since*

$$\{G, H\} = 0 \quad \Longleftrightarrow \quad \{H, G\} = 0,$$

*it follows that $G$ is conserved along the solutions to Hamilton's equations for $H$ if and only if $H$ is conserved along the solutions to Hamilton's equations for $G$.*

The previous remark is essential to the proof of the next theorem, which generalizes Proposition 1.28.

**Theorem 1.60** *If $H$ is rotationally invariant, in the sense that*

$$H\left(R\mathbf{q}_1, \ldots, R\mathbf{q}_N, R\mathbf{p}_1, \ldots, R\mathbf{p}_N\right) = H\left(\mathbf{q}_1, \ldots, \mathbf{q}_N, \mathbf{p}_1, \ldots, \mathbf{p}_N\right),$$

*for all $(\mathbf{q}, \mathbf{p})$ and all rotations $R$, then angular momentum is conserved.*

*Proof* Let $\boldsymbol{\pi}$ be angular momentum, and $\boldsymbol{\xi}$ be any non-zero vector in $\mathbb{R}^3$. For all solutions of Hamilton's equations for the Hamiltonian $G := \boldsymbol{\pi} \cdot \boldsymbol{\xi}$, each $\mathbf{q}_i(t)$ is a circular path, centred at the origin, with angular velocity $\boldsymbol{\xi}$ – see Exercise 1.9. Since $H$ is rotationally invariant, $H$ is conserved. By the previous remark, this implies that $G$ is conserved along the solutions of $H$. Since $G = \boldsymbol{\pi} \cdot \boldsymbol{\xi}$ is conserved for every $\boldsymbol{\xi}$, it follows that $\boldsymbol{\pi}$ is conserved. ■

**Theorem 1.61** *Hamilton's equations for a given* $H : \mathbb{R}^{2dN} \to \mathbb{R}$ *are equivalent to:*

$$\dot{F} = \{F, H\} \quad \textit{for all differentiable } F : \mathbb{R}^{2dN} \to \mathbb{R}\,.$$

*More precisely,* $(\mathbf{q}(t), \mathbf{p}(t))$ *is a solution of Hamilton's equations if and only if* $\frac{\mathrm{d}}{\mathrm{d}t} F(\mathbf{q}(t), \mathbf{p}(t)) = \{F, H\}$ *for all differentiable* $F : \mathbb{R}^{2dN} \to \mathbb{R}$.

*Proof* We have already observed that $\dot{F} = \{F, H\}$ along any solution of Hamilton's equations, by a calculation similar to the proof of Theorem 1.56. Now suppose that this equation is satisfied, for all $F : \mathbb{R}^{2dN} \to \mathbb{R}$. Then, taking $F(\mathbf{q}, \mathbf{p}) = \mathbf{q}_i^j$ (the $j$th coordinate of $\mathbf{q}_i$) we have

$$\frac{\mathrm{d}}{\mathrm{d}t}\mathbf{q}_i^j = \{\mathbf{q}_i^j, H\} = \frac{\partial \mathbf{q}_i^j}{\partial \mathbf{q}_i} \cdot \frac{\partial H}{\partial \mathbf{p}_i} = \frac{\partial H}{\partial \mathbf{p}_i^j}\,.$$

Similarly, taking $F(\mathbf{q}, \mathbf{p}) = \mathbf{p}_i^j$, we have

$$\frac{\mathrm{d}}{\mathrm{d}t}\mathbf{p}_i^j = \{\mathbf{q}_i^j, H\} = -\frac{\partial \mathbf{p}_i^j}{\partial \mathbf{p}_i} \cdot \frac{\partial H}{\partial \mathbf{q}_i} = -\frac{\partial H}{\partial \mathbf{q}_i^j}\,.$$

■

This description of Hamiltonian systems generalizes in a very useful way: see Chapter 4.

**Exercise 1.8**
Recall from Example 1.34 that a charged point mass moving in a static magnetic field in $\mathbb{R}^3$ has Lagrangian

$$L = \frac{1}{2}m \left\|\dot{\mathbf{q}}\right\|^2 + e\mathbf{A}(\mathbf{q}) \cdot \dot{\mathbf{q}},$$

where $\mathbf{A}(\mathbf{q})$ is a vector potential of the magnetic field $\mathbf{B}(\mathbf{q})$, i.e. $\mathbf{B}(\mathbf{q}) = \nabla \times \mathbf{A}(\mathbf{q})$. Compute the Legendre transform; invert it, and check that the Lagrangian is hyperregular; and compute the Hamiltonian.

**Exercise 1.9**
Compute Hamilton's equations in $\mathbb{R}^6$ determined by

$$H(\mathbf{q},\mathbf{p}) := \boldsymbol{\pi}\cdot\boldsymbol{\xi} = (\mathbf{q}\times\mathbf{p})\cdot\boldsymbol{\xi},$$

for a fixed non-zero $\boldsymbol{\xi}\in\mathbb{R}^3$. Verify that the solutions of these equations are circular paths, centred at the origin, with angular velocity $\boldsymbol{\xi}$.

**Exercise 1.10**
Show that the canonical Poisson bracket satisfies the ***Jacobi identity***:

$$\{F,\{G,H\}\}+\{H,\{F,G\}\}+\{G,\{H,F\}\}=0.$$

Given two constants of motion, what does the Jacobi identity imply about additional constants of motion?

## 1.5 Rigid bodies

A ***rigid body*** is a system of three or more point masses, not all collinear, constrained so that the distance between any two point masses remains constant over time. Instead of a finite number of point masses, we may consider an infinite collection of points with mass density instead of mass, forming a solid object. We will assume here that the body is solid, and note that essentially the same analysis applies to assemblies of a finite number of particles, with integrals replaced by sums. We assume that the motion is continuous, which implies that the orientation of the object is preserved. This assumption, together with the constraint that the inter-particle distances remain constant, implies that the body can only move by combinations of rotations and translations.[4]

It is often possible to deal separately with the rotational and translational motion. For example, this is the case for the motion of a spherically symmetric body (with symmetric mass distribution) in a gravitational field: gravitation determines the translational motion but has no effect on the rotation of the body. In this book, we will only deal with the rigid body's rotational motion. We will assume that one point of the body – the 'pivot

[4] More details can be found in [Arn78]

point' – remains fixed in some inertial frame, and base our coordinate systems at that point. It is common to assume that this point is the centre of mass of the body, but this is not necessary. In Section 7.4, we will consider the heavy top, in which the pivot point of the top is not at the centre of mass.

We fix an inertial coordinate system, called the ***spatial coordinate system***, with origin at the pivot point. The position and velocity of a given particle in the body at time $t$ are denoted $\mathbf{x}(t)$ and $\dot{\mathbf{x}}(t)$. Both are considered as coordinate vectors in $\mathbb{R}^3$, as defined by the spatial coordinate system. We also fix a ***reference configuration*** of the body. The position of a given particle when the body is in the reference configuration is denoted $\mathbf{X}$ and called the particle's ***label.***

The configuration of the body at time $t$ is determined by a rotation matrix $\mathbf{R}(t)$ that takes every label $\mathbf{X}$ to the position $\mathbf{x}(t)$ (in spatial coordinates) of the corresponding particle at time $t$:

$$\mathbf{x}(\mathbf{X}, t) = \mathbf{R}(t)\mathbf{X}\,.$$

Often, the dependence of $\mathbf{x}$ on $\mathbf{X}$ is suppressed and the above equation is written as

$$\mathbf{x}(t) = \mathbf{R}(t)\mathbf{X}\,. \tag{1.35}$$

At a particular time $t$, the map $\mathbf{X} \mapsto \mathbf{x} = \mathbf{R}\mathbf{X}$ is called the ***body-to-space map***; note that this map is just the rotation $\mathbf{R}$.

Recall that all rotation matrices are orthogonal, so that $\mathbf{R}^T = \mathbf{R}^{-1}$.

At every time $t$, there exists a unique ***angular velocity vector*** $\boldsymbol{\omega}(t)$ such that, for every particle in the body,

$$\dot{\mathbf{x}} = \boldsymbol{\omega} \times \mathbf{x}. \tag{1.36}$$

In this equation, all vectors are expressed in spatial coordinates, so $\boldsymbol{\omega}$ is also called the ***spatial angular velocity vector***. The existence of such a vector $\boldsymbol{\omega}$ can be derived directly from the constraint of constant inter-point distances (see [JS98]). Alternatively, it can be deduced from properties of rotation matrices, as follows. From the basic relation $\mathbf{x} = \mathbf{R}\mathbf{X}$, we deduce that

$$\dot{\mathbf{x}} = \dot{\mathbf{R}}\mathbf{X} = \dot{\mathbf{R}}\mathbf{R}^{-1}\mathbf{x} = \dot{\mathbf{R}}\mathbf{R}^T\mathbf{x}\,. \tag{1.37}$$

Since all rotation matrices are orthogonal, $\mathbf{R}(t)$ satisfies $\mathbf{R}\mathbf{R}^T = I$ for all $t$. Differentiating with respect to $t$ gives

$$\dot{\mathbf{R}}\mathbf{R}^T + \mathbf{R}\dot{\mathbf{R}}^T = 0,$$

which implies that $\dot{\mathbf{R}}\mathbf{R}^T$ is skew-symmetric, i.e. is of the form

$$\dot{\mathbf{R}}\mathbf{R}^T = \begin{bmatrix} 0 & -\omega_3 & \omega_2 \\ \omega_3 & 0 & -\omega_1 \\ -\omega_2 & \omega_1 & 0 \end{bmatrix},$$

for some $\omega_1, \omega_2, \omega_3$. Defining $\boldsymbol{\omega} := (\omega_1, \omega_2, \omega_3)$, it can be directly verified that

$$\dot{\mathbf{R}}\mathbf{R}^{-1}\mathbf{x} = \dot{\mathbf{R}}\mathbf{R}^T\mathbf{x} = \boldsymbol{\omega} \times \mathbf{x}. \tag{1.38}$$

We introduce notation for the ***hat map***

$$\hat{\boldsymbol{\omega}} := \begin{bmatrix} 0 & -\omega_3 & \omega_2 \\ \omega_3 & 0 & -\omega_1 \\ -\omega_2 & \omega_1 & 0 \end{bmatrix}, \tag{1.39}$$

so that

$$\dot{\mathbf{R}}\mathbf{R}^{-1} = \hat{\boldsymbol{\omega}}. \tag{1.40}$$

The correspondence between skew-symmetric matrices and vectors will be discussed further in Chapter 5.

Recall that the total angular momentum (around the origin) of a system of $N$ particles is $\boldsymbol{\pi} := \sum m_i \, \mathbf{x}_i \times \dot{\mathbf{x}}_i$. For a solid body, the corresponding definition is an integral over the body. Let $\rho(\mathbf{X})$ be the density of the body at the point with label $\mathbf{X}$. Let $\mathcal{B}$ be the region of space occupied by the body in its reference configuration, and note that the labels $\mathbf{X}$ range over $\mathcal{B}$. The ***(spatial) angular momentum*** of the body at any given time is

$$\boldsymbol{\pi} := \int_{\mathcal{B}} \rho(\mathbf{X}) \, \mathbf{x} \times \dot{\mathbf{x}} \, \mathrm{d}^3\mathbf{X} = \int_{\mathcal{B}} \rho(X) \, \mathbf{x} \times (\boldsymbol{\omega} \times \mathbf{x}) \, \mathrm{d}^3\mathbf{X}, \tag{1.41}$$

where $\mathbf{x} = \mathbf{R}\mathbf{X}$. By standard vector identities, $\mathbf{x} \times (\boldsymbol{\omega} \times \mathbf{x}) = (\mathbf{x} \cdot \mathbf{x}) \, \boldsymbol{\omega} - (\mathbf{x} \cdot \boldsymbol{\omega}) \, \mathbf{x}$ and $(\mathbf{x} \cdot \boldsymbol{\omega}) \, \mathbf{x} = \left(\mathbf{x}^T \boldsymbol{\omega}\right) \mathbf{x} = \mathbf{x} \left(\mathbf{x}^T \boldsymbol{\omega}\right) = \left(\mathbf{x}\mathbf{x}^T\right) \boldsymbol{\omega}$. Therefore

$$\begin{aligned} \boldsymbol{\pi} &= \int_{\mathcal{B}} \rho(\mathbf{X}) \left( \|\mathbf{x}\|^2 \mathbf{I} - \mathbf{x}\mathbf{x}^T \right) \boldsymbol{\omega} \, \mathrm{d}^3\mathbf{X} \\ &= \left( \int_{\mathcal{B}} \rho(\mathbf{X}) \left( \|\mathbf{x}\|^2 \mathbf{I} - \mathbf{x}\mathbf{x}^T \right) \mathrm{d}^3\mathbf{X} \right) \boldsymbol{\omega}, \end{aligned} \tag{1.42}$$

where $\mathbf{I}$ is the $3 \times 3$ identity matrix. The last step is valid because angular velocity $\boldsymbol{\omega}$ is a property of the entire body, independent of $\mathbf{X}$. This motivates the definition of the ***spatial moment of inertia tensor*** of the rigid body,

$$\mathbb{I}_S := \int_{\mathcal{B}} \rho(\mathbf{X}) \left( \|\mathbf{x}\|^2 \mathbf{I} - \mathbf{x}\mathbf{x}^T \right) \mathrm{d}^3\mathbf{X}.$$

Note that $\mathbb{I}_S$, $\boldsymbol{\pi}$ and $\boldsymbol{\omega}$, are all time-dependent, since $\mathbf{x}$ and $\mathbf{R}$ are. With this definition, eqn (1.42) becomes:

$$\boldsymbol{\pi} = \mathbb{I}_S\, \boldsymbol{\omega}.$$

Note the similarity to the familiar formula $\mathbf{p} = m\dot{\mathbf{x}}$ defining linear momentum. We see that the moment of inertia is the rotational analogue of mass. However, there are important difference: mass is a scalar, and it is constant and independent of the choice of coordinate system; whereas the spatial moment of inertia is a matrix with entries that depend on time and the choice of coordinate system.

The moment of inertia tensor may look more familiar if we write $\mathbf{x} = (x, y, z)$, in which case,

$$\rho(X)\left(\|\mathbf{x}\|^2\mathbf{I} - \mathbf{x}\mathbf{x}^T\right) = \rho(X)\begin{bmatrix} y^2+z^2 & -xy & -xz \\ -xy & x^2+z^2 & -yz \\ -xz & -yz & x^2+y^2 \end{bmatrix}.$$

For example, the term $\rho(X)(x^2+y^2)$ is analogous to the formula $m\left(x^2+y^2\right)$ for the scalar moment of inertia of a particle of mass $m$ moving around the $z$ axis.

By analogy with the $N$-particle total torque $\sum_i m_i\mathbf{x} \times \ddot{\mathbf{x}}$, the ***total torque*** on a rigid body is defined as

$$\int_{\mathcal{B}} \rho(X)\mathbf{x} \times \ddot{\mathbf{x}}\, \mathrm{d}^3X.$$

As for $N$-particle systems (see Proposition 1.8), when the total torque on a rigid body is zero, the total spatial angular momentum is conserved: $\dot{\boldsymbol{\pi}} = \mathbf{0}$. This can be easily checked from eqn (1.41).

Now suppose that there is no *external* torque on the rigid body, that is, the only torque (if any) arises from the forces of inter-particle interaction within the body. For $N$-particle systems, we have seen in Proposition 1.15 that forces of interaction within the system always give rise to zero total torque, and hence the conservation of spatial angular momentum. The same is true for solid bodies, a fact sometimes called ***Euler's Law***:

***In the absence of external torque, spatial angular momentum is constant, i.e.*** $\dot{\boldsymbol{\pi}} = 0$.

**Remark 1.62** *We will show later, in Chapter 8, that the conservation of spatial angular momentum can also be deduced from symmetry considerations. In particular, systems with no external forces are rotationally*

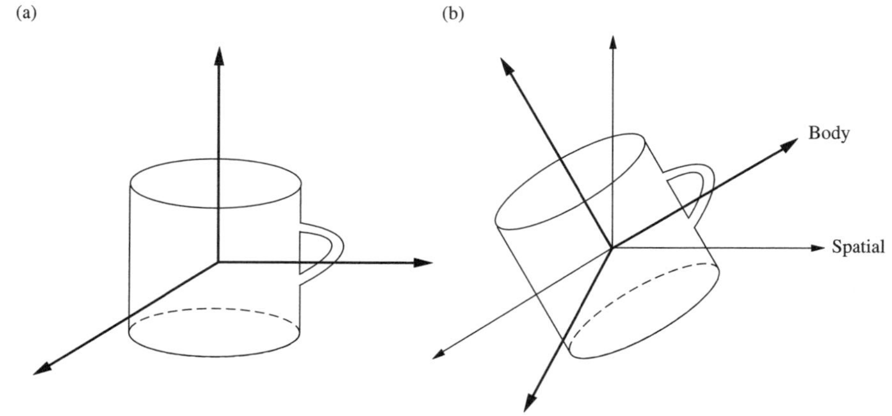


**Figure 1.1** (a) When the body is in the reference configuration, the body and spatial coordinate systems coincide. (b) The spatial coordinate system is fixed, while the body coordinate system moves with the body. The *body coordinates* $\mathbf{X}$ of a particle in the body always remain fixed, while the *spatial coordinates* $\mathbf{x}$ of the particle vary as the body moves.

*symmetric, and we will prove that this implies conservation of spatial angular momentum.*

We now express all of the quantities above in a new non-inertial coordinate system, called the ***body coordinate system***. This is defined to be the coordinate system that moves with the body but agrees with the spatial system when the body is in the reference configuration, see Figure 1.1. By definition, the position in body coordinates of a particle on the body is constant, and is the same thing as particle's label $\mathbf{X}$. The spatial position vector $\mathbf{x}$ and the label $\mathbf{X}$ are related by $\mathbf{x} = \mathbf{R}\mathbf{X}$, as in eqn (1.35). In general, left-multiplying by $\mathbf{R}$ changes a vector's coordinates from body to spatial, and vice versa, left-multiplying by $\mathbf{R}^{-1}$ changes its coordinates from spatial to body.

In particular, changing $\boldsymbol{\omega}$ and $\boldsymbol{\pi}$ into body coordinates gives the

$$\textbf{body angular velocity vector: } \boldsymbol{\Omega} := \mathbf{R}^{-1}\boldsymbol{\omega}, \qquad \text{and the}$$

$$\textbf{body angular momentum vector: } \boldsymbol{\Pi} := \mathbf{R}^{-1}\boldsymbol{\pi}.$$

We remark that the body angular velocity vector is the angular velocity vector as first defined in the spatial coordinate system and *then* converted into body coordinates. This is *not* the same as writing eqn (1.36) in body coordinates, which would be $\dot{\mathbf{X}} = \boldsymbol{\Omega} \times \mathbf{X}$ (false!), which would always imply that $\boldsymbol{\Omega} = 0$, since the labels $\mathbf{X}$ are constant over time. The difference may be clarified by thinking of a person who perceives the spinning Earth as stationary in the spinning 'body' frame yet recognizes that the Earth has an angular velocity vector parallel to its North/South axis. Note that

$$\boldsymbol{\Omega} \times \mathbf{X} = \mathbf{R}^{-1}\boldsymbol{\omega} \times \mathbf{R}^{-1}\mathbf{x} = \mathbf{R}^{-1}(\boldsymbol{\omega} \times \mathbf{x}) = \mathbf{R}^{-1}\dot{\mathbf{R}}\mathbf{R}^{-1}\mathbf{x} = \mathbf{R}^{-1}\dot{\mathbf{R}}\mathbf{X}. \quad (1.43)$$

Thus, using the notation of eqn (1.39),

$$\widehat{\boldsymbol{\Omega}} = \mathbf{R}^{-1}\dot{\mathbf{R}}. \quad (1.44)$$

It is important to remember that the body coordinate system is not (in general) inertial. Nonetheless, we can translate the formulae we have derived in spatial coordinates into body coordinates. In particular, the conservation of angular momentum $\dot{\boldsymbol{\pi}} = 0$ becomes:

$$\frac{\mathrm{d}}{\mathrm{d}t}(\mathbf{R}\boldsymbol{\Pi}) = 0,$$

which, using the product rule, is equivalent to $\dot{\mathbf{R}}\boldsymbol{\Pi} + \mathbf{R}\dot{\boldsymbol{\Pi}} = 0$ or,

$$\dot{\boldsymbol{\Pi}} = -\mathbf{R}^{-1}\dot{\mathbf{R}}\,\boldsymbol{\Pi} = -\boldsymbol{\Omega} \times \boldsymbol{\Pi} = \boldsymbol{\Pi} \times \boldsymbol{\Omega}. \quad (1.45)$$

The relation $\boldsymbol{\pi} = \mathbb{I}_S\,\boldsymbol{\omega}$ becomes $\mathbf{R}\Pi = \mathbb{I}_S\,\mathbf{R}\Omega$, or equivalently

$$\boldsymbol{\Pi} = \left(\mathbf{R}^{-1}\mathbb{I}_S\mathbf{R}\right)\boldsymbol{\Omega}. \quad (1.46)$$

We define the ***body moment of inertia*** to be $\mathbb{I} := \mathbf{R}^{-1}\mathbb{I}_S\mathbf{R}$, so that eqn (1.46) becomes

$$\boldsymbol{\Pi} = \mathbb{I}\boldsymbol{\Omega}. \quad (1.47)$$

Further, if $\mathbb{I}$ is invertible, then eqn (1.45) can be written as

$$\dot{\boldsymbol{\Pi}} = \boldsymbol{\Pi} \times \left(\mathbb{I}^{-1}\boldsymbol{\Pi}\right), \quad (1.48)$$

in which form it is called ***Euler's equation*** for a rigid body. Note that

$$\begin{aligned}
\mathbb{I} &= \mathbf{R}^{-1}\left(\int_{\mathcal{B}} \rho(\mathbf{X})\left(\|\mathbf{x}\|^2\mathbf{I} - \mathbf{x}\mathbf{x}^T\right)\mathrm{d}^3\mathbf{X}\right)\mathbf{R} && (1.49)\\
&= \int_{\mathcal{B}} \rho(\mathbf{X})\left(\|\mathbf{x}\|^2\mathbf{I} - \mathbf{R}^{-1}\mathbf{x}\mathbf{x}^T\mathbf{R}\right)\mathrm{d}^3\mathbf{X} && (1.50)\\
&= \int_{\mathcal{B}} \rho(\mathbf{X})\left(\|\mathbf{X}\|^2\,\mathbf{I} - \mathbf{X}\mathbf{X}^T\right)\mathrm{d}^3\mathbf{X}. && (1.51)
\end{aligned}$$

Thus, the ***body moment of inertia*** $\mathbb{I}$ is independent of time, which is the great advantage of working in body coordinates. We will call $\mathbb{I}$ simply the

***moment of inertia tensor*** from now on. Since $\mathbb{I}$ is constant, $\dot{\mathbf{\Pi}} = \mathbb{I}\dot{\mathbf{\Omega}}$, so Euler's equation has the following equivalent form,

$$\mathbb{I}\dot{\mathbf{\Omega}} = \mathbb{I}\mathbf{\Omega} \times \mathbf{\Omega}. \tag{1.52}$$

Since $\mathbb{I}$ is symmetric, it is diagonalizable by a rotation matrix. Its eigenvalues are called the ***principal moments of inertia*** and its eigenvectors are called the ***principal axes***. If coordinates are chosen so that $\mathbb{I}$ is diagonal, with diagonal entries $I_1, I_2, I_3$, then it is easily checked that Euler's equation is equivalent to the following set of three scalar equations,

$$\begin{aligned} I_1\dot{\Omega}_1 &= (I_2 - I_3)\Omega_2\Omega_3, \\ I_2\dot{\Omega}_2 &= (I_3 - I_1)\Omega_3\Omega_1, \\ I_3\dot{\Omega}_3 &= (I_1 - I_2)\Omega_1\Omega_2. \end{aligned}$$

We emphasize that these equations are just the conservation of spatial angular momentum, $\dot{\boldsymbol{\pi}} = 0$, written in body coordinates. They hold whenever there is no external torque on the body.

Since the moment of inertia is constant (in body coordinates), the scalars $I_1, I_2, I_3$ are constant. Thus Euler's equations form a system of first-order, linear ordinary differential equations in $\Omega_1, \Omega_2, \Omega_3$, which can in fact be integrated in closed form for all time, though doing so requires elliptic integrals (see [LL76]). Once a solution $\mathbf{\Omega}(t)$ is found, it can in principle be used to obtain the body's configuration $\mathbf{R}(t)$ by solving the so-called ***reconstruction equation***

$$\dot{\mathbf{R}} = \mathbf{R}\hat{\mathbf{\Omega}},$$

which is simply a rearrangement of eqn (1.44). However, the interpretation of this last differential equation is a bit tricky, because the dependent variable $\mathbf{R}$ is a rotation matrix, and the set of rotation matrices, denoted $SO(3)$, is not a Euclidean space. In fact, $SO(3)$ is a 3-dimensional manifold, as we will see in the next chapter. We will return to the study of rigid bodies, including the Lagrangian and Hamiltonian formulations of their equations of motion, in Chapters 7 and 9.

Though finding explicit solutions for $\mathbf{\Omega}(t)$ and $\mathbf{R}(t)$ is difficult, we can learn a lot just by considering angular momentum and energy. As we have seen, the spatial angular momentum is conserved, which implies that $\|\boldsymbol{\pi}\|$ is conserved. Since the body angular momentum vector $\mathbf{\Pi}$ satisfies $\mathbf{\Pi} = \mathbf{R}^{-1}\boldsymbol{\pi}$, it follows that $\|\mathbf{\Pi}\|$ is conserved as well. Thus, $\mathbf{\Pi}(t)$ moves on a sphere of constant radius, say

$$\|\mathbf{\Pi}\| = c.$$

The ***kinetic energy*** of the rigid body is

$$K := \frac{1}{2}\mathbf{\Omega}^T \mathbb{I}\mathbf{\Omega} = \frac{1}{2}\mathbf{\Pi}^T \mathbb{I}^{-1}\mathbf{\Pi}.$$

We will show in Chapter 7 that this formula is the natural generalization of the kinetic energy of a system of point masses. However, it suffices at this point to take it as a definition. Assuming that there are no external forces, the kinetic energy is in fact the total energy of the rigid body, which we denote by $H$. It follows from Euler's equation that $H$ is conserved:

$$\begin{aligned}\frac{\mathrm{d}H}{\mathrm{d}t} &= \frac{1}{2}\dot{\mathbf{\Pi}}^T \mathbb{I}^{-1}\mathbf{\Pi} + \frac{1}{2}\mathbf{\Pi}^T \mathbb{I}^{-1}\dot{\mathbf{\Pi}} \\ &= \frac{1}{2}\left(\mathbf{\Pi} \times \mathbb{I}^{-1}\mathbf{\Pi}\right)\cdot\left(\mathbb{I}^{-1}\mathbf{\Pi}\right) + \frac{1}{2}\left(\mathbb{I}^{-1}\mathbf{\Pi}\right)\cdot\left(\mathbf{\Pi} \times \mathbb{I}^{-1}\mathbf{\Pi}\right) = 0.\end{aligned}$$

Thus $\mathbf{\Pi}(t)$ moves on a level set of $H$, say

$$\frac{1}{2}\mathbf{\Pi}^T \mathbb{I}^{-1}\mathbf{\Pi} = h,$$

which is an ellipsoid. In summary, $\mathbf{\Pi}(t)$ must move on the intersection of an angular momentum sphere and an energy ellipsoid. As long as the principal moments of inertia (i.e. the eigenvalues of $\mathbb{I}$) are not all equal, the ellipsoid is not spherical, so the intersection of the ellipsoid and the sphere is a union of a finite number of curves and points. This is sufficient to determine the solution curve given any initial condition, though not the time-parameterization of that curve or even its orientation. Figure 1.2 shows several solution curves on a sphere of constant angular momentum, for the case when the principal moments of inertia are all distinct.

**Exercise 1.11**
Show that

$$(\mathbf{R}\boldsymbol{\omega}) \times (\mathbf{R}\mathbf{x}) = \mathbf{R}\left(\boldsymbol{\omega} \times \mathbf{x}\right)$$

for any rotation matrix $\mathbf{R}$.

**Exercise 1.12**
Suppose there are two orthogonal spatial coordinate systems, with the same origin, with a fixed change-of-coordinates matrix $P$ from the first to the second system. If $\boldsymbol{\omega}$ and $\boldsymbol{\omega}'$ are the angular velocity vectors defined with respect to the two coordinate systems, show that $\boldsymbol{\omega}' = P\boldsymbol{\omega}$.

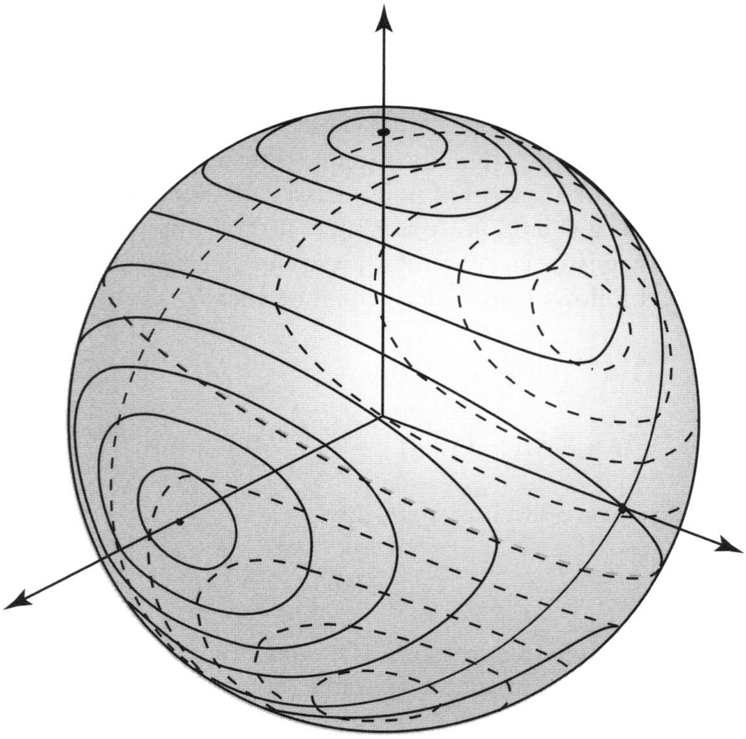

**Figure 1.2** Rigid body motion takes place on a body angular momentum sphere along the intersections of level sets of the energy.

**Exercise 1.13**
Consider an ellipsoid with axes coinciding with the $x, y, z$-axes. Show that, if the ellipsoid has semi-axes of lengths $a, b, c$, then its moment of inertia tensor is

$$\begin{bmatrix} \frac{1}{5}M(b^2+c^2) & 0 & 0 \\ 0 & \frac{1}{5}M(a^2+c^2) & 0 \\ 0 & 0 & \frac{1}{5}M(a^2+b^2) \end{bmatrix}.$$

**Exercise 1.14**
Show that if a body is two- or three-dimensional, meaning that there is no line that contains all of the mass of the body, then its moment of inertia matrix is always invertible.

**Exercise 1.15**
Check the equivalence of the vector and scalar forms of Euler's equation.

## Solutions to selected exercises

**Solution to Exercise 1.2** Let $U(r) = 1/r$. Then, the equation of motion reads

$$\frac{1}{2}\left(\dot{r}^2 + \frac{C^2}{r^2}\right) - \frac{1}{r} = E. \tag{1.53}$$

Integrating with respect to $t$ gives a complicated quadrature. Therefore, we integrate with respect to $\theta(t)$ by making the following transformation:

$$\frac{\mathrm{d}}{\mathrm{d}t} = \dot{\theta}\frac{\mathrm{d}}{\mathrm{d}\theta}.$$

Since we are studying a central force problem, angular momentum is conserved, in fact $r^2\dot{\theta} = C$. It is also convenient to introduce the coordinate change $1/r = u$. The transformation of the derivative $\dot{r}$ to $u'$, where $'$ denotes differentiation with respect to $\theta$, is given by:

$$\dot{r} = \dot{\theta}r' = \frac{Cr'}{r^2} = -Cu'.$$

Equation (1.53) transforms into an equation involving $\theta$ as follows:

$$\dot{r}^2 + \frac{C^2}{r^2} - \frac{2}{r} = C^2(u')^2 + u^2C^2 - 2u = 2E.$$

Therefore,

$$
\begin{aligned}
(u')^2 &= \frac{2E}{C^2} + \frac{2u}{C^2} - u^2 \\
&= \frac{2EC^2+1}{C^4} - \left(\frac{1}{C^4} - \frac{2u}{C^2} + u^2\right) \\
&= \frac{2EC^2+1}{C^4} - \left(u - \frac{1}{C^2}\right)^2.
\end{aligned}
$$

This leads us to introduce $v = (u - 1/C^2)$, noting that $v' = u'$. In terms of $v$, eqn (1.53) becomes

$$(v')^2 = \left(\frac{e}{C^2}\right)^2 - v^2,$$

where

$$e^2 = 2EC^2 + 1 > 0.$$

The solution to this differential equation is easily seen to be

$$v = \frac{e}{C^2}\cos(\theta - \theta_0).$$

Finally, $r$ in terms of $\theta$ is:

$$r = \frac{1}{u} = \frac{C^2}{C^2 v + 1} = \frac{C^2}{e\cos(\theta - \theta_0) + 1}.$$

This is the equation for a conic section.

**Solution to Exercise 1.4**
a) Since our system is a 1-particle system, $V = V(\boldsymbol{q})$. Translational invariance then gives $V(\boldsymbol{q}) = V(\boldsymbol{q} - \boldsymbol{q}) = V(0) =$ const.
b) For each $\boldsymbol{q}$, let $R_{\boldsymbol{q}}$ be a rotation such that $R_{\boldsymbol{q}}\widehat{\boldsymbol{q}} = \mathrm{e}$ for some constant unit vector, e. Now consider a 1-particle system with potential $V = V(\boldsymbol{q})$. Rotational invariance gives us $V(\boldsymbol{q}) = V(R_{\boldsymbol{q}}\boldsymbol{q}) = V(\|\boldsymbol{q}\|\mathrm{e})$. Therefore, we can always express the potential as $V = V(\|\boldsymbol{q}\|)$. The converse is trivial.
c) For a 2-particle system, $V = V(\boldsymbol{q}_1, \boldsymbol{q}_2)$. Translational invariance gives us $V(\boldsymbol{q}_1, \boldsymbol{q}_2) = V(0, \boldsymbol{q}_2 - \boldsymbol{q}_1)$. Therefore, $V = V(\boldsymbol{q}_2 - \boldsymbol{q}_1)$. Now applying part b) yields $V = V(\|\boldsymbol{q}_2 - \boldsymbol{q}_1\|)$. Again, the converse is trivial.

**Solution to Exercise 1.5** The problem is best set up in Cartesian coordinates $\boldsymbol{q} = (x, y, z)$ such that $\boldsymbol{B} = (0, 0, B)$. Let the initial position and velocity of the point mass be $(x_0, y_0, z_0)$ and $(v_x, v_y, v_z)$, respectively. To calculate $z(t)$ we observe that

$$\begin{aligned}\frac{\mathrm{d}\dot{z}}{\mathrm{d}t} &= \frac{\mathrm{d}}{\mathrm{d}t}\left(\dot{\boldsymbol{q}}\cdot(0,0,1)\right) = \ddot{\boldsymbol{q}}\cdot(0,0,1)\\ &= \frac{1}{m}\left(e\dot{\boldsymbol{q}}\times(0,0,B)\right)\cdot(0,0,1) = 0.\end{aligned}$$

Therefore $\dot{z} = v_z$ is constant, and $z(t) = v_z t + z_0$. Calculating the equations of motion for $\dot{x}$ and $\dot{y}$ yields

$$\begin{aligned}m\ddot{\boldsymbol{q}} &= m\left(\ddot{x},\ddot{y},0\right) = e\left(\dot{\boldsymbol{q}}\times\boldsymbol{B}\right)\\ &= eB\left(\dot{x},\dot{y},v_z\right)\times(0,0,1) = eB\left(\dot{y},-\dot{x},0\right).\end{aligned}$$

Integrating these equations gives,

$$\begin{bmatrix}\dot{x}\\ \dot{y}\end{bmatrix} = \begin{bmatrix}\cos(Ct) & \sin(Ct)\\ -\sin(Ct) & \cos(Ct)\end{bmatrix}\begin{bmatrix}v_x\\ v_y\end{bmatrix},$$

where

$$C = \frac{eB}{m}.$$

Integrating a second time gives

$$\begin{bmatrix}x\\ y\end{bmatrix} = \begin{bmatrix}x_0 + v_y/C\\ y_0 - v_x/C\end{bmatrix} + \frac{1}{C}\begin{bmatrix}\cos\left(Ct-\frac{\pi}{2}\right) & \sin\left(Ct-\frac{\pi}{2}\right)\\ -\sin\left(Ct-\frac{\pi}{2}\right) & \cos\left(Ct-\frac{\pi}{2}\right)\end{bmatrix}\begin{bmatrix}v_x\\ v_y\end{bmatrix}.$$

Putting all the solutions together yields the equation

$$\boldsymbol{q}-\boldsymbol{q}_0 = \frac{1}{C}\begin{bmatrix}v_y\\ -v_x\\ 0\end{bmatrix} = \frac{1}{C}\begin{bmatrix}\cos\left(Ct-\frac{\pi}{2}\right) & \sin\left(Ct-\frac{\pi}{2}\right) & 0\\ -\sin\left(Ct-\frac{\pi}{2}\right) & \cos\left(Ct-\frac{\pi}{2}\right) & 0\\ 0 & 0 & Ct\end{bmatrix}\begin{bmatrix}v_x\\ v_y\\ v_z\end{bmatrix},$$

which is a helix along the $z$-axis of square radius $(v_x^2 + v_y^2)/C^2$.

**Solution to Exercise 1.6** For a central force problem, $V = V(\|\boldsymbol{q}\|)$. The kinetic energy in polar coordinates $(r,\theta) \in \mathbb{R}^2$ is

$$K = \frac{m}{2}\|\dot{\boldsymbol{q}}\| = \frac{m}{2}\left(\dot{r}^2 + r^2\dot{\theta}^2\right).$$

Therefore, the Lagrangian is

$$L = K - V = \frac{m}{2}\left(\dot{r}^2 + r^2\dot{\theta}^2\right) - V(r).$$

Clearly the Lagrangian does not depend on $\theta$. The Euler–Lagrange equation for $\theta$ gives

$$\frac{\mathrm{d}}{\mathrm{d}t}\frac{\partial L}{\partial \dot{\theta}} - \frac{\partial L}{\partial \theta} = \frac{\mathrm{d}}{\mathrm{d}t}\frac{\partial L}{\partial \dot{\theta}} = \frac{\mathrm{d}}{\mathrm{d}t}\left(mr^2\dot{\theta}\right) = 0.$$

Thus $mr^2\dot{\theta}$ is a conserved quantity.

**Solution to Exercise 1.12** Let $\boldsymbol{x}' = \boldsymbol{P}\boldsymbol{x}$ for some orthogonal $3 \times 3$ matrix $\boldsymbol{P}$. The defining relationship for the angular velocity, $\boldsymbol{\omega}$, is $\dot{\boldsymbol{x}} = \boldsymbol{\omega} \times \boldsymbol{x}$. Initiating the change of coordinates using the formula derived in Exercise 1.11 gives

$$\begin{aligned}\dot{\boldsymbol{x}}' &= \boldsymbol{P}\dot{\boldsymbol{x}} = \boldsymbol{P}\left(\boldsymbol{\omega} \times \boldsymbol{x}\right) \\ &= \boldsymbol{P}\boldsymbol{\omega} \times \boldsymbol{P}\boldsymbol{x} = \boldsymbol{P}\boldsymbol{\omega} \times \boldsymbol{x}' = \boldsymbol{\omega}' \times \boldsymbol{x}'.\end{aligned}$$

Therefore, $\boldsymbol{\omega}' = \boldsymbol{P}\boldsymbol{\omega}$.

# 2 Manifolds

A *manifold* is basically a smooth curve or surface or higher-dimensional analogue. The key property of a manifold is the existence of consistent local coordinate charts, for example spherical coordinates on a sphere, which allow us to do calculus on manifolds just as in Euclidean spaces $\mathbb{R}^n$. Manifolds appear in many contexts in mechanics, including as configuration spaces. For example, the unit sphere $S^2 = \{(x, y, z) : x^2 + y^2 + z^2 = 1\}$ is the configuration space of a spherical pendulum. The sphere is a *submanifold* of $\mathbb{R}^3$. Abstract manifolds can also be defined that are not submanifolds of $\mathbb{R}^n$. Abstract manifolds are important in mechanics as *reduced spaces*, obtained from phase spaces by identifying points via a symmetry.

In Part I, we consider only finite-dimensional manifolds. Throughout this book, 'smooth' will mean $C^\infty$.

## 2.1 Submanifolds of $\mathbb{R}^n$

Recall that the ***graph*** of $y = f(x)$, for some function $f\colon \mathbb{R} \to \mathbb{R}$, is the set of points $\{(x, f(x)) : x \in \mathbb{R}\}$. The graph of $x = f(y)$ is $\{(f(y), y) : y \in \mathbb{R}\}$. Similarly, the graph of $y = f(x, z)$, for some function $f\colon A \to \mathbb{R}$, with $A$ a subset of $\mathbb{R}^2$, is

$$\{(x, f(x, z), z) : (x, z) \in A\}.$$

A general definition of 'graph' is awkward for notational reasons, but the idea should be clear from the preceding examples.

A submanifold is the union of one or more graphs of functions. The following definition ensures that these graphs fit together smoothly. Recall that a ***neighbourhood*** of a point $\mathbf{a} \in \mathbb{R}^n$ is an open set $U$ containing $\mathbf{a}$.

**Definition 2.1** *An $m$-**dimensional (embedded) submanifold** of $\mathbb{R}^n$ is a subset $M$ of $\mathbb{R}^n$ such that, for every $\mathbf{a} \in M$, there is a neighbourhood $U$ of $\mathbf{a}$ such that $M \cap U$ is the graph of some smooth function (with an open domain)*[1]

[1] We will assume that, for a function to be smooth, or even just differentiable, its domain must be an open set. This is usually part of the definition of differentiability.

*expressing $(n-m)$ of the standard coordinates in terms of the other $m$. The* **codimension** *of $M$ is $n-m$. Special cases: by definition, any non-empty open subset of $\mathbb{R}^n$ is an $n$-dimensional submanifold of $\mathbb{R}^n$; and any non-empty subset of $\mathbb{R}^n$ containing only isolated points is a 0-dimensional submanifold of $\mathbb{R}^n$.*

**Example 2.2** $S^1$ is a submanifold of $\mathbb{R}^2$, because the top and bottom semicircles are graphs of the functions $y = \pm\sqrt{1-x^2}$, with domain $(-1, 1)$, and the left and right semicircles are the graphs of $x = \pm\sqrt{1-y^2}$, with domain $(-1, 1)$.

**Example 2.3** Any line segment in any $\mathbb{R}^n$, without its endpoints, is a one-dimensional submanifold of $\mathbb{R}^n$; but the same segment *with* its endpoints is not (it is a 'manifold with boundary', which we will not define).

**Definition 2.4** *For any subset $M$ of $\mathbb{R}^n$, subsets of $M$ of the form $M \cap U$, with $U$ open, are called* **open relative to** *$M$, or just* **relatively open** *(if $M$ is understood).*

**Remark 2.5** *If $M$ is a submanifold of $\mathbb{R}^n$, and $N$ is open relative to $M$, then it follows easily from the definitions that $N$ is a submanifold of $\mathbb{R}^n$, of the same dimension as $M$.*

**Remark 2.6** *The significance of the name 'embedded submanifold' will become apparent later – see Theorem 2.32 and the remarks that precede it. There are also immersed submanifolds (see Remark 2.27). In this book 'submanifold', without qualification, means embedded submanifold.*

Manifolds are often described as level sets, for example $S^1 = \{(x, y) \in \mathbb{R}^2 : x^2 + y^2 = 1\}$, which is $f^{-1}(1)$, where $f(x, y) = x^2 + y^2$. Configuration spaces in mechanical problems are typically defined in this way, by one or more constraints: $f_1 = c_1, \ldots, f_k = c_k$, for some smooth functions $f_i : \mathbb{R}^n \to \mathbb{R}$. We now discuss conditions under which level sets are submanifolds of $\mathbb{R}^n$.

**Definition 2.7** *Given $f \colon A \to B$, The* **image** *of $D \subseteq A$ is $f(D) := \{f(x) \in B : x \in D\}$. The* **image** *of $f$, or* **range** *of $f$, is $\operatorname{Im} f := f(A)$. The* **preimage** *of $C \subseteq B$ is $f^{-1}(C) := \{x \in A : f(x) \in C\}$. (This notation does not imply that $f$ is an invertible function). For sets containing only one point, $f^{-1}(c) := f^{-1}(\{c\}) = \{x \in A : f(x) = c\}$. A* **level set** *of $f$ is $f^{-1}(c)$ for some $c \in B$. This set is also sometimes called the* **locus** *of the equation $f(\mathbf{x}) = \mathbf{c}$.*

**Example 2.8** Let $f(x, y) = xy$. The level sets $f(x, y) = c$, for $c \neq 0$, are the hyperbolae $y = c/x$, which are submanifolds. At each point $(x, y)$, the gradient vectors $\nabla f(x, y) = (y, x)$ is perpendicular to the hyperbola passing through $(x, y)$. The level set $f(x, y) = 0$ is the union of the $x$- and $y$-axes. and this set is not a submanifold, because no function has a graph that crosses itself. The 'trouble point' is the origin, because if it were removed, then each of the remaining four rays would be submanifolds. Note that this is the only point where the gradient vector, $\nabla f(x, y) = (y, x)$, equals zero.

In general, a level set $f(\mathbf{x}) = c$, for a real-valued function $f$, is a submanifold whenever the gradient vectors $\nabla f(\mathbf{x})$ are non-zero at all points $\mathbf{x}$ in the level set. This is a consequence of the Implicit Function Theorem. But before we state that theorem, we consider the general situation in which there may be more than one constraint, $f_1 = c_1, \ldots, f_k = c_k$.

**Definition 2.9** *Let $f_1, \ldots, f_k$ be differentiable functions from $A \subseteq \mathbb{R}^n$ to $\mathbb{R}$. If, for all $\mathbf{x} \in A$, the gradient vectors $\nabla f_1(\mathbf{x}), \ldots, \nabla f_k(\mathbf{x})$ are linearly independent, then $f_1, \ldots, f_k$ are said to be* ***functionally independent****.*

Any set of $k$ functions $f_1, \ldots, f_k : A \to \mathbb{R}$, taken together, define a function $f = (f_1, \ldots, f_k) : A \to \mathbb{R}^k$. Since the rows of the matrix $\mathrm{D}f(\mathbf{x})$ (the derivative, i.e. Jacobian) are $\nabla f_1(\mathbf{x}), \ldots, \nabla f_k(\mathbf{x})$, it follows that the $f_i$s are functionally independent if and only if $\mathrm{D}f(\mathbf{x})$ has rank $k$, for every $\mathbf{x}$, which is equivalent to the linear transformation $\mathrm{D}f(\mathbf{x})$ being a surjective (i.e. onto). Such functions $f$ are called *submersions*:

**Definition 2.10** *A differentiable function $f\colon (A \subseteq \mathbb{R}^n) \to \mathbb{R}^k$ is a* ***submersion at*** *$\mathbf{x} \in A$ if the derivative $\mathrm{D}f(\mathbf{x})$ is surjective. A function that satisfies this condition at all points $\mathbf{x} \in A$ is a* ***submersion****.*

**Remark 2.11** *A function $f\colon (A \subseteq \mathbb{R}^n) \to \mathbb{R}^k$ can only be a submersion if $n \geq k$.*

**Remark 2.12** *The condition that $\mathrm{D}f(\mathbf{x})$ is surjective is equivalent to each of the following conditions (with $n$ and $k$ as above): (i) $\mathrm{Im}\,\mathrm{D}f(\mathbf{x}) = \mathbb{R}^k$; (ii) $\mathrm{rank}\,\mathrm{D}f(\mathbf{x}) = k$; and (iii) $\dim\ker\mathrm{D}f(\mathbf{x}) = n - k$.*

**Remark 2.13** *A real-valued function $f$ is a submersion if and only if $\nabla f(\mathbf{x}) \neq 0$ for every $\mathbf{x}$.*

**Definition 2.14** *A* ***diffeomorphism*** *is a differentiable map with a differentiable inverse.*

**Remark 2.15** *If $f$ is a diffeomorphism, then $\mathrm{D}f(\mathbf{x})$ is invertible, for every $\mathbf{x}$. Thus, all diffeomorphisms are submersions. However, the converse is false. For example $f\colon \mathbb{R}^2 \to \mathbb{R}$, $f(x,y) = x$ is a submersion that is not a diffeomorphism.*

Now, $\mathrm{D}f(\mathbf{a})$ is surjective if and only if $k$ of the columns of $\mathrm{D}f(\mathbf{a})$ are linearly independent, say columns $i_1, \ldots, i_k$, which occurs if and only if

$$\frac{\partial\,(f_1, \ldots, f_k)}{\partial\left(x_{i_1}, \ldots, x_{i_k}\right)}(\mathbf{a}) \neq 0,$$

where this notation denotes the ***Jacobian determinant***, i.e. the determinant of the $k \times k$ matrix formed by columns $i_1, \ldots, i_k$ of $\mathrm{D}f(\mathbf{a})$. When this is true, and $f$ is smooth and $f(\mathbf{a}) = \mathbf{c}$, the Implicit Function Theorem states that, for $\mathbf{x}$ in $f^{-1}(\mathbf{c})$ and near enough to $\mathbf{a}$, it is possible to solve for $x_{i_1}, \ldots, x_{i_k}$ as a smooth function $g$ of the other $x_i$s. An equivalent conclusion is that there exists a neighbourhood $U$ of $\mathbf{a}$ such that $f^{-1}(\mathbf{c}) \cap U$ is the graph of $g$. Thus, the Implicit Function Theorem, which is commonly stated in terms of the Jacobian determinant, can be restated in the following form:

**Theorem 2.16 (Implicit Function Theorem)** *Let $A$ be an open subset of $\mathbb{R}^n$ and let $f\colon A \to \mathbb{R}^k$ be a smooth function. Let $\mathbf{a} \in A$ and let $\mathbf{c} = f(\mathbf{a})$. If $\mathrm{D}f(\mathbf{a})$ is surjective, then there exists a neighbourhood $U$ of $\mathbf{a}$ such that $f^{-1}(\mathbf{c}) \cap U$ is the graph of some smooth function expressing $k$ of the standard variables $x_1, \ldots, x_n$ in terms of the others.*

*In particular, if columns $i_1, \ldots, i_k$ of $\mathrm{D}f(\mathbf{a})$ are linearly independent, then $x_{i_1}, \ldots, x_{i_k}$ can be expressed as a smooth function of the other $x_i$s.*

Informally, we say: '$f^{-1}(\mathbf{c})$ is locally the graph of a smooth function'.

Note that the condition on $\mathrm{D}f$ in the above theorem is that $\mathrm{D}f(\mathbf{a})$ is surjective, and not that $\mathrm{D}f(\mathbf{x})$ is surjective for all $\mathbf{x} \in A$.

**Definition 2.17** *A **regular value** of $f\colon A \to \mathbb{R}^k$ is a $\mathbf{c} \in \mathbb{R}^k$ such that $\mathrm{D}f(\mathbf{x})$ is surjective for all $\mathbf{x} \in f^{-1}(\mathbf{c})$.*

**Corollary 2.18** *If $f\colon (A \subseteq \mathbb{R}^n) \to \mathbb{R}^k$ is smooth and $\mathbf{c}$ is a regular value of $f$, then $f^{-1}(\mathbf{c})$ is a submanifold of $\mathbb{R}^n$, of codimension $k$.*

**Example 2.19** Let $S^1 = f^{-1}(1)$, where $f(x,y) = x^2 + y^2$. Since $\nabla f(x,y) = (2x, 2y)$, which is non-zero when $(x,y) \neq (0,0)$, it follows $\mathrm{D}f(x,y)$ is surjective for all $(x,y) \neq (0,0)$, and in particular, 1 is a regular value of $f$. Thus, the previous corollary implies that $S^1$ is a submanifold of $\mathbb{R}^2$, of dimension 1 and codimension 1.

**Example 2.20** The unit sphere $S^2$ is a submanifold of $\mathbb{R}^3$, of dimension 2 and codimension 1, since $S^2 = g^{-1}(1)$, where $g(x,y,z) = x^2 + y^2 + z^2$, and 1 is a regular value of $g$.

Recall that, for a map to be a submersion, the dimension of its domain must be as least as large as the dimension of its codomain. The following closely related definition applies in the opposite situation.

**Definition 2.21** *A differentiable function $f\colon (A \subseteq \mathbb{R}^m) \to \mathbb{R}^n$ is an* ***immersion at*** $\mathbf{x} \in A$ *if* $\mathrm{D}f(\mathbf{x})$ *is injective (that is, one-to-one). A function that satisfies this condition at all points* $\mathbf{x} \in A$ *is an* ***immersion.*** *Equivalent definitions are, for every* $\mathbf{x} \in A$*:* $\ker \mathrm{D}f(\mathbf{x}) = \{0\}$*;* $\dim \mathrm{Im}\, \mathrm{D}f(\mathbf{x}) = m$*; and* $\operatorname{rank} \mathrm{D}f(\mathbf{x}) = m$*. Note that this is only possible when* $m \le n$*.*

**Remark 2.22** *For both submersions and immersions,* $\operatorname{rank} \mathrm{D}f(\mathbf{x})$ *is as large as possible, given the dimensions of the domain and codomain. Another way of saying this (for both submersions and immersions) is that* $\mathrm{D}f(\mathbf{x})$ *has full rank.*

**Remark 2.23** *All diffeomorphisms are immersions, but the converse is false. For example,* $f\colon \mathbb{R} \to \mathbb{R}^2$, $f(x) = (x,0)$ *is an immersion that is not a diffeomorphism.*

**Example 2.24** If $A$ is an interval in $\mathbb{R}$, then a map $f\colon A \to \mathbb{R}^n$ is a parameterized curve, i.e. a path. In this case, $f$ is an immersion if and only if $f'(t) \neq 0$ for every $t$.

**Remark 2.25** *In the special case of* $m = 2$ *and* $n = 3$*, the definition of an immersion coincides with the definition of a regular surface parameterization in the differential geometry of surfaces (except that such a parameterization might also be required to be injective, i.e. one-to-one). Indeed, using notation common in that subject, if* $\mathbf{x} : \mathbb{R}^2 \to \mathbb{R}^3$*, and the columns of* $\mathrm{D}\mathbf{x}(u,v)$ *are written* $\mathbf{x}_u(u,v)$ *and* $\mathbf{x}_v(u,v)$*, then* $\mathrm{D}\mathbf{x}$ *is injective (at* $(u,v)$*) if and only if* $\mathbf{x}_u$ *and* $\mathbf{x}_v$ *are linearly independent, which is equivalent to* $\mathbf{x}_u \times \mathbf{x}_v \neq 0$*.*

**Example 2.26** Let $\psi : \mathbb{R}^2 \to S^2 \subset \mathbb{R}^3$ be given by

$$\psi(\theta,\phi) = (\sin\phi\cos\theta, \sin\phi\sin\theta, -\cos\phi).$$

Then,

$$\mathrm{D}\psi(\theta,\phi) = \begin{bmatrix} -\sin\phi\sin\theta & \cos\phi\cos\theta \\ \sin\phi\cos\theta & \cos\phi\sin\theta \\ 0 & \sin\phi \end{bmatrix}.$$

If $\phi$ is a multiple of $\pi$ then the first column of this matrix is zero, so $D\psi(\theta, \phi)$ is not injective. Hence, $\psi$ is not an immersion. However, if $\phi$ is not a multiple of $\pi$, then the two columns are linearly independent. Therefore, the restriction of $\psi$ to the domain $\mathbb{R} \times (0, \pi)$ is an immersion. The restriction of $\psi$ to the domain $(-\pi, \pi) \times (0, \pi)$ is an injective immersion. The image of this restricted $\psi$ is $S^2$ minus the 'poles' $(0, 0, \pm 1)$ and minus the semicircle corresponding to $\phi = \pi$ (which connects the two poles). This $\psi$ defines *spherical coordinates* $(\theta, \phi)$ on $S^2$. Note that $\theta$ is 'longitude' and $\phi$ is the angle from the 'South pole' $(0, 0, -1)$.

**Remark 2.27** *The image of an immersion is a kind of submanifold, called an* ***immersed submanifold*** *(a general definition appears later in Definition 2.104). But it need not satisfy Definition 2.1, as the next example shows.*

**Example 2.28** Define $f: \mathbb{R} \to \mathbb{R}^2$ by $f(t) = (2\cos(t - \pi/2), \sin 2(t - \pi/2))$. This is an immersion – see Exercise 2.8, and its restriction to $(0, 2\pi)$ is injective. It's image $L := \operatorname{Im} f$ is called a ***lemniscate***, and is shaped like a figure '8'. The lemniscate is not a submanifold according to Definition 2.1, because its image crosses itself at the origin, and no graph of a function can cross itself. What 'goes wrong' is that $f$ maps ends of the open interval $(0, 2\pi)$ arbitrarily close to $f(\pi)$, so $f^{-1} : L \to (0, 2\pi)$ is not continuous. For example, if $x_n = \frac{1}{n}$ for all $n \in \mathbb{N}$, then $f(x_n) \to (0, 0) = f(\pi)$ but $x_n \to 0 \neq \pi$.

**Definition 2.29** *An* ***embedding*** *is an immersion $\psi : (U \subseteq \mathbb{R}^m) \to \mathbb{R}^n$ such that $\psi^{-1} : \psi(U) \to U$ is continuous.*

**Remark 2.30** *Continuity of $\psi^{-1} : \psi(U) \to U$ is defined in the usual '$\varepsilon$–$\delta$' way, or equivalently via sequences, as in the preceding and following examples. Another equivalent definition is that $\psi^{-1}$ is continuous if and only if, for every open set $W \subseteq U$, the set $\psi(W)$ is relatively open in $\psi(U)$ (see Definition 2.4 and Exercise 2.4).*

**Example 2.31** Let $\psi : \mathbb{R} \to \mathbb{R}^2$ be given by $\psi(\theta) = (\cos\theta, \sin\theta)$. Then, $\psi'(\theta) = (-\sin\theta, \cos\theta)$, which never equals $(0, 0)$, so $\psi$ is an immersion. Its image is $S^1$. Since $\psi$ is not injective, it is not an embedding. However, the restriction of $\psi$ to $(0, 2\pi)$ is injective, and its inverse $\psi^{-1} : \left(S^1 \setminus \{(1, 0)\}\right) \to (0, 2\pi)$ is continuous. Indeed, suppose that $\theta_1, \theta_2, \ldots$ is a sequence in $(0, 2\pi)$ such that $\psi(\theta_i)$ tends to $\psi(\theta_0)$, for some $\theta_0 \in (0, 2\pi)$. Then $\cos\theta_i \to \cos\theta_0$ and $\sin\theta_i \to \sin\theta_0$. Since $\theta_i \in (0, 2\pi)$ for all $i$, this implies that $\theta_i \to \theta_0$. Thus, $\psi^{-1}$ is continuous, and hence $\psi$ is an embedding.

In the following theorem, Condition 2 is essentially Definition 2.1, so the three conditions are equivalent alternative definitions of an $m$-dimensional (embedded) submanifold of $\mathbb{R}^n$. The name 'embedded submanifold' comes from the role of embeddings in Condition 3.

**Theorem 2.32 (Equivalent definitions of a submanifold of $\mathbb{R}^n$)**
*Let $M \subseteq \mathbb{R}^n$, and let $0 < m < n$ and $k = n - m$. The following are equivalent.*

1. *For every $\mathbf{a} \in M$ there exists a neighbourhood $U$ of $\mathbf{a}$, a smooth submersion $f: U \to \mathbb{R}^k$ and a $\mathbf{c} \in \mathbb{R}^k$ such that $M \cap U = f^{-1}(\mathbf{c})$.*
2. *For every $\mathbf{a} \in M$ there exists a neighbourhood $U$ of $\mathbf{a}$ such $M \cap U$ is the graph of a smooth function expressing $k$ of the standard coordinates in terms of the others.*
3. *For every $\mathbf{a} \in M$ there exists a neighbourhood $U$ of $\mathbf{a}$ and a smooth embedding $\psi : (V \subseteq \mathbb{R}^m) \to \mathbb{R}^n$ such that $\psi(V) = M \cap U$.*

**Proof** *Sketch*[2]
(1) $\Rightarrow$ (2). Apply the Implicit Function Theorem (Theorem 2.16) (the $U$s may have to shrink).
(2) $\Rightarrow$ (1) and (3). Suppose that $M \cap U$ is the graph of a smooth function $g$, and suppose, without loss of generality, that $U = U_1 \times U_2$, for some $U_1 \subseteq \mathbb{R}^m$ and $U_2 \subseteq \mathbb{R}^k$, and that $M \cap U = \{(\mathbf{x}, g(\mathbf{x})) : \mathbf{x} \in U_1\}$. Define $f: U \to \mathbb{R}^k$ by $f(\mathbf{x}, \mathbf{z}) = \mathbf{z} - g(\mathbf{x})$. Then, $f$ is a smooth submersion and $f^{-1}(0) = M \cap U$. This proves Condition 1. Let $V = U_1$ and defined $\psi : V \to \mathbb{R}^n$ by $\psi(\mathbf{x}) = (\mathbf{x}, g(\mathbf{x}))$. Then, $\psi$ is a smooth immersion and $\psi^{-1} : M \cap U \to V$ is continuous because it equals projection onto the first $m$ coordinates. This proves Condition 3.
(3) $\Rightarrow$ (2). Let $\psi : (V \subseteq \mathbb{R}^m) \to \mathbb{R}^n$ be as in Condition 3, and let $\mathbf{b} = \psi^{-1}(\mathbf{a})$. The rank of $\mathrm{D}\psi(\mathbf{b})$ is $m$, so the image of $\mathrm{D}\psi(\mathbf{b})$ cannot contain more than $m$ of the standard basis vectors in $\mathbb{R}^n$. Without loss of generality, suppose that it does not contain any of the last $k$ standard basis vectors. Writing $\pi : \mathbb{R}^n \to \mathbb{R}^m$ for projection onto the first $m$ coordinates, this means that the $\mathrm{Im}\,\mathrm{D}\psi(\mathbf{b}) \cap \ker \pi = \{0\}$. It follows from the rank-nullity theorem that the restriction of $\pi$ to $\mathrm{Im}\,\mathrm{D}\psi(\mathbf{b})$ has rank $m$. Hence, $\mathrm{D}(\pi \circ \psi)(\mathbf{b}) : \mathbb{R}^m \to \mathbb{R}^m$ has rank $m$. By the Inverse Function Theorem, there exists an open neighbourhood $W_1$ of $\mathbf{b}$ such that the restriction of $\pi \circ \psi$ to $W_1$ has a smooth inverse. Since $\psi^{-1} : M \cap U \to V$ is continuous, $\psi(W_1)$ is relatively open in $M \cap U$, so there exists a $U' \subseteq U$ such that $\psi(W_1) = M \cap U'$. Let $W_2 = \pi \circ \psi(W_1) = \pi(M \cap U')$, and let $h = \psi \circ (\pi \circ \psi)^{-1} : W_2 \to \mathbb{R}^n$. Note that $\pi \circ h$ is the identity on $W_2$, and that the components of $\pi \circ h$ are the

[2] Our proof essentially follows [BG88].

first $m$ components of $h$. Let $g : W_2 \to \mathbb{R}^k$ be the last $k$ components of $h$. Then, $g$ is smooth and its graph is $\{(\mathbf{x}, g(\mathbf{x})) : \mathbf{x} \in W_2\} = \{h(\mathbf{x}) : \mathbf{x} \in W_2\} = h(W_2) = \psi(W_1) = M \cap U'$. ∎

**Remark 2.33** *From the proof of the previous theorem, and the definitions of regular value, submersion and immersion, it follows that Conditions 1 and 3 of the theorem can be restated in the following forms which, while* a priori *weaker, are in fact equivalent:*

1′. *For every* $\mathbf{a} \in M$ *there exists a neighbourhood* $U$ *of* $\mathbf{a}$*, a smooth function* $f: U \to \mathbb{R}^k$ *and a* $\mathbf{c} \in \mathbb{R}^k$ *such that* $f$ *is a submersion at* $\mathbf{a}$ *and* $M \cap U = f^{-1}(\mathbf{c})$.
1″. *For every* $\mathbf{a} \in M$ *there exists a neighbourhood* $U$ *of* $\mathbf{a}$*, a smooth function* $f: U \to \mathbb{R}^k$ *and a* $\mathbf{c} \in \mathbb{R}^k$ *such that* $\mathbf{c}$ *is a regular value of* $f$ *and* $M \cap U = f^{-1}(\mathbf{c})$.
3′. *For every* $\mathbf{a} \in M$ *there exists a neighbourhood* $U$ *of* $\mathbf{a}$ *and a smooth map* $\psi : (V \subseteq \mathbb{R}^m) \to \mathbb{R}^n$ *such that* $\psi(V) = M \cap U$ *and* $\psi^{-1} : M \cap U \to V$ *is continuous and* $\psi$ *is an immersion at* $\psi^{-1}(\mathbf{a})$.

**Example 2.34** Consider the unit circle $S^1$ in $\mathbb{R}^2$. We verify directly that $S^1$ satisfies each of the conditions in the preceding theorem.

1. $S^1 = f^{-1}(1)$, where $f(x, y) = x^2 + y^2$. This $f$, restricted to domain $\mathbb{R}^2 \setminus \{(0, 0)\}$, is a submersion.
2. This was shown in Example 2.2.
3. The map $\psi : (0, 2\pi) \to \mathbb{R}^2$, $\psi(\theta) = (\cos\theta, \sin\theta)$, is an embedding with image $S^1$ (see Example 2.31). The same would be true if $\psi$ were given any other domain that is an open interval of length less than or equal to $2\pi$. Since $S^1$ is covered by the images of such embeddings, Conditions 3 is satisfied.

The most commonly encountered manifolds are each the level set of a single submersion, so the reader may wonder whether Conditions 1 and 2 can be made simpler by removing the '$M \cap U$' part. However, there do exist manifolds that are *not* the level set of any single submersion. These manifolds are *non-orientable*. We will not define orientability here, but a familiar example of a non-orientable manifold is the Möbius strip.

Condition 1 of Theorem 2.32, which uses level sets, is very useful for proving that a set is a submanifold, as in the proof of the next theorem. Condition 2, which concerns graphs of functions, is intuitively appealing, and is useful for proving that some sets are *not* submanifolds, as in

Example 2.8. Condition 3 is the most natural from the point of view of intrinsic differential geometry.

**Theorem 2.35 (Product manifolds)** *Let $M_1$ be an $m_1$-dimensional submanifold of $\mathbb{R}^{n_1}$, and $M_2$ be an $m_2$-dimensional submanifold of $\mathbb{R}^{n_2}$. Then, $M_1 \times M_2$ is an $(m_1 + m_2)$-dimensional submanifold of $\mathbb{R}^{n_1+n_2}$.*

*Proof* Let $(\mathbf{a}_1, \mathbf{a}_2) \in M_1 \times M_2$, and let $k_1 = n_1 - m_1$ and $k_2 = n_2 - m_2$. By Condition 1a of the preceding theorem, there exist, for $i = 1, 2$, a neighbourhood $U_i$ of $\mathbf{a}_i$, a smooth submersion $f_i : U_i \to \mathbb{R}^{k_i}$ and a $\mathbf{c}_i \in \mathbb{R}^{k_i}$ such that $M_i \cap U_i = f^{-1}(\mathbf{c}_i)$. Let $U = U_1 \times U_2$ and $\mathbf{c} = (\mathbf{c}_1, \mathbf{c}_2)$, and define $f \colon U \to \mathbb{R}^{k_1+k_2}$ by $f(\mathbf{x}_1, \mathbf{x}_2) = (f_1(\mathbf{x}_1), f_2(\mathbf{x}_2))$. It is easiliy verified that $f$ is a smooth submersion and $f^{-1}(\mathbf{c}) = (M_1 \times M_2) \cap (U_1 \times U_2)$. ■

**Example 2.36** The ***torus*** $T^2$, defined as $S^1 \times S^1$, is a 2-dimensional submanifold of $\mathbb{R}^4$. See also Exercises 2.5 and 2.38.

**Definition 2.37** *An embedding $\psi$ as in Condition 3 in the preceding theorem is a **parameterization** of the manifold $M$. If the domain and image of $\psi$ are $V$ and $M \cap U$, then the inverse $\varphi := \psi^{-1} : M \cap U \to V$ is a **coordinate chart**, and the components of $\varphi$ are called **local coordinates**. Given two different coordinate charts $\varphi_1$ and $\varphi_2$ with overlapping domains, the map $\varphi_2 \circ \varphi_1^{-1}$ is called a **transition function** or **change-of-coordinates transformation**.*

**Remark 2.38** *By definition, every coordinate chart is continuous and has a continuous inverse. A continuous map with a continuous inverse is called a **homeomorphism**. Since the domain of a coordinate chart is not an open subset of Euclidean space, the elementary definition of differentiability does not apply (but see Remark 2.45).*

**Theorem 2.39** *If $\varphi_1$ and $\varphi_2$ are coordinate charts on a submanifold $M$ of $\mathbb{R}^n$, with overlapping domains, then the transition function $\varphi_2 \circ \varphi_1^{-1}$ is a smooth diffeomorphism with a smooth inverse.*

*Proof* Let $M$ be an $m$-dimensional submanifold of $\mathbb{R}^n$. Let $\varphi_i : M \cap U_i \to (V_i \subseteq \mathbb{R}^m)$ be coordinate charts for $M$, for $i = 1, 2$, and suppose that $U_1 \cap U_2$ is non-empty. The transition function $\varphi_2 \circ \varphi_1^{-1}$ is well defined on the domain $\varphi_1(U_1 \cap U_2)$, which is open since $\varphi_1$ is the inverse of an embedding.

Let $\psi_2 = \varphi_2^{-1}$. By definition, $\psi_2 : V_2 \to M \cap U_2$ is an embedding and $\varphi_2^{-1} = \psi_2 : M \cap U_2 \to V_2$ is continuous. The last part of the proof of Theorem 2.32 shows that there is a projection $\pi : \mathbb{R}^n \to \mathbb{R}^m$ such that

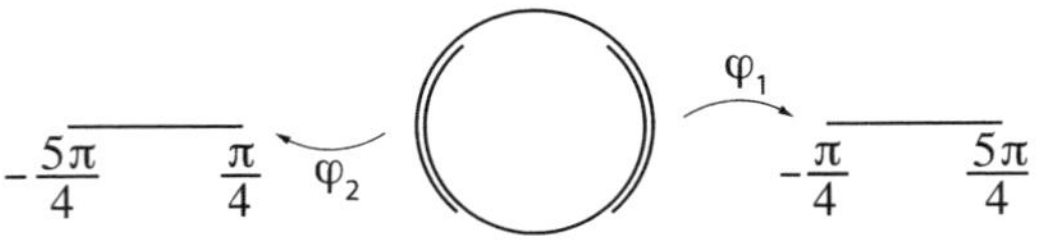


**Figure 2.1** Two coordinate charts for $S^1$

(shrinking $U_2$ and $V_2$ if necessary) $\pi \circ \psi_2$ has a smooth inverse. Now,

$$\varphi_2 \circ \varphi_1^{-1} = \psi_2^{-1} \circ \varphi_1^{-1} = (\pi \circ \psi_2)^{-1} \circ \pi \circ \varphi_1^{-1},$$

from which it follows that $\varphi_2 \circ \varphi_1^{-1}$ is smooth. The same argument, with the roles of $\varphi_1$ and $\varphi_2$ reversed, shows that $\varphi_1 \circ \varphi_2^{-1}$ is smooth, which completes the proof. ■

**Example 2.40** Consider the following two coordinate charts for the unit circle $S^1$, illustrated in Figure 2.1,

$$\varphi_1 : U_1 \longrightarrow \left(-\frac{\pi}{4}, \frac{5\pi}{4}\right), \qquad (\cos\theta, \sin\theta) \longmapsto \theta,$$

and

$$\varphi_2 : U_2 \longrightarrow \left(-\frac{5\pi}{4}, \frac{\pi}{4}\right), \qquad (\cos\theta, \sin\theta) \longmapsto \theta,$$

where $U_1 = S^1 \cap \{(x, y) : y > -|x|\}$ and $U_2 = S^1 \cap \{(x, y) : y < |x|\}$. These are coordinate charts because their inverses are parameterizations – see Example 2.34. The change of coordinates is given by the transition function $\varphi_2 \circ \varphi_1^{-1}$, which is well defined on $\varphi_1 (U_1 \cap U_2) = \left(-\frac{\pi}{4}, \frac{\pi}{4}\right) \cup \left(\frac{3\pi}{4}, \frac{5\pi}{4}\right)$. The smoothness of the transition function is guaranteed by Theorem 2.39, but can also be checked directly, since $\varphi_2 \circ \varphi_1^{-1}(\theta) = \theta$ for all $\theta \in \left(-\frac{\pi}{4}, \frac{\pi}{4}\right)$ and $\varphi_2 \circ \varphi_1^{-1}(\theta) = \theta - 2\pi$ for all $\theta \in \left(\frac{3\pi}{4}, \frac{5\pi}{4}\right)$.

**Definition 2.41** *Let $f\colon M \to N$. Given any coordinate charts $\varphi_1$ and $\varphi_2$ for $M$ and $N$, respectively, the* ***representation of $f$ in local coordinates*** *is*

$$\varphi_2 \circ f \circ \varphi_1^{-1}$$

*(with domain restricted so that this is well defined).*

Though potentially confusing, $\varphi_2 \circ f \circ \varphi_1^{-1}$ is often called just '$f$', as long as the choice of coordinates is clear from the context.

The usual definitions of 'differentiable' and 'smooth' do not apply to general maps $f$ between manifolds, since the domain of $f$ need not be an open subset of Euclidean space. One way to remedy this is to use local representations $\varphi_2 \circ f \circ \varphi_1^{-1}$, because their domains and codomains are always open subsets of Euclidean spaces, so the usual definitions *do* apply.

**Definition 2.42** *Let $M$ be a submanifold of $\mathbb{R}^p$ and $N$ a submanifold of $\mathbb{R}^s$. The map $f$ is* ***differentiable at*** $\mathbf{a} \in M$ *if for every coordinate chart $\varphi_1$ of $M$ with domain containing* $\mathbf{a}$, *and every coordinate chart $\varphi_2$ of $N$ with domain containing $f(\mathbf{a})$, the composition $\varphi_2 \circ f \circ \varphi_1^{-1}$ is differentiable at $\varphi_1(\mathbf{a})$. A map $f\colon M \to N$ is* ***differentiable*** *if it is differentiable at all points in its domain, i.e. if every representation of $f$ in local coordinates is differentiable. The definition of* ***smooth*** *is analagous. The map $f$ is a* ***diffeomorphism*** *if $f$ is differentiable and has a differentiable inverse.*

**Remark 2.43** *In fact, though the preceding definition says 'for every coordinate chart', it suffices to check just enough charts to cover the two manifolds $M$ and $N$, because of Theorem 2.39 – see Exercise 2.10.*

**Remark 2.44** *It is easily checked that the composition of two differentiable (resp. smooth) maps is differentiable (resp. smooth). Also, since every coordinate chart is continuous and has a continuous inverse, every differentiable map is continuous.*

**Remark 2.45** *Every coordinate chart is smooth in the sense of the preceding definition. Indeed, let $\varphi : M \cap U \to (V \subseteq \mathbb{R}^m)$ be a coordinate chart for $M$. The identity map $\mathrm{id}_V : V \to V$ is a coordinate chart for $V$, and the representation of $\varphi$ in coordinate charts $\varphi$ and $\mathrm{id}_V$ is $\mathrm{id}_V \circ \varphi \circ \varphi^{-1} = \mathrm{id}_V$, which is clearly smooth.*

*In conclusion, if $\psi : V \to \mathbb{R}^p$ is a parameterization of $M$ with image $M \cap U$, then both $\psi$ and $\psi^{-1} : M \cap U \to V$ are smooth. In this sense, $\psi$ is a (smooth) diffeomorphism; but since $\psi$ is not surjective, we say instead that $\psi$ is a 'diffeomorphism onto its image'. Since the image of every embedding $\psi$ is a submanifold with parameterization $\psi$, we have shown that every embedding is a diffeomorphism onto its image. Indeed, an equivalent definition of an embedding is: an immersion that is a diffeomorphism onto its image.*

**Theorem 2.46** *Let $M$ be a submanifold of $\mathbb{R}^p$ and $N$ a submanifold of $\mathbb{R}^s$. If $f\colon M \to N$ is the restriction of a differentiable (resp. smooth) map $f\colon U \to \mathbb{R}^s$, where $U$ is an open neighbourhood of $M$ in $\mathbb{R}^p$, then $f$ is differentiable (resp. smooth).*

*Proof* Let $f$ and $F$ be as in the statement of the theorem, and let $\varphi_1$ and $\varphi_2$ be coordinate charts for $M$ and $N$, respectively. Since

$$\varphi_2 \circ f \circ \varphi_1^{-1} = \varphi_2 \circ F \circ \varphi_1^{-1}$$

and $\varphi_2$ and $\varphi^{-1}$ are both smooth, the result follows. ■

**Example 2.47** Let $f\colon \mathbb{R}^3 \to \mathbb{R}^3$ be the linear transformation with matrix

$$\begin{bmatrix} \cos\alpha & -\sin\alpha & 0 \\ \sin\alpha & \cos\alpha & 0 \\ 0 & 0 & 1 \end{bmatrix},$$

which is a rotation by an angle $\alpha$ around the $z$-axis. Since $F$ is linear, it is a smooth. It is easily verified that $F$ maps every $\mathbf{x} \in S^2$ to another point in $S^2$, so $F$ can be restricted to a map $f\colon S^2 \to S^2$, which is smooth by Theorem 2.46.

We can also check directly that $f$ is smooth, using Definition 2.42. Let $\psi$ be a parameterization for $S^2$ with formula

$$\psi(\theta, \phi) = (\sin\phi\cos\theta, \sin\phi\sin\theta, -\cos\phi)$$

and a suitable domain, as in Example 2.26. For every $(\theta, \phi)$ in the domain of $\psi$, if $f(\theta, \phi)$ is also in the domain of $\psi$, then we have

$$\begin{aligned} \psi^{-1} \circ f \circ \psi(\theta, \phi) &= \psi^{-1} \circ f(\sin\phi\cos\theta, \sin\phi\sin\theta, -\cos\phi) \\ &= \psi^{-1}(\sin\phi\cos(\theta+\alpha), \sin\phi\sin(\theta+\alpha), -\cos\phi) \\ &= (\theta+\alpha, \phi). \end{aligned}$$

Thus, the expression for $f$ in the 'usual' spherical coordinates is $f(\theta, \phi) = (\theta + \alpha, \phi)$. This is clearly smooth. To finish the proof, we need to consider enough parameterizations to cover the entire sphere. Smoothness at the poles may be shown in either polar or stereographic coordinates. The latter are defined in Example 2.49 below.

**Example 2.48** The formula $\xi(x, y) = 2x/(1 - y)$ defines a map called a ***stereographic projection***, with domain $S^1$ minus the 'North pole' $(0, 1)$, and range $\mathbb{R}$. It can be easily checked that $\xi$ is the $x$-coordinate of the point on the line $y = -1$, illustrated in Figure 2.2. The map $\xi$ can be visualized as taking every point to its shadow cast by a light shining from the North pole. Note that the domain of $\xi$ is $S^1 \cap U$, where $U = \{(x, y) : y \neq 1\}$,

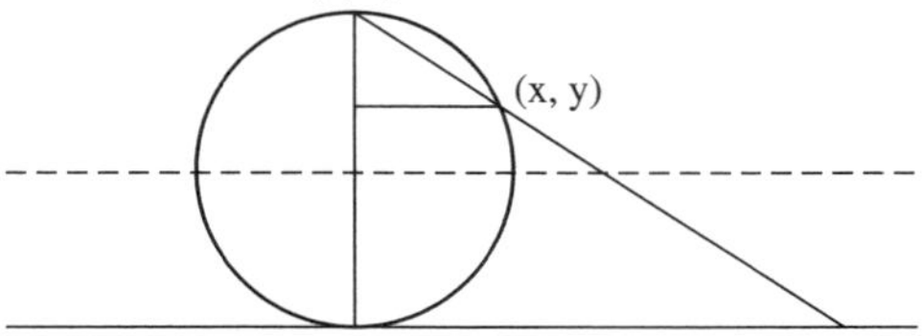


**Figure 2.2** The stereographic projection from $S^1$ onto the tangent line through the 'South pole' $(0, -1)$ takes $(x, y)$ to $(\xi, -1)$, where $\xi = 2x/(1 - y)$.

which is an open set. The formula $(x, y) \mapsto 2x/(1 - y)$ defines a smooth map from $U$ to $\mathbb{R}$, so $\xi$ is smooth.

Now let $\theta$ be the coordinate on $S^1$ defined by the parameterization $\psi$ : $\theta \mapsto (\cos\theta, \sin\theta)$, with domain $\left(-\frac{3\pi}{2}, \frac{\pi}{2}\right)$, and image equal to $S^1$ minus the North pole. The expression for $\xi$ in the coordinate $\theta$ on $S^1$ and the standard coordinate on $\mathbb{R}$ is $\xi(\theta) = \frac{2\cos\theta}{1-\sin\theta}$. Since this map has a smooth inverse (by the Inverse Function Theorem), $\xi : S^1 \setminus \{(0, 1)\} \to \mathbb{R}$ is a diffeomorphism. One consequence is that $\xi^{-1}$, viewed as a map from $\mathbb{R} \to \mathbb{R}^2$, is an embedding, so $\xi$ is a coordinate chart for $S^1$.

**Example 2.49** The 2-dimensional analogue of the previous example is the ***stereographic projection*** from the 'North pole' $(0, 0, 1)$ of $S^2$ onto the tangent plane $z = -1$. It has the formula

$$\xi = \frac{2x}{1 - z}, \quad \eta = \frac{2y}{1 - z}.$$

The domain of this map is $S^2$, minus the North pole, and the range is $\mathbb{R}^2$. It is easily checked that this map takes circles of constant latitude into circles in the plane centred on the origin. One can check, as in the previous example, that this map is a diffeomorphism and a coordinate chart for $S^2$.

**Exercise 2.1**

Consider $f: \mathbb{R}^3 \to \mathbb{R}^2$ given by $f(x, y, z) = \left(x^2 + y^2, y^2 + z^2\right)$. At which points $(x, y, z)$ is $\mathrm{D}f(x, y, z)$ surjective? Which points $\mathbf{c} \in \mathbb{R}^2$ are regular values for $f$? Sketch some of the level sets of $f$, that is, $f^{-1}(\mathbf{c})$, for various $\mathbf{c} \in \mathbb{R}^2$. Check that $f^{-1}(\mathbf{c})$ is a codimension-2 submanifold of $\mathbb{R}^3$ whenever $\mathbf{c}$ is a regular value. Are there any other values of $\mathbf{c}$ for which $f^{-1}(\mathbf{c})$ is a submanifold?

**Exercise 2.2**
Show that, if $f$ and $g$ are both submersions, and the composition $g \circ f$ is well defined, then $g \circ f$ is also a submersion. Is it true that, if $g \circ f$ is a submersion, then $g$ must be a submersion? What about $f$?

**Exercise 2.3**
Give an example of an $f: \mathbb{R}^2 \to \mathbb{R}$ such that $f^{-1}(c)$ is a submanifold of $\mathbb{R}^2$, even though $c$ is not a regular value of $f$.

**Exercise 2.4**
Let $U$ be an open subset of $\mathbb{R}^m$, and consider an injective map $\psi : U \to \mathbb{R}^n$. Prove that $\psi^{-1} : \psi(U) \to U$ is continuous in the usual '$\varepsilon$–$\delta$' sense if and only if, for every open set $W \subseteq U$, the set $\psi(W)$ is relatively open in $\psi(U)$ (see Definition 2.4).

**Exercise 2.5**
If we start with two identical circles in the $xz$-plane, of radius $r$ and centred at $x = \pm 2r$, then rotate them round the $z$-axis in $\mathbb{R}$, we get a ***torus***. Show that this torus is a submanifold. For a different kind of torus, see Example 2.36 and Exercise 2.38.

**Exercise 2.6**
Show that the preimage of a submanifold by a submersion is also a submanifold, of the same codimension as the original. Give an example. Note that a special case of this result is: the image of a submanifold by a diffeomorphism $F$ is a submanifold of the same dimension as the original. (This is true since $F^{-1}$ is a submersion.) What about the image of a submanifold by a submersion – is this always a submanifold?

**Exercise 2.7**
Show directly that $S^2$ satisfies each of the three conditions in Theorem 2.32.

**Exercise 2.8**
Show that the function $f$ defined in Example 2.28 is an immersion.

**Exercise 2.9**
Give counterexamples to show that neither the union nor the intersection of two submanifolds need be a submanifold.

**Exercise 2.10**
Show that if $\psi_2^{-1} \circ f \circ \psi_1$ is differentiable at $\psi_1^{-1}(\mathbf{a})$ for some parameterization $\psi_1$ of $M$ and some parameterization $\psi_2$ of $N$, then the same holds for every parameterization $\psi_1$ of $M$ such that $\mathbf{a} \in \operatorname{Im} \psi_1$, and every parameterization $\psi_2$ of $M$ such that $f(\mathbf{a}) \in \operatorname{Im} \psi_2$. It follows that $f$ is differentiable at $\mathbf{a}$.

## 2.2 Tangent vectors and derivatives

Consider a particle moving on a submanifold $M$ of $\mathbb{R}^n$. If $g(t)$ is the position of a particle at time $t$, then the velocity vector at time $t$ is the derivative $g'(t)$. The vector $g'(t)$ is called a *tangent vector* to $M$, based at the point $g(t) \in M$. The set of all tangent vectors based at a given point $\mathbf{x}$, corresponding to all possible paths (i.e. parameterized curves) in $M$ through $\mathbf{x}$, is the *tangent space* to $M$ at $\mathbf{x}$. This is a generalization of the tangent plane to a surface and the tangent line to a curve. There is one important difference: the tangent space is a vector space, so passes through $\mathbf{0}$, while the tangent plane or tangent line is the tangent space *translated* so that it passes through $\mathbf{x}$ – see Figure 2.4.

**Definition 2.50** *Let $M$ be a submanifold of $\mathbb{R}^n$. A* ***tangent vector*** *to $M$ is $g'(0)$ for some smooth path $g : \mathbb{R} \to M$ such that $g(0) = \mathbf{x}$. The point $\mathbf{x}$ is called the* ***base point*** *of the tangent vector. The* ***tangent space*** *to $M$ at $\mathbf{x}$ is the set of all tangent vectors based at $\mathbf{x}$, denoted $T_{\mathbf{x}}M$. Every tangent space is a vector space, as will follow from Theorem 2.53.*

A slight variation on the definition of a tangent vector is often useful: keeping track of the base point, we write $(\mathbf{x}, \mathbf{v})$ instead of just $\mathbf{v}$, so

$$T_{\mathbf{x}}M = \{(\mathbf{x}, \mathbf{v}) : \mathbf{v} = g'(0) \text{ for some path } g(t) \text{ in } M \text{ with } g(0) = \mathbf{x}\}.$$

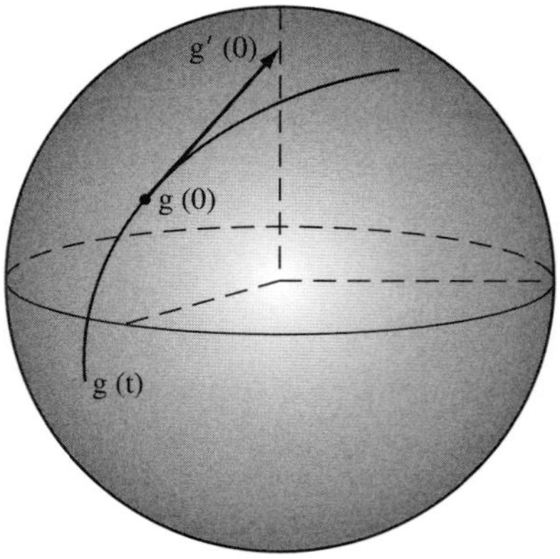


**Figure 2.3** A tangent vector to $S^2$ at $\mathbf{x}$ is $g'(0)$ for some path $g(t)$ in $S^2$ with $g(0) = \mathbf{x}$.

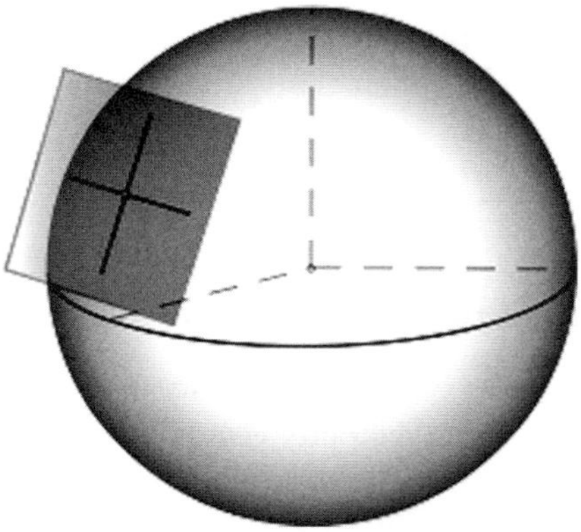

**Figure 2.4** The tangent space $T_{\mathbf{x}}S^2$ is the set of all tangent vectors to $S^2$ at $\mathbf{x}$.

The latter version of the definition, may be called the *bound vector* version, and the former the *free vector* version. We will use the free vector version when discussing only a single tangent space $T_{\mathbf{x}}M$, with $\mathbf{x}$ clear from the context. The bound vector version of the definition ensures that tangent spaces based at different points do not intersect, which is essential to the following definition.

**Definition 2.51** *The* ***tangent bundle*** *of a submanifold $M$ of $\mathbb{R}^n$, denoted by $TM$, is the union of all of the tangent spaces to $M$:*

$$TM = \bigcup_{\mathbf{x} \in M} T_{\mathbf{x}}M.$$

*The* ***tangent bundle projection*** *is the map $\pi : TM \to M$ given by $(\mathbf{x}, \mathbf{v}) \mapsto \mathbf{x}$, i.e. the projection takes any tangent vector to its base point.*

**Example 2.52** If $U \subseteq \mathbb{R}^n$ is open, then $TU = U \times \mathbb{R}^n$, because any $\mathbf{v} \in \mathbb{R}^n$ can be obtained as $\mathbf{v} = g'(0)$ for $g(t) := \mathbf{x} + t\mathbf{v}$.

There are different ways to compute tangent spaces of a submanifold, depending on how the submanifold is defined. Two ways are given in the next two theorems, and a third in Exercise 2.13.

**Theorem 2.53 (Tangent spaces of parameterized manifolds)** *Let $M$ be an $m$-dimensional submanifold of $\mathbb{R}^n$, and let $\mathbf{x} \in M$. If $M \cap U = \operatorname{Im} \psi$ for some neighbourhood $U$ of $\mathbf{x}$ and some parameterization $\psi$ with $\psi(\mathbf{a}) = \mathbf{x}$, then*

$$T_{\mathbf{x}}M = \operatorname{Im} \mathrm{D}\psi(\mathbf{a}),$$

*i.e. the image of the derivative of $\psi$ at $\mathbf{a}$. Alternatively, using the 'bound vector' definition of tangent vectors, we have $T_{\mathbf{x}}M = \{\mathbf{x}\} \times \operatorname{Im} \mathrm{D}\psi(\mathbf{a})$.*

*Proof* Let $\psi$ be as in the statement of the theorem, with domain $W$ open in $\mathbb{R}^m$. Every tangent vector to $M$ at $\mathbf{x}$ is $g'(0)$ for some path $g : I \to M \cap U$ with $g(0) = \mathbf{x}$. Let $\alpha(t) = \psi^{-1} \circ g(t)$, for all $t \in I$. Then, $\alpha(0) = \mathbf{a}$ and $g = \psi \circ \alpha$, and $g'(0) = \mathrm{D}\psi(\mathbf{a})\alpha'(0)$. Since $\alpha$ is an arbitrary path in $W$, which is open in $\mathbb{R}^m$, the vector $\alpha'(0)$ can take any value in $\mathbb{R}^m$. The result follows. ∎

**Example 2.54** Consider $S^1$ and the parameterization $\psi : (0, 2\pi) \to S^1$, $\psi(\theta) = (\cos\theta, \sin\theta)$. Let $a = \pi$ and $\mathbf{x} = \psi(\pi) = (-1, 0)$. Then,

$$\mathrm{D}\psi(\pi) = \begin{bmatrix} -\sin\theta \\ \cos\theta \end{bmatrix}\Bigg|_{\theta=\pi} = \begin{bmatrix} 0 \\ -1 \end{bmatrix}.$$

Then, $T_{\mathbf{x}}S^1 = T_{(-1,0)}S^1 = \operatorname{Im} \mathrm{D}\psi(\pi) = \operatorname{span}\{(0, -1)\} = \{0\} \times \mathbb{R}$.

At a general point $\mathbf{x} = (x, y) = \psi(\theta)$,

$$\mathrm{D}\psi(\theta) = \begin{bmatrix} -\sin\theta \\ \cos\theta \end{bmatrix} = \begin{bmatrix} -y \\ x \end{bmatrix},$$

so

$$T_{(x,y)}S^1 = \operatorname{Im} \begin{bmatrix} -y \\ x \end{bmatrix} = \{(-\lambda y, \lambda x) : \lambda \in \mathbb{R}\}.$$

At $\mathbf{x} = (1, 0)$, we must use a modified parameterization with a different domain, but the end result is the same. Therefore, the tangent bundle to $S^1$ is

$$TS^1 = \{(x, y, -\lambda y, \lambda x) : x^2 + y^2 = 1 \text{ and } \lambda \in \mathbb{R}\}.$$

There is a diffeomorphism from $TS^1$ to $S^1 \times \mathbb{R}$ given by $(x, y, -\lambda y, \lambda x) \mapsto ((x, y), \lambda)$ – see Exercise 2.11.

**Theorem 2.55 (Tangent spaces of level-set manifolds)** *Let $M$ be an $m$-dimensional submanifold of $\mathbb{R}^{m+k}$, and let $\mathbf{x} \in M$. If $M \cap U = f^{-1}(\mathbf{c})$, for some neighbourhood $U$ of $\mathbf{x}$ and for some submersion $f\colon U \to \mathbb{R}^k$, then*

$$T_{\mathbf{x}}M = \ker \mathrm{D}f(\mathbf{x}) = \left\{\mathbf{v} \in \mathbb{R}^m : \nabla f_i(\mathbf{x}) \cdot \mathbf{v} = 0,\ i = 1, \ldots, k\right\}.$$

*When using the bound vector definition, $T_{\mathbf{x}}M = \{\mathbf{x}\} \times \ker \mathrm{D}f(\mathbf{x})$.*

*Proof* Every tangent vector $\mathbf{v} \in T_{\mathbf{x}}M$ is of the form $\mathbf{v} = g'(0)$ for some path $g$ in $M \cap U$ with $g(0) = \mathbf{x}$. Since $f(g(t)) = \mathbf{c}$ for all $t$, the chain rule implies that

$$\mathrm{D}f(\mathbf{x}) \cdot \mathbf{v} = \mathrm{D}f(\mathbf{x}) \cdot g'(0) = \left.\frac{\mathrm{d}}{\mathrm{d}t}\right|_{t=0} f(g(t)) = \mathbf{0}.$$

This proves that $T_{\mathbf{x}}M \subseteq \ker \mathrm{D}f(\mathbf{x})$. For the converse, note that $\dim \ker \mathrm{D}f(\mathbf{x}) = m$ (since $f$ is a submersion), and $\dim T_{\mathbf{x}}M = m$, by Theorem 2.53. Since $T_{\mathbf{x}}M$ is contained in $\ker \mathrm{D}f(\mathbf{x})$ and has the same dimension as $\ker \mathrm{D}f(\mathbf{x})$, it must equal $\ker \mathrm{D}f(\mathbf{x})$. ■

**Example 2.56 (Tangent space to the sphere in $\mathbb{R}^3$)** The sphere $S^2$ is the set of points $(x, y, z) \in \mathbb{R}^3$ satisfying $x^2 + y^2 + z^2 = 1$. By the previous theorem, with $f(x, y, z) = x^2 + y^2 + z^2$, we find that the tangent space to the sphere a point $(x, y, z)$ on the sphere is

$$T_{(x,y,z)}S^2 = \left\{(u, v, w) \in \mathbb{R}^3 : xu + yv + zw = 0\right\}.$$

If we are using the 'bound vector' version of the definition of a tangent space, then we write $T_{(x,y,z)}S^2 = \left\{(x, y, z; u, v, w) \in \mathbb{R}^6 : xu + yv + zw = 0\right\}$.

The tangent bundle $TS^2$ of $S^2 \in \mathbb{R}^3$ is the union of the tangent spaces of $S^2$:

$$TS^2 = \left\{(x, y, z; u, v, w) \in \mathbb{R}^6 \mid x^2 + y^2 + z^2 = 1 \text{ and } xu + yv + zw = 0\right\}.$$

In contrast to $TS^1$, the tangent bundle $TS^2$ is *not* diffeomorphic to $S^2 \times \mathbb{R}^2$.

Consider a parameterization $\psi : (W \subseteq \mathbb{R}^m) \to (U \subseteq \mathbb{R}^n)$ of an $m$-dimensional submanifold $M$, and suppose $\psi(\mathbf{q}) = \mathbf{x} \in M$. Since $\psi$ is a parameterization, $\mathrm{D}\psi(\mathbf{q})$ is injective. From Theorem 2.53, the image of $\mathrm{D}\psi(\mathbf{q})$ is $T_{\mathbf{x}}M$. Therefore, $\mathrm{D}\psi(\mathbf{q})$ is an isomorphism from $\mathbb{R}^m$ to $T_{\mathbf{x}}M$.

**Definition 2.57** *In the above context, we define the following special tangent vectors at* $\mathbf{x}$,

$$\frac{\partial}{\partial q^i}(\mathbf{x}) := \frac{\partial}{\partial q^i}(\psi(\mathbf{q})) = \mathrm{D}\psi(\mathbf{q}) \cdot \mathbf{e}_i,$$

*where* $\mathbf{e}^i$ *is the* $i$*th standard basis vector for* $\mathbb{R}^m$*. Note that this is the* $i$*th column of* $\mathrm{D}\psi(\mathbf{q})$.

**Remark 2.58** *Since* $\mathrm{D}\psi(\mathbf{q}) : \mathbb{R}^m \to T_{\mathbf{x}}M$ *is an isomorphism, the vectors* $\frac{\partial}{\partial q^i}(\mathbf{x})$*, for* $i = 1, \ldots, m$*, form a basis for* $T_{\mathbf{x}}M$.

**Example 2.59** Consider a spherical parameterization of $S^2$ with formula

$$(x, y, z) = \psi(\theta, \phi) = (\sin\phi\cos\theta, \sin\phi\sin\theta, -\cos\phi),$$

where $\phi \in (0, \pi)$. We compute:

$$\frac{\partial}{\partial\theta}(x, y, z) = \mathrm{D}\psi(\theta, \phi) \cdot \mathbf{e}_1 = \begin{bmatrix} \frac{\partial x}{\partial\theta} \\ \frac{\partial y}{\partial\theta} \\ \frac{\partial z}{\partial\theta} \end{bmatrix} = \begin{bmatrix} -\sin\phi\sin\theta \\ \sin\phi\cos\theta \\ 0 \end{bmatrix} = \begin{bmatrix} -y \\ x \\ 0 \end{bmatrix},$$

$$\frac{\partial}{\partial\phi}(x, y, z) = \mathrm{D}\psi(\theta, \phi) \cdot \mathbf{e}_2 = \begin{bmatrix} \frac{\partial x}{\partial\phi} \\ \frac{\partial y}{\partial\phi} \\ \frac{\partial z}{\partial\phi} \end{bmatrix} = \begin{bmatrix} \cos\phi\cos\theta \\ \cos\phi\sin\theta \\ \sin\phi \end{bmatrix} = \frac{1}{\sqrt{1-z^2}} \begin{bmatrix} -xz \\ -yz \\ 1-z^2 \end{bmatrix}.$$

Recall from Example 2.52 that the tangent bundle of an open $W \subseteq \mathbb{R}^m$ is $TW = W \times \mathbb{R}^m$. If $\mathbf{q} = (q^1, \ldots, q^m)$ are the standard coordinates for $\mathbb{R}^m$, then the standard coordinates for $TW$ are written $(\mathbf{q}, \dot{\mathbf{q}}) = (q^1, \ldots, q^m, \dot{q}^1, \ldots, \dot{q}^m)$. Every tangent vector based at $\psi(\mathbf{q})$ can be expressed as $\mathrm{D}\psi(\mathbf{q}) \cdot \dot{\mathbf{q}}$ for a unique $\dot{\mathbf{q}} \in \mathbb{R}^m$. If $(\mathbf{x}, \mathbf{v}) = (\psi(\mathbf{q}), \mathrm{D}\psi(\mathbf{q}) \cdot \dot{\mathbf{q}})$, then $(\mathbf{x}, \mathbf{v})$ is called the tangent vector ***represented*** by $(\mathbf{q}, \dot{\mathbf{q}})$. (We often 'identify' $(\mathbf{x}, \mathbf{v})$ with $(\mathbf{q}, \dot{\mathbf{q}})$, meaning that we consider them the same, as long as the coordinate system is clear from the context.) Note that

$$\mathrm{D}\psi(\mathbf{q}) \cdot \dot{\mathbf{q}} = \mathrm{D}\psi(\mathbf{q}) \cdot \left( \sum_i \dot{q}^i \mathbf{e}_i \right) = \sum_i \dot{q}^i \left( \mathrm{D}\psi(\mathbf{q}) \cdot \mathbf{e}_i \right) = \sum_i \dot{q}^i \frac{\partial}{\partial q^i}(\mathbf{x}).$$

So $(\dot{q}^1, \ldots, \dot{q}^n)$ are also the coordinates on $T_{\mathbf{x}}M$ corresponding to the basis $\left( \frac{\partial}{\partial q^1}(\mathbf{x}), \ldots, \frac{\partial}{\partial q^n}(\mathbf{x}) \right)$.

If $\mathbf{x} = \psi(\mathbf{q})$ and If $\mathbf{v} = \mathrm{D}\psi(\mathbf{q}) \cdot \dot{\mathbf{q}}$, then $q^i$ is called the $i$th ***component*** of $\mathbf{x}$ (with respect to $\psi$), also written as $x^i$, and $\dot{q}^i$ is called the $i$th ***component*** of $\mathbf{v}$

(with respect to $\psi$), also written as $v^i$. These are also called the components of $\mathbf{x}$ and $\mathbf{v}$ with respect to the coordinate chart $\psi^{-1}$.

**Proposition 2.60** *Given a coordinate chart $\varphi : U \to \mathbb{R}^m$ for $M$, with components $(\varphi^1, \ldots, \varphi^m)$, and a tangent vector $\mathbf{v} = g'(0)$ to $M$ at $\mathbf{x} \in U$, the components of $\mathbf{x}$ with respect to $\varphi$ are $x^i = \varphi^i(\mathbf{x})$, and the components of $\mathbf{v}$ with respect to $\varphi$ are*

$$v^i := \left.\frac{\mathrm{d}}{\mathrm{d}t}\right|_{t=0} \varphi^i\left(g(t)\right).$$

*Proof* Let $\psi = \varphi^{-1}$, which is a parameterization of $M$. Let $\mathbf{q} = \varphi(\mathbf{x})$, so that $\psi(\mathbf{q}) = \mathbf{x}$. Then, the $i$th component of $\mathbf{x}$ is $x^i = q^i = \varphi^i(\mathbf{x})$. Let $c(t) = \varphi(g(t))$, so that $g(t) = \psi \circ c(t)$. By the chain rule,

$$\mathbf{v} = \left.\frac{\mathrm{d}}{\mathrm{d}t}\right|_{t=0} \psi \circ c(t) = \mathrm{D}\psi(\mathbf{q}) \cdot c'(0).$$

Let $\dot{\mathbf{q}} = c'(0)$. (Since $\mathrm{D}\psi(\mathbf{q})$ is injective, this is the unique value of $\dot{\mathbf{q}}$ such that $\mathbf{v} = \mathrm{D}\psi(\mathbf{q}) \cdot \dot{\mathbf{q}}$.) Then,

$$\dot{\mathbf{q}} = c'(0) = \left.\frac{\mathrm{d}}{\mathrm{d}t}\right|_{t=0} \varphi\left(g(t)\right).$$

The $i$th component of $\mathbf{v}$ is

$$v^i := q^i = \left.\frac{\mathrm{d}}{\mathrm{d}t}\right|_{t=0} \varphi^i\left(g(t)\right).$$

■

**Example 2.61** Consider the path $g(t) = \left(0, \cos\left(t + \frac{\pi}{4}\right), \sin\left(t + \frac{\pi}{4}\right)\right)$ in $S^2$, and let $\mathbf{x} = g(0) = \left(0, \frac{1}{\sqrt{2}}, \frac{1}{\sqrt{2}}\right)$. Then, $\mathbf{v} := g'(0) = \left(0, -\frac{1}{\sqrt{2}}, \frac{1}{\sqrt{2}}\right)$ is a tangent vector to $S^2$ at $\mathbf{x}$. The stereographic projection in Example 2.49, given by $\varphi = (\xi, \eta) = \left(\frac{x}{1-z}, \frac{y}{1-z}\right)$, is a coordinate chart for $S^2$. Since $\xi(g(t)) = 0$ for all $t$, and $\eta(g(t)) = \frac{\cos(t+\frac{\pi}{4})}{1-\sin(t+\frac{\pi}{4})}$, the components of $\mathbf{v}$ with respect to $\varphi$ are

$$v^1 = \left.\frac{\mathrm{d}}{\mathrm{d}t}\right|_{t=0} \xi(g(t)) = 0,$$

$$v^2 = \left.\frac{\mathrm{d}}{\mathrm{d}t}\right|_{t=0} \eta(g(t)) = \left.\frac{\mathrm{d}}{\mathrm{d}t}\right|_{t=0} \frac{\cos\left(t + \frac{\pi}{4}\right)}{1 - \sin\left(t + \frac{\pi}{4}\right)} = \frac{1}{1 - \sin\left(\frac{\pi}{4}\right)} = \frac{\sqrt{2}}{\sqrt{2} - 1}.$$

**Definition 2.62** *The **tangent lift** of a differentiable map $\psi : (W \subseteq \mathbb{R}^m) \to \mathbb{R}^n$ is*

$$\begin{aligned} T\psi : TW &\longrightarrow T\mathbb{R}^n, \\ (\mathbf{q}, \dot{\mathbf{q}}) &\longmapsto (\psi(\mathbf{q}), \mathrm{D}\psi(\mathbf{q}) \cdot \dot{\mathbf{q}}). \end{aligned} \tag{2.1}$$

**Theorem 2.63** *The tangent bundle $TM$ of any $m$-dimensional submanifold $M$ of $\mathbb{R}^n$ is a $2m$-dimensional submanifold of $\mathbb{R}^{2n}$.*

*Proof* By Theorem 2.32, every $\mathbf{x} \in M$ has a neighbourhood $U \subseteq \mathbb{R}^n$ such that $M \cap U = \operatorname{Im}\psi$ for some embedding $\psi : W \to U$. Consider the tangent lift $T\psi : W \times \mathbb{R}^n \to U \times \mathbb{R}^n$. It is straightforward to check that $T\psi$ is an embedding. (**Exercise**). It follows from Theorem 2.53 that $\operatorname{Im} T\psi = TM \cap (U \times \mathbb{R}^n)$. Therefore, $T\psi$ is a parameterization of $TM$, for every parameterization $\psi$ of $M$. Since every element of $TM$ is in the image of one of these maps $T\psi$, it follows that $TM$ is a submanifold of $\mathbb{R}^{2n}$ ■

If $\psi$ is a parameterization of $M \subseteq \mathbb{R}^n$ with image $M \cap U$, then $\psi^{-1} : (M \cap U) \to W$ is a coordinate chart for $M$ and $(T\psi)^{-1}$ is the corresponding coordinate chart for $TM$. The components $(\psi^{-1})^i : M \to \mathbb{R}$ are called ***coordinate functions*** on $M$. We often denote the function $\psi^{-1}$ by $\mathbf{q}$, and the coordinate functions by $q^i$. The components of $(T\psi)^{-1}$, denoted by $(\mathbf{q}, \dot{\mathbf{q}}) = \left(q^1, \ldots, q^n, \dot{q}^1, \ldots, \dot{q}^n\right)$ and defined by eqn (2.1), are coordinate functions on $TM$ called the ***tangent-lifted coordinates*** corresponding to $\mathbf{q}$.

We now turn to maps between manifolds.

**Definition 2.64** *Let $f \colon M \to N$ be differentiable. The **tangent map of** $f$ **at** $\mathbf{x}$ is the map*

$$\begin{aligned} T_{\mathbf{x}} f \colon T_{\mathbf{x}} M &\to T_{f(\mathbf{x})} N \\ \mathbf{v} &\mapsto (T_{\mathbf{x}} f)(\mathbf{v}) := \left.\frac{\mathrm{d}}{\mathrm{d}t}\right|_{t=0} f(g(t)), \end{aligned}$$

*where $g(t)$ is a path in $M$ such that $g(0) = \mathbf{x}$ and $\left.\frac{\mathrm{d}}{\mathrm{d}t}\right|_{t=0} g(t) = \mathbf{v}$. The map $T_{\mathbf{x}} f$ is often written $\mathrm{D}f(\mathbf{x})$ or $\mathrm{d}f(\mathbf{x})$ or $f_*(\mathbf{x})$, and called the **derivative of** $f$ **at** $\mathbf{x}$.*

It is not immediately obvious that the previous definition is independent of the choice of the path $g(t)$. It is, however, because $\left.\frac{\mathrm{d}}{\mathrm{d}t}\right|_{t=0} f(g(t))$ can be evaluated using the chain rule, as in eqn (2.2) or eqn (2.4) below, and only depends on $\mathbf{v}$.

**Definition 2.65** *All of the maps $T_{\mathbf{x}}f$, for all $\mathbf{x} \in M$, together define the* ***tangent map*** *of $f$, also known as the* ***tangent lift*** *of $f$,*

$$Tf\colon TM \to TN,$$

*given by $Tf(\mathbf{v}) = (T_{\mathbf{x}}f)(\mathbf{v})$ for all $\mathbf{v} \in T_{\mathbf{x}}M$.*

*If $M$ and $N$ are submanifolds and we use the 'bound vector' version of the definition of a tangent vector, then the tangent lift is written*

$$\begin{aligned} Tf\colon TM &\to TN \\ (\mathbf{x},\mathbf{v}) &\mapsto (f(\mathbf{x}), (T_{\mathbf{x}}f)(\mathbf{v}))\,. \end{aligned}$$

The reason for the term 'lift' is that tangent bundles are often visualized as sitting above their corresponding base spaces, as illustrated in the following diagram, in which $\pi_M : TM \to M$ and $\pi_N : TN \to N$ are the tangent bundle projection maps:

$$\begin{array}{ccc} TM & \xrightarrow{Tf} & TN \\ \tau_M \downarrow & & \downarrow \tau_N \\ M & \xrightarrow{f} & N\,. \end{array}$$

The diagram 'commutes', meaning that

$$\pi_N \circ Tf = f \circ \pi_M\,,$$

which can be verified from the formula $Tf(\mathbf{x},\mathbf{v}) = (f(\mathbf{x}), (T_{\mathbf{x}}f)(\mathbf{v}))$.

**Remark 2.66** *If $f$ has a differentiable inverse, then $(Tf)^{-1} = T\left(f^{-1}\right)$.*

For computations, we need to work in coordinates. We have two choices: coordinates in the 'ambient' space, i.e. the Euclidean space in which the submanifold sits; or local coordinate charts.

We begin with the former. Suppose $M$ is a submanifold of $\mathbb{R}^p$ and $N$ is a submanifold of $\mathbb{R}^s$, and $f\colon M \to N$ is the restriction of a smooth map $F\colon U \to \mathbb{R}^s$ for some $U$ open in $\mathbb{R}^p$. For any $(\mathbf{x},\mathbf{v}) \in TM$, and any path $g(t)$ in $M$ such that $g(0) = \mathbf{x}$ and $g'(0) = \mathbf{v}$, we obtain, using the chain rule,

$$(T_{\mathbf{x}}f)(\mathbf{v}) = \left.\frac{\mathrm{d}}{\mathrm{d}t}\right|_{t=0} f\,(g(t)) = \left.\frac{\mathrm{d}}{\mathrm{d}t}\right|_{t=0} F\,(g(t)) = \mathrm{D}F(\mathbf{x})\cdot g'(0) = \mathrm{D}F(\mathbf{x})\cdot\mathbf{v}\,, \tag{2.2}$$

where $\mathrm{D}F(\mathbf{x})\cdot\mathbf{v}$ means the matrix product of the Jacobian matrix (the derivative) $\mathrm{D}F(\mathbf{x})$ with the column vector $\mathbf{v}$. Thus, the tangent map $T_{\mathbf{x}}f$ is just the restriction of the derivative $\mathrm{D}F(\mathbf{x}) : \mathbb{R}^p \to \mathbb{R}^s$ to domain $T_\mathbf{x}M$ and codomain $T_{f(\mathbf{x})}N$. The tangent map $Tf\colon M \to N$ is given by

$$Tf(\mathbf{x}, \mathbf{v}) = (f(\mathbf{x}), \mathrm{D}F(\mathbf{x}) \cdot \mathbf{v}).$$

**Example 2.67** Let $f\colon \mathbb{R}^3 \to \mathbb{R}^3$ be a rotation by an angle $\alpha$ around the $z$-axis, with matrix

$$\begin{bmatrix} \cos\alpha & -\sin\alpha & 0 \\ \sin\alpha & \cos\alpha & 0 \\ 0 & 0 & 1 \end{bmatrix}.$$

Since this is a linear map, it is its own derivative, i.e. $\mathrm{D}F(\mathbf{x}) = F$, for every $\mathbf{x} \in \mathbb{R}^3$ (see Exercise 2.16). It is easily verified that $F$ maps every $\mathbf{x} \in S^2$ to another point in $S^2$, so $F$ can be restricted to a map $f\colon S^2 \to S^2$. The tangent map of $f$ at $\mathbf{x}$ is

$$\begin{aligned} T_\mathbf{x}f : T_\mathbf{x}S^2 &\longrightarrow T_\mathbf{x}S^2 \\ \mathbf{v} &\longmapsto \mathrm{D}F(\mathbf{x}) \cdot \mathbf{v} = F(\mathbf{v}). \end{aligned}$$

For example, if $\mathbf{x} = (1, 0, 0)$ then the the general form of a tangent vector to $S^2$ at $\mathbf{x}$ is $\mathbf{v} = (0, k, l)$, for some constants $k, l$ (see Example 2.56). We compute:

$$T_{(1,0,0)}f(0, k, l) = F(0, k, l) = (-k\sin\alpha, k\cos\alpha, l)\,,$$

which is a tangent vector to $S^2$ based at $F(1, 0, 0) = (\cos\alpha, \sin\alpha, 0)$. The full tangent map $Tf\colon TS^2 \to TS^2$ is

$$Tf\,(\mathbf{x}, \mathbf{v}) = (f(\mathbf{x}), (T_\mathbf{x}f)\,(\mathbf{v})) = (F(\mathbf{x}), F(\mathbf{v})).$$

(This formula is specific to linear maps $F$.)

Now suppose $\varphi$ and $\psi$ are coordinate charts for $M$ and $N$, respectively, and write $\mathbf{q} = \varphi(\mathbf{x})$, for $\mathbf{x} \in M$, and $\mathbf{r} = \psi(\mathbf{y})$, for $\mathbf{y} \in N$, as illustrated in Figure 2.5. Any $f\colon M \to N$ is expressed in coordinates $\mathbf{q}$ and $\mathbf{r}$ by $\mathbf{r} = \left(\psi \circ f \circ \varphi^{-1}\right)(\mathbf{q})$. The tangent lift $Tf$ has a corresponding expression in coordinates,

$$(\mathbf{r}, \dot{\mathbf{r}}) = T\left(\psi \circ f \circ \varphi^{-1}\right)(\mathbf{q}, \dot{\mathbf{q}}).$$

Since $\psi \circ f \circ \varphi^{-1}$ is a map between open subsets of Euclidean spaces, we have

$$(\mathbf{r}, \dot{\mathbf{r}}) = \left(\psi \circ f \circ \varphi^{-1}(\mathbf{q}), \mathrm{D}\left(\psi \circ f \circ \varphi^{-1}\right)(\mathbf{q}) \cdot \dot{\mathbf{q}}\right), \qquad (2.3)$$

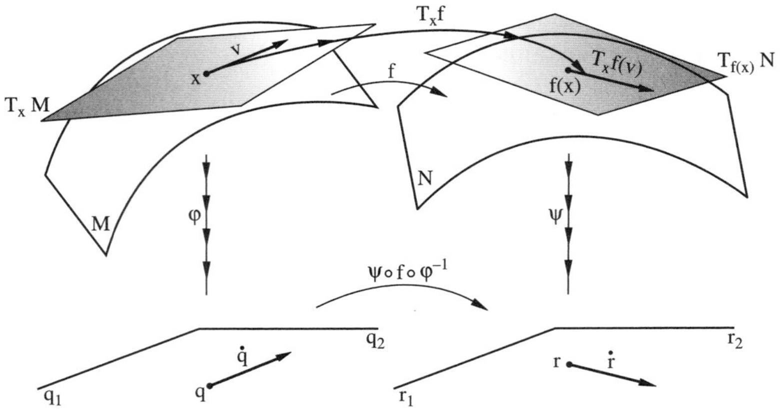


**Figure 2.5** Computing the tangent map in local coordinates: $T_{\mathbf{x}} f$, expressed in local coordinates, is the derivative $T\left(\psi \circ f \circ \varphi^{-1}\right)(\mathbf{q})$, where $\mathbf{q} = \varphi(\mathbf{x})$ and $\psi \circ f \circ \varphi^{-1}$ is the local coordinate expression for $f$.

where $\mathrm{D}\left(\psi \circ f \circ \varphi^{-1}\right)(\mathbf{q})$, for any $\mathbf{q}$, is the usual derivative, i.e. the Jacobian matrix.

**Example 2.68** Let $f\colon S^2 \to S^2$ be rotation around the $z$-axis by an angle $\alpha$, as in Example 2.67, and note that $f$ is given in spherical coordinates by $f(\theta, \phi) = (\theta + \alpha, \phi)$ (see Example 2.47). In these coordinates, the derivative at any point is the identity transformation, $\mathrm{D}f(\theta, \phi) = I$. It follows that the tangent lift of $f$ is given in spherical coordinates by

$$Tf\left(\theta, \phi, \dot{\theta}, \dot{\phi}\right) = \left(f(\theta, \phi), \mathrm{D}f(\theta, \phi)\left(\dot{\theta}, \dot{\phi}\right)\right) = \left(\theta + \alpha, \phi, \dot{\theta}, \dot{\phi}\right).$$

We can use the local coordinate expression in eqn (2.3) to give an alternative formula for $Tf$ itself. Note that the parameterizations $\varphi^{-1}$ and $\psi^{-1}$ are maps between Euclidean spaces, so can be differentiated in the usual way, but this doesn't work for the coordinate charts $\varphi$ and $\psi$ since their domains are not generally open subsets of Euclidean space. Suppose $\mathbf{v} = g'(0)$ for some path $g(t)$ with $g(0) = \mathbf{x}$. Then, $\varphi(g(t))$ is a path in $U$, and by the chain rule,

$$\mathbf{v} = \left.\frac{\mathrm{d}}{\mathrm{d}t}\right|_{t=0} g(t) = \left.\frac{\mathrm{d}}{\mathrm{d}t}\right|_{t=0} \varphi^{-1}(\varphi(g(t))) = \mathrm{D}\varphi^{-1}(\mathbf{q}) \cdot \left(\left.\frac{\mathrm{d}}{\mathrm{d}t}\right|_{t=0} \varphi(g(t))\right).$$

The representation of $\mathbf{v}$ in local coordinates is $\dot{\mathbf{q}}$ defined by $\mathrm{D}\varphi^{-1}(\mathbf{q}) \cdot \dot{\mathbf{q}} = \mathbf{v}$. Thus,

$$\dot{\mathbf{q}} = \left.\frac{\mathrm{d}}{\mathrm{d}t}\right|_{t=0} \varphi(g(t)).$$

Note that

$$f = \psi^{-1} \circ \left( \psi \circ f \circ \varphi^{-1} \right) \circ \varphi.$$

Applying the chain rule again gives,

$$\begin{aligned}
(T_{\mathbf{x}} f)(\mathbf{v}) &= \left.\frac{\mathrm{d}}{\mathrm{d}t}\right|_{t=0} f(g(t)) \\
&= \left.\frac{\mathrm{d}}{\mathrm{d}t}\right|_{t=0} \psi^{-1} \circ \left( \psi \circ f \circ \varphi^{-1} \right) \circ \varphi \circ g(t) \\
&= \mathrm{D}\psi^{-1}(\mathbf{r}) \circ \mathrm{D} \left( \psi \circ f \circ \varphi^{-1} \right)(\mathbf{q}) \cdot \left( \left.\frac{\mathrm{d}}{\mathrm{d}t}\right|_{t=0} \varphi(g(t)) \right) \\
&= \mathrm{D}\psi^{-1}(\mathbf{r}) \circ \mathrm{D} \left( \psi \circ f \circ \varphi^{-1} \right)(\mathbf{q}) \cdot \dot{\mathbf{q}}. \qquad (2.4)
\end{aligned}$$

This formula is more complicated than our original definition of the tangent map, but it does not depend on a choice of path $g(t)$ such that $g'(0) = \mathbf{v}$. This shows that the original definition of the tangent map is well defined.

The formula $\dot{\mathbf{r}} = \mathrm{D}\left(\psi \circ f \circ \varphi^{-1}\right)(\mathbf{q}) \cdot \dot{\mathbf{q}}$ in eqn (2.3) can also be written

$$\dot{r}^i = \sum_j \frac{\partial r^i}{\partial q^j} \dot{q}^j, \qquad (2.5)$$

(with the base point $\mathbf{q}$ omitted). Sometimes, if the choice of coordinates is clear from the context, the map $\psi \circ f \circ \varphi^{-1}$ (which we have called $\mathbf{r}$) will also be called $f$, in which case we write $\dot{\mathbf{r}} = \mathrm{D}f(\mathbf{q}) \cdot \dot{\mathbf{q}}$, or equivalently,

$$\dot{r}^i = \sum_j \frac{\partial f^i}{\partial q^j} \dot{q}^j. \qquad (2.6)$$

An important special case occurs when $f$ is the identity map from $M$ to itself, and $\varphi$ and $\psi$ are two overlapping coordinate charts on $M$, defining two systems of coordinates, $\mathbf{q}$ and $\mathbf{r}$. In this case, eqn (2.5) applies with $\mathbf{r} = \psi \circ \varphi^{-1}(\mathbf{q})$, and gives the change of coordinates from $\dot{\mathbf{q}}$ to $\dot{\mathbf{r}}$ corresponding to the change of coordinates from $\mathbf{q}$ to $\mathbf{r}$.

**Exercise 2.11**
Show that the map $TS^1 \to S^1 \times \mathbb{R}$, $(x, y, -\lambda y, \lambda x) \mapsto ((x, y), \lambda)$, which appeared in Example 2.54, is a diffeomorphism.

**Exercise 2.12**
Compute the tangent bundle of the cylinder $C$ given by

$$C = S^1 \times \mathbb{R} = \left\{(x, y, z) \in \mathbb{R}^3 : x^2 + y^2 = 1\right\}.$$

**Exercise 2.13 (Tangent spaces of manifolds defined as graphs)** Let $M$ be an $n$-dimensional submanifold of $\mathbb{R}^{n+k}$, let $\mathbf{x} \in M$ and let $U$ be a neighbourhood of $\mathbf{x}$. Show that, if $M \cap U$ is the graph of a smooth function $g$ of the first $n$ standard coordinates of $\mathbb{R}^{n+k}$, then

$$T_{\mathbf{x}}M = \{\mathbf{x}\} \times \{(\mathbf{v}, \mathrm{D}g\,(x_1, \ldots, x_n)\,(\mathbf{v})) : \mathbf{v} \in \mathbb{R}^n\}.$$

**Exercise 2.14**
Consider spherical coordinates on $S^2$ defined by

$$(x, y, z) = (\sin\phi\cos\theta, \sin\phi\sin\theta, -\cos\phi).$$

Let $(\mathbf{x}, \mathbf{v})$ be the tangent vector to $S^2$ given in the spherical coordinates by $(\theta, \phi, \dot{\theta}, \dot{\phi}) = \left(0, \frac{\pi}{4}, 2, -1\right)$. Compute $(\mathbf{x}, \mathbf{v})$ in the standard coordinates $(x, y, z)$ for $\mathbb{R}^3$.

**Exercise 2.15**
Let $\varphi : S^2 \to \mathbb{R}^2$ be the stereographic projection $(x, y, z) \mapsto (\xi, \eta)$ given in Example 2.49. Compute $\frac{\partial}{\partial\theta}(x, y, z)$ in $(\dot{\xi}, \dot{\eta})$ coordinates, where $\theta$ is the usual spherical coordinate, as in Exercise 2.14.

**Exercise 2.16**
Check that if $F\colon \mathbb{R}^p \to \mathbb{R}^s$ is a linear map, then $\mathrm{D}F(\mathbf{x}) = F$ for all $\mathbf{x}$.

**Exercise 2.17**
Compute the change of coordinates from $(\dot{r}, \dot{\theta})$ to $(\dot{x}, \dot{y})$ corresponding to the change from polar to Cartesian coordinates in $\mathbb{R}^2$.

**Exercise 2.18**
Consider the cylinder $C = S^1 \times \mathbb{R} = \{(x, y, z) \in \mathbb{R}^3 : x^2 + y^2 = 1\}$. Let $f \colon S^2 \to C$ be given by $f(\theta, \phi) = (\theta, -\cos\phi)$, in spherical coordinates on $S^2$ as in Exercise 2.14, and cylindrical coordinates $(\theta, z)$ on $C$ defined by $(x, y, z) = (\cos\theta, \sin\theta, z)$. Compute $Tf$ in these coordinates.

**Exercise 2.19**
Compute the tangent map of the stereographic projection $S^2 \to \mathbb{R}^2$ given in Example 2.49, in two ways: first, using Cartesian coordinates $(x, y, z)$ on $S^2 \subseteq \mathbb{R}^3$, and second, using spherical coordinates $(\theta, \phi)$ on $S^2$.

**Exercise 2.20**
The tangent bundle projection map, $\tau : TM \to M$, given by $(\mathbf{x}, \mathbf{v}) \mapsto \mathbf{x}$, is written in any tangent-lifted coordinate chart $(\mathbf{q}, \dot{\mathbf{q}})$ as $\tau(\mathbf{q}, \dot{\mathbf{q}}) = \mathbf{q}$. Write down the matrix $D\tau(\mathbf{q}, \dot{\mathbf{q}})$ in these coordinates, and check that $\tau$ is a submersion.

**Exercise 2.21**
Give an alternative proof that $TM$ is a submanifold, by describing $M$ as a union of level sets.

## 2.3 Differentials and cotangent vectors

Let $f$ be a smooth real-valued function on a manifold $M$. Note that $T_{f(\mathbf{x})}\mathbb{R} = \mathbb{R}$, for any $\mathbf{x} \in M$, so the tangent map of $f$ at $\mathbf{x}$ can be written $T_{\mathbf{x}} f \colon T_{\mathbf{x}} M \to \mathbb{R}$. In this case, $T_{\mathbf{x}} f$ is also called the ***differential*** of $f$ at

$\mathbf{x}$, and written $\mathrm{d}f(\mathbf{x}) : T_{\mathbf{x}}M \to \mathbb{R}$. From the definition of the tangent lift, we have

$$\mathrm{d}f(\mathbf{x}) \cdot \mathbf{v} = (f \circ g)'(0),$$

for any smooth path $g(t)$ in $M$ such that $g(0) = \mathbf{x}$ and $g'(0) = \mathbf{v}$. In any local coordinates $\mathbf{q} = \left(q^1, \ldots, q^n\right)$ on $M$, we have

$$\mathrm{d}f(\mathbf{q}) \cdot \dot{\mathbf{q}} = \mathrm{D}f(\mathbf{q}) \cdot \dot{\mathbf{q}} = \nabla f(\mathbf{q}) \cdot \dot{\mathbf{q}} = \sum_i \frac{\partial f}{\partial q^i} \dot{q}^i, \tag{2.7}$$

for all $\dot{\mathbf{q}} \in T_{\mathbf{q}}M$, where the partial derivatives are evaluated at the coordinate values corresponding to $\mathbf{q}$. Note that the dot in '$\mathrm{d}f(\mathbf{q}) \cdot \dot{\mathbf{q}}$' stands for the evaluation of the linear function $\mathrm{d}f(\mathbf{q})$ at $\dot{\mathbf{q}}$, while the dot in '$\nabla f(\mathbf{q}) \cdot \dot{\mathbf{q}}$' is the dot product in $\mathbb{R}^n$.

For any $i$, the local coordinate $q^i$ is itself a real-valued function defined in some neighbourhood of $x$, so it has a differential $\mathrm{d}q^i$ at every point in that neighbourhood. In local coordinates, we have

$$\mathrm{d}q^i(\mathbf{q}) \cdot \dot{\mathbf{q}} = \sum_j \frac{\partial q^i}{\partial q^j} \dot{q}^j = \dot{q}^i. \tag{2.8}$$

Since this holds for all $\mathbf{v} \in T_x M$, comparison with eqn (2.7) gives

$$\mathrm{d}f(\mathbf{q}) = \sum_i \frac{\partial f}{\partial q^i} \mathrm{d}q^i(\mathbf{q}). \tag{2.9}$$

Note that the previous equation, with the basepoint $\mathbf{q}$ omitted, is familiar from calculus, with $\mathrm{d}f$ and $\mathrm{d}q^i$ interpreted as 'infinitesimal distances'.

**Example 2.69** Let $(u, v)$ be any local coordinates on a 2-dimensional manifold $M$, and let $f$ be given in these coordinates by $f(u, v) = u^2 v$. Then, $\mathrm{d}f = 2uv\,\mathrm{d}u + u^2\,\mathrm{d}v$. (We have suppressed the base point, $(u, v)$, of all the differentials.)

If $M$ is an submanifold of $\mathbb{R}^n$ and $f$ is the restriction of some smooth $f\colon U \to \mathbb{R}$, for some neighbourhood $U$ of $M$, then there is another way to calculate $\mathrm{d}f$, in Cartesian coordinates on $\mathbb{R}^n$:

$$\mathrm{d}f(\mathbf{x}) \cdot \mathbf{v} = \mathrm{D}F(\mathbf{x}) \cdot \mathbf{v} = \sum_i \frac{\partial F}{\partial x^i} v^i. \tag{2.10}$$

Thus,

$$\mathrm{d}F(\mathbf{x}) = \sum_i \frac{\partial F}{\partial x^i}\,\mathrm{d}x^i(\mathbf{x}), \tag{2.11}$$

and $\mathrm{d}f(\mathbf{x})$ is the restriction of $\mathrm{d}F(\mathbf{x})$ to $T_{\mathbf{x}}M$.

**Example 2.70** Let $f\colon S^2 \to \mathbb{R}$ be given by $f(x,y,z) = x+y$. Exercise 2.22 concerns the calculation of $\mathrm{d}f$ in spherical coordinates on $S^2$. Another way to calculate $\mathrm{d}f$ is in Cartesian coordinates in the 'ambient' space $\mathbb{R}^3$, since $f$ is the restriction of $F\colon \mathbb{R}^3 \to \mathbb{R}$ defined by the same formula $F(x,y,z) = x+y$. The differential of $F$ at any $(x,y,z)$ is $\mathrm{d}F(x,y,z) = \mathrm{d}x(x,y,z) + \mathrm{d}y(x,y,z)$, and $\mathrm{d}f(x,y,z)$ is the restriction of $\mathrm{d}x(x,y,z) + \mathrm{d}y(x,y,z)$ to $T_{(x,y,z)}S^2$. Equivalently, $\mathrm{d}f(\mathbf{x})(\mathbf{v}) = v^1 + v^2$, for all $\mathbf{x} \in S^2$ and all $\mathbf{v} \in T_{\mathbf{x}}S^2$.

Recall that the ***dual space*** of any real vector space $V$, denoted $V^*$, is the set of linear maps from $V$ to $\mathbb{R}$, which is itself a real vector space, with the usual operations of addition and scalar mulfiplication of maps. Elements of $V^*$ are called ***covectors***. Since $\mathrm{d}f(\mathbf{x})$ is a linear map from $T_{\mathbf{x}}M$ to $\mathbb{R}$, it is an element of the dual space of $T_{\mathbf{x}}M$, which we denote $T^*_{\mathbf{x}}M$.

**Definition 2.71** *The* ***cotangent space*** *to $M$ at $\mathbf{x}$ is $T^*_{\mathbf{x}}M := (T_{\mathbf{x}}M)^*$, the dual space of $T_{\mathbf{x}}M$. Elements of $T^*_{\mathbf{x}}M$ are called* ***cotangent vectors*** *based at $\mathbf{x}$. The* ***cotangent bundle*** *$T^*M$ of $M$, by*

$$T^*M = \bigcup_{\mathbf{x}} T^*_{\mathbf{x}}M.$$

*The* ***cotangent bundle projection*** *is the map $\pi : T^*M \to M$ defined by $\pi(\mathbf{x},\mathbf{p}) = \mathbf{x}$.*

**Remark 2.72** *In the definition of the cotangent bundle, we assume that the cotangent spaces are disjoint. This means that we are using the 'bound vector' version of the definition of a tangent space as a set of pairs $(\mathbf{x},\mathbf{v})$.*

It is a standard fact of linear algebra that, for any finite-dimensional vector space $V$ and any basis $\{\mathbf{b}_1,\ldots,\mathbf{b}_n\}$ for $V$, there is a ***dual basis*** $\{\mathbf{b}^1,\ldots,\mathbf{b}^n\}$ for $V^*$ defined by $\mathbf{b}^i(\mathbf{b}_j) = \delta^i_j$, where $\delta^i_j$ is the *Kronecker delta*, defined by

$$\delta^i_j = \begin{cases} 1, & \text{if } i = j, \\ 0, & \text{otherwise.} \end{cases}$$

Recall that $\left(\frac{\partial}{\partial q^1}(\mathbf{x}), \ldots, \frac{\partial}{\partial q^n}(\mathbf{x})\right)$ is a basis for $T_{\mathbf{x}}M$, with corresponding coordinates $(\dot{q}^1, \ldots, \dot{q}^n)$. So for any $\mathbf{v} \in T_{\mathbf{x}}M$, we have $\mathbf{v} = \sum_j \dot{q}^j \frac{\partial}{\partial q^j}(\mathbf{x})$, and in local coordinates we write $\mathbf{v} = \dot{\mathbf{q}}$. Thus, eqn (2.8) can be rewritten as

$$\mathrm{d}q^i(\mathbf{x}) \cdot \left( \sum_j \dot{q}^j \frac{\partial}{\partial q^j}(\mathbf{x}) \right) = \dot{q}^i. \tag{2.12}$$

This shows that $(\mathrm{d}q^1(\mathbf{x}), \ldots, \mathrm{d}q^n(\mathbf{x}))$ is a basis for $T_{\mathbf{x}}^*M$ that is dual to the basis $\left(\frac{\partial}{\partial q^1}(\mathbf{x}), \ldots, \frac{\partial}{\partial q^n}(\mathbf{x})\right)$ for $T_{\mathbf{x}}M$.

**Remark 2.73** *For any real vector space $V$, for any $\mathbf{v} \in V$, $\boldsymbol{\alpha} \in V^*$, there is a bilinear operation $\langle \cdot, \cdot \rangle : V^* \times V \to \mathbb{R}$ called the* ***natural pairing****, defined by $\langle \boldsymbol{\alpha}, \mathbf{v} \rangle := \boldsymbol{\alpha}(\mathbf{v})$. Note that, for $V = T_{\mathbf{x}}M$, and any local coordinates $(\dot{q}^1, \ldots, \dot{q}^n)$,*

$$\langle \boldsymbol{\alpha}, \dot{\mathbf{q}} \rangle = \left\langle \sum_i \alpha_i \, \mathrm{d}q^i, \sum_i \dot{q}^i \frac{\partial}{\partial q^i} \right\rangle = \sum_i \alpha_i \dot{q}^i = \boldsymbol{\alpha} \cdot \dot{\mathbf{q}} = \boldsymbol{\alpha}\dot{\mathbf{q}},$$

*where in the last expression $\boldsymbol{\alpha}$ is considered as a row vector and and $\dot{\mathbf{q}}$ a column vector, and their product is defined by matrix multiplication. If $\boldsymbol{\alpha}$ and $\dot{\mathbf{q}}$ are both considered as column vectors, then $\langle \boldsymbol{\alpha}, \dot{\mathbf{q}} \rangle = \boldsymbol{\alpha}^T \dot{\mathbf{q}}$.*

Let $U \subseteq M$ be the domain of definition of the local coordinates $(q^1, \ldots, q^n)$. For every $\mathbf{x} \in U$, the covectors $\mathrm{d}q^i(\mathbf{x})$ form a basis for $T_{\mathbf{x}}^*M$. Let $(p_1, \ldots, p_n)$ be the coordinates corresponding to this basis, so

$$\mathbf{p} = \sum_i p_i \, \mathrm{d}q^i(\mathbf{x}), \qquad \text{for every } \mathbf{p} \in T_{\mathbf{x}}^*M.$$

Since this applies at every $\mathbf{x} \in U$, this defines the ***cotangent-lifted coordinates*** $(q^1, \ldots q^n, p_1, \ldots, p_n)$ on $T^*U \subseteq T^*M$. These local coordinate systems on $T^*M$, one for every local coordinate system on $M$, give the cotangent bundle $T^*M$ a manifold structure. The detailed proof of this fact is similar to that of Theorem 2.63 for tangent bundles.

**Theorem 2.74** *$T^*M$ is a manifold, with dimension twice that of $M$.*

**Why the distinction between subscripts and superscripts?** This is to keep track of how quantities vary if coordinates are changed. Recall eqn (2.5), repeated here:

$$\dot{r}^i = \sum_j \frac{\partial r^i}{\partial q^j} \dot{q}^j . \tag{2.13}$$

This gives the change of coordinates transformation between two sets of tangent-lifted coordinates on $TM$ induced by a given change of coordinates on $M$. If we write the original change of coordinates as $\mathbf{r} = \varphi(\mathbf{q})$, and consider $\mathbf{r}$ and $\mathbf{q}$ to be column vectors, then the above equation is equivalent to

$$\dot{\mathbf{r}} = \mathrm{D}\varphi(\mathbf{q})\, \dot{\mathbf{q}}.$$

Let $(\mathbf{q}, \mathbf{p})$ and $(\mathbf{r}, \mathbf{s})$ be the corresponding cotangent-lifted coordinates on $T^*M$. It can be shown (this is Exercise 2.26) that

$$s_i = \sum_j p_j \frac{\partial q^j}{\partial r^i} .$$

Considering $\mathbf{s}$ and $\mathbf{p}$ as row vectors, the transformation can be expressed as

$$\mathbf{s} = \mathbf{p}\, \mathrm{D}\varphi(\mathbf{q})^{-1}.$$

We now define *cotangent lifts* of maps between manifolds, which are analogous to tangent lifts.

Let $T : V \to W$ be any linear transformation. Then, for any $\boldsymbol{\beta}$ in $W^*$, the composition $\boldsymbol{\beta} \circ T$ is in $V^*$,

$$V \xrightarrow{T} W \xrightarrow{\boldsymbol{\beta}} \mathbb{R}.$$

The ***dual*** of $T$ is the transformation $T^* : W^* \to V^*$ defined by $T^*(\boldsymbol{\beta}) = \boldsymbol{\beta} \circ T$, or equivalently,

$$\langle T^*(\boldsymbol{\beta}), \mathbf{v} \rangle = \langle \boldsymbol{\beta}, T(\mathbf{v}) \rangle,$$

for all $\boldsymbol{\beta} \in W^*$ and all $\mathbf{v} \in V$. If $T$ is represented by a matrix $A$, with respect to a certain basis, then $T^*$ is represented by the transpose matrix $A^T$, with respect to the dual basis, since

$$\left\langle T^*(\boldsymbol{\beta}), \mathbf{v} \right\rangle = \langle \boldsymbol{\beta}, T(\mathbf{v}) \rangle = \boldsymbol{\beta}^T A\mathbf{v} = \left(A^T \boldsymbol{\beta}\right)^T \mathbf{v} = \left\langle \left(A^T \boldsymbol{\beta}\right), \mathbf{v} \right\rangle, \tag{2.14}$$

for all $\boldsymbol{\beta} \in W^*$ and all $\mathbf{v} \in V$, where $\boldsymbol{\beta}$ and $\mathbf{v}$ are considered as column vectors. If $\boldsymbol{\beta}$ is considered as a row vector, then $T^*(\boldsymbol{\beta}) = \boldsymbol{\beta} A$.

**Definition 2.75** *Let $f\colon M \mapsto N$. The **cotangent map** of $f$ **at** $\mathbf{x}$ is the dual of the tangent map $T_{\mathbf{x}}f$, denoted $T^*_{\mathbf{x}}f$. That is, $T^*_{\mathbf{x}}f := (T_{\mathbf{x}}f)^* : T^*_{f(\mathbf{x})}N \to T^*_{\mathbf{x}}M$ and*

$$\langle \left(T^*_{\mathbf{x}}f\right)(\boldsymbol{\beta}), \mathbf{w}\rangle = \langle \boldsymbol{\beta}, (T_{\mathbf{x}}f)(\mathbf{w})\rangle, \tag{2.15}$$

*for all $\mathbf{w} \in T_{\mathbf{x}}M$ and all $\boldsymbol{\beta} \in T^*_{f(\mathbf{x})}N$. If $f$ is invertible, then the maps $T^*_{\mathbf{x}}f$, for all $\mathbf{x} \in M$, together define the **cotangent map** of $f$,*

$$T^*f\colon T^*N \to T^*M,$$

*given by $(T^*f)(\boldsymbol{\beta}) = \left(T^*_{\mathbf{x}}f\right)(\boldsymbol{\beta}) = (T_{\mathbf{x}}f)^*(\boldsymbol{\beta})$ for all $\boldsymbol{\beta} \in T^*_{f(\mathbf{x})}N$.*

*If $f$ is a diffeomorphism, then the **cotangent lift** of $f$ at $\mathbf{x}$ is the cotangent map of $f^{-1}$ at $f(\mathbf{x})$, which is the map $T^*_{f(\mathbf{x})}\left(f^{-1}\right) : T^*_{\mathbf{x}}M \to T^*_{f(\mathbf{x})}N$ given by*

$$\langle T^*_{f(\mathbf{x})}f^{-1}(\boldsymbol{\alpha}), \mathbf{v}\rangle = \langle \boldsymbol{\alpha}, T_{f(\mathbf{x})}f^{-1}(\mathbf{v})\rangle, \tag{2.16}$$

*for all $\boldsymbol{\alpha} \in T^*_{\mathbf{x}}M$ and $\mathbf{v} \in T_{f(\mathbf{x})}N$. The maps $T^*_{f(\mathbf{x})}f^{-1}$, for all $\mathbf{x} \in M$, together define the **cotangent lift** of $f$,*

$$T^*f^{-1} : T^*M \to T^*N,$$

*given by $T^*f^{-1}(\boldsymbol{\alpha}) = T^*_{f(\mathbf{x})}f^{-1}(\boldsymbol{\alpha}) = \left(T_{f(\mathbf{x})}f^{-1}\right)^*(\boldsymbol{\alpha})$ for all $\boldsymbol{\alpha} \in T^*_{\mathbf{x}}M$.*

Tangent and cotangent maps are illustrated in Figure 2.6.

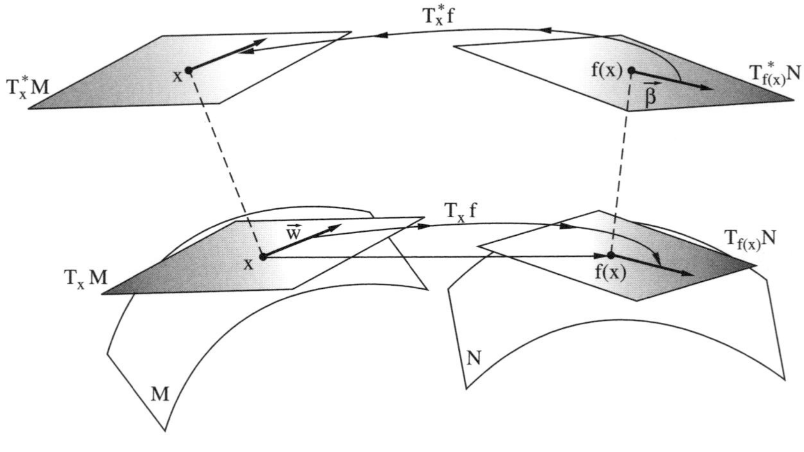


**Figure 2.6** The tangent and cotangent maps. Note that the tangent map $T_{\mathbf{x}}f$ goes in the 'same direction' as $f$, whereas the cotangent map $T^*_{\mathbf{x}}f$ goes 'backwards'.

**Remark 2.76** $(T^*f)^{-1} = T^*\left(f^{-1}\right)$, *so the notation* $T^*f^{-1}$ *is unambiguous. The base points are tricky, as* $\left(T^*_{\mathbf{x}}f\right)^{-1} = T^*_{f(\mathbf{x})}\left(f^{-1}\right)$. *Since the definition is complicated, the cotangent lift is best remembered as the most 'natural' way of getting a cotangent vector based at* $f(\mathbf{x})$ *from a cotangent vector based at* $\mathbf{x}$.

**Remark 2.77** *Note that if* $f$ *maps from* $M$ *to* $N$, *then its cotangent map* $T^*f$ *maps from* $T^*N$ *to* $T^*M$, *i.e. it 'goes backwards', as illustrated in Figure 2.6. For this reason, when* $f$ *is a diffeomorphism, we more often use the cotangent lift* $T^*f^{-1}$, *which goes 'in the same direction' as* $f$. *We say that the cotangent lift of* $f$ *'covers'* $f$.

The cotangent lift is often represented in the following diagram, in which $\pi_M : T^*M \to M$ and $\pi_N : T^*N \to N$ are the cotangent bundle projection maps:

$$\begin{array}{ccc} T^*M & \overset{(T^*f)^{-1}}{\longrightarrow} & T^*N \\ \tau_M \downarrow & & \downarrow \tau_N \\ M & \overset{f}{\longrightarrow} & N\,. \end{array}$$

The diagram commutes, which means that $\tau_N \circ (T^*f)^{-1} = f \circ \tau_M$. Indeed, if $\boldsymbol{\alpha} \in T_{\mathbf{x}}M$, then $(T^*f)^{-1}(\boldsymbol{\alpha}) \in T_{f(\mathbf{x})}N$, so

$$\tau_N \circ \left(T^*f\right)^{-1}(\boldsymbol{\alpha}) = f(\mathbf{x}) = f \circ \tau_M(\boldsymbol{\alpha}).$$

We now consider representations of cotangent lifts in local coordinates. Let $(\mathbf{q}, \dot{\mathbf{q}})$ and $(\mathbf{r}, \dot{\mathbf{r}})$ be tangent-lifted local coordinates, and let $(\mathbf{q}, \mathbf{p})$ and $(\mathbf{r}, \mathbf{s})$ be the corresponding cotangent-lifted coordinates. Recall from eqn (2.6) that the tangent lift $Tf$ can be written in coordinates as $(\mathbf{r}, \dot{\mathbf{r}}) = Tf(\mathbf{q}, \dot{\mathbf{q}})$, where

$$\dot{r}^i = \sum_j \frac{\partial f^i}{\partial q^j}\dot{q}^j, \quad \text{or } \dot{\mathbf{r}} = \mathrm{D}f(\mathbf{q})\,\dot{\mathbf{q}} \qquad \text{(tangent lift in coordinates).}$$

From this it follows that, if $(\mathbf{r}, \mathbf{s}) = (T^*f)^{-1}(\mathbf{q}, \mathbf{p})$, then

$$p_j = \sum_i s_i \frac{\partial f^i}{\partial q^j}, \quad \text{or } \mathbf{s} = \mathbf{p}\left(\mathrm{D}f(\mathbf{q})\right)^{-1} \qquad \text{(cotangent lift in coordinates),}$$

where $\mathbf{p}$ and $\mathbf{s}$ are considered as row vectors. If $\mathbf{p}$ and $\mathbf{s}$ are considered as column vectors, then $\mathbf{s} = \left((\mathrm{D}f(\mathbf{q}))^{-1}\right)^T \mathbf{p}$.

**Example 2.78** Consider the cylinder $S^1 \times \mathbb{R} = \{(x, y, z) \in \mathbb{R}^3 : x^2 + y^2 = 1\}$, with cylindrical coordinates $(\theta, z)$ defined by $(x, y, z) = (\cos\theta, \sin\theta, z)$. Let $f\colon S^2 \to S^1 \times \mathbb{R}$ be given, in spherical coordinates on $S^2$ and cylindrical coordinates on $S^1 \times \mathbb{R}$, by $f(\theta, \phi) = (\theta, -\cos\phi)$. In these coordinates,

$$\mathrm{D}f(\theta, \phi) = \begin{bmatrix} 1 & 0 \\ 0 & \sin\phi \end{bmatrix},$$

so the tangent lift of $f$ is

$$Tf(\theta, \phi, \dot{\theta}, \dot{\phi}) = \left(f(\theta, \phi), \mathrm{D}f(\theta, \phi)[\dot{\theta}, \dot{\phi}]^T\right) = \left(\theta, -\cos\phi, \dot{\theta}, \sin\phi\,\dot{\phi}\right),$$

and the cotangent lift of $f$ is

$$\begin{aligned} T^*f^{-1}(\theta, \phi, p_\theta, p_\phi) &= \left(f(\theta, \phi), [p_\theta, p_\phi]\mathrm{D}f(\theta, \phi)^{-1}\right) \\ &= \left(\theta, -\cos\phi, p_\theta, \frac{p_\phi}{\sin\phi}\right). \end{aligned}$$

**Example 2.79** Consider $f\colon \mathbb{R}^n \to \mathbb{R}^n$ defined (with respect to the standard coordinates) by $f(\mathbf{x}) = A\mathbf{x}$, for some $n \times n$ matrix $A$. Since $f$ is linear, $\mathrm{D}f(\mathbf{x}) = f$ for all $\mathbf{x}$ (see Exercise 2.16), so $\mathrm{D}f(\mathbf{x})$ has matrix $A$. The tangent lift of $f$ is given in tangent-lifted coordinates by $Tf(\mathbf{q}, \dot{\mathbf{q}}) = (A\mathbf{q}, A\dot{\mathbf{q}})$. The cotangent lift of $f$ is given in cotangent-lifted coordinates by $(T^*f)^{-1}(\mathbf{q}, \mathbf{p}) = (A\mathbf{q}, \mathbf{p}A^{-1})$ (if we think of $\mathbf{p}$ as a row vector), or $(T^*f)^{-1}(\mathbf{q}, \mathbf{p}) = (A\mathbf{q}, (A^{-1})^T\mathbf{p})$ (if we think of $\mathbf{p}$ as a column vector).

**Exercise 2.22**
Let $f\colon S^2 \to \mathbb{R}$ be given by $f(x, y, z) = x + y$. Express $f$ in spherical coordinates $(\theta, \phi)$ and then compute $\mathrm{d}f$ in these coordinates.

**Exercise 2.23**
For every $\mathbf{v} \in \mathbb{R}^n$, let $\alpha_\mathbf{v} \in (\mathbb{R}^n)^*$ be given by by $\langle \alpha_\mathbf{v}, \mathbf{w}\rangle := \mathbf{v} \cdot \mathbf{w}$. Show that the map $\mathbb{R}^n \to (\mathbb{R}^n)^*$, given by $\mathbf{v} \mapsto \alpha_\mathbf{v}$, is an isomorphism. Note that an analogous isomorphism exists for any finite-dimensional inner product space.

**Exercise 2.24**
Let $V$ be a finite-dimensional vector space. The ***double dual*** of $V$ is the dual of $V^*$, i.e. $V^{**} := (V^*)^*$. Show that the map $\Phi : V \to V^{**}$ defined by

$$\langle \Phi(v), \alpha \rangle = \langle \alpha, v \rangle$$

is an isomorphism.

**Exercise 2.25**
If $V$ is an inner product space, and $T : V \to V$, what is the relationship between the dual of $T$ and the adjoint of $T$?

**Exercise 2.26**
Show that the change of coordinates transformation between cotangent-lifted coordinates $(\mathbf{q}, \mathbf{p})$ and $(\mathbf{r}, \mathbf{s})$ on $T^*M$ is

$$s_i = \sum_j p_j \frac{\partial q^j}{\partial r^i}.$$

Show that, considering $\mathbf{s}$ and $\mathbf{p}$ as row vectors, the transformation can be expressed as

$$\mathbf{s} = \mathbf{p}\, \mathrm{D}\varphi(\mathbf{q})^{-1},$$

where $\mathbf{r} = \varphi(\mathbf{q})$. Note the equivalent formulae,

$$p_j = \sum_i s_i \frac{\partial r^i}{\partial q^j}, \qquad \mathbf{p} = \mathbf{s}\, \mathrm{D}\varphi(\mathbf{q}).$$

**Exercise 2.27**
Consider $f \colon \mathbb{R}^2 \to \mathbb{R}^2$ given by $f(x, y) = (2x + 1, \sin xy)$. Calculate the tangent and cotangent lifts of $f$, and calculate $\left(T^*_{(x,y)} f\right)^{-1} (\mathrm{d}y\,(x, y))$ in terms of $\mathrm{d}x\,(x, y)$ and $\mathrm{d}y\,(x, y)$ .

**Exercise 2.28**
If $f\colon M \to N$ and $g : N \to \mathbb{R}$ are both differentiable, show that $T_{\mathbf{x}}^* f\left(\mathrm{d}g(f(\mathbf{x}))\right) = \mathrm{d}(g \circ f)(\mathbf{x})$ for all $\mathbf{x} \in M$.

## 2.4 Matrix groups as submanifolds

The set of all $n \times n$ matrices with real entries, $\mathcal{M}(n,\mathbb{R})$, is isomorphic to $\mathbb{R}^{n^2}$. If we 'identify' $\mathcal{M}(n,\mathbb{R})$ with $\mathbb{R}^{n^2}$, i.e. consider them as the same, then the definition of a submanifold carries over to sets of matrices. Similarly, the set of all $n \times n$ matrices with complex entries, $\mathcal{M}(n,\mathbb{C})$, is isomorphic to $\mathbb{C}^{n^2}$, which is isomorphic to $\mathbb{R}^{2n^2}$, so we may identify $\mathcal{M}(n,\mathbb{C})$ with $\mathbb{R}^{2n^2}$, and consider submanifolds of this space.

It turns out that all of the classical groups of matrices, for example $SO(n)$ and $U(n)$, are submanifolds of $\mathcal{M}(n,\mathbb{R})$ or $\mathcal{M}(n,\mathbb{C})$. They are examples of *Lie groups*, which will be studied in more depth in Chapter 5.

We will now review the definitions of some of these groups, and review why they are really groups, and then show that they are submanifolds. To start with, recall the abstract definition of a group:

**Definition 2.80** *A **group** is a set of elements $G$ with a binary operation $G \times G \to G$, which we will write as $(g,h) \mapsto gh$, with the following properties:*

1. *it is **associative**, that is $(gh)k = g(hk)$ for all $g,h,k \in G$;*
2. *it has an **identity** element $e \in G$, satisfying $g = eg = ge$ for all $g \in G$; and*
3. *each $g \in G$ has an **inverse** element $g^{-1} \in G$ satisfying $gg^{-1} = g^{-1}g = e$.*

*If in addition, $gh = hg$, for all $g,h \in G$, then the group is called **commutative** or **Abelian**.*

**Definition 2.81** *A **subgroup** of a group $G$ is a subset $H$ of $G$ such that the group operation on $G$, restricted to $H$, makes $H$ a group.*

**Remark 2.82** *It is easily checked that $H$ is of a subgroup $G$ if and only if:*

1. *$H$ is closed under the group operation, i.e. $h_1h_2 \in H$ whenever $h_1$, $h_2 \in H$; and*
2. *$H$ is closed under inversion. That is, $h^{-1} \in H$ whenever $h \in H$.*

*Note that these two conditions imply that the group identity of $G$ is the group identity of $H$. Also note that the intersection of any two subgroups of $G$ is also a subgroup of $G$.*

Note that we have used multiplicative notation for the group operation and group inverse in the previous definitions. This is because we wish to consider sets of matrices as groups with respect to the matrix multiplication operation:

**Definition 2.83** *A **matrix group** is a subset of $\mathcal{M}(n,\mathbb{R})$ or $\mathcal{M}(n,\mathbb{C})$ that is a group, with the group operation being matrix multiplication.*

**Example 2.84** The ***general linear group*** $GL(n,\mathbb{R})$ is the set of all linear isomorphisms from $\mathbb{R}^n$ to itself. Every such isomorphism corresponds to a matrix in $\mathcal{M}(n,\mathbb{R})$, so we can consider $GL(n,\mathbb{R})$ to be a subset of $\mathcal{M}(n,\mathbb{R})$, which we do from now on: $GL(n,\mathbb{R})$ is the set of all invertible $n\times n$ matrices with real entries. Since the product of two invertible matrices is invertible, $GL(n,\mathbb{R})$ is closed under matrix multiplication. It is easily checked that $GL(n,\mathbb{R})$ satisfies the definition of a group, with the group operation being matrix multiplication, with the identity matrix $I$ as identity element; i.e. $GL(n,\mathbb{R})$ is a matrix group. The ***complex general linear group*** $GL(n,\mathbb{C})$ is defined similarly.

**Remark 2.85** *All other matrix groups are subgroups of $GL(n,\mathbb{R})$ and $GL(n,\mathbb{C})$, since the elements of a matrix group must be invertible. By Remark 2.82, a subset of $GL(n,\mathbb{R})$ or $GL(n,\mathbb{C})$ is a subgroup if and only if it is closed under matrix multiplication and inversion.*

**Definition 2.86** *A **symmetric** matrix is one satisfying $A^T = A$. We denote the set of real symmetric $n\times n$ matrices by $Sym(n,\mathbb{R})$.*

*A **skew-symmetric** or **anti-symmetric** matrix is one satisfying $A^T = -A$. We denote the set of real skew-symmetric $n\times n$ matrices by $Skew(n,\mathbb{R})$.*

*An **orthogonal** matrix is one satisfying $A^T A = I$, where $I$ is the $n\times n$ identity matrix (for some $n$). The group of real orthogonal $n\times n$ matrices is denoted $O(n,\mathbb{R})$ or simply $O(n)$.*

*A **unitary** matrix is a complex matrix $U$ such that $U^*U = I$, where $U^*$ is the conjugate transpose of $U$. The group of $n\times n$ unitary matrices is denoted $U(n)$.*

*A **symplectic** matrix is one satisfying $A^T J A = J$, where $J = \begin{bmatrix} 0 & I \\ -I & 0 \end{bmatrix}$. The group of real symplectic $2n\times 2n$ matrices is denoted $Sp(2n,\mathbb{R})$. (There are no odd-dimensional symplectic matrices!)*

All of the matrix groups defined above are important in mechanics. $O(n)$ is especially important as a symmetry group, while $Sp(2n, \mathbb{R})$ appears in the general setting of Hamiltonian systems on symplectic manifolds.

**Definition 2.87** *A* ***matrix Lie group*** *is a matrix group that is also a submanifold of $\mathcal{M}(n, \mathbb{R})$ (or $\mathcal{M}(n, \mathbb{C})$).*

**Remark 2.88** *All of the examples we will consider are embedded submanifolds. Some matrix Lie groups may only be immersed submanifolds (see Remark 2.27).*

**Example 2.89** $GL(n, \mathbb{R})$ is an $n^2$-dimensional submanifold of $\mathcal{M}(n, \mathbb{R})$, since it is an open subset of $\mathcal{M}(n, \mathbb{R})$. Similarly, $GL(n, \mathbb{C})$ is an open subset, and hence a submanifold, of $\mathcal{M}(n, \mathbb{C})$.

In fact, all of the matrix groups defined above are matrix Lie groups, as shown in the following proposition and Exercises 2.31 and 2.32.

**Proposition 2.90** *$O(n)$ is a matrix Lie group of dimension $n(n-1)/2$, and $T_I O(n) = Skew(n, \mathbb{R})$.*

*Proof* It is easily checked that $O(n)$ is a matrix group (see Exercise 2.29). Since $A^T A$ is always symmetric, the following map is well defined,

$$\begin{aligned} F\colon GL(n, \mathbb{R}) &\longrightarrow Sym(n, \mathbb{R}), \\ A &\longmapsto A^T A - I \,. \end{aligned}$$

Note that $GL(n, \mathbb{R})$ is an open subset of $\mathcal{M}(n, \mathbb{R})$ and $O(n) = F^{-1}(0)$, where 0 is the zero matrix. We now compute $\mathrm{D}F(A)$, for an arbitrary $A \in GL(n, \mathbb{R})$:

$$\begin{aligned} \mathrm{D}F(A)(B) &= \left.\frac{\mathrm{d}}{\mathrm{d}t}\right|_{t=0} F(A + tB) \\ &= \left.\frac{\mathrm{d}}{\mathrm{d}t}\right|_{t=0} (A + tB)^T (A + tB) - I = B^T A + A^T B \,, \end{aligned}$$

for any $B \in \mathcal{M}(n, \mathbb{R})$. To prove that $\mathrm{D}F(A)$ is surjective, we must show that, for any $n \times n$ symmetric matrix $S$, there exists a matrix $B$ such that $B^T A + A^T B = S$. But this is proven by setting

$$B = \frac{1}{2} A^{-T} S \,,$$

where $A^{-T} := \left(A^{-1}\right)^T$. Hence, $\mathrm{D}F(A)$ is surjective, for any $A$. We conclude that $F$ is a submersion. It follows that $O(n)$ is a submanifold of $\mathcal{M}(n, \mathbb{R})$.

Since $Sym(n, \mathbb{R})$ has dimension $n(n+1)/2$, the group $O(n)$ has dimension $n^2 - n(n+1)/2 = n(n-1)/2$.

Substituting $A = I$ in the above calculations gives

$$\mathrm{D}F(I)(B) = B^T + B\,.$$

Thus, using Theorem 2.55,

$$T_I O(n) = \ker \mathrm{D}F(I) = \{B \in \mathcal{M}(n, \mathbb{R}) : B^T + B = 0\} = Skew(n, \mathbb{R}).$$

■

**Definition 2.91** *The **special linear group** $SL(n, \mathbb{R})$ is the subgroup of $GL(n, \mathbb{R})$ consisting of all matrices of determinant* 1. *The **complex special linear group** $SL(n, \mathbb{C})$ is defined similarly.*

*The **special orthogonal group** is $SO(n) := O(n) \cap SL(n, \mathbb{R})$.*

*The **special unitary group** is $SU(n) := U(n) \cap SL(n, \mathbb{C})$.*

It is easily checked that these are all matrix groups.

**Proposition 2.92** *$SL(n, \mathbb{R})$ is a matrix Lie group of dimension $n^2 - 1$, i.e. codimension* 1.

*Proof* $SL(n, \mathbb{R}) = f^{-1}(1)$, where $f\colon GL(n, \mathbb{R}) \to \mathbb{R}$ is the determinant map $f(A) := \det A$. We will show that $f$ is a submersion. For any $A \in GL(n, \mathbb{R})$, define the path $\gamma$ in $GL(n, \mathbb{R})$ by $\gamma(t) = (1+t)A$. Then, $\gamma'(0) \in T_A GL(n, \mathbb{R})$ and

$$\begin{aligned} \mathrm{D}f(A)\left(\gamma'(0)\right) &= (f \circ \gamma)'(0) = \left.\frac{\mathrm{d}}{\mathrm{d}t}\right|_{t=0} \det\left((1+t)A\right) \\ &= \left.\frac{\mathrm{d}}{\mathrm{d}t}\right|_{t=0} (1+t)^n \det A = n \det A\,. \end{aligned}$$

Since $\det A \neq 0$, this shows that $\mathrm{D}f(A)$ is a non-zero linear transformation. Since the codomain of $f$ is $\mathbb{R}$, it follows that $\mathrm{D}f(A)$ is surjective. Hence, $f$ is a submersion, and therefore $SL(n, \mathbb{R})$ is a submanifold of $\mathcal{M}(n, \mathbb{R})$. Its codimension is the dimension of the range of $f$, which is 1. Since $SL(n, \mathbb{R})$ is also a matrix group, it is a matrix Lie group. ■

The situation for $SO(n)$ is rather different, since all orthogonal matrices automatically have determinant $\pm 1$, see Exercise 2.30.

**Remark 2.93** *$SO(3)$ is the group of rotations in $\mathbb{R}^3$. $O(3)$ is the group of all 'rigid linear transformations', namely rotations, reflections, and compositions of these. Rotations have determinant $+1$, while other elements* *of $O(3)$ have determinant $-1$. Both groups often appear in mechanics as symmetry groups. $SO(3)$ is the configuration space of a rigid body with one point fixed, as discussed in Section 1.5.*

**Proposition 2.94** *$SO(n)$ is a matrix Lie group of dimension $n(n-1)/2$.*

*Proof* $SO(n) = O(n) \cap U$, where $U = \det^{-1}(0, \infty)$, where $\det : \mathcal{M}(n, \mathbb{R}) \to \mathbb{R}$ is the determinant map. Since det is continuous (in fact, it is a polynomial), $U$ is open. Thus, $SO(n)$ is a non-empty relatively open subset of $O(n)$. Since $O(n)$ is a submanifold of $\mathcal{M}(n, \mathbb{R})$, by Proposition 2.90, it follows that $SO(n)$ is also a submanifold of $\mathcal{M}(n, \mathbb{R})$, of the same dimension as $O(n)$. Since $SO(n)$ is a matrix group, the result follows. ■

**Remark 2.95** *In fact, all of the matrix groups we have seen so far are connected, except for $O(n)$ and $U(n)$. For proofs, see [MR02].*

**Proposition 2.96** *$O(n)$ has two connected components: $SO(n)$ (which contains the identity matrix) and another component diffeomorphic to $SO(n)$.*

*Proof* We saw in Proposition 2.94 that $SO(n)$ is open relative to $O(n)$. Recall that all elements of $O(n)$ have determinant $\pm 1$. If $A$ is any element of $O(n)$ with $\det A = -1$, then the map $B \mapsto BA$ is a diffeomorphism that maps $SO(n)$ onto $O(n) \setminus SO(n)$. It follows that $O(n) \setminus SO(n)$ is also open relative to $O(n)$, and therefore $O(n)$ is not connected. Since $SO(n)$ is connected (see Exercise 2.34 and [MR02]), and hence $O(n) \setminus SO(n)$ is connected, it follows that $SO(n)$ and $O(n) \setminus SO(n)$ are the connected components of $O(n)$. ■

**Exercise 2.29**
Prove that all of the groups defined in Definition 2.86 really are matrix groups. Prove that $Sym(n, \mathbb{R})$ and $Skew(n, \mathbb{R})$ are vector spaces and are not matrix groups.

**Exercise 2.30**
Show that all orthogonal matrices and all symplectic matrices have determinant $\pm 1$. (In fact, all symplectic matrices have determinant 1, but this is harder to prove – see [MS95].)

**Exercise 2.31**
Show that $Sp(2n, \mathbb{R})$ is a matrix Lie group, of dimension $2n^2 + n$, and

$$T_I Sp(2n, \mathbb{R}) = \{B \in \mathcal{M}(n, \mathbb{R}) : B^T J + JB = 0\}.$$

HINT: Modify the definition of $F$ in the proof of Proposition 2.90, and replace $Sym(n, \mathbb{R})$ with $Skew(n, \mathbb{R})$.

**Exercise 2.32**
Modify the proof of Proposition 2.90 to prove that $U(n)$ is a submanifold of $\mathcal{M}(n, \mathbb{C})$. What is its (real) dimension? Calculate $T_I U(n)$.

**Exercise 2.33**
Prove that $SL(n, \mathbb{C})$ is a matrix Lie group of (real) dimension $2(n^2 - 1)$.

**Exercise 2.34**
Convince yourself that $SO(3)$ is path-connected, and hence connected. HINT: a rotation is completely determined by how it transforms the standard orthonormal frame. Starting with the standard frame, how can you continuously move it so as to coincide with an arbitrary given right-handed orthonormal frame? This defines a path in $SO(3)$.

**Exercise 2.35**
Prove that $SU(n)$ is a matrix Lie group, with (real) dimension $n^2 - 1$. This is considerably more difficult than for $SO(n)$, since $SU(n)$ is of codimension 1 in $U(n)$.

## 2.5 Abstract manifolds

A manifold is a set covered by consistent local coordinate charts; the precise definition appears below. All submanifolds of Euclidean spaces are manifolds, but there are other kinds of manifold that satisfy the general 'abstract' definition, notably quotient spaces (see Definition 2.101). These appear in mechanics in many ways. However, abstract manifolds do not play a central role in this book, so this section could be skipped on first reading.

To motivate the definition, recall that whenever $\varphi_1$ and $\varphi_2$ are coordinate charts for a submanifold of $\mathbb{R}^n$, with overlapping domains, the *transition function* $\varphi_2 \circ \varphi_1^{-1}$ (also called the *change-of-coordinates transformation*) is smooth (see Theorem 2.39). For the abstract definition of a manifold, we keep only this requirement that transition functions $\varphi_2 \circ \varphi_1^{-1}$ be smooth, without requiring anything of the maps $\varphi_1$ and $\varphi_2$ individually.

**Definition 2.97** *A* ***coordinate chart*** *(or* ***local coordinate system****) for a set $M$ is a 1-to-1 mapping $\varphi : U \to \mathbb{R}^n$, for some subset $U$ of $M$ and some $n$, such that $\varphi(U)$ is an open subset of $\mathbb{R}^n$. The set $U$ is called a* ***coordinate patch****. The components of $\varphi$ are called* ***local coordinates****. The inverse mapping $\varphi^{-1}$ (with domain $\varphi(U)$) is called a* ***parameterization****.*

*Two coordinate charts $\varphi_1 : U_1 \to \mathbb{R}^n$ and $\varphi_2 : U_2 \to \mathbb{R}^n$ for $M$ are* ***compatible*** *if the map $\varphi_2 \circ \varphi_1^{-1}$, with domain $\varphi_1(U_1 \cap U_2)$, is a smooth diffeomorphism with a smooth inverse. (Implicit in this definition is the assumption that $\varphi_1(U_1 \cap U_2)$ and $\varphi_2(U_1 \cap U_2)$ are open subsets of $\mathbb{R}^n$.) The map $\varphi_2 \circ \varphi_1^{-1}$ is called a* ***transition function*** *or* ***change-of-coordinates transformation****.*

*An n-dimensional* ***atlas*** *for a set $M$ is a family of mutually compatible n-dimensional charts covering $M$, i.e. a family $\{\varphi_i : U_i \to \mathbb{R}^n\}_{i \in J}$, for some index set $J$, such that $\bigcup_{i \in J} U_i = M$.*

*Two atlases for $M$ are* ***equivalent*** *if all of their charts are mutually compatible. A* ***smooth structure*** *(or* ***differential structure****) on $M$ is an equivalence class of atlases for $M$; each of the atlases in the class is said to determine the smooth structure. A* ***smooth manifold*** *is a set $M$ with a smooth structure. (Informally, we usually say 'M is a manifold', if the smooth structure is clear from the context.)*

**Example 2.98** The two charts for $S^1$ in Example 2.40 form an atlas for $S^1$. But we could equally well have chosen two different compatible charts that covered $S^1$, with the same formula but different domains and codomains. These two charts would define a second atlas for $S^1$, equivalent to the first. Both atlases determine the same smooth structure for $S^1$.

**Theorem 2.99** *Every submanifold of $\mathbb{R}^N$ is a manifold, with a smooth structure determined by the coordinate charts given in Definition 2.37.*

*Proof* The smoothness of transition functions is guaranteed by Theorem 2.39. ∎

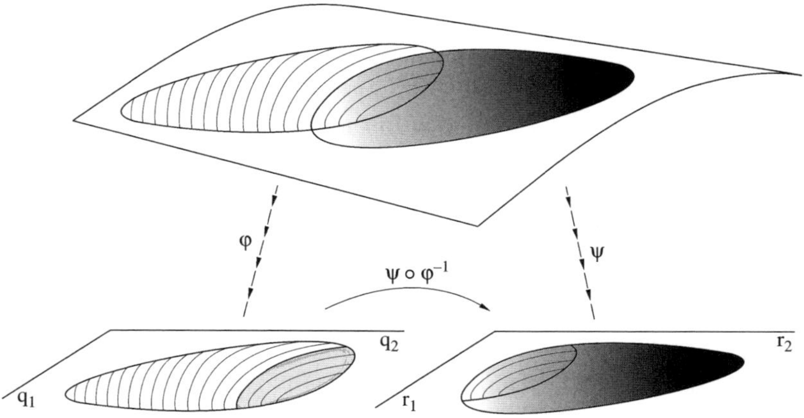


**Figure 2.7** Two coordinate charts for an $n$-dimensional manifold.

**Example 2.100** Consider the equivalence relation on $\mathbb{R}$ given by

$$x \sim y \iff x - y \in \mathbb{Z},$$

and denote by $[x]$ the equivalence class containing $x$. The set of equivalence classes is denoted $\mathbb{R}/\mathbb{Z}$ and called the *quotient* of $\mathbb{R}$ by $\mathbb{Z}$. Let $V_1 = (0,1)$ and $V_2 = (-1/2, 1/2)$. For each $i$, let $U_i = \{[x] : x \in V_i\}$ define $\varphi_i : U_i \to V_i$ by $[x] \mapsto x$ for all $x \in V_i$. Since $\varphi_i$ is injective and its image is an open subset of $\mathbb{R}$, it is a coordinate chart in the sense of Definition 2.97. Note that $U_1 \cup U_2 = \mathbb{R}/\mathbb{Z}$. The transition map $\varphi_2 \circ \varphi_1^{-1}$ is defined on the domain $(0, 1/2) \cup (1/2, 1)$ and is given by:

$$x \mapsto \begin{cases} x, & \text{if } x \in (0, 1/2); \\ x - 1, & \text{if } x \in (1/2, 1). \end{cases}$$

This map is clearly smooth. This proves that $\mathbb{R}/\mathbb{Z}$ is a 1-dimensional manifold. Note that $\mathbb{R}/\mathbb{Z}$ inherits a group structure, defined by $[x] + [y] = [x + y]$.

**Definition 2.101** *An* ***identification space*** *is the set of all equivalence classes for some equivalence relation. An important kind of identification space is the* ***quotient space*** *of a manifold $M$ by a group of transformations $G$, written $M/G$. Two elements in $M$ are considered equivalent if there exists a transformation in $G$ that takes the first element onto the second. In Example 2.100, we can take $G = \mathbb{Z}$, with every element $n \in \mathbb{Z}$ interpreted as the map $r \mapsto r + n$. Quotient spaces will be studied in Chapter 6.*

A smooth structure is exactly what is needed to define what it means for a map between manifolds to be smooth. Indeed, given any local coordinate systems $\varphi$ and $\psi$ for manifolds $M$ and $N$, respectively, any map $f: M \to N$ between manifolds has a ***representation in local coordinates*** $\psi \circ f \circ \varphi^{-1}$ (we assume that the domain of definition of this map is non-empty). Since this is a map between subsets of Euclidean spaces, we can apply existing definitions from calculus and Section 2.1. In examples, this is very natural, since maps are typically defined in terms of local coordinates in the first place.

**Definition 2.102** *A map $f: M \to N$ between manifolds is* ***smooth*** *if, for any choice of coordinate charts $\varphi$ for $M$ and $\psi$ for $N$, the composition $\psi \circ f \circ \varphi^{-1}$ (if it has a non-empty domain) is smooth. Note that it is implicit in the definition of smoothness that the domain of each local representative $\psi \circ f \circ \varphi^{-1}$ is an open set. The definitions of* ***differentiable, diffeomorphism, continuous, submersion, immersion*** *and* ***embedding*** *are analogous.*

In fact, though the definition of smoothness says 'for any choice of coordinate charts', it suffices to check just enough charts to cover the two manifolds $M$ and $N$ – this is Exercise 2.42.

**Example 2.103 (Circle group)** The identification space $\mathbb{R}/2\pi\mathbb{Z}$ is defined by the equivalence

$$x \sim y \iff x - y \in 2\pi\mathbb{Z}.$$

It has a manifold structure analogous to that of $\mathbb{R}/\mathbb{Z}$ in Example 2.100. The map

$$f: \mathbb{R}/\mathbb{Z} \to \mathbb{R}/2\pi\mathbb{Z}, \qquad f([x]) = [2\pi x],$$

is a diffeomorphism, Indeed, if $\varphi : [x] \to x$ (with a suitable domain) is a chart for $\mathbb{R}/\mathbb{Z}$ and $\psi : [x] \to x$ is a chart for $\mathbb{R}/2\pi\mathbb{Z}$, then the domain of $\psi \circ f \circ \varphi^{-1}$ is a union of open intervals, and on each of these intervals $\psi \circ f \circ \varphi^{-1}(x) = 2\pi x + \text{constant}$, which is smooth. The proof that $f^{-1}$ is smooth is similar.

The spaces $\mathbb{R}/\mathbb{Z}$ and $\mathbb{R}/2\pi\mathbb{Z}$ are both diffeomorphic to the unit circle $S^1 \subset \mathbb{R}^2$. Indeed, consider the map $f: \mathbb{R}/2\pi\mathbb{Z} \to S^1$ given by $f([x]) = (\cos x, \sin x)$. This map is well defined since $x \sim y$ implies that $x - y$ is a multiple of $2\pi$. It is also invertible. If $\varphi : [x] \to x$ (with a suitable domain) is a chart for $\mathbb{R}/2\pi\mathbb{Z}$ and $\psi : (\cos\theta, \sin\theta) \mapsto \theta$ is a chart for $S^1$, then the domain of $\psi \circ f \circ \varphi^{-1}$ is a union of open intervals, and on each of these intervals $\psi \circ f \circ \varphi^{-1}(x) = x + \text{constant}$, which is smooth. The proof that $f^{-1}$ is smooth is similar. Thus, $f$ is a diffeomorphism. The three spaces in this example are all commonly called 'the circle' and denoted by $S^1$. Since $S^1$ has a group operation, as defined in Example 2.100, it is called the ***circle group***.

**Definition 2.104** *An **immersed submanifold** of a manifold $N$ is the image of an immersion $\psi : M \to N$ for some manifold $M$.*

*An **embedded submanifold** of a manifold $N$ is the image of an embedding $\psi : M \to N$ for some manifold $M$.*

Recall that an immersion need not be injective, but an embedding must be.

**Remark 2.105** *Note how simple the definition of an embedded submanifold is, compared to the definitions in Section 2.1! The reason is that, in the new definition, the domain of $\psi$ can be any manifold, not necessarily an open subset of a Euclidean space, so one embedding suffices to cover the entire manifold.*

**Example 2.106** Let $S^2 \subset \mathbb{R}^3$ be the unit sphere. The map $f\colon S^2 \to \mathbb{R}^3$, $f(\mathbf{x}) = 2\mathbf{x}$, is an embedding. Therefore, the image of $f$, which is an ellipsoid, is an embedded submanifold.

In a certain sense, all manifolds 'are' submanifolds of some Euclidean space:

**Theorem 2.107 (Whitney embedding theorem)** *Every $m$-dimensional manifold can be embedded in $\mathbb{R}^{2m}$.*

However, the proof of this theorem does not show how to construct such embeddings explicitly.

**Remark 2.108** *Recall that if $M$ is a subset of $\mathbb{R}^N$, then a subset of $M$ is open relative to $M$ if it is of the form $M \cap U$ for some neighbourhood $U$ of $\mathbb{R}^N$. This specification of which subsets of $M$ are considered to be open is called the relative topology on $M$. A manifold structure on $M$ is **compatible** with the relative topology if the notion of openness given by the coordinate charts is consistent with the relative topology, in the following sense: for every $V \subseteq M$, the set $V$ is relatively open if and only if, for every coordinate chart $\varphi : U \to \mathbb{R}^n$ for $M$, $\varphi(U \cap V)$ is open. An equivalent requirement is that every coordinate chart is a **homeomorphism**, i.e. a continuous map with a continuous inverse. The question 'is $X$ a manifold?' usually means: does $X$ have a manifold structure compatible with a given topology[3]?*

[3] A ***topology*** for $X$ is a family of subsets of $X$, called the *open* sets, that contains at least the empty set and $X$ itself, and such that any union of open sets, and any finite intersection of open sets, is open.

**Example 2.109** Let $C$ be the graph of $z = \sqrt{x^2 + y^2}$, which is a cone. This is not a submanifold of $\mathbb{R}^3$. Yet the map $\varphi : \mathbb{R}^2 \to C$ given by $\varphi(x, y) = (x, y, \sqrt{x^2 + y^2})$ is a homeomorphism. Since the image of this map is the whole of $C$, this single map constitutes an atlas for $C$ and defines a manifold structure on $C$. Since $\varphi$ is a homeomorphism, this manifold structure is compatible with the relative topology on $C$.

**Example 2.110** Let $M = \{(x, y) : x = 0 \text{ or } y = 0\}$. We have seen in Example 2.8 that this set is not a submanifold of $\mathbb{R}^2$. In fact, it does not even have any manifold structure compatible with the relative topology, in the sense explained in the preceding remark. Indeed, note that $(1, 2) \times \{0\}$ is open relative to $M$ and homeomorphic with $(1, 2)$. So if $M$ were a manifold, it would have to be 1-dimensional.[4] Now suppose that there exists a coordinate chart $\varphi : U \to \mathbb{R}$ for some neighbourhood $U$ of $(0, 0)$. For this chart to be compatible with the given topology, $\varphi$ must be a homeomorphism. By shrinking $U$ if necessary, we can assume that $U$ and $\varphi(U)$ are connected. Removing the point $(0, 0)$ from $U$ leaves 4 connected components, whereas removing one point from an interval leaves 2 connected components. This contradicts $U$ being a homeomorphism.

We now define tangent vectors, spaces and bundles for abstract manifolds. Let $\gamma$ be a differentiable path in a manifold $M$. As before, we wish to consider $\gamma'(t)$ as a tangent vector based at $\gamma(t)$, but now there is no ambient space $\mathbb{R}^N$, containing $M$, in which to calculate $\gamma'(t)$. In order to calculate $\gamma'(t)$, we need to use a coordinate chart $\varphi$ for $M$, and compute $(\varphi \circ \gamma)'(t)$. Of course, using different coordinate charts gives different answers, and there is no preferred coordinate chart. To get around this, we declare that the tangent vector is actually an *equivalence class* of paths, in the following sense.

**Definition 2.111** *Two paths $\gamma_1(t)$ and $\gamma_2(t)$ in $M$ are called* ***equivalent at*** *$x \in M$ if*

$$\gamma_1(0) = \gamma_2(0) = x \text{ and } (\varphi \circ \gamma_1)'(0) = (\varphi \circ \gamma_2)'(0)$$

*for some coordinate chart $\varphi$.*

Though the preceding definition says 'for some coordinate chart', it would be equivalent to say 'for any coordinate chart' – see Exercise 2.43.

---

[4] Euclidean spaces of different dimensions are not homeomorphic, a topological fact that is not difficult to prove, though we do not do so here.

**Definition 2.112** *A **tangent vector** to $M$ at $x$ is an equivalence class of paths at $x$, in the sense defined above. For any tangent vector and any path $\gamma(t)$ in the equivalence class (note that this implies $\gamma(0) = x$), we say that the tangent vector $v$ is **represented** by $\gamma(t)$, and we write $v = \gamma'(0)$. The point $x$ is called the **base point** of the tangent vector. The **tangent space** to $M$ at $x$, denoted $T_xM$, is the set of all tangent vectors at $x$. The **tangent bundle** of $M$ is*

$$TM := \bigcup_{x \in M} T_xM.$$

*The **tangent bundle projection** is the map $\tau : TM \to M$ taking every tangent vector to its base point.*

The reader should check that this definition generalizes the one given earlier for submanifolds of $\mathbb{R}^N$ – this is Exercise 2.44.

We can now define derivatives and tangent maps exactly as for submanifolds of $\mathbb{R}^n$. The only difference is that we don't need the 'bound version' of the definition of a tangent vector '$(\mathbf{x}, \mathbf{v})$', since our definition of a tangent vector at $x$ for abstract manifolds already implicitly includes the base point $x$.

**Definition 2.113** *The **tangent map of $f$ at** $x$ is the map*

$$\begin{aligned} T_x f \colon T_xM &\to T_{f(x)}N \\ v \mapsto (T_x f)(v) &:= \left.\frac{\mathrm{d}}{\mathrm{d}t}\right|_{t=0} f(g(t)), \end{aligned}$$

*where $g(t)$ is a path in $M$ such that $g(0) = x$ and $\left.\frac{\mathrm{d}}{\mathrm{d}t}\right|_{t=0} g(t) = v$. The map $T_x f$ is often written $\mathrm{D}f(x)$ or $\mathrm{d}f(x)$ or $f_*(x)$, and called the **derivative of** $f$ **at** $x$. The **tangent map** of $f$ is $Tf \colon TM \to TN$ given by $Tf(v) = (T_x f)(v)$ for all $v \in T_xM$.*

**Definition 2.114** *Let $M$ be a manifold. Given any coordinate chart $\varphi : U \to V \subseteq \mathbb{R}^n$ for $M$, the associated **tangent-lifted coordinate chart** for $TM$ is $T\varphi : TU \to TV = V \times \mathbb{R}^n$. If the components of $\varphi$ are denoted $\mathbf{q} = (q^1, \ldots, q^n)$, then the components of $T\varphi$ are denoted $(\mathbf{q}, \dot{\mathbf{q}}) = (q^1, \ldots, q^n, \dot{q}^1, \ldots, \dot{q}^n)$, and called the **tangent-lifted coordinates** corresponding to $\mathbf{q}$.*

If $f\colon M \to N$ is expressed in local coordinates $\mathbf{q}$ on $M$ and $\mathbf{r}$ on $N$, then $Tf$ is expressed in the corresponding tangent-lifted coordinates as

$$(\mathbf{r}, \dot{\mathbf{r}}) = (f(\mathbf{q}), \mathrm{D}f(\mathbf{q}) \cdot \dot{\mathbf{q}}) .$$

**Theorem 2.115** *The tangent bundle $TM$ of any manifold $M$ is a manifold of twice the dimension of $M$, with coordinate charts $T\varphi$ for any coordinate chart $\varphi$ of $M$, If $N$ is another manifold and $f\colon M \to N$ is smooth, then $Tf\colon TM \to TN$ is also smooth.*

*Proof* Exercise. ■

**Remark 2.116** *The tangent bundle is more than just a manifold: it is an example of a* ***vector bundle****, which we will now describe by highlighting and generalizing certain properties of $TM$. For any coordinate chart $\varphi : U \to V \subseteq \mathbb{R}^n$, the tangent map $T\varphi : TU \to V \times \mathbb{R}^n$ is a diffeomorphism. Also, $TU = \pi^{-1}(U)$, where where $\pi : TM \to M$ is the tangent bundle projection. Hence, we have the diffeomorphisms,*

$$\pi^{-1}(U) = TU \cong V \times \mathbb{R}^n \cong U \times \mathbb{R}^n .$$

*Since $M$ is covered by an atlas of coordinate charts, every point in $M$ has a neighbourhood $U$ such that*

$$\pi^{-1}(U) \cong U \times \mathbb{R}^n .$$

*This property is informally described as '$TM$ locally looks like $M \times \mathbb{R}^n$'. In general, an n-dimensional vector bundle is a base manifold $M$, a total manifold $P$ (generalizing $TM$), and a submersion $\pi : P \to M$, such that '$P$ locally looks like $M \times \mathbb{R}^n$', meaning that, for every open subset $U$ of $M$, there is a diffeomorphism from $\pi^{-1}(U)$ to $U \times \mathbb{R}^n$. For any $x \in M$, the preimage $\pi^{-1}(x) \in P$ is called the* ***fibre*** *over $x$ (it is diffeomorphic to $\mathbb{R}^n$). For the tangent bundle, the fibre over $x$ is the tangent space $T_xM$.*

**Exercise 2.36**
Consider the ***stereographic projection*** of $S^1$ given by $\xi_N(x, y) = \frac{2x}{1-y}$. This is a diffeomorphism from $S^1$, excluding the 'North pole' $(0, 1)$, to $\mathbb{R}$, as shown in Example 2.48. The map $\xi_S(x, y) = \frac{2x}{1+y}$ has a similar effect, using lines through the South pole instead of the North. Check that the two maps $\xi_N$ and $\xi_S$ form an atlas for $S^1$, making $S^1$ a manifold. Prove that this manifold structure is compatible with the structure on

$S^1$ defined by the parameterizations $\varphi_1$ and $\varphi_2$ given at the beginning of this section.

**Exercise 2.37**
Let $(\xi_N, \eta_N)$ and $(\xi_S, \eta_S)$ be the stereographic projections of $S^2$ defined in Example 2.49. Show that the transition from coordinates $(\xi_N, \eta_N)$ to $(\xi_S, \eta_S)$ is a diffeomorphism, thus showing that these two projections are coordinate charts defining a manifold structure on $S^2$. (HINT: $(1 + z)(1 - z) = 1 - z^2 = x^2 + y^2$ on $S^2$.) This exercise generalizes in the obvious way to $S^n \subseteq \mathbb{R}^n$ for any $n$.

**Exercise 2.38**
The symbol '$T^2$', which denotes a 2-dimensional torus, has two standard interpretations: (i) $S^1 \times S^1 \subseteq \mathbb{R}^4$, see Example 2.36; and (ii) a submanifold of $\mathbb{R}^3$ shaped like the surface of an idealized doughnut (with a hole), see Exercise 2.5. Show that these are diffeomorphic.

**Exercise 2.39**
Show that the ray $[0, \infty)$ is not a manifold, in the sense explained in Remark 2.108. The ray is an example of what is called a 'manifold with boundary'.

**Exercise 2.40**
If $M$ is a set and $N$ is a manifold, and $\varphi : M \to N$ is a bijection, show that the manifold structure on $N$ can be used to define a manifold structure on $M$ such that $\varphi$ is a diffeomorphism, and that this is the unique manifold structure on $M$ with this property. This manifold structure on $M$ is called the structure ***induced*** by $\varphi$, or the ***pull-back*** by $\varphi$ of the manifold structure on $N$.

**Exercise 2.41**
If we begin with a figure eight in the $xz$-plane, along the $x$-axis and centred at the origin, and spin it round the $z$-axis in $\mathbb{R}^3$, we get a 'pinched

surface' that looks like a sphere that has been 'pinched' so that the North and South poles touch. Prove that this is not a manifold.

**Exercise 2.42**
Let $F: M \to N$. Let $\varphi_1$ and $\varphi_2$ be two coordinate charts for $M$ with the same domain, and let $\psi_1$ and $\psi_2$ two coordinate charts for $N$ with the same domain. Show that if $\psi_1 \circ F \circ \varphi_1^{-1}$ is smooth then $\psi_2 \circ F \circ \varphi_2^{-1}$ is as well. Conclude that, to check smoothness, it suffices to consider only one atlas of charts for $M$ and one atlas of charts for $N$.

**Exercise 2.43**
Check that the definition if a tangent vector in Definition 2.111 is independent of the choice of chart $\varphi$. More precisely, check that if $\varphi$ and $\psi$ are two charts (with $x$ in the domain of both), then $(\varphi \circ \gamma_1)'(0) = (\varphi \circ \gamma_2)'(0)$ if and only if $(\psi \circ \gamma_1)'(0) = (\psi \circ \gamma_2)'(0)$.

**Exercise 2.44**
Explain why Definition 2.112 generalizes the definitions of tangent vectors, spaces and bundles given earlier for submanifolds of $\mathbb{R}^N$.

## Solutions to selected exercises

**Solution to Exercise 2.1** Let $f(x, y, z) = (x^2 + y^2, y^2 + z^2)$ then,

$$Df(x, y, z) = 2\begin{bmatrix} x & y & 0 \\ 0 & y & z \end{bmatrix}.$$

This map is onto $\mathbb{R}^2$ when its rank is 2. Therefore, $Df(x, y, z)$ is surjective for all $(x, y, z)$ except on the axes. A point $(a, b) \in \mathbb{R}^2$ is a regular value of $f$ if and only if the level set $f^{-1}\{(a, b)\}$ does not intersect the axes. Thus, $(a, b)$ is a regular value of $f$ if and only if $a$ and $b$ are unequal and non-zero.

The level sets satisfy $x^2 + y^2 = a$ and $y^2 + z^2 = b$. Therefore, if $a, b > 0$, the levels sets are intersections of cylinders about the $z$-axis and cylinders about the $x$-axis. If $a \neq b$, then the level set is the union of two disjoint closed curves, which is a 1-dimensional submanifold. This verifies that the preimage of any regular value is a codimension-2 submanifold of $\mathbb{R}^3$. If $a = b$, then the cylinders are of equal radius and their intersection is the union of two circles that intersect on the $y$-axis; this is not a submanifold. If one or both of $a$ and $b$ is zero, then $f^{-1}\{(a,b)\}$ consists of either one or two points, and is thus a 0-dimensional manifold. These are all submanifolds that are preimages of non-regular values.

**Solution to Exercise 2.2** Suppose $g \circ f$ is well defined. By the chain rule, $\mathrm{D}(g \circ f)(z) = \mathrm{D}g(f(z))\mathrm{D}f(z)$ for every $z$ in the domain of $f$. If $f$ and $g$ are both submersions, then $\mathrm{D}f(z)$ and $\mathrm{D}g(f(z))$ are both surjective, which implies that $\mathrm{D}g(f(z))\mathrm{D}f(z)$ is surjective, for all $z$ in the domain of $f$, which is also the domain of $g \circ f$. Hence, $g \circ f$ is a submersion.

If $g \circ f$ is a submersion, then $\mathrm{D}g(f(z))\mathrm{D}f(z)$ is surjective, for every $z$ in the domain of $f$. This does not imply that $\mathrm{D}f(z)$ is surjective, so $f$ need not be a submersion. It does imply that $\mathrm{D}g(f(z))$ is surjective for every $y = f(z)$ in the image of $f$, but not necessarily for every $y$ in the domain of $g$, so $g$ need not be a submersion (even though it is a submersion *at* $f(z)$, for every $z$). For example, if $f\colon (0,\infty) \to \mathbb{R}^2$ and $g : \mathbb{R}^2 \to \mathbb{R}$ are given by

$$g(x,y) = x^3, \quad f(z) = (z,0),$$

then

$$T_{(x,y)}g = \begin{bmatrix} 3x^2 & 0 \end{bmatrix}, \quad T_z f = \begin{bmatrix} 1 \\ 0 \end{bmatrix}, \quad T_z(g \circ f) = \begin{bmatrix} 3z^2 \end{bmatrix}.$$

Since $T_z f$ is never surjective, $f$ is not a submersion. Since $T_{(x,y)}g$ is not surjective at points with $x = 0$, the map $g$ is not a submersion either. The domain of $g \circ f$ is $(0,\infty)$ and $T_z(g \circ f)$ is surjective at all $z \in (0,\infty)$, therefore $g \circ f$ is a submersion.

**Solution to Exercise 2.3** Let $f\colon \mathbb{R}^2 \to \mathbb{R}^1$ be given by $f(x,y) = 1$. Then, $\mathrm{D}f = 0$ so $\mathrm{D}f$ is not surjective, but $f^{-1}(1) = \mathbb{R}^2$, which is a submanifold of $\mathbb{R}^2$.

**Solution to Exercise 2.5** The relation

$$\left(2r - \sqrt{x^2 + y^2}\right)^2 + z^2 = r^2,$$

relates the Cartesian coordinates, $(x,y,z) \in \mathbb{R}^3$, to the radius, $r \in \mathbb{R}$, of a torus. We want to solve this relation for $r$ to obtain a submersion. Expanding the relation gives:

$$3r^2 - 4r\sqrt{x^2 + y^2} + z^2 = 0,$$

and solving for $r$ gives:

$$r = \frac{2}{3}\left(\sqrt{x^2+y^2} \pm \sqrt{4\left(x^2+y^2\right) - 3z^2}\right).$$

The restriction $x^2 + y^2 > \frac{3}{4}z^2$ is required for a real solution. This is acceptable since it gives an open subset of $\mathbb{R}^3$ that contains the torus we want to describe. Now let

$$f(x,y,z) = \frac{2}{3}\left(\sqrt{x^2+y^2} \pm \sqrt{4\left(x^2+y^2\right) - 3z^2}\right), \quad x^2+y^2 > \frac{3}{4}z^2.$$

Then, $f$ is a smooth map and, for $x^2 + y^2 > \frac{3}{4}z^2$,

$$\mathrm{D}f(x,y,z) = \frac{2}{3}\left(\frac{1}{\sqrt{x^2+y^2}}\begin{bmatrix} x & y & 0 \end{bmatrix} + \frac{1}{\sqrt{4\left(x^2+y^2\right) - 3z^2}}\begin{bmatrix} 4x 4y - 3z \end{bmatrix}\right),$$

which is surjective. Therefore, $f$ is a submersion such that $f^{-1}(r)$ is the torus of radius $r$. Consequently, the torus of radius $r$ is a submanifold of $\mathbb{R}^3$.

**Solution to Exercise 2.6** Let $f\colon \mathbb{R}^n \to \mathbb{R}^p$ be a submersion and let $M$ be a submanifold of $\mathbb{R}^p$ of codimension $k$. For every $z \in f^{-1}(M)$, since $M$ is a submanifold, there exists a neighbourhood $U$ of $f(z)$ and a smooth submersion $g : U \to \mathbb{R}^k$ such that $M \cap U = g^{-1}(\mathbf{c})$ for some $\mathbf{c} \in \mathbb{R}^p$. Let $V = f^{-1}(U)$, and note that $z \in V$, and $V$ is open because $f$ is continuous. Let $h = g \circ f\colon V \to \mathbb{R}^k$, which is a smooth submersion since both $f$ and $g$ are smooth submersions (see the solution to Exercise 2.2). Then,

$$\begin{aligned} f^{-1}(M) \cap V &= f^{-1}(M) \cap f^{-1}(U) \\ &= f^{-1}(M \cap U) = f^{-1}\left(g^{-1}(\mathbf{c})\right) = (g \circ f)^{-1}(\mathbf{c}). \end{aligned}$$

Since this is true at every $z \in f^{-1}(M)$, it follows that $f^{-1}(M)$ is a submanifold of $\mathbb{R}^n$. Since $g \circ f$ has codomain $\mathbb{R}^k$, the codimension of $f^{-1}(M)$ is $k$, which is the codimension of $M$.

For example, let $M = S^1$, which is a codimension-1 submanifold of $\mathbb{R}^2$, and consider $G : \mathbb{R}^3 \to \mathbb{R}^2$ defined by $G(x,y,z) = (x,y)$. Then, $G^{-1}(M) = S^1 \times \mathbb{R}$, a cylinder, which is a codimension-1 submanifold of $\mathbb{R}^3$.

The image of a submanifold by a submersion is not always a submanifold. Consider for example, $f(x,y,z) = (x,y)$. It is easy to construct a path $c(t)$ in $\mathbb{R}^3$ such that its image is a submanifold of $\mathbb{R}^3$ but $f(c(t))$ is a path in $\mathbb{R}^2$ that crosses itself. The image of such a path cannot be a submanifold.

**Solution to Exercise 2.12** The cylinder is described by the constraint $x^2 + y^2 = c$ in $\mathbb{R}^3$. Differentiating this relationship we find

$$2\,(x\dot{x} + y\dot{y}) = 0.$$

This is the condition on $T\mathbb{R}^3$ that describes the tangent bundle to the level sets of the submersion $f\colon \mathbb{R}^3 \to \mathbb{R}$, $f(x,y,z) = x^2 + y^2$. Note that $f^{-1}(c) = S^1 \times \mathbb{R}$. Therefore, the tangent bundle of the cylinder is

$$T(S^1 \times \mathbb{R}) = \left\{ (x, y, z, \dot{x}, \dot{y}, \dot{z}) | x^2 + y^2 = c,\ x\dot{x} + y\dot{y} = 0 \right\}.$$

**Solution to Exercise 2.14** The parameterization of the sphere in $\mathbb{R}^3$ by standard spherical coordinates is given by

$$f(\theta, \phi) = (\sin\phi\cos\theta, \sin\phi\sin\theta, -\cos\phi) \in \mathbb{R}^3.$$

The derivative of $f$ is given by:

$$\mathrm{D}f(\theta,\phi) = \begin{bmatrix} -\sin\phi\sin\theta & \cos\phi\cos\theta \\ \sin\phi\cos\theta & \cos\phi\sin\theta \\ 0 & \sin\phi \end{bmatrix}.$$

Therefore,

$$\begin{bmatrix} \dot{x} \\ \dot{y} \\ \dot{z} \end{bmatrix} = \mathrm{D}f(0, \pi/4) \cdot \begin{bmatrix} 2 \\ -1 \end{bmatrix} = \begin{bmatrix} 0 & \frac{1}{\sqrt{2}} \\ \frac{1}{\sqrt{2}} & 0 \\ 0 & \frac{1}{\sqrt{2}} \end{bmatrix} \begin{bmatrix} 2 \\ -1 \end{bmatrix} = \begin{bmatrix} -\frac{1}{\sqrt{2}} \\ \sqrt{2} \\ -\frac{1}{\sqrt{2}} \end{bmatrix}.$$

Thus, in Cartesian coordinates in $\mathbb{R}^3$,

$$(\mathbf{x}, \mathbf{v}) = (x, y, z, \dot{x}, \dot{y}, \dot{z}) = \frac{1}{\sqrt{2}}(1, 0, -1, -1, 2, -1).$$

**Solution to Exercise 2.15** The stereographic projection is given by

$$\xi = \frac{2x}{1-z}, \qquad \eta = \frac{2y}{1-z}.$$

Therefore, using the standard parameterization of the sphere in $\mathbb{R}^3$,

$$\xi = \frac{2\sin\phi\cos\theta}{1+\cos\phi}, \qquad \eta = \frac{2\sin\phi\sin\theta}{1+\cos\phi}.$$

Consequently,

$$\frac{\partial}{\partial\theta}\begin{bmatrix}\xi\\ \eta\end{bmatrix}=\begin{bmatrix}-\dfrac{2\sin\phi\sin\theta}{1+\cos\phi}\\ \dfrac{2\sin\phi\cos\theta}{1+\cos\phi}\end{bmatrix}=\begin{bmatrix}-\eta\\ \xi\end{bmatrix}.$$

**Solution to Exercise 2.18** Given a map $f\colon S^2\to S^1\times\mathbb{R}$, $Tf$ is computed in the given coordinates as

$$Tf(\theta,\phi)\cdot\begin{bmatrix}\dot\theta\\ \dot\phi\end{bmatrix}=\left(f(\theta,\phi),\mathrm{D}f(\theta,\phi)\cdot\begin{bmatrix}\dot\theta\\ \dot\phi\end{bmatrix}\right).$$

Here substituting $f(\theta,\phi)=(\theta,\cos\phi)$ yields,

$$Tf(\theta,\phi)\cdot\begin{bmatrix}\dot\theta\\ \dot\phi\end{bmatrix}=\left(\theta,\cos\phi,\dot\theta,-\sin\phi\,\dot\phi\right).$$

**Solution to Exercise 2.23** Suppose $V$ is a finite-dimensional inner product space with inner product $\langle\langle\cdot\,,\,\cdot\rangle\rangle : V\times V\to\mathbb{R}$. Consider the map:

$$f\colon V\to V^*,\quad f(\boldsymbol{v})=\alpha_{\boldsymbol{v}}:=\langle\langle\boldsymbol{v}\,,\,\cdot\rangle\rangle.$$

The claim is that $f$ is an isomorphism. Clearly $f$ is linear since the inner product is bilinear. It is injective since $f(\boldsymbol{v})=0$ implies $\langle\langle\boldsymbol{v}\,,\,\boldsymbol{w}\rangle\rangle=0$ for all $\boldsymbol{w}\in V$, which, by non-degeneracy of the inner product, implies that $\boldsymbol{v}=0$. To prove surjectivity, let $\{\boldsymbol{e}_1,\ldots,\boldsymbol{e}_n\}$ be an orthonormal basis for $V$. For every $\alpha\in V^*$, set

$$\boldsymbol{v}=\sum_{i=1}^{n}\alpha(\boldsymbol{e}_i)\boldsymbol{e}_i,$$

so that $v^i=\alpha(\boldsymbol{e}_i)$ for $i=1,\ldots,n$. Then,

$$\langle\langle\boldsymbol{v}\,,\,\boldsymbol{w}\rangle\rangle=\left\langle\!\!\left\langle v^i\boldsymbol{e}_i\,,\,w^j\boldsymbol{e}_j\right\rangle\!\!\right\rangle=v^iw^j\langle\langle\boldsymbol{e}_i\,,\,\boldsymbol{e}_j\rangle\rangle=\alpha(\boldsymbol{e}_i)w^j\delta^i_j=\alpha\left(\boldsymbol{w}\right).$$

Since $f(\boldsymbol{v})\left(\boldsymbol{w}\right)=\langle\langle\boldsymbol{v}\,,\,\boldsymbol{w}\rangle\rangle$ for all $\boldsymbol{w}\in V$, by definition of $f$, this shows that $f(\boldsymbol{v})=\alpha$. Therefore, $f$ is surjective, and hence an isomorphism.

**Solution to Exercise 2.26** Changing the form coordinates by a push-forward gives

$$s_i=p_j\frac{\partial q^j}{\partial r^i}.$$

The form transforms as

$$dr^i = \frac{\partial r^i}{\partial q^j} dq^j.$$

Combining these two expressions yields

$$s_i dr^i = p_j \frac{\partial q^j}{\partial r^i} \frac{\partial r^i}{\partial q^j} dq^j = p_j dq^j.$$

**Solution to Exercise 2.30** The defining relationship for orthogonal matrices is $\boldsymbol{Q}^T \boldsymbol{Q} = \boldsymbol{I}$. Taking the determinant of this relationship gives,

$$1 = \det\left(\boldsymbol{Q}^T \boldsymbol{Q}\right) = \det \boldsymbol{Q}^T \det \boldsymbol{Q} = \left(\det \boldsymbol{Q}\right)^2.$$

Thus, orthogonality requires $\det \boldsymbol{Q} = \pm 1$. Similarly, symplectic matrices have the defining relationship $\boldsymbol{Q}^T \boldsymbol{J} \boldsymbol{Q} = \boldsymbol{J}$. Again, taking determinants gives

$$\begin{aligned} 1 = \det \boldsymbol{J} &= \det\left(\boldsymbol{Q}^T \boldsymbol{J} \boldsymbol{Q}\right) \\ &= \det \boldsymbol{Q}^T \det \boldsymbol{J} \det \boldsymbol{Q} = \left(\det \boldsymbol{Q}\right)^2. \end{aligned}$$

Therefore, symplecticity also requires $\det \boldsymbol{Q} = \pm 1$.

**Solution to Exercise 2.31** It is easy to show $Sp(2n, \mathbb{R})$ is a matrix subgroup of $GL(2n, \mathbb{R})$. Consider the map

$$F\colon GL(2n, \mathbb{R}) \to Skew(2n, \mathbb{R}), \quad F(\boldsymbol{A}) = \boldsymbol{A}^T \boldsymbol{J} \boldsymbol{A} - \boldsymbol{J}.$$

Note that $F^{-1}(0) = Sp(2n, \mathbb{R})$. Now $DF(\boldsymbol{A}) \cdot \boldsymbol{B}$ is computed for some $\boldsymbol{B} \in M(2n, \mathbb{R})$ as follows,

$$\begin{aligned} DF(\boldsymbol{A}) \cdot \boldsymbol{B} &= \left.\frac{d}{dt}\right|_{t=0} (\boldsymbol{A} + t\boldsymbol{B})^T \boldsymbol{J} (\boldsymbol{A} + t\boldsymbol{B}) - \boldsymbol{J} \\ &= \boldsymbol{B}^T \boldsymbol{J} \boldsymbol{A} + \boldsymbol{A}^T \boldsymbol{J} \boldsymbol{B}. \end{aligned}$$

Suppose $\boldsymbol{A} \in Sp(2n, \mathbb{R})$. To prove that $DF(\boldsymbol{A})$ is surjective it suffices to show that for any $\boldsymbol{S} \in Skew(2n, \mathbb{R})$ there is a $\boldsymbol{B} \in M(2n, \mathbb{R})$ such that $\boldsymbol{S} = \boldsymbol{B}^T \boldsymbol{J} \boldsymbol{A} + \boldsymbol{A}^T \boldsymbol{J} \boldsymbol{B}$. Such a $\boldsymbol{B}$ is given by

$$\boldsymbol{B} = -\frac{1}{2} \boldsymbol{J} \boldsymbol{A}^{-T} \boldsymbol{S}.$$

This is verified as follows,

$$B^T J A + A^T J B = -\left(\frac{1}{2} J A^{-T} S\right)^T J A - \frac{1}{2} A^T J^2 A^{-T} S$$
$$= -\frac{1}{2} S A^{-1} J^2 A + \frac{1}{2} S = S$$

(since $J^2 = -I$). Thus, $F$ is a submersion and $F^{-1}(0) = Sp(2n, \mathbb{R})$ is a submanifold of $GL(2n, \mathbb{R})$. Therefore, $Sp(2n, \mathbb{R})$ is a matrix Lie group. The dimension of $Skew(2n, \mathbb{R})$ is $2n(2n - 1)/2$. Therefore, the dimension of $Sp(2n, \mathbb{R})$ is given by

$$4n^2 - \frac{4n^2 - 2n}{2} = 2n^2 + n.$$

Also, observe that substituting $A = I$ into our expression for D$F$ yields

$$T_I Sp(2n, \mathbb{R}) = \ker \mathrm{D}F(I) = \left\{ B \in M(2n, \mathbb{R}) | B^T J + J B = 0 \right\}.$$

**Solution to Exercise 2.33** It is easy to show $SL(n, \mathbb{C})$ is a matrix subgroup of $GL(n, \mathbb{C})$. Consider the map

$$F \colon GL(2n, \mathbb{C}) \to \mathbb{C}, \quad F(A) = \det A - 1.$$

Note that $F^{-1}(0) = SL(n, \mathbb{C})$. Since det is a submersion it follows that $F$ is also a submersion. Therefore, $SL(n, \mathbb{C})$ is a Lie subgroup of $GL(n, \mathbb{C})$. The dimension of $\mathbb{C}$ is 2, so $SL(n, \mathbb{C})$ has dimension $2(n^2 - 1)$.

**Solution to Exercise 2.35** The defining relationship of $U(n)$ is $Q^\dagger Q = I$ for $Q \in \mathbb{C}$, so that $\det Q = e^{i\phi}$ for some $\phi \in \mathbb{R}$. Therefore, consider the map

$$F \colon U(n) \to \mathbb{R}, \quad F(A) = -\mathrm{i} \log \det A.$$

Note that $F^{-1}(0) = SU(n)$ and compute $\mathrm{D}F(A) \cdot B$ for some $B \in T_A U(n)$ as

$$\mathrm{D}F(A) \cdot B = \left.\frac{\mathrm{d}}{\mathrm{d}t}\right|_{t=0} - \mathrm{i} \log \det A = -\mathrm{i} \frac{\mathrm{D} \det(A) \cdot B}{\det A}.$$

Since det is a submersion, D det$(A)$ is surjective, so $F$ is also a submersion. Thus, $SU(n)$ a Lie subgroup of $U(n)$ of codimension 1. Therefore, $SU(n)$ has dimension $n^2 - 1$.

# 3 Geometry on manifolds

***Differential geometry*** is geometry studied via advanced calculus. It applies not just in Euclidean space but on general smooth manifolds.

***Riemannian geometry*** is the oldest branch of differential geometry, and the branch that most directly generalizes Euclidean geometry: it is concerned with lengths and angles. In mechanics, Riemannian geometry is used to describe kinetic energy; and also the paths followed by particles in the absence of external forces, which are *geodesics*.

***Symplectic geometry*** is very different: it is not concerned with lengths or angles, but it does generalize area. It is used in mechanics to generalize Hamilton's equations of motion. These can be further generalized using *Poisson brackets*.

***Vector fields*** are essentially ordinary differential equations. Vector fields, general ***tensors*** and ***differential forms*** are used in all branches of differential geometry.

We begin with vector fields and differential 1-forms, then introduce general tensor fields. The chapter concludes with brief introductions to Riemannian and symplectic geometry.

## 3.1 Vector fields

A vector field is a collection of tangent vectors, one at every point in the manifold:

**Definition 3.1** *A **vector field** on a manifold $M$ is a map $X : M \to TM$ such that $X(z) \in T_zM$ for all $z \in M$. Vector fields can be added to each other, and multiplied by functions $k : M \to \mathbb{R}$ (called **scalar fields**), as follows:*

$$(X_1 + X_2)(z) := X_1(z) + X_2(z), \quad (kX)(z) := k(z)X(z).$$

*A **smooth** (resp. **differentiable**) vector field is one that is a smooth (resp. differentiable) map from $M$ to $TM$. The set of all smooth vector fields on $M$ is written $\mathfrak{X}(M)$. The set of all smooth scalar fields is denoted $\mathcal{F}(M)$.*

Given any tangent-lifted local coordinates $(q^1, \ldots, q^n, \dot{q}^1, \ldots, \dot{q}^n)$ defined on $TU \subseteq TM$, the $i$th ***component*** of a vector field $X$ is the scalar field

$X^i : U \to \mathbb{R}$ such that $X^i(z)$ is the $\dot{q}^i$ coordinate of $X(z)$ for all $z \in U$. Equivalently,

$$X = X^i \frac{\partial}{\partial q^i}.$$

We write $X = \left(X^1, \ldots, X^n\right)$. By the definition of the manifold structure on $TM$, a vector field $X$ is smooth if and only if all of its components are smooth, in any coordinates.

**Definition 3.2** *An* ***integral curve*** *of $X$ is a diffentiable map $c : I \to M$, where $I$ is an open interval in $\mathbb{R}$, such that $c'(t) = X(c(t))$ for all $t \in I$. Integral curves are also called* ***solution curves, particular solutions*** *or* ***trajectories*** *of the corresponding vector field.*

**Remark 3.3** *When written in coordinates, vector fields represent differential equations.*

**Example 3.4** The vector field $X$ on $\mathbb{R}^2$ defined by $X(x, y) = (-y, x) \in T_{(x,y)}\mathbb{R}^2$ corresponds to the system of differential equations,

$$\begin{cases} \dot{x} = -y, \\ \dot{y} = x. \end{cases}$$

It has solution curves of the form $(x(t), y(t)) = (r\cos(t + \omega), r\sin(t + \omega))$. Note that, for every $(x, y) \in S^1$, the vector $X(x, y)$ is tangent to $S^1$. Therefore, the restriction of $X$ to the unit circle $S^1$ is a vector field on $S^1$.

**Example 3.5** The vector field $X(x, y, z) = (-y, x, 0)$ on $\mathbb{R}^3$ corresponds to the system of differential equations

$$\begin{cases} \dot{x} = -y, \\ \dot{y} = x, \\ \dot{z} = 0. \end{cases}$$

Since $(-y, x, 0) \cdot (x, y, z) = 0$ always, the restriction of $X$ to $S^2$ is a vector field on $S^2$ (see Example 2.56). The restriction of $X$ to $S^2$ has solution curves of the form

$$(x(t),\ y(t),\ z(t)) = (\sin\phi\cos(t + \omega), \sin\phi\sin(t + \omega), -\cos\phi),$$

for any constants $\phi$ and $\omega$. (When $\phi$ is a multiple of $\pi$, the 'curves' are just points at the North or South pole.)

The same vector field can be written in spherical coordinates as $X(\theta,\phi) = (1,0)$. To verify this, let $\psi(\theta,\phi) = (x,y,z) = (\sin\phi\cos\theta, \sin\phi\sin\theta, -\cos\phi)$ and calculate:

$$\mathrm{D}\psi(\theta,\phi)\begin{bmatrix}1\\0\end{bmatrix} = \begin{bmatrix}\partial x/\partial\theta\\ \partial y/\partial\theta\\ \partial z/\partial\theta\end{bmatrix} = \begin{bmatrix}-\sin\phi\sin\theta\\ \sin\phi\cos\theta\\ 0\end{bmatrix} = \begin{bmatrix}-y\\ x\\ 0\end{bmatrix} = X(x,y,z).$$

In spherical coordinates, $X$ represents the system of differential equations

$$\begin{cases}\dot{\theta} = 1,\\ \dot{\varphi} = 0.\end{cases}$$

**Example 3.6** Let $M$ be a manifold. Since $TM$ is also a manifold, we can consider vector fields on it. In tangent-lifted coordinates, every vector field on $TM$ has the form

$$X = \sum_i a^i \frac{\partial}{\partial q^i} + b^i \frac{\partial}{\partial \dot{q}^i},$$

where the $a^i$ and $b^i$ are functions of $q$ and $\dot{q}$. Note that the same symbol $q^i$ has two interpretations: as a coordinate on $TM$ and as a coordinate on $M$, so $\partial/\partial q^i$ can mean a vector field $TM$ (as above) or on $M$. Similarly, in cotangent-lifted coordinates, every vector field on $T^*M$ has the form

$$X = \sum_i a^i \frac{\partial}{\partial q^i} + b^i \frac{\partial}{\partial p_i}.$$

Using the technique of ***reduction of order*** for ordinary differential equations (ODEs), any ODE of order two or above may be transformed into an equivalent first-order ODE on a higher-dimensional space. This implies that every ODE corresponds to a vector field, so vector fields are equivalent to ODEs. The difference is that the definition of a vector field is coordinate independent, whereas writing down the corresponding ODE requires a choice of local coordinates.

The following theorem says that integral curves exist and are unique. Further, the integral curves $\Phi_z(t)$ with initial condition $z$ at $t = 0$ can be pieced together into a smooth ***flow*** $\Phi(z,t) := \Phi_z(t)$. In ODE terminology, a flow is a ***general solution***. In the simplest case, the flow is a map from $M \times \mathbb{R}$ to $M$, but the domain of $\Phi$ may be restricted, because the integral curves may have limited domains, and the domains can be different for different $z$ values.

**Definition 3.7** *Let $X$ be a differentiable vector field on a manifold $M$. A **flow** of $X$ is a differentiable map $\Phi : U \times I \to M$, where $I \subseteq \mathbb{R}$ is an interval containing 0 and $U$ is an open subset $M$, such that, for any $z \in U$, the map $\Phi_z(t) := \Phi(z,t)$ is an integral curve of $X$ with $\Phi_z(0) = z$.*

**Theorem 3.8** *Let $X$ be a differentiable vector field on a manifold $M$. Then, integral curves of $X$ exist, with any given initial condition, though the domain of the integral curve need not be all of $\mathbb{R}$. On any domain on which an integral curve is defined, it is defined uniquely. For every $z \in M$ there exists a flow $\Phi : U \times I \to M$, where $U$ is a neighbourhood of $z$. If $X$ is smooth, then its integral curves and flows are smooth.*

For a proof of this and the following theorem, and more details, see [AM78].

One way of thinking about a flow is to fix a 'time' $t$, giving a map $\Phi_t(z) := \Phi(z,t)$, called a ***time-t flow*** of $X$.

**Theorem 3.9** *Let $\Phi$ be a flow of a differentiable vector field $X$ on a manifold $M$. Then:*

- $\Phi_0 = \mathrm{Id}$, *(the identity function on the domain of $\Phi_0$), i.e. $\Phi_0(z) = z$ for all $z$;*
- $\Phi_{t+s} = \Phi_t \circ \Phi_s$; *(the **flow property**)*
- $\Phi_t$ *is a diffeomorphism onto its image, for every $t$.*

*If $\Phi$ is defined on all of $M \times \mathbb{R}$, and $\Phi_t : M \to M$ is surjective for all $t$, then the above three properties imply that the set of all of the time-$t$ maps $\Phi_t$ forms a group, with the group operation being composition.*

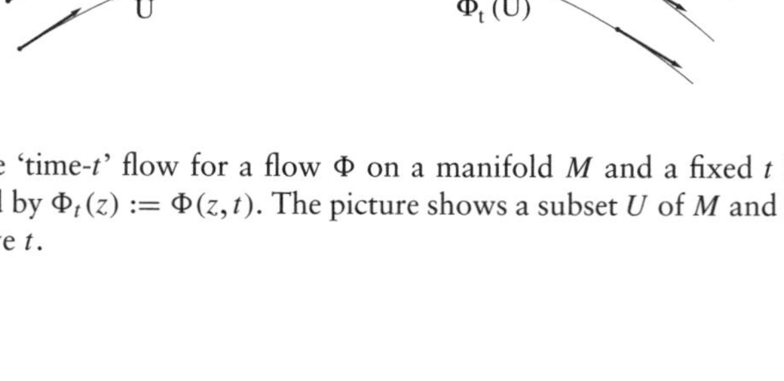

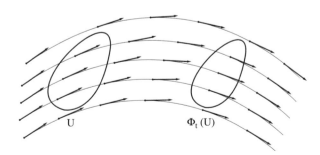

**Figure 3.1** The 'time-$t$' flow for a flow $\Phi$ on a manifold $M$ and a fixed $t$ is the map $\Phi_t : M \to M$ defined by $\Phi_t(z) := \Phi(z,t)$. The picture shows a subset $U$ of $M$ and its image $\Phi_t(U)$ for some positive $t$.

We will usually assume that vector fields are complete, i.e. they have a globally defined flow $\Phi : M \times \mathbb{R} \to M$. We will also assume that each $\Phi_t$ is a diffeomorphism. When we say 'the' flow of $X$, we are assuming that a globally defined flow exists, but the reader will be able to supply slight generalizations of our statements for locally defined flows.

We now define the 'push-forward' and 'pull-back' of a vector field, which are the natural transformations of a vector field associated to a diffeomorphism on the base manifold. The push-forward operation is essentially a tangent lift, but applied to all vectors in a vector field simultaneously. A pull-back is the opposite of a push-forward.

**Definition 3.10** *Let $\varphi : M \to N$ be differentiable and invertible. The* ***push-forward*** *of a vector field $X$ on $M$ by $\varphi$ is the vector field $\varphi_* X$ on $N$ defined by*

$$\varphi_* X = T\varphi \circ X \circ \varphi^{-1};$$

$$\text{i.e.} \qquad (\varphi_* X)(\varphi(z)) = T_z\varphi\,(X(z)), \qquad \text{for all } z \in M.$$

*The push-forward is illustrated in the following commuting diagram:*

$$\begin{array}{ccc} TM & \xrightarrow{T\varphi} & TN \\ X \uparrow & & \uparrow \varphi_* X \\ M & \xrightarrow{\varphi} & N. \end{array}$$

*In local coordinates, writing $\mathbf{r} = \varphi(\mathbf{q})$,*

$$(\varphi_* X)(\mathbf{r}) = \mathrm{D}\varphi(\mathbf{q}) \cdot X(\mathbf{q}) = \frac{\mathrm{d}\mathbf{r}}{\mathrm{d}\mathbf{q}} \cdot X(\mathbf{q}),$$

*which is equivalent to*

$$(\varphi_* X)^i = \frac{\partial r^i}{\partial q^j} X^j. \tag{3.1}$$

*The* ***pull-back*** *of a vector field $Y$ on $N$ by a diffeomorphism $\varphi : M \to N$ is the vector field $\varphi^* Y$ on $M$ defined by*

$$\varphi^* Y = \left(\varphi^{-1}\right)_* Y = T\varphi^{-1} \circ Y \circ \varphi.$$

*In local coordinates, writing* $\mathbf{r} = \varphi(\mathbf{q})$ *and applying the Inverse Function Theorem,*

$$\begin{aligned}\left(\varphi^*Y\right)(\mathbf{q}) &= \mathrm{D}\left(\varphi^{-1}\right)(\mathbf{r})\cdot Y(\mathbf{r})\\ &= (\mathrm{D}\varphi(\mathbf{q}))^{-1}\cdot Y(\mathbf{r})\\ &= \frac{\mathrm{d}\mathbf{q}}{\mathrm{d}\mathbf{r}}\cdot Y(\mathbf{r}).\end{aligned} \tag{3.2}$$

**Remark 3.11** *If* $M = N$*, and* $\varphi$ *is interpreted as a change of coordinates, then eqn (3.1) is just the familiar formula for the tangent lift of a change of coordinates, seen earlier in eqn (2.5).*

**Example 3.12** Define $\psi\colon \left(U \subset \mathbb{R}^2\right) \to \left(V \subset \mathbb{R}^2\right)$ by $\psi(r,\theta) = (r\cos\theta, r\sin\theta)$, where $U = \mathbb{R}_+ \times (0, 2\pi)$ and $V = \{(x,y) : y \neq 0 \text{ or } x < 0\}$. Define $X$ on $U$ by $X(r,\theta) = (0,1)$. Then, for any $(r,\theta)$,

$$\begin{aligned}(\psi_*X)(r\cos\theta, r\sin\theta) &= \left(T\psi\circ X\circ\psi^{-1}\right)(\psi(r,\theta)) = (T\psi\circ X)(r,\theta)\\ &= \mathrm{D}\psi(r,\theta)\cdot(0,1)\\ &= \begin{bmatrix}\cos\theta & -r\sin\theta\\ \sin\theta & r\cos\theta\end{bmatrix}\begin{bmatrix}0\\1\end{bmatrix} = \begin{bmatrix}-r\sin\theta\\ r\cos\theta\end{bmatrix}.\end{aligned}$$

Note that if we consider $\psi$ as a parameterization of $\mathbb{R}^2$ by polar coordinates, then $\psi_*X$ is the representation in Cartesian coordinates of the vector field with polar representation $X$.

**Remark 3.13** *If* $\Phi$ *is the flow of* $X$ *and* $\Psi$ *is the flow of a push-forward* $\varphi_*X$*, then* $\Psi$ *is a kind of 'push-forward' of* $\Phi$*, in the sense that (see Exercise 3.3),*

$$\Psi_t(z) = \varphi\circ\Phi_t(z), \quad \textit{for all } t.$$

**Example 3.14** Figure 3.2 shows the pull-back of a vector field by the stereographic projection map defined in Example 2.48.

We now study differentiation along vector fields. Let $f$ be a smooth scalar field on $M$, i.e. a smooth map $f : M \to \mathbb{R}$. For any $z \in M$, and any $v \in T_zM$, we can compute a ***directional derivative***[1] of $f$ at $z$ along $v$, namely $\mathrm{d}f(z)\cdot v$.

[1] Some definitions of directional derivative require $v$ to have unit length.

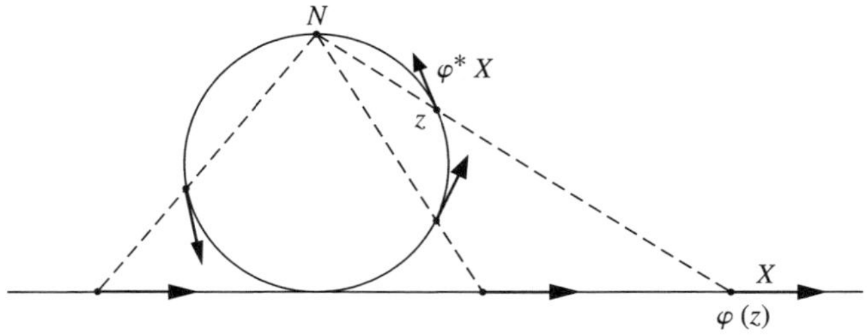


**Figure 3.2** The pull-back of a vector field by the stereographic projection map. The dotted lines show the correspondence between points on the circle and points on the line given by the stereographic projection map $\varphi : S^1\backslash\{N\} \to \mathbb{R}$. The vector field $X$ defined on the line has a corresponding vector field $\varphi^* X$ on $S^1\backslash\{N\}$, called the pull-back of $X$ by $\varphi$.

In any local coordinates $z^1, \ldots, z^n$,

$$\mathrm{d}f(z) \cdot v = \nabla f(z) \cdot v = \sum_i \frac{\partial f}{\partial z^i} v^i.$$

For any curve $c(t)$ with $c(0) = z$ and $c'(0) = v$, the chain rule implies that

$$\mathrm{d}f(z) \cdot v = \left.\frac{\mathrm{d}}{\mathrm{d}t}\right|_{t=0} f \circ c(t).$$

Now, let $X$ be a smooth vector field on a manifold $M$, with flow $\Phi$. For any $z \in M$, since $\Phi_z(0) = z$ and $\Phi_z'(0) = X(z)$, we can take $v = X(z)$ and $c = \Phi_z$ in the above discussion, giving the following 'directional derivative of $f$ along $X$':

**Definition 3.15** *The* ***Lie derivative*** *of $f$ along $X$ is the scalar field $\mathcal{L}_X f$ defined by*

$$(\mathcal{L}_X f)(z) := \mathrm{d}f(z) \cdot X(z) = \left.\frac{\mathrm{d}}{\mathrm{d}t}\right|_{t=0} f \circ \Phi_t(z),$$

*where $\Phi$ is the flow of $X$. In many books, $\mathcal{L}_X f$ is written $Xf$ or $X[f]$. In a context where $X$ is understood, we write $\dot{f} = \mathcal{L}_X f$.*

Note that $(\mathcal{L}_X f)(z)$ depends on $X(z)$ but not on the rest of $X$.

**Example 3.16** Let $X(x, y) = (x, -y^2)$ and $f(x, y) = xy$. Then,

$$(\mathcal{L}_X f)(x, y) := \mathrm{d}f(x, y) \cdot (X(x, y)) = (y, x) \cdot (x, -y^2) = xy - xy^2.$$

We now introduce some terminology that will allow us to draw parallels with the push-forward and pull-back introduced earlier for vector fields.

**Definition 3.17** *Let $\varphi : M \to N$. For any scalar field $f$ on $N$, the* **pull-back** *of $f$ by $\varphi$ is $\varphi^* f := f \circ \varphi$, which is a scalar field on $M$.*
*If $\varphi$ is invertible and $g$ is any scalar field on $M$, then the* **push-forward** *of $g$ by $\varphi$ is $\varphi_* g := g \circ \varphi^{-1}$, which is a scalar field on $N$.*

**Remark 3.18** *When we express a function in different coordinates, we are in fact computing a pull-back or push-forward. For example, the function $f(x,y) = 2x$, when expressed in polar coordinates is $f(r,\theta) = 2r\cos\theta$, i.e. $\varphi^* f$, where $\varphi(r,\theta) = (r\cos\theta, r\sin\theta)$. From a pure mathematical point of view, we should not use the same letter '$f$' for both $f$ and $\varphi^* f$ because they are really different functions, but this is common.*

**Remark 3.19** *The Lie derivative can be expressed in terms of pull-backs as*

$$\mathcal{L}_X f = \left.\frac{\mathrm{d}}{\mathrm{d}t}\right|_{t=0} \Phi_t^* f\,; \quad i.e.\ (\mathcal{L}_X f)(z) = \left.\frac{\mathrm{d}}{\mathrm{d}t}\right|_{t=0} \left(\Phi_t^* f\right)(z).$$

Now, let $X$ and $Y$ be two vector fields on the same manifold $M$. We would like to define a 'derivative of $Y$ along $X$'. A reasonable first thought is to write $Y$ in local coordinates $\left(\dot{q}^1, \ldots, \dot{q}^n\right)$ as $Y = \left(Y^1, \ldots, Y^n\right)$, where each $Y^i$ a scalar field, and define the derivative to be

$$\left(\mathcal{L}_X Y^i\right) \frac{\partial}{\partial q^i}. \tag{3.3}$$

The problem with this is that we can get different answers in different coordinate systems, as in the following example.

**Example 3.20** Consider the following vector fields on $\mathbb{R}^2$, both defined in Cartesian coordinates:

$$X(x,y) = (x,y) \text{ and } Y(x,y) = (1,0).$$

In these coordinates, $Y$ is constant ($Y^1 = 1$ and $Y^2 = 0$), so $\mathcal{L}_X Y^i = 0$ for $i = 1,2$. We now express both $X$ and $Y$ in polar coordinates, using the map $\psi(r,\theta) = (r\cos\theta, r\sin\theta)$. Since

$$(\mathrm{D}\psi(r,\theta))^{-1} = \begin{bmatrix} \cos\theta & -r\sin\theta \\ \sin\theta & r\cos\theta \end{bmatrix}^{-1} = \begin{bmatrix} \cos\theta & \sin\theta \\ -\left(\dfrac{1}{r}\right)\sin\theta & \left(\dfrac{1}{r}\right)\cos\theta \end{bmatrix},$$

we have

$$
\begin{aligned}
&(\mathrm{D}\psi(r,\theta))^{-1}\cdot X(r\cos\theta, r\sin\theta)\\
&= \begin{bmatrix} \cos\theta & \sin\theta \\ -\left(\dfrac{1}{r}\right)\sin\theta & \left(\dfrac{1}{r}\right)\cos\theta \end{bmatrix} \begin{bmatrix} r\cos\theta \\ r\sin\theta \end{bmatrix} = \begin{bmatrix} r \\ 0 \end{bmatrix};\\
&(\mathrm{D}\psi(r,\theta))^{-1}\cdot Y(r\cos\theta, r\sin\theta)\\
&= \begin{bmatrix} \cos\theta & \sin\theta \\ -\left(\dfrac{1}{r}\right)\sin\theta & \left(\dfrac{1}{r}\right)\cos\theta \end{bmatrix} \begin{bmatrix} 1 \\ 0 \end{bmatrix} = \begin{bmatrix} \cos\theta \\ -\left(\frac{1}{r}\right)\sin\theta \end{bmatrix}.
\end{aligned}
$$

Thus, in polar coordinates, $X(r,\theta) = (r,0)$ and $Y(r,\theta) = \left(\cos\theta, -\frac{1}{r}\sin\theta\right)$. Then,

$$\mathcal{L}_X Y^1(r,\theta) = \mathrm{d}Y^1(r,\theta)\cdot(r,0) = (0, -\sin\theta)\cdot(r,0) = 0;$$

but $$\mathcal{L}_X Y^2(r,\theta) = \mathrm{d}Y^2(r,\theta)\cdot(r,0) = \left(\frac{1}{r^2}, \cos\theta)\right)\cdot(r,0) = \frac{1}{r}.$$

To summarize this example, in Cartesian coordinates we have $\mathcal{L}_X Y^1 = \mathcal{L}_X Y^2 = 0$, while in polar coordinates we have $\mathcal{L}_X Y^1 = 0$ and $\mathcal{L}_X Y^2 = 1/r$. So the expression in eqn (3.3) is not coordinate independent.

We now give a coordinate-free definition of a 'derivative of $Y$ along $X$', similar to the definition of $\mathcal{L}_X f$ in Remark 3.19. See Figure 3.3.

**Definition 3.21** *The* ***Lie derivative*** *of $Y$ along $X$ is*

$$\mathcal{L}_X Y \equiv \left.\frac{\mathrm{d}}{\mathrm{d}t}\right|_{t=0} \Phi_t^* Y,$$

*where $\Phi$ is the flow of $X$. By eqn (3.2) we have, for every $z$,*

$$(\mathcal{L}_X Y)(z) = \left.\frac{\mathrm{d}}{\mathrm{d}t}\right|_{t=0} (\Phi_t^* Y)(z) = \left.\frac{\mathrm{d}}{\mathrm{d}t}\right|_{t=0} (\mathrm{D}\Phi_t(z))^{-1}\cdot Y(\Phi_t(z)).$$

**Example 3.22** Consider the same vector fields as in Example 3.20, both defined in Cartesian coordinates $\mathbb{R}^2$:

$$X(x,y) = (x,y) \text{ and } Y(x,y) = (1,0).$$

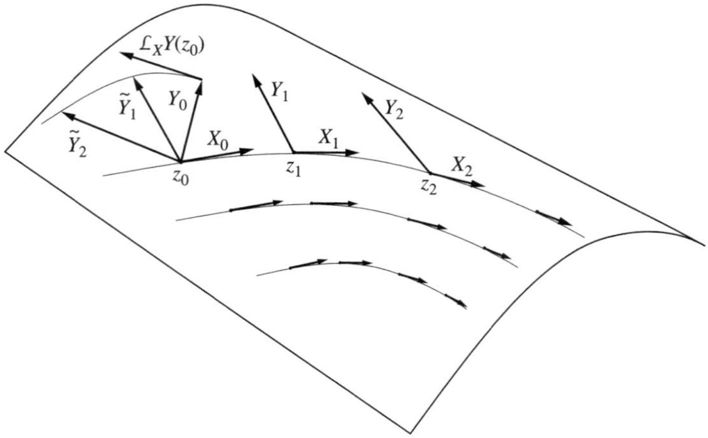


**Figure 3.3** The Lie derivative $\mathcal{L}_X Y$ is the derivative of the vector field $Y$ along the flow of the vector field $X$. To calculate $\mathcal{L}_X Y(z_0)$, the vectors $Y(z)$ at points $z$ on the integral curve of $X$ through $z_0$ are pulled back to vectors $\tilde{Y}$ based at $z_0$. The Lie derivative $\mathcal{L}_X Y(z_0)$ is the rate of change of the $\tilde{Y}$ vectors.

The time-$t$ flow of $X$ is $\Phi_t(x, y) = \left(\mathrm{e}^t x, \mathrm{e}^t y\right)$, the derivative of which is

$$\mathrm{D}\Phi_t(x, y) = \begin{bmatrix} \mathrm{e}^t & 0 \\ 0 & \mathrm{e}^t \end{bmatrix}.$$

Using the definition of the Lie derivative, we have

$$\begin{aligned} \mathcal{L}_X Y(x, y) &= \left.\frac{\mathrm{d}}{\mathrm{d}t}\right|_{t=0} \left(\Phi_t^* Y\right)(x, y) \\ &= \left.\frac{\mathrm{d}}{\mathrm{d}t}\right|_{t=0} (\mathrm{D}\Phi_t(x, y))^{-1} \cdot Y(\Phi_t(x, y)) \\ &= \left.\frac{\mathrm{d}}{\mathrm{d}t}\right|_{t=0} \begin{bmatrix} \mathrm{e}^{-t} & 0 \\ 0 & \mathrm{e}^{-t} \end{bmatrix} \begin{bmatrix} 1 \\ 0 \end{bmatrix} \\ &= \begin{bmatrix} -1 \\ 0 \end{bmatrix}. \end{aligned}$$

We now define the ***Jacobi–Lie bracket*** operation on vector fields. In local coordinates on an $n$-dimensional manifold $M$, a vector field $X$ on $M$ can be expressed as $X(z) = \left(X^1(z), \ldots, X^n(z)\right)$ and, in the domain of the coordinate system, the base points $z$ can be identified with elements of $\mathbb{R}^n$, so $X$ can be considered (locally) as a map from $\mathbb{R}^n$ to $\mathbb{R}^n$. Let $\mathrm{D}X(z)$ denote the derivative of this map at $z$.

**Definition 3.23** *The* ***Jacobi–Lie bracket*** *operation on* $\mathfrak{X}(M)$ *is defined in local coordinates by*

$$[X, Y] := (\mathrm{D}Y) \cdot X - (\mathrm{D}X) \cdot Y.$$

*More precisely,* $[X, Y]$ *is the vector field defined in local coordinates by*

$$[X, Y](z) := \mathrm{D}Y(z) \cdot X(z) - \mathrm{D}X(z) \cdot Y(z).$$

Since the rows of $\mathrm{D}X(z)$ are the gradient vectors $\nabla X^i(z)$,

$$(\mathrm{D}X(z) \cdot Y(z))^i = \nabla X^i(z) \cdot Y(z),$$

for every $\mathbf{z} \in M$. For this reason, the Jacobi–Lie bracket is often written as

$$[X, Y] = X \cdot \nabla Y - Y \cdot \nabla X.$$

A priori, the above definition is coordinate dependent. However, it is actually coordinate *independent*, a fact that follows from the next theorem, since the definition of the Lie derivative is coordinate-free.

**Theorem 3.24** $\mathcal{L}_X Y = [X, Y]$

The proof of this theorem uses the identity

$$\left(A^{-1}\right)' = -A^{-1} A' A^{-1}, \tag{3.4}$$

the proof of which is left as Exercise 3.12.

*Proof* In the following calculation, we work in local coordinates. Thus, we may consider everything as matrices, which allows us to use the product rule, the rule in eqn (3.4), and the following consequence of the equality of mixed second partial derivatives: $\frac{\mathrm{d}}{\mathrm{d}t}(\mathrm{D}\Phi_t(x)) = \mathrm{D}\left(\frac{\mathrm{d}}{\mathrm{d}t}\Phi_t\right)(x)$.

For all $x$,

$$
\begin{aligned}
\mathcal{L}_X Y(x) &= \left.\frac{\mathrm{d}}{\mathrm{d}t}\right|_{t=0} \Phi_t^* Y(x) \\
&= \left.\frac{\mathrm{d}}{\mathrm{d}t}\right|_{t=0} (\mathrm{D}\Phi_t(x))^{-1} \cdot Y(\Phi_t(x)) \\
&= \Bigg[\left(\frac{\mathrm{d}}{\mathrm{d}t}(\mathrm{D}\Phi_t(x))^{-1}\right) \cdot Y(\Phi_t(x)) \\
&\qquad + (\mathrm{D}\Phi_t(x))^{-1} \cdot \left(\frac{\mathrm{d}}{\mathrm{d}t} Y(\Phi_t(x))\right)\Bigg]_{t=0} \\
&= \Bigg[-(\mathrm{D}\Phi_t(x))^{-1}\left(\frac{\mathrm{d}}{\mathrm{d}t}\mathrm{D}\Phi_t(x)\right)(\mathrm{D}\Phi_t(x))^{-1} \cdot Y(\Phi_t(x)) \\
&\qquad + (\mathrm{D}\Phi_t(x))^{-1} \cdot \left(\frac{\mathrm{d}}{\mathrm{d}t} Y(\Phi_t(x))\right)\Bigg]_{t=0} \\
&= -\left(\left.\frac{\mathrm{d}}{\mathrm{d}t}\right|_{t=0} \mathrm{D}\Phi_t(x)\right) \cdot Y(x) + \left.\frac{\mathrm{d}}{\mathrm{d}t}\right|_{t=0} Y(\Phi_t(x)) \\
&= -\mathrm{D}\left(\left.\frac{\mathrm{d}}{\mathrm{d}t}\right|_{t=0} \Phi_t(x)\right) \cdot Y(x) + \mathrm{D}Y(x) \cdot \left(\left.\frac{\mathrm{d}}{\mathrm{d}t}\right|_{t=0} \Phi_t(x)\right) \\
&= -\mathrm{D}X(x) \cdot Y(x) + \mathrm{D}Y(x) \cdot X(x) \\
&= [X, Y](x).
\end{aligned}
$$

Therefore, $\mathcal{L}_X Y = [X, Y]$. ■

**Theorem 3.25** $\phi_*[X, Y] = [\phi_* X, \phi_* Y]$ *for any diffeomorphism* $\phi : M \to M$.

*Proof* Exercise. ■

**Theorem 3.26** $[X, [Y, Z]] + [Y, [Z, X]] + [Z, [X, Y]] = 0$, *for all* $X, Y, Z$.

*Proof* Exercise. ■

**Exercise 3.1**
Verify Theorem 3.9 for solutions to a linear vector field $X(z) = Az$, defined on $\mathbb{R}^n$, with $A$ a constant matrix.

**Exercise 3.2**
Prove Theorem 3.9.

**Exercise 3.3**
Prove the formula in Remark 3.13 regarding the flow of the push-forward of a vector field.

**Exercise 3.4**
Let $(\theta, \phi)$ be the spherical coordinates on $S^2$ defined by

$$(x, y, z) = (\sin\phi\cos\theta, \sin\phi\sin\theta, -\cos\phi),$$

for $\theta \in (-\pi, \pi), \phi \in (0, \pi)$. With $f$ as in Exercise 2.18, and $X$ on $S^2$ defined in spherical coordinates by $(\dot{\theta}, \dot{\phi}) = X(\theta, \phi) = (1, 1)$, calculate $f_* X$. Sketch $X$ and $f_* X$.

**Exercise 3.5**
Sketch the vector field $X$ defined in spherical coordinates by $X(\theta, \phi) = (1, \sin\phi)$. Express $X$ in the stereographic coordinates $(\xi, \eta)$ defined in Example 2.49.

**Exercise 3.6**
If $X(x, y) = (-y, x)$ and $H(x, y) = \frac{1}{2}\left(x^2 + y^2\right)$, compute $\mathcal{L}_X H$.

**Exercise 3.7**
Prove the ***Leibniz rule*** (a.k.a. product rule): $\mathcal{L}_X(fg) = (\mathcal{L}_X f)\, g + f\,(\mathcal{L}_X g)$.

**Exercise 3.8**
Calculate $[X, Y]$ for $X$ and $Y$ as in Example 3.22, and check that your answer equals $\mathcal{L}_X Y$ as calculated directly in that example.

**Exercise 3.9**
Let $X(x, y, z) = (-y, x, 1)$ and $Y(x, y, z) = (x, y, z)$. Calculate $\mathcal{L}_X Y$ in two ways: directly from Definition 3.21, and as $[X, Y]$.

**Exercise 3.10**
Show that, if $X = X^i \frac{\partial}{\partial q^i}$, then $\mathcal{L}_X q^i = X^i$.

**Exercise 3.11**
Let $\Phi$ be the flow of $X$. Show that $\mathcal{L}_X Y = -\left.\frac{\mathrm{d}}{\mathrm{d}t}\right|_{t=0} (\Phi_t)_* Y$.

**Exercise 3.12**
Let $A(t)$ be a path in $M(n, \mathbb{R})$, the set of $n \times n$ matrices. By differentiating the equation $A^{-1}A = I$ with respect to $t$, show that $(A^{-1})' = -A^{-1}A'A^{-1}$.

## 3.2 Differential 1-forms

Just as a vector field is a 'field' of tangent vectors, a differential 1-form is a 'field' of cotangent vectors, one for every base point. Recall that a cotangent vector based at $z \in M$ is a linear map from $T_zM$ to $\mathbb{R}$, and the set of all such maps is the cotangent space $T_z^*M$, which is the dual to the tangent space $T_zM$. Differential 1-forms are often called just 1-forms.

**Definition 3.27** *A (**differential) 1-form** on a manifold $M$ is a map $\theta: M \to T^*M$ such that $\theta(z) \in T_z^*M$ for every $z \in M$. Differential 1-forms can be added together, and multiplied by scalar fields $k : M \to \mathbb{R}$, as follows:*

$$(\alpha + \beta)(z) := \alpha(z) + \beta(z), \quad (k\theta)(z) := k(z)\theta(z).$$

*A **smooth** (resp. **differentiable**) **1-form** is one that is a smooth (resp. differentiable) map from $M \to T^*M$.*

The most important example of a 1-form is the differential of a scalar field, the definition of which appeared in Chapter 2 and is repeated here.

**Definition 3.28** *The **differential** of a scalar field $f : M \to \mathbb{R}$ is the 1-form $\mathrm{d}f$ such that, for any $z \in M$, $\mathrm{d}f(z) : T_zM \to \mathbb{R}$ is the derivative, i.e. the tangent map, of $f$ at $z$.*

Not all 1-forms are differentials, as the next example shows.

**Example 3.29** Using standard coordinates on $\mathbb{R}^2$, let $\alpha = \mathrm{d}x + x\,\mathrm{d}y$. Suppose there exists a smooth scalar field $f : U \to \mathbb{R}$, for some open $U \subseteq \mathbb{R}^2$ such that $\mathrm{d}f(x,y) = \alpha(x,y)$ for all $(x,y) \in U$. Then, since $\mathrm{d}f = \frac{\partial f}{\partial x}\mathrm{d}x + \frac{\partial f}{\partial y}\mathrm{d}y$, we must have $\frac{\partial f}{\partial x} \equiv 1$ and $\frac{\partial f}{\partial y} \equiv x$ (where '$\equiv$' means: at all $(x,y) \in U$). But this implies $\frac{\partial^2 f}{\partial y \partial x} \equiv 0$ and $\frac{\partial^2 f}{\partial x \partial y} \equiv 1$, which contradicts the equality of mixed second partials.

**Definition 3.30** *If $\alpha = \mathrm{d}f$ for some smooth $f$, then $\alpha$ is **exact**.*

Note that this is consistent with the meaning of exactness in the study of differential equations. Indeed, consider the ODE

$$M(x,y)\mathrm{d}x + N(x,y)\mathrm{d}y = 0,$$

which is *exact* if there exists a $\Phi(x,y)$ such that $\partial\Phi/\partial x = M$ and $\partial\Phi/\partial y = N$. If $\alpha$ is the 1-form $M(x,y)\mathrm{d}x + N(x,y)\mathrm{d}y$, then this condition is equivalent to $\alpha = \mathrm{d}\Phi$.

Though not all 1-forms are differentials, all 1-forms can be expressed as linear combinations of differentials, with the coefficients being scalar fields that depend on the base point. Indeed, given any cotangent-lifted local coordinates $(q^1, \ldots, q^n, p_1, \ldots, p_n)$ defined on $T^*U \subseteq T^*M$, the $i$th ***component*** of a 1-form $\alpha$ is the scalar field $\alpha_i : U \to \mathbb{R}$ such that $\alpha_i(z)$ is the $p_i$ coordinate of $\alpha(z)$ for all $z \in U$. Then, by definition of the $p_i$ coordinates, and using the summation convention,[2]

$$\alpha = \alpha_i \,\mathrm{d}q^i.$$

By the definition of the manifold structure on $T^*M$, a 1-form is smooth if and only if all of its components are smooth, in any coordinates.

Just as cotangent vectors are naturally paired with tangent vectors, 1-forms are naturally paired with vector fields, by applying the tangent–cotangent pairing at every base point:

$$\langle \alpha, X \rangle (z) := \langle \alpha(z), X(z) \rangle .$$

[2] The ***Einstein summation convention*** is: whenever an index appears twice in the same expression, once as a superscript and once as a subscript, summation over that index is assumed. For the purposes of this convention, a superscript in a denominator is assumed to be a subscript, and a subscript in a denominator is considered to be a superscript.

Note that $\langle \alpha, X \rangle$ is a scalar field. In coordinates,

$$\left\langle \alpha_i \, \mathrm{d}q^i, X^i \frac{\partial}{\partial q^i} \right\rangle (z) = \left\langle \alpha_i(z) \, \mathrm{d}q^i(z), X^i(z) \frac{\partial}{\partial q^i}(z) \right\rangle = \alpha_i(z) X^i(z).$$

The natural pairing is often called ***contraction*** and written $X \lrcorner \alpha := \langle \alpha, X \rangle$.

**Example 3.31** If $X = 2x \frac{\partial}{\partial x} - \frac{\partial}{\partial y}$ and $\alpha = y^2 \, \mathrm{d}x + y \, \mathrm{d}y$, then

$$X \lrcorner \alpha = \langle \alpha, X \rangle = 2xy^2 - y.$$

Pull-backs, push-forwards and Lie derivatives of 1-forms can be defined, as was done for vector fields in the previous section, but we leave these definitions to Section 3.3, where they will appear in a more general context.

For any manifold $M$, the cotangent bundle $T^*M$ is also a manifold. The 1-forms on $T^*M$ are smooth maps from $T^*M$ to $T^*(T^*M)$. In cotangent-lifted local coordinates $(q^1, \ldots, q^n, p_1, \ldots, p_n)$, the general 1-form on $T^*M$ has the form $a_i \, \mathrm{d}q^i + b_i \, \mathrm{d}p_i$, where $a_i$ and $b_i$ are functions of $(q, p)$.

The following 1-form on $T^*M$ is particularly important in symplectic geometry and mechanics. We first define it in terms of local coordinates, and then show that the definition is independent of the choice of local coordinates and in fact can be rewritten in a coordinate-free way.

**Definition 3.32** *The **Liouville 1-form** or **canonical 1-form** on $T^*M$ is defined, in cotangent-lifted local coordinates, by*

$$\theta = p_i \, \mathrm{d}q^i, \tag{3.5}$$

*and also written in the short form* $\mathbf{p} \cdot \mathbf{dq}$ *or* $\mathbf{p} \, \mathbf{dq}$.

At first sight, this definition seems to depend on a choice of coordinates. However, one can check directly that this is not the case – see Exercise 3.15. Further, it is possible to state an equivalent version of the definition that does not even use coordinates, as we now show.

Note that the $q^i$ in the definition of the Liouville 1-form are considered to be coordinates on $T^*M$, so that $\mathrm{d}q^i$ is the 1-form on $T^*M$ given by

$$\left\langle \mathrm{d}q^i, a^j \frac{\partial}{\partial q^j} + b^j \frac{\partial}{\partial p_j} \right\rangle = a^i. \tag{3.6}$$

Recall the cotangent bundle projection map $\pi : T^*M \to M$, which is defined in cotangent-lifted local coordinates by $\pi(\mathbf{q},\mathbf{p}) = \mathbf{q}$. The tangent lift of this projection is $T\pi : TT^*M \to TM$, given in local coordinates by

$$T\pi\left(a^j\frac{\partial}{\partial q^j} + b^j\frac{\partial}{\partial p_j}\right) = a^j\frac{\partial}{\partial q^j}.$$

Comparison of this formula with eqn (3.6) shows that

$$\left\langle \mathrm{d}q^i, a^j\frac{\partial}{\partial q^j} + b^j\frac{\partial}{\partial p_j}\right\rangle = \left\langle \mathrm{d}q^i, T\pi\left(a^j\frac{\partial}{\partial q^j} + b^j\frac{\partial}{\partial p_j}\right)\right\rangle,$$

for all scalar fields $a^i$ and $b^i$, where the $q^i$ on the left is a coordinate on $T^*M$ and the $q^i$ on the right is a coordinate on $M$. If $q^i$ is a coordinate on $M$, then $p_i \mathrm{d}q^i = \mathbf{p}$, by definition of the coordinates $p_i$. It follows that,

$$\left\langle p_i\,\mathrm{d}q^i, a^j\frac{\partial}{\partial q^j} + b^j\frac{\partial}{\partial p_j}\right\rangle = \left\langle \mathbf{p}, T\pi\left(a^j\frac{\partial}{\partial q^j} + b^j\frac{\partial}{\partial p_j}\right)\right\rangle.$$

Thus, the Liouville 1-form $\theta$ has an equivalent coordinate-free definition,

$$\langle\theta(\mathbf{q},\mathbf{p}),\mathbf{v}\rangle = \langle\mathbf{p}, T\pi(\mathbf{v})\rangle, \qquad \text{for all } \mathbf{v}\in T_{(\mathbf{q},\mathbf{p})}T^*M,$$

see Figure 3.4. This implies that the expression in eqn (3.5) is coordinate independent.

**Proposition 3.33** *The cotangent lift of any diffeomorphism $\varphi : M \to N$ preserves the Liouville 1-forms on $T^*M$ and $T^*N$, that is,*

$$\left(T^*\varphi^{-1}\right)^*\theta_N = \theta_M.$$

*Proof* See Exercise 3.16 ■

**Exercise 3.13**

Given $z_0 \in M$, show that there exists a scalar field $f$, defined on some neighbourhood of $z_0$, such that $\alpha(z_0) = \mathrm{d}f(z_0)$. In fact, making a mild topological assumption on $M$ (paracompactness – see for example [AM78]), this $f$ can be extended to a scalar field defined on all of $M$. Note, however, that it is *not* always possible to find a single smooth $f$ such that $\alpha(z) = \mathrm{d}f(z)$ for *all* points $z \in M$ (see Exercise 3.35), or even all points $z$ in some neighbourhood of $z_0$ (see Exercise 3.14).

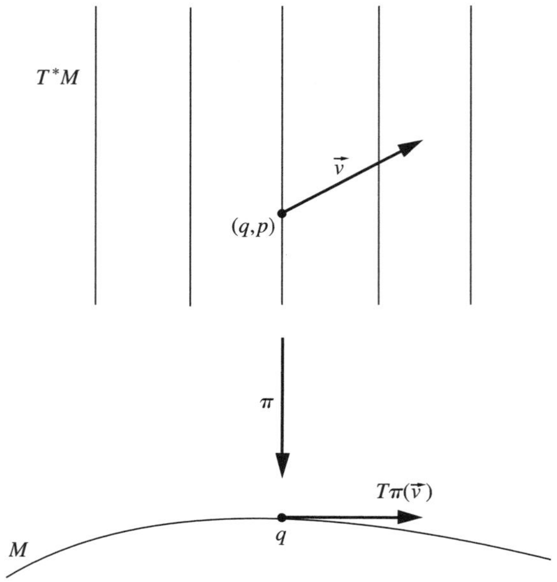


**Figure 3.4** The tangent map of the cotangent bundle projection $\pi : T^*M \to M$ is used in the definition of the Liouville 1-form. The vector $\mathbf{v} \in T_{(q,p)}T^*M$ is projected onto the vector $T\pi(\mathbf{v}) \in T_qM$.

**Exercise 3.14**
For what constant values of $a$ and $b$ is the 1-form $\omega = a\, y\, \mathrm{d}x - b\, x\, \mathrm{d}y$ exact?

**Exercise 3.15**
Show that the Liouville 1-form, defined by $p_i\, \mathrm{d}q^i$, is independent of the choice of cotangent-lifted coordinates. That is, show that if $\left(q^1, \ldots, q^n, p_1, \ldots, p_n\right)$ and $\left(r^1, \ldots, r^n, s_1, \ldots, s_n\right)$ are two sets of cotangent-lifted coordinates, then $p_i\, \mathrm{d}q^i = s_i\, \mathrm{d}r^i$.

**Exercise 3.16**
Prove Proposition 3.16: cotangent lifts preserve Liouville 1-forms. This can be done using the coordinate-free definition; or in local coordinates, using Exercise 3.15.

## 3.3 Tensors

Let $V$ be a finite-dimensional real vector space.

**Definition 3.34** *A* ***covariant*** $k$***-tensor*** *on* $V$ *is a multilinear map*

$$V^k \to \mathbb{R},$$

*where* $V^k = V \times \cdots \times V$, *with* $k$ *copies of* $V$, *and 'multilinear' means that the map is linear in each of its variables separately, holding the others fixed.* *A* ***contravariant*** $k$***-tensor*** *on* $V$ *is a multilinear map*

$$(V^*)^k \to \mathbb{R}.$$

*In both cases, the* ***rank*** *of the tensor is* $k$. *A* 0***-tensor*** *is just a real number. Addition and scalar multiplication of tensors are defined as for general maps.*

Some examples of tensors are:

- Covariant 2-tensors are bilinear forms.
- Covariant 1-tensors are linear maps from $V$ to $\mathbb{R}$, i.e. elements of $V^*$, i.e. covectors.
- Contravariant 1-tensors are vectors. This is because linear maps from $V^*$ to $\mathbb{R}$ may be naturally identified with elements of $V$ (i.e. $V^{**} \simeq V$), assuming $V$ is finite-dimensional (see Exercise 2.24).

In mechanics,

- If the configuration space is $V = \mathbb{R}^n$, then position vectors $\mathbf{q}$ are contravariant 1-tensors.
- For a general configuration space $Q$, let $q \in Q$. Taking $V = T_q Q$, a velocity vector $v \in T_q Q$ is a contravariant 1-tensor, while a conjugate momentum vector $p \in T_q^* Q$ is a covariant 1-tensor.

**Definition 3.35** *A tensor is* ***symmetric*** *if its value is unchanged whenever the order of the variables is changed.*

*A tensor is* ***skew-symmetric*** *(or* ***anti-symmetric***, *or* ***alternating****) if it changes sign whenever any two variables are interchanged.*

*A tensor is* ***positive-definite*** *if its values are always non-negative, and are zero if and only if all of its arguments are zero.*

For covariant 2-tensors, these definitions agree with the usual ones for bilinear forms. An ***inner product*** on $V$ is a positive-definite, symmetric covariant 2-tensor.

The remainder of this section focuses on covariant tensors. The theory for contravariant tensors is analogous.

If $V$ has dimension $n$, then any covariant tensor of rank 2 can be described, relative to a given basis $(\mathbf{e}_1, \ldots, \mathbf{e}_n)$, by the $n \times n$ matrix $M$ defined by

$$T(\mathbf{v}, \mathbf{w}) = \mathbf{v}^T M \mathbf{w},$$

for all $\mathbf{v}, \mathbf{w} \in V$ (written as column vectors). The $(i, j)$th entry of $M$ is called the ***tensor coefficient*** $T_{ij}$, which satisfies

$$T_{ij} = T\left(\mathbf{e}_i, \mathbf{e}_j\right).$$

Using bilinearity, it follows that, for any $\mathbf{v}, \mathbf{w} \in V$,

$$T\left(\mathbf{v}, \mathbf{w}\right) = T_{ij} \mathbf{v}^i \mathbf{w}^j,$$

where $\mathbf{v} = \mathbf{v}^i \mathbf{e}_i$ and we are using the summation convention.

A 2-tensor is ***non-degenerate*** if its matrix is invertible. Note that every positive-definite (or negative-definite) matrix is invertible, so all inner products are non-degenerate.

**Example 3.36** The inner product on $\mathbb{R}^2$ defined by $T(\mathbf{u}, \mathbf{v}) = \langle\langle \mathbf{u}, \mathbf{v} \rangle\rangle := u^1 v^1 + 3v^1 v^2$ is a covariant 2-tensor with matrix $\begin{bmatrix} 1 & 0 \\ 0 & 3 \end{bmatrix}$ with respect to the standard basis. It is non-degenerate, and in fact positive-definite (like all inner products). Its tensor coefficients, with respect to the same basis, are $T_{11} = 1, T_{12} = T_{21} = 0, T_{22} = 3$.

**Remark 3.37** *Let $T$ be a covariant 2-tensor on $V$, and let $\mathcal{B}$ and $\mathcal{C}$ be bases for $V$. Let $M$ be the matrix for $T$ with respect to $\mathcal{B}$, and $P$ the change-of-basis matrix from $\mathcal{C}$ to $\mathcal{B}$. Writing $[\mathbf{v}]_{\mathcal{B}}$ for the component vector of $\mathbf{v}$ with respect to basis $\mathcal{B}$ (and similarly for $\mathcal{C}$), we have*

$$T(\mathbf{v}, \mathbf{w}) = [\mathbf{v}]_{\mathcal{B}}^T M \, [\mathbf{w}]_{\mathcal{B}} = [\mathbf{v}]_{\mathcal{C}}^T \, P^T M P \, [\mathbf{w}]_{\mathcal{C}},$$

*so $T$ has matrix $P^T M P$ with respect to $\mathcal{C}$.*

Tensors are often described as quantities that transform in certain ways. What this means for 2-tensors is that, when changing bases, the coefficients of *covariant* tensors transform as in the previous remark, while the coefficients of *contravariant* tensors transform as in Exercise 3.19.

For general $k$, covariant tensors of rank $k$ can be described, with respect to a given basis $(\mathbf{e}_1, \ldots, \mathbf{e}_n)$ for $V$, by the $n^k$ coefficients,

$$T_{j_1 \cdots j_k} := T\left(\mathbf{e}_{j_1}, \ldots, \mathbf{e}_{j_k}\right).$$

Thus, tensors can be thought of as '$k$-dimensional matrices'. For any vectors $\mathbf{v}_1, \ldots, \mathbf{v}_k \in V$,

$$T(\mathbf{v}_1, \ldots, \mathbf{v}_k) = T_{j_1 \cdots j_k} \mathbf{v}_1^{j_1} \cdots \mathbf{v}_k^{j_k},$$

where $\mathbf{v}_a = \mathbf{v}_a^j \mathbf{e}_j$.

Given a choice of basis for $V$, the tensor is determined by its tensor coefficients. A logical notation would be $T = \left(T_{j_1 \cdots j_k}\right)$, but the more common notation is just $T = T_{j_1 \cdots j_k}$.

**Definition 3.38** *The* ***tensor product*** *of a covariant k-tensor S and a covariant l-tensor T is the covariant $(k+l)$-tensor $S \otimes T$ defined by*

$$(S \otimes T)(v_1, \ldots, v_{k+l}) = S(v_1, \ldots, v_k)\, T(v_{k+1}, \ldots, v_{k+l}),$$

*i.e. the first k vectors get 'plugged into' S, and the rest get 'plugged into' T.*

It is easily checked that tensor product is an associative bilinear operation.

**Example 3.39** Let $\left(\mathbf{e}^1, \mathbf{e}^2, \mathbf{e}^3\right)$ be the dual basis to the standard basis on $\mathbb{R}^3$; so for example, $\mathbf{e}^1(\mathbf{u}) = u^1$ for all $\mathbf{u} = (u^1, u^2, u^3) \in \mathbb{R}^3$. Note that each $\mathbf{e}^i$ is a map from $\mathbb{R}^3$ to $\mathbb{R}$, so it is a covariant 1-tensor on $\mathbb{R}^3$. Then, $\left(\mathbf{e}^1 \otimes \mathbf{e}^2\right)(\mathbf{u}, \mathbf{v}) = u^1 v^2$. The bilinear form $T$ defined by $T(\mathbf{u}, \mathbf{v}) = 2u^1 v^3 - 3u^2 v^1$ can be written as $T = 2\mathbf{e}^1 \otimes \mathbf{e}^3 - 3\mathbf{e}^2 \otimes \mathbf{e}^1$.

The next proposition, which generalizes the previous example, is a consequence of the multilinearity of tensors and the definition of tensor coefficients.

**Proposition 3.40** *Given a basis $\left(\mathbf{e}^1, \ldots, \mathbf{e}^n\right)$ for $V^*$, any covariant k-tensor T on V satisfies $T = T_{i_1 \ldots i_k} \mathbf{e}^{i_1} \otimes \cdots \otimes \mathbf{e}^{i_k}$, where the sum is over all possible choices of the k indices $i_1, \ldots, i_k$. This expression uniquely defines the tensor coefficients, so the set $\{\mathbf{e}^{i_1} \otimes \cdots \otimes \mathbf{e}^{i_k}\}$, of all possible kth-order tensor products of the basis elements, forms a basis for $T_k^0 V$.*

**Definition 3.41** *A* ***tensor field*** *on a manifold M is a smoothly varying family of tensors $\{T(z)\}$, all of the same rank, such that $T(z)$ is a tensor on $T_z M$, for each $z \in M$. The tensor coefficients now depend on z, so they are* *scalar fields. They also depend on a choice of local coordinate system, and are defined by*

$$T_{j_1\ldots j_k}(z) := T\left(\frac{\partial}{\partial q^{j_1}}(z), \ldots, \frac{\partial}{\partial q^{j_k}}(z)\right).$$

*The tensor field is* ***smooth*** *if all of the tensor coefficients are smooth (in any coordinate system).*

The set of all tensor fields on $M$ of a given rank is called a ***tensor bundle*** over $M$, generalizing the tangent and cotangent bundles. It is a manifold.

A tensor field $T$ is ***symmetric*** (resp. ***skew-symmetric***, ***non-degenerate***, ***positive-definite***) if $T(z)$ is symmetric (resp. skew-symmetric, non-degenerate, positive-definite), for every $z \in M$.
Some examples of tensor fields are:

- a 0-tensor field is a scalar field;
- a covariant 1-tensor field is a differential 1-form;
- a contravariant 1-tensor field is a vector field;
- a ***Riemannian metric*** on $M$ is a positive-definite, symmetric covariant 2-tensor field on $M$, i.e. a tensor $g$ such that $g(z)$ is an inner product on $T_zM$, for every $z \in M$. See Section 3.4.

Covariant tensor fields of rank 2 are represented by matrices with entries that depend on the base point, $z$, as in the following example.

**Example 3.42** Let $(\theta, \phi)$ be spherical coordinates on the unit sphere $S^2$, and suppose that $T$ is a covariant 2-tensor field on $S^2$ with matrix representation

$$T = \begin{bmatrix} 0 & \sin\phi \\ -\sin\phi & 0 \end{bmatrix}.$$

If $\mathbf{v}$ and $\mathbf{w}$ are tangent vectors to $S^2$ based at the same point $(\theta, \phi)$, then

$$T(\mathbf{v}, \mathbf{w}) = \begin{bmatrix} v^\theta & v^\phi \end{bmatrix} \begin{bmatrix} 0 & \sin\phi \\ -\sin\phi & 0 \end{bmatrix} \begin{bmatrix} w^\theta \\ w^\phi \end{bmatrix} = \sin\phi \left(v^\theta w^\phi - v^\phi w^\theta\right).$$

Note that $T$ is skew-symmetric. In fact, $T$ is the 'area form' for $S^2$, as explained later in Examples 3.52 and 3.53.

**Remark 3.43 (Change of coordinates)** *If $\psi$ is a coordinate transformation from 'new' to 'old' coordinates on $M$, then $D\psi(z)$ is the corresponding linear change-of-basis transformation on $T_zM$ (for every $z$). If $M$ is the matrix representation of a covariant 2-tensor $T(z)$ with respect to the old* *coordinates, then, by Remark 3.37, the matrix representation of $T(z)$ in the new coordinates is $(D\psi(z))^T M D\psi(z)$.*

**Definition 3.44** *The **tensor product** of two tensor fields is defined 'pointwise':*

$$(\alpha \otimes \beta)(z) = \alpha(z) \otimes \beta(z), \qquad \textit{for every } z.$$

**Example 3.45** For any smooth $f, g : M \to R$, the covariant 2-tensor field $\mathrm{d}f \otimes \mathrm{d}g$ is defined by

$$\left(\mathrm{d}f \otimes \mathrm{d}g\right)(z)(v, w) = \left(\mathrm{d}f(z) \cdot v\right)\left(\mathrm{d}f(z) \cdot w\right)$$

for any $z \in M$ and any $v, w \in T_z M$.

**Proposition 3.46** *Given any local coordinates $q^1, \ldots, q^n$ on $M$, any covariant $k$-tensor field $T$ on $M$ can be locally expressed as*

$$T = T_{i_1 \ldots i_k} \mathrm{d}q^{i_1} \otimes \cdots \otimes \mathrm{d}q^{i_k}.$$

*Proof* This follows from Proposition 3.40 and the fact that, for every $z$, the covectors $\left(\mathrm{d}q^{i_1}(z), \ldots, \mathrm{d}q^{i_k}(z)\right)$ form a basis for $T_z M$. ■

**Definition 3.47** *The **symmetric product** of two covectors (i.e. covariant 1-tensors) $\alpha$ and $\beta$ is the symmetric covariant 2-tensor $\alpha\beta$ defined by*

$$\alpha\beta = \frac{1}{2}(\alpha \otimes \beta + \beta \otimes \alpha), \quad \textit{i.e.} \tag{3.7}$$

$$(\alpha\beta)(v, w) = \frac{1}{2}(\alpha(v)\beta(w) + \beta(v)\alpha(w)).$$

*The definition applies 'pointwise' to tensor fields.*

**Example 3.48** If $g$ is a symmetric covariant 2-tensor field then $g_{ij} = g_{ji}$ for all $i, j$, so (using the summation convention)

$$\begin{aligned} g = g_{ij}\,\mathrm{d}x^i \otimes \mathrm{d}x^j &= \frac{1}{2}\left(g_{ij} + g_{ji}\right)\mathrm{d}x^i \otimes \mathrm{d}x^j \\ &= \frac{1}{2} g_{ij}\left(\mathrm{d}x^i \otimes \mathrm{d}x^j + \mathrm{d}x^j \otimes \mathrm{d}x^i\right) = g_{ij}\mathrm{d}x^i \mathrm{d}x^j. \end{aligned}$$

**Remark 3.49** *The following notation is widely used, despite being potentially ambiguous:* $\mathrm{d}x^2 := (\mathrm{d}x)^2 := \mathrm{d}x\,\mathrm{d}x$.

**Definition 3.50** *The **wedge product** of two covectors $\alpha$ and $\beta$ is the skew-symmetric covariant 2-tensor $\alpha \wedge \beta$ defined by*

$$\alpha \wedge \beta = \alpha \otimes \beta - \beta \otimes \alpha, \quad i.e. \tag{3.8}$$
$$(\alpha \wedge \beta)(v, w) = \alpha(v)\beta(w) - \beta(v)\alpha(w).$$

*The definition applies 'pointwise' to tensor fields.*

**Example 3.51** $(\mathrm{d}q^1 \wedge \mathrm{d}q^2)(z)\left(2\frac{\partial}{\partial q^1} + 3\frac{\partial}{\partial q^2}, 5\frac{\partial}{\partial q^1}\right) = -15.$

**Example 3.52** The ***standard area form*** on $\mathbb{R}^2$ is $\omega := \mathrm{d}x \wedge \mathrm{d}y$. It measures signed area on tangent spaces, since $\omega(x, y)((a, b), (c, d)) = ad - bc$, which is the signed area of the parallelogram spanned by the vectors $(a, b)$ and $(c, d)$. More generally, recalling that the area of a surface $S$ in $\mathbb{R}^3$ parameterized by $(u, v)$ is $\int_S \left\| \frac{\partial}{\partial u} \times \frac{\partial}{\partial v} \right\| \, \mathrm{d}u \, \mathrm{d}v$, we define the ***area form*** on $S$ to be

$$\omega := \left\| \frac{\partial}{\partial u} \times \frac{\partial}{\partial v} \right\| \mathrm{d}u \wedge \mathrm{d}v,$$

where $\partial/\partial u$ and $\partial/\partial v$ are expressed in Cartesian coordinates in $\mathbb{R}^3$ before taking the cross-product. The area form is often denoted by $\mathrm{d}A$.

**Example 3.53** Consider the cylinder $C = \{(x, y, z) \in \mathbb{R}^3 : x^2 + y^2 = 1\}$, with cylindrical coordinates $(\theta, z)$ defined by $(x, y, z) = (\cos\theta, \sin\theta, z)$. The area form on the cylinder is

$$\omega_C := \left\| \frac{\partial}{\partial \theta} \times \frac{\partial}{\partial z} \right\| d\theta \wedge dz = \left\| \begin{bmatrix} -\sin\theta \\ \cos\theta \\ 0 \end{bmatrix} \times \begin{bmatrix} 0 \\ 0 \\ 1 \end{bmatrix} \right\| d\theta \wedge dz = d\theta \wedge dz.$$

Similarly, the area form on $S^2$, in spherical coordinates defined by $(x, y, z) = (\sin\phi\cos\theta, \sin\phi\sin\theta, -\cos\phi)$, may be calculated as follows:

$$\begin{aligned} \omega_S &:= \left\| \frac{\partial}{\partial \theta} \times \frac{\partial}{\partial \phi} \right\| \mathrm{d}\theta \wedge \mathrm{d}\phi \\ &= \left\| \begin{bmatrix} -\sin\phi\sin\theta \\ \sin\phi\cos\theta \\ 0 \end{bmatrix} \times \begin{bmatrix} \cos\phi\cos\theta \\ \cos\phi\sin\theta \\ \sin\phi \end{bmatrix} \right\| \mathrm{d}\theta \wedge \mathrm{d}\phi \\ &= \sin\phi \, \mathrm{d}\theta \wedge \mathrm{d}\phi. \end{aligned}$$

The definitions of symmetric product and wedge product can be generalized to tensors of arbitrary rank. We will do this for the wedge product in Section 3.5. The general symmetric product will be not be needed in this book.

We now define pull-backs and push-forwards, which are the transformations of tensors (or tensor fields) that are naturally associated with a given map between vector spaces (or manifolds). We define them only for covariant tensors. The definitions for contravariant tensors are similar, and generalize the definition given for vector fields in Section 3.1.

**Definition 3.54** *Let $T$ be a covariant $k$-tensor on a vector space $W$. The **pull-back** of $T$ by a linear map $L : V \to W$ is the covariant $k$-tensor $L^*T$ on $V$ defined by*

$$\left(L^*T\right)(v_1, \ldots, v_k) = T\left(L(v_1), \ldots, L(v_k)\right).$$

***Push-forward** is the opposite of pull-back: the push-forward of a covariant $k$-tensor $S$ on $V$ by an invertible linear map $L$ is the covariant $k$-tensor field $L_*S$ on $W$ defined by $L_*S = \left(L^{-1}\right)^* S$, that is,*

$$\left(L_*S\right)(w_1, \ldots, w_k) = S\left(L^{-1}(w_1), \ldots, L^{-1}(w_k)\right).$$

**Definition 3.55** *Let $\omega$ be a covariant $k$-tensor field on a manifold $N$. The **pull-back** of $\omega$ by a smooth map $\varphi : M \to N$ is the covariant $k$-tensor field $\varphi^*\omega$ on $M$ defined by*

$$\begin{aligned}\left(\varphi^*\omega\right)(z)(v_1, \ldots, v_k) &= \omega\left(\varphi(z)\right)\left(T\varphi(v_1), \ldots, T\varphi(v_k)\right)\\ &= \omega\left(\varphi(z)\right)\left(\mathrm{D}\varphi(z)\cdot v_1, \ldots, \mathrm{D}\varphi(z)\cdot v_k\right).\end{aligned}$$

*Let $\alpha$ be a covariant $k$-tensor field on $M$. The **push-forward** of $\alpha$ by a diffeomorphism $\varphi$ is the covariant $k$-tensor field $\varphi_*\alpha$ on $N$ defined by $\varphi_*\alpha = \left(\varphi^{-1}\right)^*\alpha$, that is,*

$$\begin{aligned}\left(\varphi_*\alpha\right)(\varphi(x))(w_1, \ldots, w_k) &= \alpha(x)\left(T\varphi^{-1}(w_1), \ldots, T\varphi^{-1}(w_k)\right)\\ &= \alpha(x)\left(\mathrm{D}\varphi^{-1}(\varphi(x))\cdot w_1, \ldots, \mathrm{D}\varphi^{-1}(\varphi(x))\cdot w_k\right)\\ &= \alpha(x)\left((\mathrm{D}\varphi(x))^{-1}\cdot w_1, \ldots, (\mathrm{D}\varphi(x))^{-1}\cdot w_k\right).\end{aligned}$$

**Proposition 3.56** *If $f$ is a scalar field on $N$ and $\alpha$ and $\beta$ are 1-forms on $N$, and $\varphi : M \to N$ is smooth, then*

1. $\varphi^*(\mathrm{d}f) = \mathrm{d}(\varphi^* f) = \mathrm{d}(f \circ \varphi)$;
2. $\varphi^*(f\alpha) = (\varphi^* f)(\varphi^*\alpha) = (f \circ \varphi)(\varphi^*\alpha)$;

3. $\varphi^* (\alpha \otimes \beta) = (\varphi^* \alpha) \otimes (\varphi^* \beta)$;
4. $\varphi^* (\alpha\beta) = (\varphi^* \alpha) (\varphi^* \beta)$;
5. $\varphi^* (\alpha \wedge \beta) = (\varphi^* \alpha) \wedge (\varphi^* \beta)$.

*Proof* Claim 1 follows from the chain rule:

$$\left(\varphi^*(\mathrm{d}f)\right)(z) \cdot v = (\mathrm{d}f)(\varphi(z)) \cdot (\mathrm{D}\varphi(z) \cdot v) = \mathrm{d}(f \circ \varphi)(z) \cdot v,$$

for all $v \in T_z M$. Recall that $f \circ \varphi = \varphi^* f$ by definition. Claims 2 and 3 follow directly from the definition of pull-back (**Exercise**). For Claim 5, let $v, w \in T_z M$, so $T_z\varphi(v), T_z\varphi(w) \in T_{\varphi(z)}N$. In the following calculation, we suppress the base points for readability.

$$\begin{aligned}\left(\varphi^* (\alpha \wedge \beta)\right)(v, w) &= (\alpha \wedge \beta)\left(T_z\varphi(v), T_z\varphi(w)\right)\\ &= \alpha(T_z\varphi(v))\, \beta(T_z\varphi(w)) - \beta(T_z\varphi(v))\, \alpha(T_z\varphi(w))\\ &= \left(\varphi^*\alpha\right)(v)\; \left(\varphi^*\beta\right)(w) - \left(\varphi^*\beta\right)(v)\; \left(\varphi^*\alpha\right)(w)\\ &= \left(\left(\varphi^*\alpha\right) \wedge \left(\varphi^*\beta\right)\right)(v, w).\end{aligned}$$

The proof for symmetric product (Claim 4) is similar. ■

**Example 3.57** Let $D = S^2 \setminus \{(0, 0, \pm 1)\}$ and let $f : D \to C := S^1 \times \mathbb{R}$ be given, in spherical coordinates on $S^2$ and cylindrical coordinates on $C$, by $f(\theta, \phi) = (\theta, -\cos\phi)$. The area forms on $S^2$ and $C$ are $\omega_S = \sin\phi \, \mathrm{d}\theta \wedge \mathrm{d}\phi$ and $\omega_C = \mathrm{d}\theta \wedge \mathrm{d}z$ (see Example 3.53). We will show that $f^*\omega_C = \omega_S$ (with both forms restricted to the domain $D$). This shows that $f$ is area preserving – see Exercise 3.21. In fact, the area forms measure *signed area*, so the fact that $f$ preserves them (without a sign change) also shows that $f$ is orientation preserving.

We illustrate two methods. The first calculation uses the definition of pull-back directly:

$$\mathrm{D}f(\theta, \phi) = \begin{bmatrix} 1 & 0 \\ 0 & \sin\phi \end{bmatrix}, \quad \text{and} \quad \mathrm{D}f(\theta, \phi) \cdot (\dot\theta, \dot\phi) = \left(\dot\theta, \sin\phi\, \dot\phi\right),$$

so

$$\begin{aligned}&\left(f^*\omega_C\right)(\theta, \phi)\left((\dot\theta_1, \dot\phi_1), (\dot\theta_2, \dot\phi_2)\right)\\ &\quad = \omega_C\,(\theta, -\cos\phi)\left(\left(\dot\theta_1, \sin\phi\, \dot\phi_1\right), \left(\dot\theta_2, \sin\phi\, \dot\phi_2\right)\right)\\ &\quad = \sin\phi\, \dot\theta_1 \dot\phi_2 - \sin\phi\, \dot\theta_2 \dot\phi_1\\ &\quad = \omega_S(\theta, \phi)\left((\dot\theta_1, \dot\phi_1), (\dot\theta_2, \dot\phi_2)\right).\end{aligned}$$

The second method uses Proposition 3.56:

$$\begin{aligned}\varphi^*\omega_C &= \varphi^*\left(\mathrm{d}\theta \wedge \mathrm{d}z\right) = \left(\varphi^*(\mathrm{d}\theta)\right) \wedge \left(\varphi^*(\mathrm{d}z)\right)\\ &= \left(\mathrm{d}(\theta \circ \varphi)\right) \wedge \left(\mathrm{d}(z \circ \varphi)\right) = \left(\mathrm{d}\theta\right) \wedge \left(\mathrm{d}(-\cos\phi)\right)\\ &= \left(\mathrm{d}\theta\right) \wedge \left(\sin\phi\, \mathrm{d}\phi\right) = \sin\phi\, \mathrm{d}\theta \wedge \mathrm{d}\phi = \omega_S.\end{aligned}$$

**Example 3.58** Let $(x, y) = \psi(r, \theta) = (r\cos\theta, r\sin\theta)$, which is a diffeomorphism when restricted to a suitable domain and codomain. We calculate $\psi_*\mathrm{d}\theta$, which equals $(\psi^{-1})^*\,\mathrm{d}\theta$. Applying Proposition 3.56,

$$\left(\psi^{-1}\right)^*\mathrm{d}\theta = \mathrm{d}\left(\theta \circ \psi^{-1}\right) = \frac{\partial(\theta \circ \psi^{-1})}{\partial x}\,\mathrm{d}x + \frac{\partial(\theta \circ \psi^{-1})}{\partial y}\,\mathrm{d}y.$$

Now, $\theta \circ \psi^{-1}(x, y) = \arctan(y/x) + \text{constant}$ whenever $x \neq 0$, and $\theta \circ \psi^{-1}(x, y) = \operatorname{arccot}(x/y) + \text{constant}$ whenever $y \neq 0$. Both of these formulae imply:

$$\left(\psi^{-1}\right)^*\mathrm{d}\theta = -\frac{y}{x^2 + y^2}\,\mathrm{d}x + \frac{x}{x^2 + y^2}\,\mathrm{d}y.$$

**Remark 3.59** *Note also that the $\psi$ in the previous example may be viewed as a change of coordinates on $\mathbb{R}^2$, in which case push-forward by $\psi$ is the corresponding change of coordinates for 1-forms. Generalizing the above calculations, we see that the general 1-form $p_r\,\mathrm{d}r + p_\theta\,\mathrm{d}\theta$ is written in Cartesian coordinates as $p_r\frac{x}{\sqrt{x^2+y^2}}\,\mathrm{d}x + p_r\frac{y}{\sqrt{x^2+y^2}}\,\mathrm{d}y - p_\theta\frac{y}{x^2+y^2}\,\mathrm{d}x + p_\theta\frac{x}{x^2+y^2}\,\mathrm{d}y$. If we equate this expression with $s_x\,\mathrm{d}x + s_y\,\mathrm{d}y$, and write both $\mathbf{p} := (p_r, p_\theta)$ and $\mathbf{s} := (s_x, s_y)$ as row vectors, then*

$$\mathbf{s} = \mathbf{p}\,(D\psi(r,\theta))^{-1}.$$

*This is an example of the general formula in Exercise 2.26.*

**Definition 3.60** *The **Lie derivative** of a tensor $\alpha$ with respect to a vector field $X$ is defined as*

$$\mathcal{L}_X\,\alpha = \left.\frac{\mathrm{d}}{\mathrm{d}t}\right|_{t=0} \Phi_t^*\alpha,$$

*where $\Phi_t$ is the time-$t$ flow of $X$.*

**Example 3.61** Let $X(x, y) = (x, y)$ and $\alpha = (x^2 + y^2)\,dx$. The time-$t$ flow of $X$ is $\Phi_t(x, y) = (e^t x, e^t y)$, so

$$\begin{aligned}\left(\Phi_t^*\alpha\right)(x, y) &= \left((e^t x)^2 + (e^t y)^2\right)\Phi_t^* dx \\ &= e^{2t}\left(x^2 + y^2\right) d(x \circ \Phi_t) \\ &= e^{2t}\left(x^2 + y^2\right)\left(e^t dx\right) \\ &= e^{3t}\alpha(x, y).\end{aligned}$$

Therefore,

$$\mathcal{L}_X \alpha = \left.\frac{d}{dt}\right|_{t=0} e^{3t}\alpha = 3\alpha.$$

**Proposition 3.62** *Let $X$ be a vector field, $\alpha$ and $\beta$ covariant $k$-tensors, and $f$ a scalar field.*

1. $\mathcal{L}_X(f\alpha + \beta) = (\mathcal{L}_X f)\alpha + f(\mathcal{L}_X \alpha) + \mathcal{L}_X \beta$;
2. $\mathcal{L}_X(df) = d(\mathcal{L}_X f)$;
3. *If $\alpha$ is a 1-form, and in local coordinates, $\alpha = \alpha_i\, dq^i$, then*

$$\mathcal{L}_X \alpha = \left(X^i \frac{\partial \alpha_j}{\partial q^i} + \frac{\partial X^i}{\partial q^j}\alpha_i\right) dq^j.$$

*Proof* For properties 1 and 2, see Exercise 3.23. If $\alpha$ is a 1-form then, applying the first two properties,

$$\begin{aligned}\mathcal{L}_X \alpha = \mathcal{L}_X\left(\alpha_i dq^i\right) &= (\mathcal{L}_X \alpha_i)\, dq^i + \alpha_i\, \mathcal{L}_X\, dq^i \\ &= X^j \frac{\partial \alpha_i}{\partial q^j}\, dq^i + \alpha_i d(\mathcal{L}_X q^i) \\ &= X^j \frac{\partial \alpha_i}{\partial q^j}\, dq^i + \alpha_i dX^i \\ &= X^j \frac{\partial \alpha_i}{\partial q^j}\, dq^i + \alpha_i \frac{\partial X^i}{\partial q^j}\, dq^j.\end{aligned}$$

This third property is a special case of ***Cartan's formula***. ∎

**Exercise 3.17**
For each of the following tensors: write down its tensor coefficients, with respect to the standard basis; and, if it's a rank 2 tensor, write down its matrix.
a) the skew-symmetric bilinear form on $\mathbb{R}^3$ defined by

$$T(\mathbf{u}, \mathbf{v}) = 3u^1v^2 - 3u^2v^1 - 4u^1v^3 + 4u^3v^1;$$

b) the dot product in $\mathbb{R}^n$;
c) the scalar triple product in $\mathbb{R}^3$.

**Exercise 3.18**
Express each of the tensors in Exercise 3.17 as linear combinations of tensor products of standard basis vectors.

**Exercise 3.19**
(See Remark 3.37). Let $\mathcal{B}$ and $\mathcal{C}$ be bases for $V$, and let $P$ be the change-of-basis matrix from $\mathcal{C}$ to $\mathcal{B}$. Let $\mathcal{B}^*$ and $\mathcal{C}^*$ be the dual bases of $\mathcal{B}$ and $\mathcal{C}$, respectively. Show that the change-of-basis matrix from $\mathcal{C}^*$ to $\mathcal{B}^*$ is $P^{-T}$ (inverse transpose). Defining the matrix of a contravariant 2-tensor by analogy with the covariant case, show that, if $T$ has matrix $M$ with respect to $\mathcal{B}^*$ then it has matrix $P^{-1}MP^{-T}$ with respect to $\mathcal{C}^*$. Compare this with Exercise 2.26.

**Exercise 3.20**
Determine how tensor coefficients transform when changing bases, by generalizing Example 3.37 and Exercise 3.19.

**Exercise 3.21**
Let $f$ be as in the Example 3.57. Using the standard change-of-variables rule from multivariable calculus, check that $f$ is area-preserving, meaning that, for any region $B \subseteq S^2$,

$$\int_B \sin\phi \, \mathrm{d}\theta \, \mathrm{d}\phi = \int_{f(B)} \mathrm{d}\theta \, \mathrm{d}z.$$

Compare your calculations with those in Example 3.57.

**Exercise 3.22**
If $X(x, y) = (-y, x)$ and $\omega = dx \wedge dy$, show that $\mathcal{L}_X \omega = 0$.

**Exercise 3.23**
Prove Proposition 3.62. HINT: The first property follows from linearity of derivatives and the product rule. For the second, use the chain rule and equality of mixed partials.

## 3.4 Riemannian geometry

A ***Riemannian metric*** on a manifold $Q$ is a smoothly varying family of inner products on each of the tangent spaces $T_q Q$. A formal definition appears below. Recall that an ***inner product*** on a vector space is a positive-definite, symmetric bilinear form. It is represented, with respect to any basis, by a positive-definite symmetric matrix. Two vectors $\mathbf{v}$ and $\mathbf{w}$ are ***orthogonal*** with respect to a given inner product if and only if $\langle\langle \mathbf{v}, \mathbf{w} \rangle\rangle = 0$. Given an inner product $\langle\langle \cdot, \cdot \rangle\rangle$, the ***associated norm*** is given by $\|\mathbf{v}\| := \langle\langle \mathbf{v}, \mathbf{v} \rangle\rangle^{1/2}$, and $\|\mathbf{v}\|$ is called the ***norm*** of $\mathbf{v}$ or the ***length*** of $\mathbf{v}$.

**Example 3.63** The ***Euclidean inner product*** on $\mathbb{R}^n$ is the bilinear form that, when written in standard Cartesian coordinates, equals the dot product:

$$\langle\langle \mathbf{v}, \mathbf{w} \rangle\rangle := \mathbf{v} \cdot \mathbf{w} = \sum_i v^i w^i .$$

It is represented, with respect to the standard basis, by the identity matrix $I$, because $\langle\langle \mathbf{v}, \mathbf{w} \rangle\rangle = \mathbf{v}^T \mathbf{w} = \mathbf{v}^T I \mathbf{w}$. The associated norm is called the ***Euclidean norm***. The ***speed*** of a particle with velocity $\dot{\mathbf{q}} \in \mathbb{R}^n$ is $\|\dot{\mathbf{q}}\|$.

**Example 3.64** Recall that the ***kinetic energy*** of an $N$-particle Newtonian system in $\mathbb{R}^3$ is $K = \frac{1}{2} \sum_i m_i \|\dot{\mathbf{q}}^i\|^2$, with $\|\cdot\|$ being the Euclidean norm.

Writing $\dot{\mathrm{q}} = (\dot{\mathrm{q}}^1, \ldots, \dot{\mathrm{q}}^N)$ and defining

$$\mathbb{M} := \operatorname{diag}(m_1, m_1, m_1, m_2, m_2, m_2, \ldots, m_N, m_N, m_N) \tag{3.9}$$

(the diagonal matrix with these entries along the diagonal), we have

$$K = \frac{1}{2}\dot{\mathrm{q}}^T \mathbb{M}\dot{\mathrm{q}}.$$

Let $\langle\langle\cdot,\cdot\rangle\rangle_{\mathbb{M}}$ be the inner product on on $\mathbb{R}^{3N}$ with matrix $\mathbb{M}$ with respect to the standard basis, and let $\|\cdot\|_{\mathbb{M}}$ be the associated norm. Then,

$$K = \frac{1}{2}\|\dot{\mathrm{q}}\|_{\mathbb{M}}^2 = \frac{1}{2}\langle\langle\dot{\mathrm{q}}, \dot{\mathrm{q}}\rangle\rangle_{\mathbb{M}}.$$

Now, consider a particle moving on a general manifold $Q$, with trajectory $q(t)$. Its velocity, $\dot{q}(t)$, is a tangent vector based at $q(t)$. So, for a general definition of speed or kinetic energy of such particles, we need a norm on $T_q Q$, for every $q$. For a system of particles, the same remarks apply, except that each point in the configuration space $Q$ represents the positions of all particles, and $\dot{q}(t)$ describes the velocity of all particles. These are motivations for the following definition.

**Definition 3.65** *A* ***Riemannian metric*** *on a manifold $Q$ is a smooth symmetric positive-definite covariant 2-tensor field $g$ on $Q$. This means that $g(q)$ is an inner product on $T_q Q$, for every $q \in Q$, and that the tensor coefficients $g_{ij}$, in any choice of local coordinates, are smooth functions of $q$.*

*Notation: $g(q)$ is also written as $g_q$ or $\langle\langle\cdot,\cdot\rangle\rangle_q$, and $g$ is written as $\langle\langle\cdot,\cdot\rangle\rangle$. Using the notation for symmetric products introduced in the previous section, $g = g_{ij}\mathrm{d}q^i \mathrm{d}q^j$; recall that $\mathrm{d}q^i \mathrm{d}q^i$ is also written as $(\mathrm{d}q^i)^2$, with the parentheses often omitted, for example $\mathrm{d}x^2 := \mathrm{d}x\,\mathrm{d}x$. The pair $(Q, g)$ is called a* ***Riemannian manifold.***

*Given $g$, the* ***associated norm*** *(really a family of norms, one at each $q \in Q$) is defined as follows: the norm of a tangent vector $v \in T_q Q$ is $\|v\|_q$ defined by $\|v\|_q^2 := \langle\langle v, v\rangle\rangle_q = g(q)(v, v) = g_{ij}(q)\, v^i v^j$ (the latter expression is in local coordinates). Two tangent vectors $v, w \in T_q Q$, based at the same point $q \in Q$, are* ***orthogonal*** *if $\langle\langle v, w\rangle\rangle_q = 0$, i.e. $g_{ij}(q)\, v^i w^j = 0$.*

The simplest manifolds are vector spaces. Any inner product on a vector space $V$ defines a *constant* Riemannian metric $g$ on $V$ such that $g(q)$ equals the given inner product, for every $q \in V$. The following example is of this type. There also exist non-constant Riemannian metrics on vector spaces.

**Example 3.66** The ***Euclidean metric*** on $\mathbb{R}^n$ is the constant Riemannian metric such that $g(q)$ is the Euclidean inner product, for every $q \in \mathbb{R}^n$. In Cartesian coordinates, each $g(q)$ has matrix $I$ (the identity matrix), so

$$g_{ij} = \delta_{ij} := \begin{cases} 0, & i \neq j \\ 1, & i = j \end{cases}$$

and $g = \mathrm{d}x_1^2 + \cdots + \mathrm{d}x_n^2$.

In $\mathbb{R}^2$, in Cartesian coordinates, the Euclidean metric is $g = \mathrm{d}x^2 + \mathrm{d}y^2$. In polar coordinates, defined by $(x, y) = \psi(r, \theta) = (r\cos\theta, r\sin\theta)$, the matrix of $g(r, \theta)$ is

$$\begin{aligned}(\mathrm{D}\psi(r,\theta))^T I\, \mathrm{D}\psi(r,\theta) &= \begin{bmatrix} \cos\theta & -r\sin\theta \\ \sin\theta & r\cos\theta \end{bmatrix}^T \begin{bmatrix} \cos\theta & -r\sin\theta \\ \sin\theta & r\cos\theta \end{bmatrix} \\ &= \begin{bmatrix} \cos\theta & \sin\theta \\ -r\sin\theta & r\cos\theta \end{bmatrix} \begin{bmatrix} \cos\theta & -r\sin\theta \\ \sin\theta & r\cos\theta \end{bmatrix} \\ &= \begin{bmatrix} 1 & 0 \\ 0 & r^2 \end{bmatrix}.\end{aligned}$$

Thus, in polar coordinates (excluding the origin), the Euclidean metric is $g = \mathrm{d}r^2 + r^2\mathrm{d}\theta^2$. An alternative method is:

$$\begin{aligned} g &= (\mathrm{d}x)^2 + (\mathrm{d}y)^2 \\ &= \left(\cos\theta\, \mathrm{d}r - r\sin\theta\, \mathrm{d}\theta\right)^2 + \left(\sin\theta\, \mathrm{d}r + r\cos\theta\, \mathrm{d}\theta\right)^2 \\ &= (\mathrm{d}r)^2 + r^2(\mathrm{d}\theta)^2. \end{aligned}$$

**Proposition 3.67** *If $M$ is a manifold, $(N, g)$ is a Riemannian manifold and $\varphi : M \to N$ is a smooth immersion, then $\varphi^* g$, the pull-back of $g$ by $\varphi$, is a Riemannian metric on $M$, called the metric on $M$* ***induced*** *by $g$ via $\varphi$.*

*Proof* By definition, $(\varphi^* g)(x)(v, w) = g(\varphi(x))(\mathrm{D}\varphi(x) \cdot v, \mathrm{D}\varphi(x) \cdot w)$. It is clear that $(\varphi^* g)(x)$ is a symmetric bilinear form on $T_x M$, for every $x$, and that in any local coordinates, the coefficients of $\varphi^* g$ vary smoothly, so $\varphi^* g$ is a smooth symmetric covariant 2-tensor. For any $v \in T_x M$, if $(\varphi^* g)(x)(v, v) = 0$ then $\mathrm{D}\varphi(x) \cdot v = 0$, by the positive-definiteness of $g$. Since $\varphi$ is an immersion, this implies $v = 0$. This proves that $\varphi^* g$ is positive-definite. ∎

**Definition 3.68** *If $M$ is a submanifold of $\mathbb{R}^n$, then the metric on $M$ induced by the Euclidean metric via the inclusion map $i : M \to \mathbb{R}^n$ is called the* ***first fundamental form*** *on $M$. Equivalently, the first fundamental form is the*

*Euclidean metric restricted to tangent spaces of $M$, i.e. $g(\mathbf{x})(\mathbf{v}, \mathbf{w}) = \mathbf{v} \cdot \mathbf{w}$ for all $\mathbf{v}, \mathbf{w} \in T_{\mathbf{x}}M$.*

The first fundamental form is the metric studied in the classical Riemannian geometry of curves and surfaces. Note that the name is historical, and in particular, the first fundamental form is *not* a differential form (defined in the next section).

**Example 3.69** We calculate the first fundamental form on the cylinder $x^2 + y^2 = R^2$. First, we change to cylindrical coordinates on $\mathbb{R}^3$, so the Euclidean metric becomes $\mathrm{d}r^2 + r^2\,\mathrm{d}\theta^2 + \mathrm{d}z^2$, as for polar coordinates in the previous example. At points on the cylinder, this is $\mathrm{d}r^2 + R^2\,\mathrm{d}\theta^2 + \mathrm{d}z^2$. The first fundamental form is the restriction of this metric to tangent spaces to the cylinder. Since tangent vectors to the cylinder are all of the form $(0, \dot{\theta}, \dot{z})$, the restricted metric is $R^2\,\mathrm{d}\theta^2 + \mathrm{d}z^2$.

**Remark 3.70** *Let $g$ be the first fundamental form on a submanifold $Q$ of $\mathbb{R}^n$, with $n = 2$ or $3$. The norm of a tangent vector $v \in T_qQ$, as defined in Definition 3.65, is its length. Two vectors $v, w \in T_qQ$ are orthogonal, as defined in Definition 3.65, if and only if they are perpendicular, in the usual sense. The angle between $v$ and $w$ is the unique $\theta \in [0, \pi]$ such that $\cos\theta\,\|v\|\,\|w\| = \langle\langle v, w\rangle\rangle_q$. The speed of a particle with trajectory $\mathbf{q}(t)$ in $Q$ is $\|\dot{q}(t)\|_{q(t)}$.*

Note however that, for a general Riemannian metric, the associated norm will *not* generally correspond to length, two orthogonal vectors need *not* be perpendicular in the usual sense, and ‘angle’ as defined in the previous remark need *not* have its usual interpretation – see Exercise 3.25.

**Remark 3.71** *In a given mechanics problem, there are often two Riemannian metrics involved: the first determining speed, and the second determining kinetic energy. Specifically, if $\mathbf{q}(t)$ is a trajectory, then $\|\dot{\mathbf{q}}(t)\|$ with respect to the first metric is speed, while $\frac{1}{2}\|\dot{\mathbf{q}}(t)\|^2$ with respect to the second metric is kinetic energy, as in Example 3.64. The first metric is often a first fundamental form, while the second differs from it by some mass or inertia constants.*

**Definition 3.72** *Let $(Q, g)$ be a Riemannian manifold, and let $f$ be a scalar field on $Q$. The* ***gradient*** *of $f$ is the vector field $\nabla f$ on $Q$ defined, at any $z \in Q$ by*

$$\mathrm{d}f(z)(v) = g_z\left(\nabla f(z), v\right), \quad \text{for all } v \in T_zQ, \tag{3.10}$$

*or equivalently, in more compact notation,*

$$\mathrm{d}f(z) = g_z(\nabla f(z), \cdot).$$

*This is well defined because $g_z$ is non-degenerate, for all $z$. In local coordinates, if $\mathrm{d}f(z)$ is written as a row vector, $\nabla f(z)$ a column vector and $g_z$ a matrix, then $\mathrm{d}f(z) = \nabla f(z)^T g(z)$, so*

$$\nabla f(z) = g_z^{-1}\, \mathrm{d}f(z)^T.$$

**Example 3.73** In $\mathbb{R}^n$ with the Euclidean metric, the gradient of a scalar field $f$ satisfies $\mathrm{d}f(\mathbf{z}) \cdot \mathbf{v} = \nabla f(\mathbf{z}) \cdot \mathbf{v}$, for every $\mathbf{v} \in \mathbb{R}^n$, so $\nabla f$ is the usual gradient vector field from calculus.

The pairing defined in the following example will be used several times in this book. It is essential for certain computations in mechanics on matrix Lie groups.

**Example 3.74 (The trace pairing)** The following real-valued bilinear operation on $\mathcal{M}(n,\mathbb{R})$ is called the ***trace pairing***:

$$\langle \mathbf{C}, \mathbf{D} \rangle := \mathrm{tr}(\mathbf{C}\mathbf{D}^T) = \sum_i \sum_j C_{ij} D_{ij} = \mathrm{tr}\left(\mathbf{C}^T \mathbf{D}\right).$$

(Recall the identities $\mathrm{tr}(\mathbf{U}^T) = \mathrm{tr}(\mathbf{U})$ and $\mathrm{tr}(\mathbf{U}\mathbf{V}) = \mathrm{tr}(\mathbf{V}\mathbf{U})$.) This pairing clearly corresponds to the Euclidean inner product on $\mathbb{R}^{n^2}$, via any of the standard identifications of $\mathcal{M}(n,\mathbb{R})$ with $\mathbb{R}^{n^2}$. Thus, the trace pairing is an inner product, and therefore $g_{\mathbf{R}}(\mathbf{C}, \mathbf{D}) := \mathrm{tr}\left(\mathbf{C}\mathbf{D}^T\right)$ defines a Riemannian metric on $\mathcal{M}(n,\mathbb{R})$. (It is the metric induced from the Euclidean metric on $\mathbb{R}^{n^2}$.) For any matrix group $G$, the restriction of this metric to the tangent spaces of $G$ is a Riemannian metric on $G$. It is the metric on $G$ induced by $g$ via the inclusion map.

Let $f$ be a scalar field on $\mathcal{M}(n,\mathbb{R}) \cong \mathbb{R}^{n^2}$. We compute the gradient of $f$, with respect to the metric $g$. Since $g$ is essentially the Euclidean metric on $\mathbb{R}^{n^2}$, the matrix $\nabla f(\mathbf{A})$ has $(i,j)$th entry $\partial f / \partial A_{ij}$. It can also be computed from eqn (3.10) as follows,

$$\mathrm{d}f(\mathbf{A})(\mathbf{C}) = g_{\mathbf{A}}(\nabla f(\mathbf{A}), \mathbf{C}) = \mathrm{tr}\left(\left(\nabla f(\mathbf{A})\right) \mathbf{C}^T\right) \quad \text{for all } \mathbf{C} \in \mathcal{M}(n,\mathbb{R}). \tag{3.11}$$

In particular, if $f(\mathbf{A}) = \operatorname{tr}(\mathbf{AB}^T)$, then, for all $\mathbf{C} \in \mathcal{M}(n, \mathbb{R})$,

$$\mathrm{d}f(\mathbf{A})(\mathbf{C}) = \left.\frac{\mathrm{d}}{\mathrm{d}t}\right|_{t=0} \operatorname{tr}\left((\mathbf{A} + t\mathbf{C})\mathbf{B}^T\right) = \operatorname{tr}\left(\mathbf{CB}^T\right) = \operatorname{tr}\left(\mathbf{BC}^T\right),$$

and therefore $\nabla f(\mathbf{A}) = \mathbf{B}$. In alternative notation,

$$\frac{\partial \operatorname{tr}(\mathbf{AB}^T)}{\partial \mathbf{A}} = \mathbf{B}. \tag{3.12}$$

Since $\operatorname{tr}(\mathbf{AB}^T) = \operatorname{tr}(\mathbf{BA}^T) = \operatorname{tr}(\mathbf{A}^T\mathbf{B})$, we also have

$$\frac{\partial \operatorname{tr}(\mathbf{A}^T\mathbf{B})}{\partial \mathbf{A}} = \mathbf{B}. \tag{3.13}$$

Now, consider a scalar field $f$ on a matrix Lie group $G$. The gradient 'vector' $\nabla f(\mathbf{A})$ is defined to be the unique element of $T_{\mathbf{A}}G$ such that

$$\mathrm{d}f(\mathbf{A})(\mathbf{C}) = g_{\mathbf{A}}\left(\nabla f(\mathbf{A}), \mathbf{C}\right) = \operatorname{tr}\left(\left(\nabla f(\mathbf{A})\right)\mathbf{C}^T\right) \quad \text{for all } \mathbf{C} \in T_{\mathbf{A}}G.$$

This is almost the same definition as in eqn (3.11), except that $\nabla f(\mathbf{A})$ and $\mathbf{C}$ are both required to be in $T_{\mathbf{A}}G$. Beware that the formulae in eqns (3.12) and (3.13) can't be applied directly, since the matrix $\mathbf{B}$ in these formulae are not necessarily elements of $T_{\mathbf{A}}G$.

**Remark 3.75** *Since all cotangent vectors in $T_z^*Q$ are of the form $df(z)$ for some scalar field $f$ on $Q$, the map $\mathrm{d}f(z) \to \nabla f(z)$ determines an isomorphism from $T_z^*Q$ to $T_zQ$. In particular, given a matrix Lie group $G$, its tangent space at the identity $\mathfrak{g} := T_IG$ is isomorphic to its dual $\mathfrak{g}^* := T_I^*G$, and the isomorphism is defined via the trace pairing introduced above.*

**Definition 3.76** *The* ***length*** *(or* ***arc length****) of a smooth path $q : [a, b] \to Q$ in a manifold $Q$ with Riemannian metric $g$ is*

$$\int_a^b \|\dot{q}(t)\|_{q(t)} \,\mathrm{d}t.$$

*The* ***distance*** *between two points in $Q$ is the length of the shortest smooth path connecting them. A map between Riemannian manifolds, $\varphi : (Q, g) \to (R, h)$,* ***preserves distances*** *if the distance between $\varphi(a)$ and $\varphi(b)$ equals the distance between $a$ and $b$, for all $a, b \in Q$.*

**Remark 3.77** *If $q : [a, b] \to Q$ and $r : [c, d] \to Q$ parameterize the same curve, and $q = r \circ \varphi$, for some smooth $\varphi$ with $\varphi(a) = c$ and $\varphi(b) = d$, then it is straightforward to check that $\int_a^b \|\dot{q}(t)\| \,\mathrm{d}t = \int_c^d \|\dot{r}(s)\| \,\mathrm{d}s$. This almost* *proves that the formula above defines the length of a curve, independent of the choice of parameterization. What remains to be shown is that, given any two smooth parameterizations* $q$ *and* $r$ *with the same image, there exists a* $\varphi$ *as above. This is true if* $r$ *is an immersion and* $r^{-1} : \operatorname{Im} r \to [c, d]$ *is continuous, by Theorem 2.39.*

**Remark 3.78** *Distance as defined above satisfies the axioms of a metric, namely:* $d(x, y) \geq 0$; $d(x, y) = 0 \Leftrightarrow x = y$; $d(x, y) = d(y, x)$; *and* $d(x, z) \leq d(x, y) + d(y, z)$.

In $\mathbb{R}^n$, the shortest curve joining two points is a line segment. On a sphere, the shortest curve joining two points is an arc of a great circle. These curves are examples of ***geodesics***. In general, a curve is a geodesic if it 'locally minimizes length', meaning that no 'small' change in the curve could produce a shorter curve connecting the same two points $q(a)$ and $q(b)$. The following formal definition uses the calculus of variations (introduced in Section 1.2).

**Definition 3.79** *A smooth path* $q : [a, b] \to Q$ *in a Riemannian manifold is a* ***geodesic*** *if it has constant non-zero speed* $\|\dot{\mathrm{q}}(t)\| = c$ *and it is a stationary point of the* ***length functional***

$$\mathcal{S}_l[\mathrm{q}(\cdot)] := \int_a^b \|\dot{\mathrm{q}}(t)\|_{\mathrm{q}(t)} \, \mathrm{d}t, \tag{3.14}$$

*with respect to variations among paths with fixed endpoints. A path defined on an unbounded interval is a geodesic if every finite piece of it is a geodesic.*

When we introduced the calculus of variations, it was for curves in $\mathbb{R}^n$, not general manifolds. But this is no obstacle because a curve is a stationary point of the length functional if and only if each piece of the curve has the same property (see Section 4.1). It follows that, to check whether a curve is a geodesic, we can cover the manifold in coordinate charts, and use local coordinates to check pieces of the curve. Note that the definition specifies a *stationary point* rather than a local minimum. However, it can be proven that, for this functional, the stationary points are always local minima.

**Remark 3.80** *A geodesic need not globally minimize length. For example, a trajectory that follows three quarters of a great circle on a sphere is a geodesic, because any small change to it would produce a longer curve. Yet there does exist a shorter curve connecting the two endpoints, namely the arc that 'goes the other way' around the great circle.*

In general, to find geodesics, we have to find stationary points of the length functional. But we already have a great tool for doing this: the Euler–Lagrange equations. Indeed, the length functional is the action functional for the Lagrangian $l(\mathbf{q}, \dot{\mathbf{q}}) := \|\dot{\mathbf{q}}\|_{\mathbf{q}}$, so by Theorem 1.38, the stationary points of the length functional are the solutions of the Euler–Lagrange equations,

$$\frac{d}{dt}\left(\frac{\partial l}{\partial \dot{q}^i}\right) - \frac{\partial l}{\partial q^i} = 0, \quad \text{for } i = 1, \ldots, n. \tag{3.15}$$

In local coordinates, $l(\mathbf{q}, \dot{\mathbf{q}}) = \|\dot{\mathbf{q}}\|_{\mathbf{q}} = \sqrt{g_{ij}(\mathbf{q})\, \dot{q}^i \dot{q}^j}$, and eqn (3.15) becomes, for $m = 1, \ldots, n$ (changing $i$ to $m$),

$$\frac{d}{dt}\left[\left(\frac{1}{2\,\|\dot{\mathbf{q}}\|_{\mathbf{q}}}\right)\frac{\partial}{\partial \dot{q}^m}\left(g_{ij}(\mathbf{q})\, \dot{q}^i \dot{q}^j\right)\right] = \left(\frac{1}{2\,\|\dot{\mathbf{q}}\|_{\mathbf{q}}}\right)\frac{\partial}{\partial q^m}\left(g_{ij}(\mathbf{q})\, \dot{q}^i \dot{q}^j\right).$$

If $\mathbf{q}(t)$ is a constant-speed parameterization then $\frac{1}{2\|\dot{\mathbf{q}}\|_{\mathbf{q}}}$ is constant and so the Euler–Lagrange equations simplify to

$$\frac{d}{dt}\left[\frac{\partial}{\partial \dot{q}^m}\left(g_{ij}(\mathbf{q})\, \dot{q}^i \dot{q}^j\right)\right] = \frac{\partial}{\partial q^m}\left(g_{ij}(\mathbf{q})\, \dot{q}^i \dot{q}^j\right), \quad \text{for } m = 1, \ldots, n, \tag{3.16}$$

or equivalently,

$$\frac{d}{dt}\left[g_{mj}\dot{q}^j + g_{im}\dot{q}^i\right] = \frac{\partial g_{ij}}{\partial q^m}\, \dot{q}^i \dot{q}^j, \quad \text{for } m = 1, \ldots, n. \tag{3.17}$$

Further calculations lead to the form given in Exercise 3.31, involving the *Christoffel symbols.*

**Example 3.81** For the Euclidean metric, $g_{ij} = \delta_{ij}$, so the geodesic equations (in eqn (3.17)) reduce to: $\ddot{\mathbf{q}} = \mathbf{0}$. This proves that the geodesics for this metric are straight lines, or segments of them.

**Theorem 3.82** *Given a Riemannian manifold, let $L$ be the Lagrangian*

$$L(\mathbf{q}, \dot{\mathbf{q}}) = \frac{1}{2}\|\dot{\mathbf{q}}\|_{\mathbf{q}}^2 = \frac{1}{2} g_{ij}(\mathbf{q})\, \dot{q}^i \dot{q}^j.$$

*Then, $\mathbf{q}(t)$ is a geodesic if and only if it is a solution of the Euler–Lagrange equations for $L$, with non-zero speed.*

*Proof* We have seen that a geodesic is a path that satisfies eqn (3.16) and has constant non-zero speed. It is easily checked that eqn (3.16) are the Euler–Lagrange equations for $L$. Recall that solutions of the Euler–Lagrange equations always conserve the energy function: this was shown for $\mathbb{R}^n$ in Theorem 1.37, and the proof applies on general manifolds. For the present Lagrangian $L$, the corresponding energy function is $L$ itself. Hence, $L$ is conserved, which implies that speed $\|\dot{\mathbf{q}}\|$ is also automatically conserved by all solutions. Therefore, all solutions of the Euler–Lagrange equations for $L$ with non-zero speed are geodesics. ■

**Remark 3.83** *In many examples, the Lagrangian* $L = \frac{1}{2}\|\dot{\mathbf{q}}\|^2$ *can be interpreted as kinetic energy. Thus, the previous theorem generalizes the fact that, in Newtonian mechanics, in the absence of forces, particles move in straight lines.*

**Definition 3.84** *An* ***isometry*** *is a diffeomorphism* $\varphi$ *from one Riemannian manifold* $(Q, g)$ *to another* $(R, h)$*, such that* $\varphi^* h = g$*. If such a* $\varphi$ *exists, we say that* $(Q, g)$ *and* $(R, h)$ *are* ***isometric.***

**Theorem 3.85** *Let* $\varphi : (Q, g) \to (R, h)$ *be a diffeomorphism between Riemannian manifolds. Then, the folllowing are equivalent:*

1. $\varphi$ *is an isometry;*
2. $\varphi$ *preserves lengths of tangent vectors, i.e.* $\|T\varphi(v)\|_h = \|v\|_g$ *for all* $v \in TQ$.
3. $\varphi$ *preserves arc length;*
4. $\varphi$ *preserves distances;*
5. $\varphi$ *preserves geodesics.*

*Proof* See Exercise 3.27. ■

**Example 3.86** Let $D = S^2 \backslash \{(0, 0, \pm 1)\}$ and $C = S^1 \times \mathbb{R}$. Let $f : D \to C$ be given, in spherical coordinates $(\theta, \phi)$ on $S^2$ and cylindrical coordinates $(\theta, z)$ on $C$, by $f(\theta, \phi) = (\theta, -\cos\phi)$. From Example 3.69 we know that the first fundamental form on the cylinder is $g_C = \mathrm{d}\theta^2 + \mathrm{d}z^2$. The first fundamental form on the unit sphere is $g_S = \sin^2\phi\, \mathrm{d}\theta^2 + \mathrm{d}\phi^2$ (this is Exercise 3.24). The derivative of $f$ is

$$\mathrm{D}f(\theta, \phi) \cdot (\dot{\theta}, \dot{\phi}) = \left(\dot{\theta}, \sin\phi\, \dot{\phi}\right).$$

By the Inverse Function Theorem, $f : D \to f(D)$ is a diffeomorphism. The pull-back of $g_C$ by $f$ is,

$$\left(f^* g_C\right)(\theta, \phi)\left((\dot\theta_1, \dot\phi_1), (\dot\theta_2, \dot\phi_2)\right) = g_C(\theta, -\cos\phi)\left((\dot\theta_1, \sin\phi\,\dot\phi_1), (\dot\theta_2, \sin\phi\,\dot\phi_2)\right)$$
$$= \dot\theta_1\dot\theta_2 + \sin^2\phi\,\dot\phi_1\dot\phi_2.$$

Therefore, $f^* g_C = \mathrm{d}\theta^2 + \sin^2\phi\,\mathrm{d}\phi^2$. Note that this does not equal $g_S$. Therefore, $f$ is not an isometry, i.e. $f$ is not distance preserving. Recall, however, from Example 3.57, that $f$ is area preserving. In cartography, this $f$ is known as the *Lambert cylindrical equal-area projection*.

**Example 3.87** The unit sphere $S^2$ minus any one point is diffeomorphic to $\mathbb{R}^2$ (for example via stereographic projection, see Example 2.49). However, there is no isometry between these two spaces. One way to see this is that the distance between two points in $S^2$ is at most $\pi$, whereas distances between points in $\mathbb{R}^2$ are unbounded; and isometries must preserve distances.

Further, there is no open subset of $S^2$ that is isometric to any open subset of $\mathbb{R}^2$. To see this, consider circles in both spaces. In any 2-dimensional Riemannian manifold, a ***circle*** is the set of points at distance $r$ from $\mathbf{c}$, for a given radius $r$ and centre $\mathbf{c}$. Since this definition depends only on the Riemannian metric, it is easily checked that circles are preserved by isometries, i.e. the image of a circle by an isometry is another circle. In $\mathbb{R}^2$, the perimeter of a circle with radius $r$ is $2\pi r$. Since isometries preserve distances, this must be true of circles in any space that is isometric to an open subset of $\mathbb{R}^2$. But a circle of radius $r$ on the unit sphere (with radius measured as an arc length in $S^2$) always has perimeter less than $r$. Indeed, if $\mathbf{x}$ is a point on a circle with centre $\mathbf{c}$ and radius $r$ (measured in $S^2$), then the angle (in $\mathbb{R}^3$) between the vectors $\mathbf{x}$ and $\mathbf{c}$ is $r$ (in radians), so the radius of the circle with respect to the Euclidean metric on $\mathbb{R}^3$ is $2\sin(r/2)$ and the perimeter of the circle has length $4\pi\sin(r/2)$.

The following theorem, not proven here, is an extension of the Whitney embedding theorem (Theorem 2.107) for general smooth manifolds.

**Theorem 3.88 (Nash embedding theorem)** *For every Riemannian manifold $(M, g)$ there exists an embedding $\varphi : M \to \mathbb{R}^n$, for some $n$, such that the pull-back of the Euclidean metric by $\varphi$ equals $g$. Thus, every Riemannian metric arises as a first fundamental form.*

**Exercise 3.24**
Show that the first fundamental form on the unit sphere $S^2$, expressed in the spherical coordinates defined by $(x, y, z) = (\sin\phi\cos\theta, \sin\phi\sin\theta, -\cos\phi)$, is $g_S := \sin^2\phi\, \mathrm{d}\theta^2 + \mathrm{d}\phi^2$. Express $g_S$ in stereographic projection coordinates.

**Exercise 3.25**
Show that $g := (x^2+1)\,\mathrm{d}x^2 + \mathrm{d}x\,\mathrm{d}y + \mathrm{d}y^2$ defines a Riemannian metric on $\mathbb{R}^2$. Show that the tangent vectors $\mathbf{v} = (3, 4)$ and $\mathbf{w} = (4, -3)$, both based at $(1, 1)$, are not orthogonal with respect to $g$. Calculate the norms of these vectors.

**Exercise 3.26**
Show that if $A$ is symmetric and $B$ skew-symmetric then $\operatorname{tr}\left(AB^T\right) = 0$.

**Exercise 3.27**
Prove Theorem 3.85. HINT FOR PART 2: Recall that an inner product can be recovered from its associated norm by the formula

$$\langle\langle v, w\rangle\rangle = \frac{1}{2}\left(\|v+w\|^2 - \|v\|^2 - \|w\|^2\right).$$

**Exercise 3.28**
Find an isometry between some relatively open subset of the cone $z = \sqrt{3(x^2+y^2)}$ and an open subset of the $xy$-plane, both equipped with the first fundamental form.

HINT: Think of unrolling a paper cone.

**Exercise 3.29**
Find the geodesics on a cylinder, with respect to the first fundamental form.

**Exercise 3.30**
Find the geodesics on a cone, with respect to the first fundamental form. HINT: use Exercise 3.28 and the fact that the geodesics for the Euclidean metric in $\mathbb{R}^n$ are straight lines (see Example 3.81).

**Exercise 3.31**
Show that the Euler–Lagrange equations for geodesics, eqn (3.17), are equivalent to the ***geodesic equations***,

$$\ddot{q}^m + \Gamma^m_{ij}\dot{q}^i\dot{q}^j = 0, \tag{3.18}$$

where

$$\Gamma^m_{ij} := \frac{1}{2}g^{km}\left(\frac{\partial g_{ik}}{\partial q^j} + \frac{\partial g_{jk}}{\partial q^i} - \frac{\partial g_{ij}}{\partial q^k}\right).$$

By $g^{km}(\mathbf{q})$ we mean the $(k,m)$th entry of the inverse of the matrix $[g_{ij}(\mathbf{q})]$. An equivalent definition is: $g_{ik}g^{km} = \delta_i^m$, for all $i,m$. The $\Gamma^m_{ij}$ are called the ***Christoffel symbols***, of the second kind, of the Levi–Civita connection on $(Q,g)$.

**Exercise 3.32**
Check that great circles on $S^2$ satisfy the geodesic equations for the first fundamental form.

HINT: without loss of generality, it suffices to consider 'the Equator'.

## 3.5 Symplectic geometry

Symplectic geometry has played a central role in the mathematical development of Hamiltonian mechanics. For simplicity, we have chosen not to emphasize it in this book. Instead, we present Hamiltonian mechanics via Poisson brackets, which generalize symplectic forms and are somewhat easier to define. This section is included mostly for general knowledge.

Recall that a bilinear form is ***non-degenerate*** if its matrix is non-singular.

**Definition 3.89** *A **symplectic bilinear form** is one that is skew-symmetric and non-degenerate. A vector space with an associated symplectic bilinear form is called a **symplectic vector space**.*

**Example 3.90** The ***canonical symplectic bilinear form*** on $\mathbb{R}^{2n}$ is

$$\omega\left((\mathbf{x}_1,\mathbf{y}_1),(\mathbf{x}_2,\mathbf{y}_2)\right) := \mathbf{x}_1 \cdot \mathbf{y}_2 - \mathbf{x}_2 \cdot \mathbf{y}_1.$$

Its matrix, with respect to the standard basis, is

$$J := \begin{bmatrix} 0 & I \\ -I & 0 \end{bmatrix}, \tag{3.19}$$

where $I$ is the $n \times n$ identity matrix. Note that, in $\mathbb{R}^2$, the bilinear form $\omega$ is the standard area form (see Example 3.52).

**Remark 3.91** *If $\mathbb{R}^{2n}$ is identified with $\mathbb{C}^n$ via*

$$\left(x^1,\ldots,x^n,y^1,\ldots,y^n\right) \leftrightarrow \left(x^1+iy^1,\ldots,x^n+iy^n\right),$$

*then the matrix of the canonical symplectic form can be written as* $-\mathrm{i}I$ *(an $n\times n$ complex diagonal matrix). Etymological note: 'symplectic' is the greek translation of the latin 'complex'.*

In order to define symplectic forms on *manifolds*, we first need to introduce some of the theory of differential forms.

**Definition 3.92** *A **differential form** (or simply **form**) is a skew-symmetric covariant tensor field.[3] An n-**form** is a form of rank n.*

Some examples are:

- a scalar field is a 0-form;
- differential 1-forms satisfy the general definition because all 1-tensors are automatically skew-symmetric;
- the area form on any surface in $\mathbb{R}^3$, defined in Example 3.52, is a differential 2-form.
- the ***volume form*** $\mu$ in $\mathbb{R}^3$, defined by $\mu(\mathbf{z})(\mathbf{u},\mathbf{v},\mathbf{w}) := (\mathbf{u}\times\mathbf{v})\cdot\mathbf{w}$, is a differential 3-form.

[3] Confusingly, bilinear forms can be skew-symmetric, symmetric or neither, while *differential* forms are assumed to be skew-symmetric.

One of the main uses of differential forms is to define integration on manifolds. We will not develop the theory of differential forms here. Instead, we will give the bare minimum to allow us to define symplectic forms, in particular what it means for a form to be *closed*. We need to define the exterior derivative operation, which in turn can be described in terms of the wedge product.

Recall from Section 3.3 the wedge product of two 1-forms,

$$\alpha \wedge \beta = \alpha \otimes \beta - \beta \otimes \alpha, \quad \text{i.e.}$$
$$(\alpha \wedge \beta)\,(z)(v, w) = \alpha(z)(v)\,\beta(z)(w) - \alpha(z)(w)\beta(z)(v).$$

**Example 3.93** If $x^1, x^2$ are coordinates, then

$$\left(\mathrm{d}x^1 \wedge \mathrm{d}x^2\right)(\mathbf{x})(\mathbf{u}, \mathbf{v}) = u^1 v^2 - v^1 u^2.$$

The next two definitions generalize the wedge product.

**Definition 3.94** *The* ***wedge product*** *of 1-forms $\alpha_1, \ldots, \alpha_r$ on a given manifold is*

$$\alpha_1 \wedge \cdots \wedge \alpha_r := \sum_{\sigma \in S_r} \operatorname{sign}(\sigma)\, \alpha_{\sigma(1)} \otimes \cdots \otimes \alpha_{\sigma(r)},$$

*where the sum runs over all permutations $\sigma$ of the indices $\{1, \ldots, r\}$, and* $\operatorname{sign}(\sigma)$ *is $+1$ if $\sigma$ is an even permutation and $-1$ if it is odd.*

**Example 3.95** Let $x, y, z$ be the standard coordinates on $\mathbb{R}^3$.

$$\begin{aligned}
&(\mathrm{d}x \wedge \mathrm{d}y \wedge \mathrm{d}z)\,(\mathbf{x})(\mathbf{u}, \mathbf{v}, \mathbf{w}) \\
&\quad = (\mathrm{d}x \otimes \mathrm{d}y \otimes \mathrm{d}z + \mathrm{d}y \otimes \mathrm{d}z \otimes \mathrm{d}x + \mathrm{d}z \otimes \mathrm{d}x \otimes \mathrm{d}y - \mathrm{d}y \otimes \mathrm{d}x \otimes \mathrm{d}z \\
&\qquad - \mathrm{d}x \otimes \mathrm{d}z \otimes \mathrm{d}y - \mathrm{d}z \otimes \mathrm{d}y \otimes \mathrm{d}x)\,(\mathbf{x})(\mathbf{u}, \mathbf{v}, \mathbf{w}) \\
&\quad = u^1 v^2 w^3 + u^2 v^3 w^1 + u^3 v^1 w^2 - u^2 v^1 w^3 - u^1 v^3 w^2 - u^3 v^2 w^1 \\
&\quad = (\mathbf{u} \times \mathbf{v}) \cdot \mathbf{w}.
\end{aligned}$$

Thus, $\mathrm{d}x \wedge \mathrm{d}y \wedge \mathrm{d}z = \mu$, the volume form on $\mathbb{R}^3$.

**Remark 3.96** *It follows easily from the previous definition that interchanging the order of any two of the arguments in the wedge product of 1-forms multiplies the result by $-1$. Thus, if any two arguments are identical, the wedge product must be zero, for example $\mathrm{d}x \wedge \mathrm{d}y \wedge \mathrm{d}x = 0$. A further implication is that $\alpha_1 \wedge \cdots \wedge \alpha_r$ is a differential $r$-form.*

**Proposition 3.97** *Every r-form α can be expressed as*

$$\alpha_{i_1 \dots i_r} \, \mathrm{d}x^{i_1} \wedge \dots \wedge \mathrm{d}x^{i_r},$$

*for some scalar fields $\alpha_{i_1 \dots i_r}$ (there is an implied summation).*

*Sketch of proof* This follows from the skew-symmetry property of differential forms, together with the general local coordinate expression for covariant tensor fields given in Proposition 3.46. ■

**Definition 3.98** *The* ***wedge product*** *(or* ***exterior product****) of two differential forms $\alpha = \alpha_{i_1 \dots i_r} \, \mathrm{d}x^{i_1} \wedge \dots \wedge \mathrm{d}x^{i_r}$ and $\beta = \beta_{j_1 \dots j_s} \, \mathrm{d}x^{j_1} \wedge \dots \wedge \mathrm{d}x^{j_s}$ is*

$$\alpha \wedge \beta = \alpha_{i_1 \dots i_r} \beta_{j_1 \dots j_r} \, \mathrm{d}x^{i_1} \wedge \dots \wedge \mathrm{d}x^{i_r} \wedge \mathrm{d}x^{j_1} \wedge \dots \wedge \mathrm{d}x^{j_s}.$$

**Example 3.99**

$$\begin{aligned}
&(y \, \mathrm{d}x \wedge \mathrm{d}y) \wedge (x^2 \mathrm{d}x + \mathrm{d}z) \\
&\quad = x^2 y \, \mathrm{d}x \wedge \mathrm{d}y \wedge \mathrm{d}x + y \, \mathrm{d}x \wedge \mathrm{d}y \wedge \mathrm{d}z = y \, \mathrm{d}x \wedge \mathrm{d}y \wedge \mathrm{d}z.
\end{aligned}$$

**Proposition 3.100** *Pull-back commutes with wedge product:*

$$\varphi^* (\alpha \wedge \beta) = (\varphi^* \alpha) \wedge (\varphi^* \beta).$$

*Proof* The proof is similar to that of Proposition 3.56, the only difficulties being notational. ■

**Definition 3.101** *The* ***exterior derivative*** *of a scalar field f is its differential,*

$$\mathrm{d}f = \frac{\partial f}{\partial x^i} \, \mathrm{d}x^i.$$

*The* ***exterior derivative*** *of a k-form, for $k > 0$, is defined by*

$$\mathrm{d}\left(\alpha_{j_1 \dots j_k} \, \mathrm{d}x^{j_1} \wedge \dots \wedge \mathrm{d}x^{j_k}\right) = \frac{\partial \alpha_{j_1 \dots j_k}}{\partial x^i} \, \mathrm{d}x^i \wedge \mathrm{d}x^{j_1} \wedge \dots \wedge \mathrm{d}x^{j_k}.$$

*Note that if α is a k-form then dα is a $(k+1)$-form.*

Some work is needed to show that the previous definition is coordinate independent. See for example [Spi65].

**Example 3.102**

1. $\mathrm{d}(\sin x\, \mathrm{d}x) = \cos x\, \mathrm{d}x \wedge \mathrm{d}x = 0$;
2. $\mathrm{d}(\sin y\, \mathrm{d}x) = \cos y\, \mathrm{d}y \wedge \mathrm{d}x = -\cos y\, \mathrm{d}x \wedge \mathrm{d}y$;
3. $\mathrm{d}\left(\mathrm{d}x \wedge \mathrm{d}y\right) = 0$;
4. $\mathrm{d}(x^2 z \mathrm{d}x \wedge \mathrm{d}y + z \mathrm{d}x \wedge \mathrm{d}y)$

$$= 2xz\, \mathrm{d}x \wedge \mathrm{d}x \wedge \mathrm{d}y + x^2\, \mathrm{d}z \wedge \mathrm{d}x \wedge \mathrm{d}y + \mathrm{d}z \wedge \mathrm{d}x \wedge \mathrm{d}y$$

$$= (x^2 + 1)\, \mathrm{d}x \wedge \mathrm{d}y \wedge \mathrm{d}z.$$

**Proposition 3.103** *Let $\alpha$ and $\beta$ be differential forms on $N$, with $\alpha$ of rank $k$, and let $a, b \in \mathbb{R}$. Let $\varphi : M \to N$ be smooth. Then:*

1. $\mathrm{d}(\alpha \wedge \beta) = \mathrm{d}\alpha \wedge \beta + (-1)^k \alpha \wedge \mathrm{d}\beta$;
2. $\mathrm{d}(a\alpha + b\beta) = a\, \mathrm{d}\alpha + b\, \mathrm{d}\beta$;
3. $\mathrm{d}\left(\mathrm{d}\alpha\right) = 0$;
4. $\mathrm{d}(\varphi^*\omega) = \varphi^*\left(\mathrm{d}\omega\right)$.

*Proof* Exercise. ■

**Definition 3.104** *A differential form $\alpha$ is **closed** if $\mathrm{d}\alpha = 0$, and it is **exact** if $\alpha = \mathrm{d}\beta$ for some form $\beta$.*

**Remark 3.105** *By Exercise 3.34, on an $n$-dimensional manifold, there are no non-zero $(n + 1)$-forms, so every $n$-form is closed.*

All exact forms are closed, since $\mathrm{d}\left(\mathrm{d}\alpha\right) = 0$ for all $\alpha$, but the converse is false (see Exercise 3.35). However, there is a partial converse:

**Proposition 3.106 (Poincaré Lemma)** *All closed forms are locally exact; that is, for every closed form $\omega$ on a manifold $M$, and every $z \in M$, there exists a neighbourhood $U$ of $z$ and a form $\alpha$ on $U$ such that $\mathrm{d}\alpha$ equals the restriction of $\omega$ to $U$.*

*Proof* See for example [Spi65]. ■

Recall that a 2-tensor $\omega$ on $M$ is ***non-degenerate*** if and only if, at every point $z \in M$, the matrix representation of $\omega(z)$ is non-singular (in any local coordinates).

**Definition 3.107** *A **symplectic form** is a closed non-degenerate 2-form. If $\omega$ is a symplectic form on a manifold $M$, then the pair $(M, \omega)$ is called a **symplectic manifold**.*

Note that if $\omega$ is a symplectic form then $\omega(z)$ is a symplectic bilinear form, for every $z \in M$.

**Example 3.108** Let $\left(x^1, \ldots, x^n, y^1, \ldots, y^n\right)$ be the standard coordinates on $\mathbb{R}^{2n}$. The ***canonical symplectic form*** on $\mathbb{R}^{2n}$ is

$$\omega_0 = \mathrm{d}\mathbf{x} \wedge \mathrm{d}\mathbf{y} := \mathrm{d}x^1 \wedge \mathrm{d}y^1 + \cdots + \mathrm{d}x^n \wedge \mathrm{d}y^n. \tag{3.20}$$

It is easily checked that this is a symplectic form, and that $\omega_0(\mathbf{z})$ is the canonical symplectic bilinear form, for all $\mathbf{z} \in \mathbb{R}^{2n}$. Note that the canonical symplectic form on $\mathbb{R}^2$ is the standard area form, see Example 3.52.

Compare a symplectic form with a Riemannian metric: both are 2-tensors, but a symplectic form (like all differential forms) is skew-symmetric, while a Riemannian metric is symmetric. A Riemannian metric is positive-definite, which implies non-degeneracy, but is stronger. 'Closed' means $\mathrm{d}\omega = 0$, where d is exterior derivative; since this is only well defined for forms, it doesn't apply to Riemannian metrics.

**Definition 3.109** *A map is **symplectic** if it preserves the symplectic forms on its domain and codomain. More precisely:*

1. *A linear map $L : (V_1, \omega_1) \to (V_2, \omega_2)$ between symplectic vector spaces is symplectic if $\omega_1 = L^* \omega_2$, that is*

$$\omega_1(v, w) = \omega_2(L(v), L(w)), \quad \textit{for all } v, w \in V_1.$$

2. *A differentiable map $\varphi : (M_1, \omega_1) \to (M_2, \omega_2)$ between symplectic manifolds is symplectic if $\omega_1 = \varphi^* \omega_2$, that is*

$$\omega_1(z)(v, w) = \omega_2(\varphi(z))(\mathrm{D}\varphi(z) \cdot v, \mathrm{D}\varphi(z) \cdot w),$$

*for all $v, w \in T_z M_1$, for all $z \in M_1$.*

**Example 3.110** A linear map $A : \mathbb{R}^{2n} \to \mathbb{R}^{2n}$ is symplectic with respect to the canonical symplectic bilinear form (on both domain and codomain) if and only if, considering $A$ as a matrix,

$$\mathbf{v}^T J \mathbf{w} = (A\mathbf{v})^T J (A\mathbf{w}) = \mathbf{v}^T A^T J A \mathbf{w}, \quad \text{for all } \mathbf{v}, \mathbf{w} \in \mathbb{R}^{2n},$$

where $J$ is the matrix defined in eqn (3.19). This is equivalent to $\mathrm{J} = \mathrm{A}^{\mathrm{T}}\mathrm{J}\mathrm{A}$. Thus, A is symplectic if and only if it is in the symplectic group $Sp(2n, \mathbb{R})$ (introduced in Definition 2.86).

A differentiable map $\varphi : \mathbb{R}^{2n} \to \mathbb{R}^{2n}$ is symplectic with respect to the canonical symplectic form $\omega$ if and only if

$$\omega(z)(v, w) = \omega(\varphi(z))(\mathrm{D}\varphi(z) \cdot v, \mathrm{D}\varphi(z) \cdot w),$$

for all $z, v, w \in \mathbb{R}^{2n}$. It follows that $\varphi$ is symplectic if and only if $\mathrm{D}\varphi(z)$ is symplectic with respect to the canonical symplectic bilinear form, for every $z$. As shown above, this is equivalent to $D\varphi(z) \in Sp(2n, \mathbb{R})$, for every $z$.

It can be shown that, for any symplectic bilinear form, there exists a basis with respect to which $\omega$ has matrix $J$, as defined in eqn (3.19). Such a basis (which is not unique) is called a ***symplectic basis***. Note that this result implies that symplectic bilinear forms only exist on even-dimensional spaces. What about symplectic forms on manifolds? The following local answer is fundamental to symplectic geometry (for proof see, for instance, [AM78] or [MS95]).

**Theorem 3.111 (Darboux's Theorem)** *Every symplectic manifold $(M, \omega)$ is locally symplectically diffeomorphic to $(\mathbb{R}^{2n}, \omega_0)$, for some $n$, where $\omega_0$ is the canonical symplectic form; that is, given any $z \in M$, there exists a diffeomorphism $\varphi : (U \subseteq M) \to (V \subseteq \mathbb{R}^{2n})$, for some neighbourhood $U$ of $z$, that is symplectic with respect to $\omega$ and $\omega_0$.*

Compare this with the situation in Riemannian geometry. At the linear level, things are similar: since all symmetric matrices are diagonalizable, all inner products have matrix I (the identity matrix) with respect to some basis; this is analogous to the fact that all symplectic bilinear forms have matrix $J$ with respect to some basis. But for manifolds, even locally, things are very different: in symplectic geometry, all manifolds are locally symplectically diffeomorphic; but in Riemannian geometry, it is not true that all manifolds are locally isometric (see Example 3.87).

**Remark 3.112** *In the context of the statement of the previous theorem, let $(x^1, \ldots, x^n, y^1, \ldots, y^n)$ be the standard coordinates on $\mathbb{R}^{2n}$, and consider the local coordinates $(q^1, \ldots, q^n, p^1, \cdots p^n)$ on $M$ defined by $q^i = x^i \circ \varphi$ and $p^i = y^i \circ \varphi$ (i.e. the pulled-back coordinates). It follows from the chain rule and the definition of a symplectic map that the matrix of $\omega(z)$ with respect to these local coordinates is $J$, for any $z \in U$. Any coordinates such* *that $\omega(z)$ has matrix $J$, for all $z$ in the domain of the coordinates, are called* ***symplectic local coordinates.***

Cotangent bundles are a very important class of symplectic manifolds, especially in Hamiltonian mechanics.

**Definition 3.113** *In any cotangent-lifted coordinates, the* ***canonical symplectic form*** *on $T^*Q$ is $\omega = \mathrm{d}\mathbf{q} \wedge \mathrm{d}\mathbf{p} := \mathrm{d}q^i \wedge \mathrm{d}p_i$.*

A priori, this definition depends on a choice of local coordinates. However, it is in fact coordinate independent. To prove this, one either has to do a direct change-of-basis calculation, or give an equivalent coordinate-free definition, as we will now do. Recall the Liouville 1-form, introduced in Section 3.2:

**Definition 3.114** *Let $Q$ be a manifold, and let $\pi : T^*Q \to Q$ be the standard cotangent bundle projection, given $\pi(\alpha) = q$ for all $\alpha \in T_q^*Q$. The* ***Liouville 1-form*** *on $T^*Q$ is $\theta$ defined by*

$$\theta\,(\alpha)\,(w) := \langle \alpha, T\pi(w)\rangle \quad \text{for all } \alpha \in T^*Q \text{ and } w \in T_\alpha T^*Q. \tag{3.21}$$

*In any cotangent-lifted coordinates $\left(q^1, \ldots, q^n, p_1, \ldots, p_n\right)$, since $\pi(\mathbf{q}, \mathbf{p}) = \mathbf{q}$, we have*

$$\theta(\mathbf{q}, \mathbf{p}) = \mathbf{p}\,\mathrm{d}\mathbf{q} := p_i \mathrm{d}q^i.$$

**Remark 3.115** *The canonical symplectic form on a cotangent bundle $T^*Q$ satisfies*

$$\omega := -\mathrm{d}\theta,$$

*where $\theta$ is the Liouville 1-form. This is an alternative, coordinate-free definition of $\omega$.*

**Remark 3.116** *It is easy to check that the canonical symplectic form on $T^*Q$ is closed, because it is exact: $\mathrm{d}\omega = -\mathrm{d}(\mathrm{d}\theta) = 0$. By Darboux's Theorem (the proof of which uses the fact that $\omega$ is closed), all symplectic forms are locally equivalent to the canonical symplectic form.*

**Proposition 3.117** *The cotangent lift of any diffeomorphism $\varphi : M \to N$ is symplectic with respect to the canonical symplectic forms on $T^*M$ and $T^*N$.*

*Proof* Let $\theta_M$ and $\theta_N$ be the Liouville 1-forms on $M$ and $N$, and let $\omega_M = -\mathrm{d}\theta_M$ and $\omega_N = -\mathrm{d}\theta_N$ be corresponding canonical sympectic forms. Let

$$\psi = T^*\varphi^{-1} : T^*M \to T^*N.$$

We need to show that $\psi^*\omega_N = \omega_M$, or equivalently,

$$\psi^*\left(\mathrm{d}\theta_N\right) = \mathrm{d}\theta_M.$$

Since exterior derivative commutes with pull-back, this is equivalent to

$$\mathrm{d}\left(\psi^*\theta_N\right) = \mathrm{d}\theta_M.$$

But Proposition 3.33 shows that $\psi^*\theta_N = \theta_M$, so we are done. ■

**Exercise 3.33**
Show that if $\alpha$ is an $r$-form and $\beta$ an $s$-form, then

$$\beta \wedge \alpha = (-1)^{r+s}\alpha \wedge \beta.$$

**Exercise 3.34**
Show that there are no non-zero $(n+1)$-forms on an $n$-dimensional manifold.

**Exercise 3.35**
Consider the 1-form ‘$\mathrm{d}\theta$’ on $S^1$, where $\theta$ is ‘the angle variable’. This definition doesn’t quite make sense, because there is no globally defined coordinate $\theta$ on $S^1$. Indeed, all coordinate patches exclude at least one point of $S^1$. So what exactly is meant by $\mathrm{d}\theta$ in this context? Prove that $\mathrm{d}\theta$ is closed but not exact.

**Exercise 3.36**
Verify that the canonical symplectic form is skew-symmetric, non-degenerate and closed. Use the definition of the wedge product to show that, if $\omega_0$ is the canonical symplectic form defined in eqn (3.20), then $\omega_0(\mathbf{z})$ is the canonical bilinear symplectic form defined in eqn (3.19), for every $\mathbf{z} \in \mathbb{R}^{2n}$.

## Solutions to selected exercises

**Solution to Exercise 3.3** Differentiating $\varphi \circ \Phi_t(z)$ with respect to $t$ yields

$$\begin{aligned}\frac{\mathrm{d}}{\mathrm{d}t}(\varphi \circ \Phi_t(z)) &= \left(T_{\Phi_t(z)}\varphi\right) \circ X(\Phi_t(z)) \\ &= \left(T_{\Phi_t(z)}\varphi\right) \circ X \circ \varphi^{-1}(\varphi \circ \Phi_t(z)) \\ &= (\varphi_* X)\,(\varphi \circ \Phi_t(z)).\end{aligned}$$

Therefore, both $\Psi_t$ and $\varphi \circ \Phi_t$ are flows of $\varphi_* X$. By uniqueness of solutions of first-order ODEs,

$$\Psi_t(z) = \varphi \circ \Phi_t(z).$$

**Solution to Exercise 3.4** Recall that $f(\theta, \varphi) = (\theta, \cos\varphi)$. Therefore,

$$\mathrm{D}f(\theta,\varphi) = \begin{bmatrix} 1 & 0 \\ 0 & -\sin\varphi \end{bmatrix}.$$

Consequently,

$$\begin{aligned}f_* X(\theta, \cos\varphi) &= (\theta, \cos\varphi, \mathrm{D}f(\theta,\varphi)\cdot X(\theta,\varphi)) \\ &= (\theta, \cos\varphi, 1, -\sin\varphi))\,.\end{aligned}$$

**Solution to Exercise 3.5** The stereographic projection is given by

$$\xi = \frac{2\cos\theta\sin\phi}{1-\cos\phi}, \qquad \eta = \frac{2\sin\theta\sin\phi}{1-\cos\phi}.$$

Setting the change of coordinate map to be $f(\theta,\phi) = (\xi, \eta)$ gives

$$\mathrm{D}f(\theta,\phi) = \begin{bmatrix} -\dfrac{2\sin\theta\sin\phi}{1-\cos\phi} & \dfrac{2\cos\theta}{1-\cos\phi}\left(\cos\phi - \dfrac{\sin^2\phi}{1-\cos\phi}\right) \\ \dfrac{2\cos\theta\sin\phi}{1-\cos\phi} & \dfrac{2\sin\theta}{1-\cos\phi}\left(\cos\phi - \dfrac{\sin^2\phi}{1-\cos\phi}\right) \end{bmatrix}$$

$$= \begin{bmatrix} -\eta & \dfrac{2\cos\theta}{1-\cos\phi}(\cos\phi - (1+\cos\phi)) \\ \xi & \dfrac{2\sin\theta}{1-\cos\phi}(\cos\phi - (1+\cos\phi)) \end{bmatrix}$$

$$= \begin{bmatrix} -\eta & -\dfrac{\xi}{\sin\phi} \\ \xi & -\dfrac{\eta}{\sin\phi} \end{bmatrix}.$$

Thus,

$$f_* X(\xi,\eta) = Tf(\theta,\phi)\cdot(\theta,\phi,1,\sin\phi) = (\xi,\eta,-\eta-\xi,\xi-\eta)\,.$$

**Solution to Exercise 3.6** Let $H(x,y) = 1/2\left(x^2+y^2\right)$ and $X = (-y,x)$. Then,

$$\mathcal{L}_X H = \mathrm{d}H\cdot X = (x,y)\cdot(-y,x) = 0.$$

**Solution to Exercise 3.9** Let $Y(x,y,z) = (x,y,z)$ and $X(x,y,z) = (-y,x,1)$. Then,

$$\mathrm{D}Y(x,y,z) = \begin{bmatrix} 1 & 0 & 0 \\ 0 & 1 & 0 \\ 0 & 0 & 1 \end{bmatrix}, \quad \mathrm{D}X(x,y,z) = \begin{bmatrix} 0 & -1 & 0 \\ 1 & 0 & 0 \\ 0 & 0 & 0 \end{bmatrix}.$$

Therefore,

$$\begin{aligned} \mathcal{L}_X Y &= [X,Y] \\ &= \mathrm{D}Y\cdot X - \mathrm{D}X\cdot Y \\ &= \begin{bmatrix} 1 & 0 & 0 \\ 0 & 1 & 0 \\ 0 & 0 & 1 \end{bmatrix}\begin{bmatrix} -y \\ x \\ 1 \end{bmatrix} - \begin{bmatrix} 0 & -1 & 0 \\ 1 & 0 & 0 \\ 0 & 0 & 0 \end{bmatrix}\begin{bmatrix} x \\ y \\ z \end{bmatrix} \\ &= \begin{bmatrix} -y \\ x \\ 1 \end{bmatrix} - \begin{bmatrix} -y \\ x \\ 0 \end{bmatrix} = \begin{bmatrix} 0 \\ 0 \\ 1 \end{bmatrix}. \end{aligned}$$

To use the pull-back method, note that the flow of $X$ is given by:

$$\varphi_t(\boldsymbol{x}) = \begin{bmatrix} x\cos t - y\sin t \\ x\sin t + y\cos t \\ t+z \end{bmatrix}, \quad \mathrm{D}\varphi_t(x) = \begin{bmatrix} \cos t & -\sin t & 0 \\ \sin t & \cos t & 0 \\ 0 & 0 & 1 \end{bmatrix}.$$

Therefore, the pull-back $\varphi_t^* Y$ is given by:

$$\varphi_t^* Y(z) = \mathrm{D}\varphi_t^{-1} \circ Y \circ \phi_t(z),$$
$$= \begin{bmatrix} \cos t & \sin t & 0 \\ -\sin t & \cos t & 0 \\ 0 & 0 & 1 \end{bmatrix} \begin{bmatrix} x\cos t - y\sin t \\ x\sin t + y\cos t \\ t+z \end{bmatrix} = \begin{bmatrix} x \\ y \\ t+z \end{bmatrix}.$$

Finally, the Lie derivative is computed as

$$\mathcal{L}_X Y(z) = \left.\frac{\mathrm{d}}{\mathrm{d}t}\right|_{t=0} \varphi_t^* Y(z) = \left.\frac{\mathrm{d}}{\mathrm{d}t}\right|_{t=0} \begin{bmatrix} x \\ y \\ t+z \end{bmatrix} = \begin{bmatrix} 0 \\ 0 \\ 1 \end{bmatrix}.$$

This is the same result as the Lie bracket calculation.

**Solution to Exercise 3.14** $\omega$ is exact if and only if there exists an $f : \mathbb{R}^2 \to \mathbb{R}$ such that

$$\frac{\partial f}{\partial x} = ay, \quad \frac{\partial f}{\partial y} = -bx.$$

Solving these equations yields

$$f = ayx + \phi(y), \quad f = -bxy + \psi(x).$$

If $a = -b$, then we can take $f(x) = ayx$; otherwise, the system has no solution. Therefore, $\omega$ is exact if and only if $a = -b$.

**Solution to Exercise 3.17**

a) $T = \begin{bmatrix} 0 & 3 & -4 \\ -3 & 0 & 0 \\ 4 & 0 & 0 \end{bmatrix}$

b) $\delta_{ij} = \begin{cases} 0, & i \neq j, \\ 1, & i = j \end{cases}$

c) $\varepsilon_{ijk} = \begin{cases} 0, & i = j, j = k, k = i, \\ 1, & \mathrm{sgn}(ijk) = 1, \\ -1, & \mathrm{sgn}(ijk) = -1, \end{cases}$

where $\mathrm{sgn}(ijk)$ is the signature of the permutation $(ijk) \in S_3$, the group of permutations of 3 elements.

**Solution to Exercise 3.19** Since $P$ is the change of basis matrix from $\mathcal{C}$ to $\mathcal{B}$,

$$\boldsymbol{P}[\mathbf{v}]_{\mathcal{C}} = [\mathbf{v}]_{\mathcal{B}}.$$

Therefore, pairing with $[\mathbf{w}]_{\mathcal{B}^*}$ gives

$$[\mathbf{w}]_{\mathcal{B}^*}^T[\mathbf{v}]_{\mathcal{B}} = [\mathbf{w}]_{\mathcal{B}^*}^T\boldsymbol{P}[\mathbf{v}]_{\mathcal{C}} = \left(\boldsymbol{P}^T[\mathbf{w}]_{\mathcal{B}^*}\right)^T[\mathbf{v}]_{\mathcal{C}}.$$

Therefore, $[\mathbf{w}]_{\mathcal{C}^*} = \boldsymbol{P}^T[\mathbf{w}]_{\mathcal{B}^*}$ and $\boldsymbol{P}^{-T}[\mathbf{w}]_{\mathcal{C}^*} = [\mathbf{w}]_{\mathcal{B}^*}$. So, the transformation matrix from $\mathcal{C}^*$ to $\mathcal{B}^*$ is $\boldsymbol{P}^{-T}$.

Suppose now that a contravariant 2-tensor $\boldsymbol{M}$ has coordinate representations $[\boldsymbol{M}]_{\mathcal{B}^*}$ and $[\boldsymbol{M}]_{\mathcal{C}^*}$, respectively, in the $\mathcal{B}^*$ and $\mathcal{C}^*$ coordinate bases. Contracting $\boldsymbol{M}$ with $[\mathbf{w}]_{\mathcal{B}^*}$ and $[\mathbf{u}]_{\mathcal{B}^*}$ yields

$$\begin{aligned}
[\mathbf{w}]_{\mathcal{B}^*}^T[\boldsymbol{M}]_{\mathcal{B}^*}[\mathbf{u}]_{\mathcal{B}^*} &= \left(\boldsymbol{P}^{-T}[\mathbf{w}]_{\mathcal{C}^*}\right)^T[\boldsymbol{M}]_{\mathcal{B}^*}\left(\boldsymbol{P}^{-T}[\mathbf{u}]_{\mathcal{C}^*}\right)\\
&= [\mathbf{w}]_{\mathcal{C}^*}^T\left(\boldsymbol{P}^{-1}[\boldsymbol{M}]_{\mathcal{B}^*}\boldsymbol{P}^{-T}\right)[\mathbf{u}]_{\mathcal{C}^*}\\
&= [\mathbf{w}]_{\mathcal{C}^*}^T[\boldsymbol{M}]_{\mathcal{C}^*}[\mathbf{u}]_{\mathcal{C}^*}.
\end{aligned}$$

Therefore, the transformation law for contravariant 2-tensors is

$$[\boldsymbol{M}]_{\mathcal{C}^*} = \boldsymbol{P}^{-1}[\boldsymbol{M}]_{\mathcal{B}^*}\boldsymbol{P}^{-T}.$$

**Solution to Exercise 3.22** Let $X(x, y) = (-y, x)$ and $\omega = \mathrm{d}x \wedge \mathrm{d}y$. The flow of $X$ is given by

$$\varphi_t(x, y) = \begin{bmatrix} x\cos t - y\sin t\\ x\sin t + y\cos t\end{bmatrix}.$$

The pull-back $\varphi_t^*\omega$ is given by

$$\begin{aligned}
\varphi_t^*\omega &= \left(\varphi_t^*(\mathrm{d}x)\right)\wedge\left(\varphi_t^*(\mathrm{d}y)\right)\\
&= \mathrm{d}\left(\varphi_t^*x\right)\wedge\mathrm{d}\left(\varphi_t^*y\right)\\
&= \mathrm{d}\left(x\cos t - y\sin t\right)\wedge\mathrm{d}\left(x\sin t + y\cos t\right)\\
&= \cos^2 t\,\mathrm{d}x\wedge\mathrm{d}y + \sin^2 t\,\mathrm{d}x\wedge\mathrm{d}y\\
&= \omega.
\end{aligned}$$

Thus, $\omega$ is conserved by the flow and

$$\mathcal{L}_X\omega = \left.\frac{\mathrm{d}}{\mathrm{d}t}\right|_{t=0} \varphi_t^*\omega = \left.\frac{\mathrm{d}}{\mathrm{d}t}\right|_{t=0} \omega = 0.$$

**Solution to Exercise 3.25** Let $g(x,y) = (x^2+1)\mathrm{d}x^2 + \mathrm{d}x\,\mathrm{d}y + \mathrm{d}y^2$. Clearly $g$ is a smooth symmetric covariant 2-tensor field. For every $\mathbf{v} = (v_x, v_y) \in \mathbb{R}^2$,

$$g(x,y)(\mathbf{v},\mathbf{v}) = (x^2+1)v_x^2 + v_xv_y + v_y^2 = x^2v_x^2 + (v_x + \frac{1}{2}v_y)^2 + \frac{3}{4}v_y^2.$$

The expression on the right is never negative and is only zero if $\mathbf{v} = 0$. Thus, $g$ is positive-definite, and hence a Riemannian metric. Now,

$$\begin{aligned} g(1,1)((3,4),(4,-3)) &= \begin{bmatrix} 3 & 4 \end{bmatrix} \begin{bmatrix} 2 & \frac{1}{2} \\ \frac{1}{2} & 1 \end{bmatrix} \begin{bmatrix} 4 \\ -3 \end{bmatrix} \\ &= \begin{bmatrix} 3 \\ 4 \end{bmatrix} \begin{bmatrix} 6.5 \\ -1 \end{bmatrix} \\ &= 15.5. \end{aligned}$$

Consequently, $(3,4)$ and $(4,-3)$ based at $(1,1)$ are not orthogonal.

**Solution to Exercise 3.28** The cone is parameterized in cylindrical coordinates by $z = \sqrt{3}r$. Therefore, the cone can be described by the embedding

$$f(a,\theta) = \left(a, \theta, \sqrt{3}a\right).$$

Now, consider 'unrolling' the cone onto the plane. This unrolling is given by the map

$$g(a,\theta) = \left(2a, \frac{\theta}{2}, 0\right),$$

in the sense that $g \circ f^{-1}$, restricted to $\theta \in (-\pi, \pi)$ for example, is a smooth map from (most of) the cone to the plane.

The Euclidean metric in cylindrical coordinates is $\mathrm{d}r^2 + r^2\mathrm{d}\theta^2 + \mathrm{d}z^2$. Thus, the first fundamental form of the cone, in coordinates $(a,\theta)$ given by $f$, is

$$\mathrm{d}a^2 + a^2\,\mathrm{d}\theta^2 + \mathrm{d}(\sqrt{3}a)^2 = 4\,\mathrm{d}a^2 + a^2\,\mathrm{d}\theta^2;$$

and the first fundamental form of the plane, in coordinates $(a, \theta)$ given by $g$, is

$$\mathrm{d}(2a)^2 + (2a)^2\,\mathrm{d}\left(\frac{\theta}{2}\right)^2 = 4\,\mathrm{d}a^2 + a^2\,\mathrm{d}\theta^2.$$

from the second. Since these are identical, the map $g \circ f^{-1}$ is an isometry.

**Solution to Exercise 3.32** The first fundamental form on $S^2$ is given by

$$\mathrm{d}\phi^2 + \sin^2\phi\,\mathrm{d}\theta^2.$$

Therefore, the Lagrangian for geodesic equations is $L = \frac{1}{2}\left(\dot{\phi}^2 + \sin^2\phi\,\dot{\theta}^2\right)$. The geodesic equations are the corresponding Euler–Lagrange equations:

$$\ddot{\phi} - \sin\phi\cos\phi\,\dot{\theta}^2 = 0,$$
$$\ddot{\theta} + 2\cot\phi\,\dot{\theta}\dot{\phi} = 0.$$

The equator can be described by $\theta$ = const. and $\phi = At + B$. Therefore, $\ddot{\theta} = \dot{\theta} = \ddot{\phi} = 0$ and the great circles satisfy the geodesic equations.

The values of the Christoffel symbols can be read off from the geodesic equations. The only non-zero terms are

$$\Gamma^{\phi}_{\theta\theta} = -\sin\phi\cos\phi,$$
$$\Gamma^{\theta}_{\phi\theta} = \cot\phi,$$
$$\Gamma^{\theta}_{\theta\phi} = \cot\phi.$$

**Solution to Exercise 3.34** All $(n+1)$-forms can be expressed as a sum of terms of the form

$$\alpha_{i_1\ldots i_r}\,\mathrm{d}x^{i_1} \wedge \cdots \wedge \mathrm{d}x^{i_{n+1}}.$$

But on an $n$-dimensional manifold there are only $n$ coordinates $x^i$, so at least one $\mathrm{d}x^i$ must appear twice in every term. Since forms are skew-symmetric, this implies that $\mathrm{d}x^{i_1} \wedge \cdots \wedge \mathrm{d}x^{i_{n+1}} = 0$. Therefore, all $(n+1)$-forms are zero.

**Solution to Exercise 3.35** An angular coordinate chart $\theta_U : U \subseteq S^1 \to \mathbb{R}$ is a smooth function such that $(\cos(\theta_U(x,y)), \sin(\theta_U(x,y))) = (x,y)$ for all $(x,y) \in U$. For any such chart, the 1-form $\mathrm{d}\theta_U$ is defined in the usual way. The 1-form $\mathrm{d}\theta$ is defined by $\mathrm{d}\theta = \mathrm{d}\theta_U$ for any angular coordinate chart $\theta_U$. To check that this definition is consistent, consider two angular coordinate charts $\theta_U : U \subseteq S^1 \to \mathbb{R}$ and $\theta_V : V \subseteq S^1 \to \mathbb{R}$. If $W$ is a connected subset of $U \cap V$, then $\theta_V - \theta_U = 2n\pi$ for some $n \in \mathbb{N}$ is constant on $W$. It follows that $\mathrm{d}\theta_U = \mathrm{d}\theta_V$ on $U \cap V$. Hence, $\mathrm{d}\theta$ is well defined.

Since $S^1$ is one-dimensional, all two forms are equal to zero, so $\mathrm{d}\theta$ is closed. However, $\mathrm{d}\theta$ is not exact, since

$$\int_{S^1} \mathrm{d}\theta = 2\pi.$$

Indeed, if $\mathrm{d}\theta$ were equal to $\mathrm{d}f$ for some $f : S^1 \to \mathbb{R}$ then we would have

$$\int_{S^1} \mathrm{d}\theta = f(2\pi) - f(0) = 0,$$

which is a contradiction.

# 4 Mechanics on manifolds

Manifolds are locally equivalent to Euclidean spaces. The Lagrangian and Hamiltonian formulations of mechanics on Euclidean spaces introduced in Chapter 1 provide the *local* mathematical formulations of mechanics on manifolds. However, global issues may still arise from the need to use multiple coordinate charts.

This chapter reformulates some key definitions from Chapter 1 in a coordinate-free manner. This reformulation emphasizes the fundamental concepts while suppressing inessential details. However, a return to coordinate expressions is still necessary for doing calculations in examples.

We assume that all manifolds are finite-dimensional and all Lagrangians are smooth.

## 4.1 Lagrangian mechanics on manifolds

In Chapter 1, working in Euclidean space, we saw that the ***Euler–Lagrange equations***:

$$\frac{\mathrm{d}}{\mathrm{d}t}\left(\frac{\partial L}{\partial \dot{\mathbf{q}}}\right) = \frac{\partial L}{\partial \mathbf{q}}, \tag{4.1}$$

are equivalent to ***Hamilton's principle of stationary action***: $\delta S = 0$, for the ***action functional***

$$S[\mathbf{q}(\cdot)] := \int_a^b L(\mathbf{q}(t), \dot{\mathbf{q}}(t))\, \mathrm{d}t, \tag{4.2}$$

with respect to variations among paths with fixed endpoints. Recall that an equivalent statement of Hamilton's principle is:

$$\frac{\partial}{\partial s} \int_a^b L(\mathbf{q}(t,s), \dot{\mathbf{q}}(t,s))\, \mathrm{d}t \bigg|_{s=0} = 0, \tag{4.3}$$

for all deformations $\mathbf{q}(t,s)$ of $\mathbf{q}(t)$ leaving the endpoints fixed.

In general, the configuration space is a manifold $Q$. Recall from Chapter 2 that every local coordinate system $(q^1, \ldots, q^n)$ on $Q$ induces a tangent-lifted

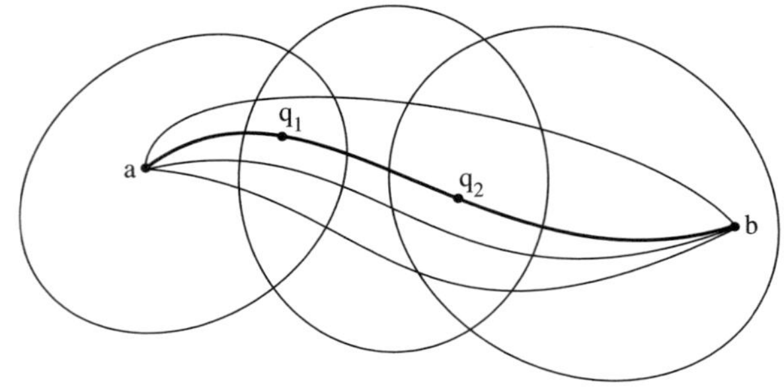


**Figure 4.1** In general, a path need not lie in a single coordinate patch, though it can always be covered with a finite number of coordinate patches. If a deformation with fixed endpoints is broken up into deformations of subpaths, the latter need not have fixed endpoints.

local coordinate system $(q^1, \ldots, q^n, \dot{q}^1, \ldots, \dot{q}^n)$ on $TQ$, which we write in shorthand as $(\mathbf{q}, \dot{\mathbf{q}})$. The Lagrangian is a map $L : TQ \to \mathbb{R}$, which in local coordinates becomes a function of $(\mathbf{q}, \dot{\mathbf{q}})$ as in Chapter 1. The Euler–Lagrange equations are defined as usual, in terms of local coordinates on $Q$. In Hamilton's principle, we must now consider paths $\mathbf{q}(t)$ and deformations $\mathbf{q}(t, s)$ that are not necessarily contained in a single coordinate patch, as illustrated in Figure 4.1. The interpretation of Hamilton's principle in terms of deformations remains the same, but extra care is needed in demonstrating the equivalence with the Euler–Lagrange equations. The main issue is that, though paths can be broken up into subpaths each lying in a single coordinate patch, and deformations can be broken up into deformations of subpaths, the resulting deformations of subpaths will not necessarily leave the endpoints of the subpaths fixed (see Figure 4.1).

For this reason, the equivalence of the Euler–Lagrange equations with Hamilton's principle does not follow trivially from the equivalence in Euclidean space. However, this can be resolved without difficulty, since the proof of Theorem 1.38 includes an intermediate result that applies all deformations, not necessarily with fixed endpoints, namely that $\delta \mathcal{S} = 0$ is equivalent to

$$\int_a^b \left( \frac{\mathrm{d}}{\mathrm{d}t} \left( \frac{\partial L}{\partial \dot{\mathbf{q}}} \right) - \frac{\partial L}{\partial \mathbf{q}} \right) \cdot \delta \mathbf{q} \, \mathrm{d}t = \frac{\partial L}{\partial \dot{\mathbf{q}}} \cdot \delta \mathbf{q} \bigg|_a^b . \tag{4.4}$$

This is the key to the proof of the following theorem.

**Theorem 4.1 (Euler–Lagrange equations)** *Let $Q$ be a manifold. For any smooth $L : TQ \to \mathbb{R}$ (called the **Lagrangian**), a path $\mathbf{q}(t)$ in $Q$ satisfies Hamilton's principle if and only it satisfies the Euler-Lagrange equations in every local coordinate system.*

*Proof* Let $\{t_0 = a, t_1, \ldots, t_r = b\}$ be a partition of $[a,b]$ such that every subpath from $t_{i-1}$ to $t_i$ is contained in a single coordinate patch, as illustrated in Figure 4.1.

First suppose that $\mathbf{q}(t)$ satisfies the Euler–Lagrange equations everywhere. Then Theorem 1.38 applies to every subpath of $\mathbf{q}(t)$, as does the intermediate result in eqn (4.4), giving

$$\int_{t_{i-1}}^{t_i} \left( \frac{\mathrm{d}}{\mathrm{d}t}\left(\frac{\partial L}{\partial \dot{\mathbf{q}}}\right) - \frac{\partial L}{\partial \mathbf{q}} \right) \cdot \delta \mathbf{q}\, \mathrm{d}t = \frac{\partial L}{\partial \dot{\mathbf{q}}} \cdot \delta \mathbf{q} \Big|_{t_{i-1}}^{t_i}, \tag{4.5}$$

for any variation $\delta\mathbf{q}$ of the subpath. Now, let $\delta\mathbf{q}$ be a variation of the entire path, such that $\delta\mathbf{q}(a) = \delta\mathbf{q}(b) = 0$. Then, the restriction of $\delta\mathbf{q}$ to any subinterval $[t_{i-1}, t_i]$ is a variation of the $i$th subpath, so the equation above holds for every subpath. When copies of this equation, for $i = 1, \ldots, r$, are added together, the terms $\frac{\partial L}{\partial \dot{\mathbf{q}}} \cdot \delta\mathbf{q}\Big|_{t_{i-1}}^{t_i}$ cancel, leaving $\frac{\partial L}{\partial \dot{\mathbf{q}}} \cdot \delta\mathbf{q}\Big|_a^b$, which equals zero because of the assumption that $\delta\mathbf{q}(a) = \delta\mathbf{q}(b) = 0$.

Conversely, suppose that $\mathbf{q}(t)$ satisfies Hamilton's principle. Then eqn (4.3) holds for any variation $\delta\mathbf{q}(t)$ that vanishes at the endpoints. For any $i$, any variation of the $i$th subpath such that $\delta\mathbf{q}(t_{i-1}) = \delta\mathbf{q}(t_i) = \mathbf{0}$ can be extended to a variation of the entire path that is trivial for $t$ outside of $[t_{i-1}, t_i]$, i.e. such that $\delta\mathbf{q}(t) = \mathbf{0}$ for $t$ outside of $[t_{i-1}, t_i]$. Unfortunately, this extension is not necessarily smooth, but we can approximate it arbitrarily closely by smooth variations $\delta\mathbf{q}(t)$ that are trivial for $t$ outside of $[t_{i-1}, t_i]$. Thus,

$$\int_{t_{i-1}}^{t_i} \left( \frac{\mathrm{d}}{\mathrm{d}t}\left(\frac{\partial L}{\partial \dot{\mathbf{q}}}\right) - \frac{\partial L}{\partial \mathbf{q}} \right) \cdot \delta \mathbf{q}\, \mathrm{d}t \simeq \int_a^b \left( \frac{d}{dt}\left(\frac{\partial L}{\partial \dot{\mathbf{q}}}\right) - \frac{\partial L}{\partial \mathbf{q}} \right) \cdot \delta \mathbf{q}\, \mathrm{d}t = 0, \tag{4.6}$$

where the $\delta\mathbf{q}$ on the left is the original variation on the subpath, and the $\delta\mathbf{q}$ on the right is a smooth approximation to the extension defined above. It follows, with a little analytical work, that in fact the left-hand side equals zero. Thus Hamilton's principle holds for the subpath from $t_{i-1}$ to $t_i$, and thus by Theorem 1.38, the Euler–Lagrange equations are satisfied by this subpath. Since this is true for all subpaths, the result follows. ■

We now examine the Euler–Lagrange equations more closely. Expanding the time derivative gives equivalent equations,

$$\frac{\partial^2 L}{\partial \dot{\mathbf{q}}^2}\ddot{\mathbf{q}} + \frac{\partial^2 L}{\partial \dot{\mathbf{q}}\, \partial \mathbf{q}}\dot{\mathbf{q}} = \frac{\partial L}{\partial \mathbf{q}}. \tag{4.7}$$

Thus, the Euler–Lagrange equations are second-order equations in $q^1, \ldots, q^n$.

**Definition 4.2** *A Lagrangian L is* ***regular*** *(or* ***non-degenerate****) if the $n \times n$ Hessian matrix $\frac{\partial^2 L}{\partial \dot{q}^2}$ is invertible, for every* $\mathbf{q}$ *and* $\dot{\mathbf{q}}$*. Equivalently,*

$$\det \frac{\partial^2 L}{\partial \dot{\mathbf{q}}^2} \neq 0.$$

If $L$ is regular, then the equations (4.7) can be solved for $\ddot{\mathbf{q}}$ as follows:

$$\ddot{\mathbf{q}} = \left( \frac{\partial^2 L}{\partial \dot{\mathbf{q}}^2} \right)^{-1} \left( \frac{\partial L}{\partial \mathbf{q}} - \frac{\partial^2 L}{\partial \dot{\mathbf{q}}\, \partial \mathbf{q}} \dot{\mathbf{q}} \right). \tag{4.8}$$

This system of ODEs is equivalent to:

$$\begin{aligned} \frac{\mathrm{d}}{\mathrm{d}t}\mathbf{q} &= \dot{\mathbf{q}} \\ \frac{\mathrm{d}}{\mathrm{d}t}\dot{\mathbf{q}} &= \left( \frac{\partial^2 L}{\partial \dot{\mathbf{q}}^2} \right)^{-1} \left( \frac{\partial L}{\partial \mathbf{q}} - \frac{\partial^2 L}{\partial \dot{\mathbf{q}}\, \partial \mathbf{q}} \dot{\mathbf{q}} \right). \end{aligned} \tag{4.9}$$

This is just the standard reduction of order procedure for differential equations, with $\dot{q}^1, \ldots, \dot{q}^n$ introduced as independent variables.

Since the Euler–Lagrange equations are coordinate independent (see Proposition 1.31 and the following remark), it follows that the form of system (4.9) is coordinate independent. Therefore, this system defines a vector field on $TQ$:

**Definition 4.3** *The* ***Lagrangian vector field*** *$Z_L$ on $TQ$ is defined, in tangent-lifted local coordinates, by the system of equations (4.9).*

In summary, we have shown,

**Theorem 4.4** *If L is regular, then a path* $\mathbf{q}(t)$ *in Q satisfies the Euler–Lagrange equations if and only if its natural lift* $(\mathbf{q}(t), \dot{q}(t)) := \left(\mathbf{q}(t), \frac{\mathrm{d}}{\mathrm{d}t}\mathbf{q}(t)\right)$ *is a solution of the* ***Lagrangian vector field*** *on $TQ$.*

**Remark 4.5** *Since L is assumed to be smooth, a standard result from the theory of ordinary differential equations guarantees the local existence and* *uniqueness of solutions to the Lagrangian vector field, and hence to the Euler–Lagrange equations.*

Whenever one encounters definitions in terms of local coordinates that later turn out to be coordinate independent, it is natural to look for alternative definitions that do not make direct use of local coordinates. In fact the regularity of a Lagrangian and the Lagrangian vector field are concepts that can be defined in a coordinate-free manner, as we will see Section 4.2.

**Exercise 4.1**
Check that Definition 4.2 is coordinate independent. This will also follow from Remark 4.7.

**Exercise 4.2**
Consider the Lagrangian $L(q,\dot{q}) = \frac{1}{2}m\dot{q}^2 + \frac{m}{|q|}$, on the configuration space $Q = \mathbb{R} \setminus \{0\}$, which is the Lagrangian for a single particle of mass $m$ moving on a line in a central gravitational field. Check that $L$ is regular (if $m \neq 0$), but that the corresponding Euler–Lagrange equation does not have any solution valid for all time $t \in \mathbb{R}$.

**Exercise 4.3**
"$L$ is regular" does not mean "all values of $L$ are regular" (see Definition 2.17), and in fact neither statement implies the other. Find examples of functions $L$ to illustrate this.

**Exercise 4.4**
[Gauge invariance] Show that the Euler–Lagrange equations are unchanged under

$$L(\mathbf{q}(t), \dot{\mathbf{q}}(t)) \to L' = L + \frac{\mathrm{d}}{\mathrm{d}t}\gamma(\mathbf{q}(t)), \tag{4.10}$$

for any function $\gamma : Q \to \mathbb{R}$.

## 4.2 The Legendre transform and Hamilton's equations

Recall from Chapter 1 that the ***Legendre transform*** for a Lagrangian $L(\mathbf{q}, \dot{\mathbf{q}})$ is the map

$$(\mathbf{q}, \dot{\mathbf{q}}) \mapsto (\mathbf{q}, \frac{\partial L}{\partial \dot{\mathbf{q}}}) \tag{4.11}$$

(see Definition 1.44). For a general configuration manifold $Q$, the same definition holds with $(\mathbf{q}, \dot{\mathbf{q}})$ being tangent-lifted local coordinates, but now $\frac{\partial L}{\partial \dot{\mathbf{q}}}$ is interpreted as a cotangent vector based at $\mathbf{q}$. Here is an equivalent coordinate-free definition of this transform:

**Definition 4.6** *Given a smooth $L : TQ \to \mathbb{R}$, the corresponding* ***Legendre transform*** *is the smooth map $\mathbb{F}L : TQ \to T^*Q$ defined by*

$$\langle \mathbb{F}L(v_q), w_q \rangle := \left.\frac{d}{ds}\right|_{s=0} L(v_q + s w_q),$$

*for all $q \in Q$ and all $v_q, w_q \in T_q Q$.*

In local tangent-lifted coordinates, $\mathbb{F}L(\mathbf{q}, \dot{\mathbf{q}}) = \left(\mathbf{q}, \frac{\partial L}{\partial \dot{\mathbf{q}}}\right)$, since

$$\langle \mathbb{F}L(\mathbf{q}, \dot{\mathbf{q}}_1), (\mathbf{q}, \dot{\mathbf{q}}_2) \rangle = \left(\mathbf{q}, \left.\frac{\mathrm{d}}{\mathrm{d}s}\right|_{s=0} L(\mathbf{q}, \dot{\mathbf{q}}_1 + s\dot{\mathbf{q}}_2)\right) = \left(\mathbf{q}, \frac{\partial L}{\partial \dot{\mathbf{q}}}(\mathbf{q}, \dot{\mathbf{q}}_1) \cdot \dot{\mathbf{q}}_2\right).$$

The map $\mathbb{F}L$ is also called the ***fibre derivative*** of $L$.

**Remark 4.7** *The derivative of $\mathbb{F}L$, in local coordinates $(\mathbf{q}, \dot{\mathbf{q}})$, has matrix*

$$D(\mathbb{F}L)(\mathbf{q}, \dot{\mathbf{q}}) = \begin{bmatrix} I & 0 \\ \dfrac{\partial^2 L}{\partial \mathbf{q} \partial \dot{\mathbf{q}}} & \dfrac{\partial^2 L}{\partial \dot{\mathbf{q}}^2} \end{bmatrix}. \tag{4.12}$$

*Thus $D(\mathbb{F}L)$ is invertible at a certain point in $TQ$ if and only if $\frac{\partial^2 L}{\partial \dot{\mathbf{q}}^2}$ is invertible at that point. It follows that $L$ is regular if and only if $D(\mathbb{F}L)$ is invertible at all points in $TQ$.*

**Definition 4.8** *A Lagrangian $L$ is* ***hyperregular*** *if $\mathbb{F}L$ is a diffeomorphism.*

Hyperregularity implies regularity. We will assume from now on that all of our Lagrangians are hyperregular.

**Definition 4.9** *The* ***energy function*** *for a Lagrangian* $L : TQ \to \mathbb{R}$ *is* $E : TQ \to \mathbb{R}$ *defined by*

$$E(v) := \langle \mathbb{F}L(v), v\rangle - L.$$

*In local coordinates,*

$$E(\mathbf{q}, \dot{\mathbf{q}}) := \frac{\partial L}{\partial \dot{\mathbf{q}}} \cdot \dot{\mathbf{q}} - L(\mathbf{q}, \dot{\mathbf{q}}).$$

**Proposition 4.10** *Let* $E$ *be the energy function corresponding to a Lagrangian* $L$*. Then energy* $E$ *is conserved along any solution to the Euler-Lagrange equations for* $L$*.*

*Proof* The calculation in the proof of Theorem 1.37 applies here as well. ■

**Definition 4.11** *The* ***Hamiltonian corresponding*** *to* $L$ *is* $H : T^*Q \to \mathbb{R}$ *defined by*

$$H := E \circ (\mathbb{F}L)^{-1},$$

*where* $E$ *is the energy function for* $L$*. In local coordinates,* $H$ *is just* $E$ *expressed as a function of* $(\mathbf{q}, \mathbf{p})$ *instead of* $(\mathbf{q}, \dot{\mathbf{q}})$*, where* $(\mathbf{q}, \mathbf{p}) = \mathbb{F}L(\mathbf{q}, \dot{\mathbf{q}})$*, i.e.,*

$$H(\mathbf{q}, \mathbf{p}) = \mathbf{p} \cdot \dot{\mathbf{q}}(\mathbf{q}, \mathbf{p}) - L(\mathbf{q}, \dot{\mathbf{q}}(\mathbf{q}, \mathbf{p})), \tag{4.13}$$

**Theorem 4.12** *For any hyperregular Lagrangian* $L : TQ \to \mathbb{R}$*, let* $H$ *be the corresponding Hamiltonian as defined above. The Euler–Lagrange equations, in any tangent-lifted local coordinates* $(\mathbf{q}, \dot{\mathbf{q}})$ *on* $TQ$*, are equivalent to* ***Hamilton's equations of motion,***

$$\frac{d\mathbf{q}}{dt} = \frac{\partial H}{\partial \mathbf{p}}, \quad \frac{d\mathbf{p}}{dt} = -\frac{\partial H}{\partial \mathbf{q}}, \tag{4.14}$$

*in the corresponding cotangent-lifted coordinates* $(\mathbf{q}, \mathbf{p})$ *on* $T^*Q$*. These equations define a vector field* $X_H$ *on* $T^*Q$*, called the* ***Hamiltonian vector field,*** *which is actually the push-forward of the* ***Lagrangian vector field*** *by the Legendre transform, i.e.* $H = (\mathbb{F}L)_* Z_L$*.*

*Proof* The Euler–Lagrange equations, together with the equations $\frac{d\mathbf{q}}{dt} = \dot{\mathbf{q}}$, form a system of implicit first order differential equations in the variables $(\mathbf{q}, \dot{\mathbf{q}})$. Applying the Legendre transform gives the equivalent system

$$\frac{d\mathbf{p}}{dt} - \frac{\partial L}{\partial \mathbf{q}} = 0, \quad \frac{d\mathbf{q}}{dt} = \dot{\mathbf{q}}(\mathbf{q}, \mathbf{p}). \tag{4.15}$$

In fact the vector field defined by these equations is the push-forward of the Lagrangian vector field for $L$, as can be verified from eqns (4.8) and (4.12) (see Exercise 4.5). The resulting vector field is called $X_H$. We now calculate the partial derivatives of $H$:

$$\frac{\partial H}{\partial \mathbf{q}} = \mathbf{p} \cdot \frac{\partial \dot{\mathbf{q}}(\mathbf{q}, \mathbf{p})}{\partial \mathbf{q}} - \frac{\partial L}{\partial \mathbf{q}} - \frac{\partial L}{\partial \dot{\mathbf{q}}} \cdot \frac{\partial \dot{\mathbf{q}}(\mathbf{q}, \mathbf{p})}{\partial \mathbf{q}} = -\frac{\partial L}{\partial \mathbf{q}}$$

$$\frac{\partial H}{\partial \mathbf{p}} = \dot{\mathbf{q}}(\mathbf{q}, \mathbf{p}) + \mathbf{p} \cdot \frac{\partial \dot{\mathbf{q}}(\mathbf{q}, \mathbf{p})}{\partial \mathbf{q}} - \frac{\partial L}{\partial \dot{\mathbf{q}}} \cdot \frac{\partial \dot{\mathbf{q}}(\mathbf{q}, \mathbf{p})}{\partial \mathbf{p}} = \dot{\mathbf{q}}(\mathbf{q}, \mathbf{p}).$$

Substituting these into eqn (4.15) gives Hamilton's equations. These are the equations for the vector field $X_H$ in local coordinates. ■

Solutions to a Lagrangian system correspond to solutions to the associated Hamiltonian system. Specifically, if $\mathbf{c}(t)$ is a path in $TQ$ that is a solution to a Lagrangian vector field, then $\mathbb{F}L \circ \mathbf{c}(t)$ is a path in $T^*Q$ that is a solution to the the associated Hamiltonian vector field. Both paths project down to the same path in $Q$, that is, they describe the same physical motion. In local coordinates, if $\mathbf{c}(t) = (\mathbf{q}(t), \dot{\mathbf{q}}(t))$ then $\mathbb{F}L \circ \mathbf{c} = (\mathbf{q}(t), \mathbf{p}(t))$, with $\mathbf{p}(t) = \frac{\partial L}{\partial \dot{\mathbf{q}}}(\mathbf{q}(t), \dot{\mathbf{q}}(t))$.

Hamilton's equations can be studied for *any* $H : T^*Q \to \mathbb{R}$, not just those corresponding to Lagrangians:

**Definition 4.13** *Let $H : T^*Q \to \mathbb{R}$ be smooth. The* ***Hamiltonian vector field*** *corresponding to $H$ is the vector field $X_H$ on $T^*Q$ that is given, in any cotangent-lifted coordinates, by Hamilton's equations.*

For this definition to make sense, it is necessary to verify that it is coordinate independent. This is a direct consequence of the definition of cotangent-lifted coordinates, and is left to Exercise 4.6. The result should be very plausible, since we have already seen that it is true for Hamiltonians that correspond to Lagrangians. Indeed, in this case, Hamilton's equations are equivalent to the Euler–Lagrange equations, which are coordinate-independent.

**Remark 4.14** *Consider a Hamiltonian function $H$. If $H$ is hyperregular (the definition is analagous to the definition for Lagrangians), then it is* *possible to define a Lagrangian $L$ in terms of $H$ such that solutions to the Lagrangian vector field $Z_L$ correspond to solutions to the Hamiltonian vector field $X_H$. Further, the two constructions: $H$ from $L$, and $L$ from $H$, are inverse to each other. Thus, there is a one-to-one correspondence between hyperregular Lagrangians and hyperregular Hamiltonians. For details, see [MR02].*

**Theorem 4.15** *The Hamiltonian is a conserved quantity of the motion defined by Hamilton's equations.*

*Proof* This can be verified directly by a simple calculation, which has already appeared in the proof of Theorem 1.56. ■

Another important property of Hamiltonian systems is conservation of phase space volume. For systems with one degree of freedom, i.e. a two-dimensional phase space, 'phase-space volume' is just area. In general, the *phase-space volume* of a region $D \subseteq T^*Q$ is the integral

$$\mathrm{vol}(D) := \int_D \mathrm{d}q^1 \mathrm{d}p_1 \cdots \mathrm{d}q^n \mathrm{d}p_n.$$

This definition is independent of the choice of cotangent-lifted coordinates (see Exercise 4.8), and it can be extended to domains $D$ that do not lie in a single coordinate patch. The flow $\Phi$ of a vector field is *volume preserving* if every time-$t$ map $\Phi_t$ preserves phase space volume, that is, $\mathrm{vol}(D) = \mathrm{vol}\,(\Phi_t(D))$, for every subset $D$ of the phase space.

**Theorem 4.16** *The flow of a Hamiltonian vector field is volume-preserving.*

*Sketch of proof* The divergence of a Hamiltonian vector field is zero:

$$\nabla \cdot X_H = \left( \frac{\partial}{\partial q^i} \frac{\partial H}{\partial p_i} \right) + \frac{\partial}{\partial q^i} \left( -\frac{\partial H}{\partial p_i} \right) = 0.$$

In two (or three) dimensions, it is intuitively clear that the flow of a divergence-free vector field preserves area (or volume). This is proven in vector calculus texts, e.g., [Spi65], and the proof generalizes to arbitrary manifolds [Spi79]. ■

For Hamiltonian systems with one degree of freedom, the previous theorem states that the flow of the system is area preserving. This is illustrated for a simple example in Figure 4.2.

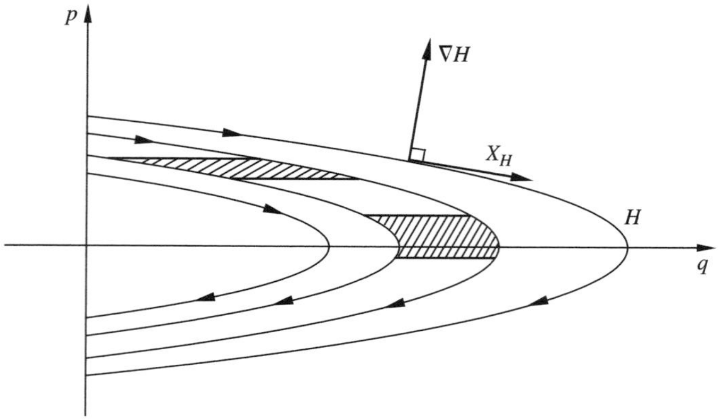


**Figure 4.2** Phase space diagram for $H = \frac{1}{2}p^2 + q$, illustrating area preservation and the fact that $X_H$ is tangent to level sets of $H$ and orthogonal to $\nabla H$.

The preservation of phase-space volume has important dynamical consequences, including:

**Corollary 4.17** *Hamiltonian systems cannot have any fixed points or periodic orbits that are either asymptotically stable or asymptotically unstable.*

We have seen that the definition of a Hamiltonian vector field is coordinate independent. There are two equivalent coordinate-free definitions of Hamiltonian vector fields on $T^*Q$. Both use differential geometry: one is in terms of the canonical Poisson bracket, as we have seen already in Chapter 1; and the other is in terms of the canonical symplectic form, which has been defined in Chapter 3. Both definitions can be generalized to phase spaces other than $T^*Q$. Poisson brackets will be presented in Section 4.3. We now make some brief comments about the symplectic approach.

Consider a Hamiltonian system on $T^*Q$. Hamilton's equations can be written in matrix form:

$$\begin{bmatrix} \dot{\mathbf{q}} \\ \dot{\mathbf{p}} \end{bmatrix} = \begin{bmatrix} 0 & I \\ -I & 0 \end{bmatrix} \begin{bmatrix} \dfrac{\partial H}{\partial \mathbf{q}} \\ \dfrac{\partial H}{\partial \mathbf{p}} \end{bmatrix}, \tag{4.16}$$

where $(\mathbf{q}, \mathbf{p})$ are cotangent-lifted local coordinates, $I$ is the $n \times n$ identity matrix, and $n$ is the dimension of $Q$. Defining

$$J := \begin{bmatrix} 0 & I \\ -I & 0 \end{bmatrix}, \tag{4.17}$$

the matrix form of Hamilton's equations can be re-expressed as

$$X_H = J\nabla H. \tag{4.18}$$

Recall from Chapter 3 the canonical symplectic form $\omega := -\mathbf{d}\theta$, where $\theta$ is the canonical 1-form. In cotangent lifted coordinates,

$$\omega = \mathbf{d}q^i \wedge \mathbf{d}p_i.$$

The matrix of $\omega$ in these coordinates is constant and equals $J$ as defined above. It follows that, for any $\mathbf{v}, \mathbf{w} \in \mathbb{R}^{2n}$,

$$\omega(\mathbf{q}, \mathbf{p})(\mathbf{v}, \mathbf{w}) = \mathbf{v}^T J\mathbf{w}.$$

In particular, since $J^{-1} = -J$,

$$\omega(X_H(\mathbf{q}, \mathbf{p}), \mathbf{w}) = X_H^T(\mathbf{q}, \mathbf{p})\, J\, \mathbf{w} = (\nabla H(\mathbf{q}, \mathbf{p}))^T\, \mathbf{w} = dH(\mathbf{q}, \mathbf{p})(\mathbf{w}),$$

for all $\mathbf{w}$. Thus,

$$\omega(X_H, \cdot) = dH. \tag{4.19}$$

This is the coordinate-independent definition of $X_H$. More common ways of writing the left-hand side in differential geometry are $X_H \,\lrcorner\, \omega$ or $\mathbf{i}_{X_H}\omega$.

The coordinate-free definition of $X_H$ in eqn (4.19) is valid on any manifold with a symplectic form. Thus, Hamiltonian systems can be defined and studied on phase spaces that are not cotangent bundles. An example is given in Exercise 4.9.

**Exercise 4.5**
Given the formula for the Lagrangian vector field $Z_L$ in local coordinates in eqn (4.8), and the formula for the matrix of $\mathrm{D}(\mathbb{F}L)(\mathbf{q}, \dot{\mathbf{q}})$ in eqn (4.12), calculate the push-forward of $Z_L$ by $\mathbf{F}L$, and show that the result is the system in Equation 4.15.

**Exercise 4.6**
Show that Hamilton's equations are coordinate independent, so they uniquely define a Hamiltonian vector field. In other words, show that if $(\mathbf{q}, \mathbf{p})$ and $(\mathbf{r}, \mathbf{s})$ are two systems of cotangent-lifted local coordinates for the same domain, then a trajectory satisfies Hamilton's equations for $(\mathbf{q}, \mathbf{p})$ if and only if it satisfies Hamilton's equations for $(\mathbf{r}, \mathbf{s})$.

**Exercise 4.7**
[Spherical pendulum] The spherical pendulum, defined in Exercise 1.42, has configuration space $S^2$. Write down its Lagrangian and the equations of motion, in spherical coordinates. Show explicitly that the Lagrangian is hyperregular. Use the Legendre transformation to convert the equations to Hamiltonian form. Find the conservation law corresponding to angular momentum about the axis of gravity by "bare hands" methods.

**Exercise 4.8**
Show that if $(\mathbf{q}, \mathbf{p})$ and $(\mathbf{r}, \mathbf{s})$ are two sets of cotangent-lifted coordinates with the same domain, and $D$ is a subset of this domain, then the phase-space volume of $D$ is the same in both coordinate systems, i.e. if $D_1$ is the domain expressed in $(\mathbf{q}, \mathbf{p})$ coordinates and $D_2$ is the same domain expressed in $(\mathbf{r}, \mathbf{s})$ coordinates, then $\int_{D_1} \mathrm{d}q^1 \mathrm{d}p_1 \cdots \mathrm{d}q^n \mathrm{d}p_n = \int_{D_2} \mathrm{d}r^1 \mathrm{d}s_1 \cdots \mathrm{d}r^n \mathrm{d}s_n$.

**Exercise 4.9**
Let $\omega$ be the area form on $S^2$, which is a symplectic form. Let $H(x, y, z) = z$. Calculate $X_H$.

## 4.3 Hamiltonian mechanics on Poisson manifolds

Recall the definition of the ***canonical Poisson bracket*** from Chapter 1, which we can rewrite, using coordinates $\left(q^1, \ldots, q^n, p_1, \ldots, p_n\right)$, as

$$\{F, G\} := \frac{\partial F}{\partial q^i}\frac{\partial G}{\partial p_i} - \frac{\partial F}{\partial p_i}\frac{\partial G}{\partial q^i}. \tag{4.20}$$

The same definition applies to cotangent bundles $T^*Q$, for arbitrary manifolds $Q$, using cotangent-lifted local coordinates. The definition is independent of the choice of cotangent-lifted coordinates (this is Exercise 4.12), so it can be extended to an operation on scalar fields $F, G : T^*Q \to \mathbb{R}$, even when $T^*Q$ cannot be covered by a single coordinate patch. Notice that eqn (4.20) can be written in matrix form as

$$\{F, G\}(z) := \mathrm{d}F(z) \begin{bmatrix} 0 & I \\ -I & 0 \end{bmatrix} \left(\mathrm{d}G(z)\right)^T , \tag{4.21}$$

where $I$ is the $n \times n$ identity matrix, and $\mathrm{d}F(z)$ and $\mathrm{d}G(z)$ are represented by row vectors, e.g. $\mathrm{d}F(z) = \nabla F(z)^T$. The reason we write the equation in terms of $\mathrm{d}F$ and $\mathrm{d}G$ instead of $\nabla F$ and $\nabla G$ is that differentials are intrinsically defined quantities, unlike gradient vectors, which depend on a choice of coordinates or Riemannian metric. The matrix in the equation above is the matrix of the ***Poisson tensor*** for the canonical Poisson bracket, which is a contravariant 2-tensor.

The Poisson bracket is preserved by cotangent-lifted transformations, as the next theorem shows:

**Theorem 4.18** *Let $\Phi : T^*Q \to T^*Q$ be the cotangent lift of some diffeomorphism $\Psi : Q \to Q$, that is $\Phi = T^*\Psi$. For any smooth $F, G$: $T^*Q \to \mathbb{R}$,*

$$\{\Phi \circ F, \Phi \circ G\} = \Phi \circ \{F, G\},$$

*where $\{\cdot,\cdot\}$ is the canonical Poisson bracket on $T^*Q$.*

The properties in the following theorem can all be verified by direct calculation. The Jacobi identity was the subject of Exercise 1.10.

**Theorem 4.19** *The canonical Poisson bracket has the following properties:*

- $\{F + G, H\} = \{F, H\} + \{G, H\}$ *and* $\{F, G + H\} = \{F, H\} + \{F, G\}$ *(**bilinearity**);*
- $\{F, G\} = -\{G, F\}$ *(**skew-symmetry**);*
- $\{F, \{G, H\}\} + \{H, \{F, G\}\} + \{G, \{H, F\}\} = 0$ *(**Jacobi identity**);*
- $\{FG, H\} = F\{G, H\} + \{F, H\}G$ *(**Leibniz identity**).*

The relation of the canonical Poisson bracket to Hamilton's equations was given in Theorem 1.61. The result generalizes to arbitrary configuration manifolds $Q$, as stated in the next theorem. The proof is the same as for Theorem 1.61, except that local coordinates are used.

**Theorem 4.20** *Let $\{\cdot,\cdot\}$ be the canonical Poisson bracket on $T^*Q$. Hamilton's equations for a given $H : T^*Q \to \mathbb{R}$ are equivalent to:*

$$\dot{F} = \{F, H\} \quad \textit{for all differentiable } F : T^*Q \to \mathbb{R}, \tag{4.22}$$

*along all integral curves. More precisely,* $(\mathbf{q}(t), \mathbf{p}(t))$ *is a solution of Hamilton's equations if and only if* $\frac{\mathrm{d}}{\mathrm{d}t} F(\mathbf{q}(t), \mathbf{p}(t)) = \{F, H\}(\mathbf{q}(t), \mathbf{p}(t))$ *for all differentiable* $F : T^*Q \to \mathbb{R}$.

Note that, if $(\mathbf{q}(t), \mathbf{p}(t))$ is a solution of a vector field $X$, then

$$\dot{F}(\mathbf{q}(t), \mathbf{p}(t)) = \left\langle \mathrm{d}F(\mathbf{q}(t), \mathbf{p}(t)), \frac{\mathrm{d}}{\mathrm{d}t}(\mathbf{q}(t), \mathbf{p}(t)) \right\rangle = \mathcal{L}_X F(\mathbf{q}(t), \mathbf{p}(t)).$$

So the condition $\dot{F} = \{F, H\}$, for all integral curves of $X$, is equivalent to $\mathcal{L}_X F = \{F, H\}$. Thus, the previous theorem can be restated as: the Hamiltonian vector field $X_H$, defined by Hamilton's equations, is uniquely determined by

$$\mathcal{L}_{X_H} F = \{F, H\} \quad \text{for all differentiable } F : T^*Q \to \mathbb{R}. \tag{4.23}$$

Since $\mathcal{L}_{X_H} F = \langle dF, X_H \rangle$, the Hamiltonian vector field $X_H$ is uniquely determined by the previous equation for arbitrary $F$.

Due to the skew-symmetry property of the bracket, an immediate corollary of the previous theorem is that $\dot{H} = 0$ along solutions, that is:

**Corollary 4.21** *The Hamiltonian* $H$ *is a conserved quantity of the flow of the corresponding Hamiltonian vector field* $X_H$.

The following definition abstracts the four properties of the canonical Poisson bracket given in Theorem 4.19.

**Definition 4.22** *A* ***Poisson bracket*** *on a manifold* $P$ *is a bilinear, skew-symmetric operator* $\{\cdot, \cdot\}$ *on* $C^\infty(P, \mathbb{R})$ *satisfying the Jacobi identity and the Leibniz identity, as defined above. The pair* $(P, \{\cdot, \cdot\})$ *is called a* ***Poisson manifold.*** *Given any smooth* $H : P \to \mathbb{R}$*, the corresponding* ***Hamiltonian vector field,*** *denoted* $X_H$*, is uniquely determined by the relation*

$$\mathcal{L}_{X_H} F = \{F, H\}, \quad \textit{for all smooth } F : P \to \mathbb{R}, \tag{4.24}$$

*i.e.,*

$$\dot{F} = \{F, H\}, \quad \textit{along all solutions to } X_H. \tag{4.25}$$

**Remark 4.23** *The preceding definition extends Hamiltonian dynamics to phase spaces that are not cotangent bundles.*

**Example 4.24 (Rigid body bracket)** The formula

$$\{F, G\}(\mathbf{\Pi}) := -\mathbf{\Pi} \cdot (\nabla F(\mathbf{\Pi}) \times \nabla G(\mathbf{\Pi}))$$

defines a Poisson bracket on $\mathbb{R}^3$ (see Exercise 4.13.) Let

$$H(\mathbf{\Pi}) = \frac{1}{2}\left(\frac{\Pi_1^2}{I_1} + \frac{\Pi_2^2}{I_2} + \frac{\Pi_3^2}{I_3}\right),$$

where $I_i > 0$, $i = 1, 2, 3$, and denote $\mathbb{I} = \operatorname{diag}(I_1, I_2, I_3)$, so that

$$H(\mathbf{\Pi}) = \frac{1}{2}\left\langle \mathbf{\Pi}, \mathbb{I}^{-1}\mathbf{\Pi}\right\rangle.$$

Note that $\nabla H(\mathbf{\Pi}) = \mathbb{I}^{-1}\mathbf{\Pi}$. Using eqn (4.25), and setting $F = \Pi_i$, we obtain three scalar equations of motion, which together have the vector form,

$$\dot{\mathbf{\Pi}} = \mathbf{\Pi} \times \mathbb{I}^{-1}\mathbf{\Pi}. \tag{4.26}$$

This is Euler's equation, which was first derived in Section 1.5.

**Definition 4.25** *Let $(P, \omega)$ be a symplectic manifold. The* ***Poisson bracket associated with*** *$\omega$ is defined by*

$$\{F, G\}(z) := \omega(z)(X_F(z), X_G(z)),$$

*for any smooth $F, G \in \mathcal{F}(P)$, where $X_F$ and $X_G$ are the Hamiltonian vector fields corresponding to $F$ and $G$, respectively, as defined in eqn (4.19). Alternative equivalent notation is: $\{F, G\} = X_G \lrcorner (X_F \lrcorner \omega)$.*

**Definition 4.26** *For any Poisson bracket on a manifold $P$, the associated* ***Poisson tensor*** *is the contravariant 2-tensor $B$ defined by*

$$\{F, H\}(z) = B(z)(\mathrm{d}F(z), \mathrm{d}H(z)),$$

*for every $F, H \in \mathcal{F}(P)$, for every $z \in P$. (Showing that the Poisson tensor is well-defined is non-trivial – see [MR02][GS84].) In any local coordinates $x^1, \ldots, x^n$ on $P$, the tensor $B(z)$ is the matrix with entries*

$$B_{ij}(z) = \{x^i, x^j\}(z).$$

**Remark 4.27** *In local coordinates, $B(z)$ is a matrix, $\mathrm{d}F(z)$ is a row vector, and*

$$\{F, H\}(z) = \mathrm{d}F(z)B(z)\mathrm{d}H(z)^T.$$

*Since the column vector $X_H(z)$ is defined by*

$$\{F, H\}(z) = \mathcal{L}_{X_H} F(z) = \mathrm{d}F(z)X_H(z),$$

*the Hamiltonian vector field $X_H$ is determined by*

$$X_H(z) = B(z)\mathrm{d}H(z)^T = B(z)\nabla H(z).$$

**Example 4.28** Let $\omega$ be a symplectic form on $P$. The Poisson tensor of the Poisson bracket associated with $\omega$ is $B$ defined by

$$B(z)(\mathrm{d}F(z), \mathrm{d}H(z)) = \omega(z)(X_F(z), X_G(z)).$$

In any local coordinates, this equation has an equivalent matrix form,

$$\mathrm{d}F(z)B(z)\mathrm{d}H(z)^T = X_F(z)^T\omega(z)X_H(z). \tag{4.27}$$

The definition of $X_F$, in eqn (4.19), has matrix form $X_F(z)^T\omega(z) = \mathrm{d}F(z)$, and similarly for $X_H$. Subsituting this into eqn (4.27) shows that

$$\omega(z)B(z)\omega(z)^T = \omega(z).$$

Since $\omega(z)$ is skew-symmetric and non-singular, this implies that $B(z) = -\omega(z)^{-1}$. In particular, in symplectic local coordinates, $\omega(z)$ and $B(z)$ both have matrix $J = \begin{bmatrix} 0 & I \\ -I & 0 \end{bmatrix}$.

In general, if $B$ is a Poisson tensor on $P$ then, for every $z \in P$, the matrix of $B(z)$ is skew-symmetric and may be singular. By a process similar to Gram–Schmidt orthogonalization, any skew-symmetric matrix can be brought into the following form by a change of basis:

$$\begin{bmatrix} 0 & I & 0 \\ -I & 0 & 0 \\ 0 & 0 & 0 \end{bmatrix}, \tag{4.28}$$

where $I$ is a $k \times k$ identity matrix, for some $2k \leq \dim P$ (see [AM78]). Thus, for every $z$, it is possible to choose a basis with respect to which the matrix of $B(z)$ has the form above. The rank of $B(z)$ is $2k$, and may vary as $z$ varies. Note that, when $B(z)$ has full rank, the matrix given above equals the matrix of the canonical Poisson bracket, given in eqn (4.21).

**Definition 4.29** *Let $(P, \{\cdot,\cdot\})$ be a Poisson manifold. Two functions $F, G : P \to \mathbb{R}$ are in* ***involution****, or* ***Poisson commute****, if $\{F, G\} = 0$. A function that Poisson commutes with all functions is called a* ***Casimir function*** *(or just a* ***Casimir****).*

**Remark 4.30** *If $B$ is a Poisson tensor, then $C$ is a Casimir for the corresponding Poisson bracket if and only if $\mathrm{d}C(z)$ is in the kernel of $B(z)$, for all $z$.*

**Example 4.31** (**$\mathbb{R}^3$ Poisson brackets.**) The rigid body Poisson bracket in Example 4.24 is part of a class of Poisson brackets on $\mathbb{R}^3$, each associated with a smooth function $C : \mathbb{R}^3 \to \mathbb{R}$ and defined by

$$\{F, G\} = -\nabla C \cdot \nabla F \times \nabla G. \tag{4.29}$$

Not all functions $C$ are allowable – see Exercise 4.14. Note that every allowed $C$ is a Casimir for the corresponding bracket. A bracket of this form, together with a Hamiltonian $H$, generates the motion

$$\dot{x}^i = \{x^i, H\} = (\nabla C \times \nabla H)^i \,, \tag{4.30}$$

that is,

$$\dot{\mathbf{x}} = \nabla C \times \nabla H. \tag{4.31}$$

Since $C$ and $H$ are both conserved quantities, the motion takes place along the intersections of level surfaces of the functions $C$ and $H$ in $\mathbb{R}^3$. This is a generalization of the observation made at the end of Chapter 1, namely that the motion of the body angular momentum vector $\boldsymbol{\Pi}$ for the rigid body takes place along intersections of angular momentum spheres and energy ellipsoids.

**Definition 4.32** *Let $(P_1, \{\cdot,\cdot\}_1)$ and $(P_2, \{\cdot,\cdot\}_2)$ be Poisson manifolds. A map $\varphi : P_1 \to P_2$ is a* ***Poisson map*** *if it preserves the brackets, meaning that, for every smooth $F, G : P_2 \to \mathbb{R}$,*

$$\{F \circ \varphi, G \circ \varphi\}_1 = \{F, G\}_2 \circ \varphi.$$

*The equivalent condition on the associated Poisson tensors is $\varphi_* B_1 = B_2$ (see Exercise 4.21).*

**Proposition 4.33** *Every symplectic map is Poisson, with respect to the associated Poisson brackets.*

*Proof* See Exercise 4.22. ■

**Proposition 4.34** *Any cotangent lift of smooth function $f : Q_1 \to Q_2$ is a Poisson map with respect to the canonical Poisson brackets on $T^*Q_1$ and $T^*Q_2$.*

*Proof* This follows from Propositions 3.117 and 4.33. ■

**Exercise 4.10**
Show that the Poisson bracket associated with the canonical symplectic form $\omega = \mathrm{d}q^i \wedge \mathrm{d}p_i$ is the canonical Poisson bracket.

**Exercise 4.11**
Let $B$ be the constant Poisson tensor on $\mathbb{R}^3$ with matrix $\begin{bmatrix} 0 & 1 & -2 \\ -1 & 0 & 0 \\ 2 & 0 & 0 \end{bmatrix}$, and $\{\cdot,\cdot\}$ the associated Poisson bracket. If $F(x,y,z) = x \sin z$ and $G(x,y,z) = y^2$, compute $\{F,G\}$ and $X_F$.

**Exercise 4.12**
Show that the definition of the canonical Poisson bracket is independent of the choice of cotangent-lifted coordinates. Specifically, show that if $(\mathbf{q},\mathbf{p})$ and $(\mathbf{r},\mathbf{s})$ are two sets of cotangent-lifted coordinates with the same domain, with $(\mathbf{r},\mathbf{s}) = T^*\Psi^{-1}(\mathbf{q},\mathbf{p}) = (\Psi(\mathbf{q}), \left(\mathrm{D}\Psi^{-1}(\mathbf{q})\right)^* (\mathbf{p}))$, then

$$\left(\sum_{i=1}^{n} \frac{\partial F}{\partial q^i}\frac{\partial G}{\partial p_i} - \frac{\partial F}{\partial p_i}\frac{\partial G}{\partial q^i}\right)(\mathbf{q},\mathbf{p}) = \left(\sum_{j=1}^{n} \frac{\partial F}{\partial r^i}\frac{\partial G}{\partial s_i} - \frac{\partial F}{\partial s_i}\frac{\partial G}{\partial r^i}\right)(\mathbf{r},\mathbf{s}). \tag{4.32}$$

Note that if we reinterpret the same calculations with $\Psi$ being a map from $Q$ to $Q$ and $(\mathbf{r},\mathbf{s})$ being the same coordinates as $(\mathbf{q},\mathbf{p})$, then the same result proves Theorem 4.18, i.e. shows that the canonical Poisson bracket is preserved by cotangent lifts.

**Exercise 4.13 (Rigid body bracket)**
Show that the formula $\{F, G\}(\Pi) := -\Pi \cdot (\nabla F(\Pi) \times \nabla G(\Pi))$ defines a Poisson bracket on $\mathbb{R}^3$. With $H$ defined as in Example 4.24, show that the Hamiltonian vector field is given by Euler's equation (4.26).

**Exercise 4.14**
Find conditions on the function $C(\mathbf{x})$ so that the $\mathbb{R}^3$ bracket

$$\{F, G\} := -\nabla C \cdot \nabla F \times \nabla G \tag{4.33}$$

satisfies the defining properties of a Poisson bracket. Note that taking $C(\Pi) = \frac{1}{2}\|\Pi\|^2$ gives the rigid body bracket in Example 4.24.

**Exercise 4.15**
Show that the motion equation

$$\dot{\mathbf{x}} = \nabla C \times \nabla H$$

for the $\mathbb{R}^3$ bracket (4.33) is invariant under certain linear transformations of the functions $C$ and $H$. Interpret this invariance geometrically.

**Exercise 4.16**
Let $F$ and $G$ be two smooth scalar fields on a manifold $M$.
a) Show that, if $M$ is 2-dimensional, with local coordinates $x_1, x_2$, then

$$\mathrm{d}F \wedge \mathrm{d}G = \left(\frac{\partial F}{\partial x_1}\frac{\partial G}{\partial x_2} - \frac{\partial G}{\partial x_1}\frac{\partial F}{\partial x_2}\right)\mathrm{d}x_1 \wedge \mathrm{d}x_2.$$

b) Find the corresponding formula for $\mathrm{d}F \wedge \mathrm{d}G$ if $M$ is 3-dimensional.

**Exercise 4.17**
How is the $\mathbb{R}^3$ bracket (defined in eqn (4.33)) related to the canonical Poisson bracket? HINT: restrict to level surfaces of the function $C(\mathbf{x})$, and apply Exercise 4.16.

**Exercise 4.18 (Casimirs of the $\mathbb{R}^3$ bracket)**
Suppose the function $C(\mathbf{x})$ is chosen so that the $\mathbb{R}^3$ bracket (4.33) satisfies the defining properties of a Poisson bracket. What are the Casimirs for the $\mathbb{R}^3$ bracket (4.33)?

**Exercise 4.19**
Show that the composition of two Poisson maps is a Poisson map.

**Exercise 4.20 (Euler angles)**
The classical way of studying the rigid body is with Euler angle coordinates on $SO(3)$. In the corresponding cotangent-lifted coordinates on $T^*SO(3)$, the angular momentum vector of the rigid body is

$$\Pi = \begin{bmatrix} \left(\left(p_\varphi - p_\psi \cos\theta\right)\sin\psi + p_\theta \sin\theta\cos\psi\right)/\sin\theta \\ \left(\left(p_\varphi - p_\psi \cos\theta\right)\cos\psi + p_\theta \sin\theta\sin\psi\right)/\sin\theta \\ p_\psi \end{bmatrix}.$$

Consider the canonical Poisson bracket on $T^*SO\,(3)$, i.e., in Euler angle coordinates,

$$\{F,G\} = \frac{\partial F}{\partial \varphi}\frac{\partial G}{\partial p_\varphi} - \frac{\partial F}{\partial p_\varphi}\frac{\partial G}{\partial \varphi} + \frac{\partial F}{\partial \psi}\frac{\partial G}{\partial p_\psi} - \frac{\partial F}{\partial p_\psi}\frac{\partial G}{\partial \psi} + \frac{\partial F}{\partial \theta}\frac{\partial G}{\partial p_\theta} - \frac{\partial F}{\partial p_\theta}\frac{\partial G}{\partial \theta};$$

and the Poisson bracket on $\mathbb{R}^3$ given in Exercise 4.13. Considering $\Pi$ as a function from $T^*SO\,(3)$ to $\mathbb{R}^3$, show by direct calculation that $\Pi$ is a Poisson map, i.e. for every $F, G : \mathbb{R}^3 \to \mathbb{R}$ and every $m \in T^*SO\,(3)$,

$$\{F \circ \Pi, G \circ \Pi\}\,(m) = -\Pi\,(m) \cdot \left(\nabla F\,(\Pi\,(m)) \times \nabla G\,(\Pi\,(m))\right).$$

The theory behind why this is a Poisson map will appear in Chapter 9.

**Exercise 4.21**
Show that $\varphi$ is a Poisson map with respect to $\{\cdot,\cdot\}_1$ and $\{\cdot,\cdot\}_2$ if and only if $\varphi_* B_1 = B_2$, where $B_i$ is the Poisson tensor associated to $\{\cdot,\cdot\}_i$.

**Exercise 4.22**
Prove Proposition 4.33: every symplectic map is Poisson, with respect to the associated Poisson brackets. HINT: one way to do this is using Poisson tensors and Example 4.28.

## 4.4 A brief look at symmetry, reduction and conserved quantities

Consider the Euler–Lagrange equations,

$$\frac{\mathrm{d}}{\mathrm{d}t}\left(\frac{\partial L}{\partial \dot{q}^i}\right) = \frac{\partial L}{\partial q^i}, \quad i = 1, \ldots, n.$$

**Definition 4.35** *A coordinate $q^i$ is* ***cyclic*** *if $L$ is independent of it.*

If $L$ is independent of $q^i$, then $\frac{\partial L}{\partial q^i} = 0$, so clearly $\frac{\partial L}{\partial \dot{q}^i}$, for the same $i$, is constant. Therefore:

**Proposition 4.36** *If $q^i$ is a cyclic coordinate for a Lagrangian $L$, then $\frac{\partial L}{\partial \dot{q}^i}$ is a first integral, i.e. a conserved quantity.*

**Example 4.37** Consider a Lagrangian on configuration space $\mathbb{R}^3$, expressed in standard Cartesian coordinates. The Lagrangian $L$ is ***translationally invariant*** it is independent of $x, y, z$, i.e. if it depends only on $\dot{x}, \dot{y}, \dot{z}$. In this case, $x, y, z$ are all cyclic variables, and by the previous proposition, all of the conjugate momenta $\dfrac{\partial L}{\partial \dot{x}}, \dfrac{\partial L}{\partial \dot{y}}, \dfrac{\partial L}{\partial \dot{z}}$ are conserved. For example, if $L = \frac{1}{2}m\left(\dot{x}^2 + \dot{y}^2 + \dot{z}^2\right)$, then the conjugate momenta are

$$\frac{\partial L}{\partial \dot{x}} = m\dot{x}, \quad \frac{\partial L}{\partial \dot{y}} = m\dot{y}, \quad \frac{\partial L}{\partial \dot{z}} = m\dot{z},$$

i.e. the components of linear momentum.

**Example 4.38** Consider a Lagrangian on configuration space $\mathbb{R}^2$, expressed in polar coordinates. $L$ is ***rotationally invariant*** if it is independent of $\theta$. By the previous proposition, the conjugate momentum $\frac{\partial L}{\partial \dot{\theta}}$ is conserved. For example, if $L = \frac{1}{2}m\left(\dot{r}^2 + r^2\dot{\theta}^2\right) - V(r)$ for some $V$, then $\partial L/\partial \dot{\theta} = mr^2\dot{\theta}$, which is the formula for angular momentum.

**Definition 4.39** *For a Hamiltonian system, a coordinate is called* ***cyclic*** *if the Hamiltonian is independent of it.*

**Remark 4.40** *If $H$ is the Hamiltonian corresponding to a Lagrangian $L$, as in eqn (4.13), then every coordinate that is cyclic for $L$ is cyclic for $H$.*

Recall Hamilton's equations,

$$\frac{\mathrm{d}}{\mathrm{d}t}q^i = \frac{\partial H}{\partial p_i}, \quad \frac{\mathrm{d}}{\mathrm{d}t}p_i = -\frac{\partial H}{\partial q^i} \quad i = 1, \ldots, n.$$

The second of these equations directly implies the following:

**Remark 4.41** *If $q^i$ is a cyclic coordinate for a Hamiltonian $H$, then the conjugate momentum $p_i$ is a first integral, i.e. a conserved quantity.*

Suppose for simplicity that $q^1$ is cyclic. Along any particular solution, $p_1 = c$, for some constant $c$ determined by the initial conditions. Hamilton's equations become

$$\begin{aligned} \frac{\mathrm{d}q^1}{\mathrm{d}t} &= \frac{\partial H}{\partial p_1}, \quad p_1 = c, \\ \frac{\mathrm{d}q^i}{\mathrm{d}t} &= \frac{\partial H}{\partial p^i}, \quad \frac{\mathrm{d}p_i}{\mathrm{d}t} = -\frac{\partial H}{\partial q^i} \quad i = 2, \ldots, n. \end{aligned}$$

Since $H$ is independent of $q^1$, and $p_1$ is constant, the last $2n - 2$ equations decouple from the first two. The last $2n - 2$ equations comprise the ***reduced*** Hamiltonian system. Note that they are Hamilton's equations for the ***reduced Hamiltonian***,

$$H_{red}(q^2, \ldots, q^n, p_2, \ldots, p_n) := H(q^1, \ldots, q^n, c, p_2, \ldots, p_n), \tag{4.34}$$

where the value of $q^1$ is immaterial.

In summary, we have:

**Proposition 4.42 (Naive reduction)** *If $q^1$ is a cyclic coordinate for a Hamiltonian $H$, then the solutions of Hamilton's equations are completely determined by the solutions to Hamilton's equations for the reduced Hamiltonian $H_{red}$ defined in eqn (4.34). Specifically, $p_1$ has a constant value $c$ given by the initial conditions; $q^i(t)$ and $p_i(t)$, for $i = 2, \ldots, n$, $j \neq i$ are obtained by solving Hamilton's equations for $H_{red}$; and then $q^1(t)$ is given* *by integrating*

$$\frac{dq^1}{dt} = \frac{\partial}{\partial p^1} H(q^1, q^2(t), \ldots, q^n(t), c, p_2(t), \ldots, p_n(t)).$$

*(This last equation is called the **reconstruction equation**.)*

When naive reduction is possible, it reduces the number of degrees of freedom in the system, i.e. reduces the size of the system of ODEs to be solved. This is of great practical importance.[1] In addition, naive reduction serves as an introduction to more general reduction theorems, which are central to geometric mechanics. Reduction theorems are ways of taking advantages of ***symmetry transformations***, which are transformations that leave $L$ or $H$ invariant. Reduction in general has two facets: 'ignoring' some variables of which $L$ or $H$ are independent (e.g. $q^i$ in naive reduction); and noting that some quantities are conserved (e.g. $p_i$ in naive reduction). In this section we concentrate on conserved quantities.

The concept of a cyclic variable is limited by its dependence on the choice of a particular coordinate system. To start with, that coordinate system may not cover the entire configuration manifold, and the conservation of the corresponding conjugate momentum would only apply to trajectories as long as they stay within that coordinate patch. Another issue is that invariance properties of $L$ or $H$ are not always naturally expressed in terms of some variables being cyclic.

These considerations point to the need for a coordinate-free concept of an invariance property, and a general way of finding conserved quantities associated with them. Recall that a flow of a vector field $X$ on a manifold $M$ is its general solution $\Phi : M \times \mathbb{R} \to M$. The concept of a flow generalizes as follows:

**Definition 4.43** *A **flow** on a manifold $M$ is a smooth map $\Phi : M \times \mathbb{R} \to M$ such that, the 'time-$t$' maps $\Phi_t : M \to M$ defined, for every $t \in \mathbb{R}$, by $\Phi_t(\cdot) = \Phi(\cdot, t)$, satisfy the three properties in Theorem 3.9.*
*The **infinitesimal flow** corresponding to $\Phi$ is the vector field $X$ defined by*

$$X(z) = \left.\frac{\mathrm{d}}{\mathrm{d}s}\right|_{s=0} \Phi(z, s).$$

There is a one-to-one correspondence between flows and complete vector fields.[2]

---

[1] In fact V.I. Arnold [Arn78] wrote, 'Almost all the solved problems in mechanics have been solved by [this method].'

[2] Assuming that all of them are sufficiently smooth. See Theorem 3.9.

**Definition 4.44** *A **symmetry transformation** (or just a **symmetry**) of a function $F : M \to \mathbb{R}$ is a map $\varphi : M \to M$ such that $F \circ \varphi = F$.*
*The function $F$ is **invariant** (or **symmetric**) with respect to the flow $\Phi$ if each time-$t$ map $\Phi_t$ is a symmetry of $F$, that is, $F \circ \Phi_t = F$ for all $t$.*

The next examples show that translational and rotational invariance can be expressed in terms of flows. Note that we are identifying $T\mathbb{R}^{dN}$ with $\mathbb{R}^{dN} \times \mathbb{R}^{dN}$.

**Example 4.45** For a given $\mathbf{v} \in \mathbb{R}^3$, Define $\Psi^{\mathbf{v}} : \mathbb{R}^{6N} \times \mathbb{R} \to \mathbb{R}^{6N}$ by

$$\Psi^{\mathbf{v}}\left((\mathbf{q}_1, \ldots, \mathbf{q}_N, \dot{\mathbf{q}}_1, \ldots, \dot{\mathbf{q}}_N), t\right) = (\mathbf{q}_1 + t\mathbf{v}, \ldots, \mathbf{q}_N + t\mathbf{v}, \dot{\mathbf{q}}_1, \ldots, \dot{\mathbf{q}}_N).$$

Thus, for every $t$, the map $\Psi^{\mathbf{v}}_t$ translates every particle by $t\mathbf{v}$, leaving the velocities unchanged. It follows directly from the definitions that $L : \mathbb{R}^{6N} \to \mathbb{R}$ is translationally invariant if and only if it is symmetric with respect to all of the flows $\Psi^{\mathbf{v}}$, for all $\mathbf{v} \in \mathbb{R}^3$.

**Example 4.46** Consider $N$-particle systems in the plane. Let $\Psi$ be the flow on $T\mathbb{R}^{2N}$ given in polar coordinates by

$$\begin{aligned} &\Psi\left((r_1, \theta_1, \ldots, r_N, \theta_N, \dot{r}_1, \dot{\theta}_1, \ldots, \dot{r}_N, \dot{\theta}_N), t\right) \\ &\qquad = (r_1, \theta_1 + t, \ldots, r_N, \theta_N + t, \dot{r}_1, \dot{\theta}_1, \ldots, \dot{r}_N, \dot{\theta}_N). \end{aligned}$$

Thus, for every $t$, the map $\Psi_t$ rotates all particles by an angle $t$, leaving their angular and radial velocities unchanged. Then $L$ is rotationally invariant if and only if it is symmetric with respect to the flow $\Psi$.

A diffeomorphism from a configuration space $Q$ to itself is called a ***point transformation***. If $\Phi$ is a flow on configuration space, then each of the time-$t$ maps $\Phi_t$ is a point transformation. The tangent lifts of these point transformations, taken together, define a flow on the phase space $TQ$:

**Definition 4.47** *The **tangent lift** of a flow $\Phi$ on $Q$ is the flow $\Psi$ on $TQ$ defined by $\Psi_t = T\Phi_t$, for all $t \in \mathbb{R}$.*

**Example 4.48** In the Example 4.45, the point transformations are

$$\Phi^{\mathbf{v}}_t(\mathbf{q}_1, \ldots, \mathbf{q}_N) := (\mathbf{q}_1 + \mathbf{v}, \ldots, \mathbf{q}_N + \mathbf{v}).$$

Since $D\Phi_t^{\mathrm{v}}(\mathbf{q}_1, \ldots, \mathbf{q}_N)$, when computed in Cartesian coordinates, is always the identity map, the transformations $\Psi_t$ in Example 4.45 are the tangent lifts of the maps $\Phi_t$.

Whenever a Lagrangian is symmetric with respect to a tangent-lifted flow, there is an associated first integral, i.e. conserved quantity. This is a famous fact proven by Emmy Noether[3].

**Theorem 4.49 (Noether's Theorem, Lagrangian version)**
*If $L : TQ \to \mathbb{R}$ is invariant with respect to the tangent lift of a flow $\Phi : Q \times \mathbb{R} \to Q$, then $\langle \theta_L, \delta\mathbf{q} \rangle$ is a first integral, where $\theta_L(\mathbf{q}, \dot{\mathbf{q}}) = \frac{\partial L}{\partial \dot{q}^i} dq^i$ and $\delta\mathbf{q}$ is the infinitestimal flow of $\Phi$. In local coordinates, this first integral is*

$$\frac{\partial L}{\partial \dot{\mathbf{q}}} \cdot \delta\mathbf{q} = \frac{\partial L}{\partial \dot{\mathbf{q}}}(\mathbf{q}, \dot{\mathbf{q}}) \cdot \left. \frac{\partial \Phi_s(\mathbf{q})}{\partial s} \right|_{s=0}.$$

*Proof* Let $L$ be a Lagrangian on $TQ$ that is invariant with respect to the tangent lift of a flow $\Phi$ on $Q$. Given any path $\mathbf{q} : [a, b] \to Q$, we define the deformation $\mathbf{q}(\cdot, \cdot)$ by

$$\mathbf{q}(t, s) = \Phi_s(\mathbf{q}(t)).$$

Note that $\mathbf{q}(\cdot, 0) = \mathbf{q}(\cdot)$, but that $\mathbf{q}(\cdot, \cdot)$ does not necessarily leave the endpoints fixed. An example is sketched in Figure 4.3. Writing

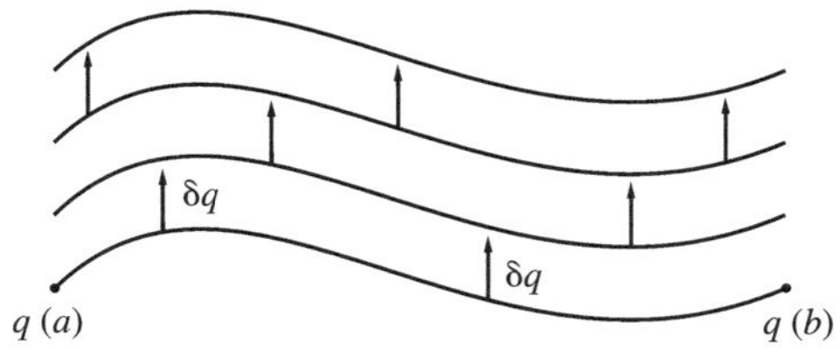


**Figure 4.3** A deformation corresponding to a flow by translations: each curve $\mathbf{q}(\cdot, s)$ is a translation of of the curve $\mathbf{q}(\cdot)$. The arrows indicate the corresponding infinitesimal flow.

---

3 This result first appeared in Noether [Noe72] . In fact, the result in [Noe72] is more general than this. In particular, in the PDE (partial differential equation) setting one must also include the transformation of the volume element in the action principle. See, for example, [Olv00] for good discussions of the history, framework and applications of Noether's theorem. See [Hol08] for a treatment of Noether's theorem for space-time symmetries, discussed along the same lines as here. Allowing time-translations enables energy conservation to be interpreted as yet another consequence of Noether's theorem.

$\dot{\mathbf{q}}(t,s) = \frac{\partial}{\partial t}\mathbf{q}(t,s)$, we have, for every $t, s$,

$$\begin{aligned} L(\mathbf{q}(t,s), \dot{\mathbf{q}}(t,s)) &= L(\Phi_s(\mathbf{q}(t)), \frac{\partial}{\partial t}\Phi_s(\mathbf{q}(t))) \\ &= L(\Phi_s(\mathbf{q}(t)), \mathrm{D}\Phi_s(\mathbf{q}(t)) \cdot \dot{\mathbf{q}}(t)) \\ &= L(T\Phi_s(\mathbf{q}(t), \dot{\mathbf{q}}(t)) \\ &= L(\mathbf{q}(t), \dot{\mathbf{q}}(t)), \end{aligned}$$

where in the last line we have used the assumption that $L$ is invariant with respect to the tangent lift of $\Psi$. It follows that

$$\left.\frac{\partial}{\partial s}\right|_{s=0} \int_a^b L(\mathbf{q}(t,s), \dot{\mathbf{q}}(t,s))dt = 0.$$

(In fact, the equation holds for all $s$, but we will only need the case $s = 0$.) As we have seen earlier in eqn (4.4), the previous equation is equivalent to

$$\int_a^b \left( \frac{\mathrm{d}}{\mathrm{d}t}\left(\frac{\partial L}{\partial \dot{\mathbf{q}}}\right) - \frac{\partial L}{\partial \mathbf{q}} \right) \cdot \delta\mathbf{q}\, \mathrm{d}t = \left.\frac{\partial L}{\partial \dot{\mathbf{q}}} \cdot \delta\mathbf{q}\right|_a^b,$$

where $\delta\mathbf{q}(t,s) := \left.\frac{\partial}{\partial s}\right|_{s=0} \mathbf{q}(t,s)$. Note that $\delta\mathbf{q}(t,s)$ is the infinitesimal flow of $\Psi$ at $(t,s)$. Now if $\mathbf{q}(\cdot)$ is a trajectory satisfying the Euler–Lagrange equations for $L$, then the expression in parentheses in the previous equation vanishes, leaving

$$\left.\frac{\partial L}{\partial \dot{\mathbf{q}}} \cdot \delta\mathbf{q}\right|_a^b = 0,$$

i.e. the value of $\frac{\partial L}{\partial \dot{\mathbf{q}}} \cdot \delta\mathbf{q}$ is the same at $(\mathbf{q}(a), \dot{\mathbf{q}}(a))$ as $(\mathbf{q}(b), \dot{\mathbf{q}}(b))$. Since this is true for all solutions to the Euler–Lagrange equations, $\frac{\partial L}{\partial \dot{\mathbf{q}}} \cdot \delta\mathbf{q}$ is a first integral, i.e. a conserved quantity. This quantity can also be expressed in terms the ***Lagrangian*** *1-**form*** on $TQ$, which is defined in local coordinates as

$$\theta_L\,(\mathbf{q}, \dot{\mathbf{q}}) := \frac{\partial L}{\partial \dot{q}^i}\,\mathrm{d}q^i.$$

A straightforward calculation shows that this definition is coordinate independent. This fact also follows from the relationship between $\theta_L$ and the canonical 1-form $\theta = p_i \mathrm{d}q^i$ on $T^*Q$, which was introduced in Chapter 3, namely,

$$\theta_L = (\mathbf{F}L)^*\,\theta.$$

If we view the vector field $\delta\mathbf{q}$ as a vector field on $TQ$ with zero $\dot{\mathbf{q}}$ component, then

$$\frac{\partial L}{\partial \dot{\mathbf{q}}} \cdot \delta\mathbf{q} = \langle \theta_L, \delta\mathbf{q} \rangle .$$

The notation on the right-hand side means that the 1-form $\theta_L$ has been paired with the vector field $\delta\mathbf{q}$ to give a scalar field on $TQ$; and the value of this scalar field, at any $(\mathbf{q}, \dot{\mathbf{q}})$, is $\langle \theta_L(\mathbf{q}, \dot{\mathbf{q}}), \delta\mathbf{q}(\mathbf{q}) \rangle$. ■

Noether's Theorem has an analogue in Hamiltonian mechanics. To state it, we need the following analogue of the tangent lift of a flow.

**Definition 4.50** *The* ***cotangent lift*** *of a flow $\Phi$ on $Q$ is the flow $\Psi$ on $TQ$ defined by $\Psi_t = T^*\Phi_t^{-1}$, for all $t \in \mathbb{R}$.*

**Theorem 4.51 (Noether's Theorem, first Hamiltonian version)**
*If $H$: $T^*Q \to \mathbb{R}$ is invariant with respect to the cotangent lift of a flow $\Phi : Q \times \mathbb{R} \to Q$ then $\langle \theta, \delta\mathbf{q} \rangle$ is a conserved quantity, where $\theta$ is the canonical 1-form and $\delta\mathbf{q}$ is the infinitestimal flow of $\Phi$. In local cotangent-lifted coordinates, this first integral is*

$$\mathbf{p} \cdot \delta\mathbf{q} = \mathbf{p} \cdot \left. \frac{\partial \Phi_s(\mathbf{q})}{\partial s} \right|_{s=0} .$$

For Hamiltonians that correspond to hyperregular Lagrangrians (in the sense of Definition 4.9), the previous theorem is a direct consequence of the Lagrangian version of Noether's Theorem and the following proposition.

**Proposition 4.52** *Let $\Phi : Q \times \mathbb{R} \to Q$ be a flow. Let $L : TQ \to \mathbb{R}$ be a hyperregular Lagrangian, with corresponding Hamiltonian $H : T^*Q \to \mathbb{R}$. Then $L$ is invariant with respect to the tangent lift of the flow $\Phi$ if and only if $H$ is invariant with respect to the cotangent lift of $\Phi$.*

*Proof* Suppose that $L$ is invariant with respect to the tangent lift of $\Phi$. A direct calculation shows that $\left(T^*\Phi_s^{-1}\right) \circ \mathbb{F}L = \mathbb{F}L \circ T\Phi_s$, for every $s$, (see Exercise 4.24). This implies that the energy function $E$ is invariant with respect to the tangent lift of $\Phi$ (see Exercise 4.25). Since $H = E \circ (\mathbb{F}L)^{-1}$ by definition, it follows that,

$$H \circ \left(T^*\Phi_s^{-1}\right) \circ \mathbb{F}L = E \circ (\mathbb{F}L)^{-1} \circ \mathbb{F}L \circ T\Phi_s = E \circ T\Phi_s = E,$$

and hence that $H \circ \left(T^*\Phi^{-1}\right) = H$. Hence $H$ is invariant under the cotangent lift of $\Phi$. The converse is similar. ■

For general Hamiltonians, Noether's Theorem is still valid, but a different method of proof is required. We defer this until Chapter 8, where a version of the theorem is given that applies to a broader class of symmetries called *Lie group actions*.

**Exercise 4.23**
Let $L = K - V$, where $K = \frac{1}{2}\left(m_1\|\dot{\mathbf{q}}_1\|^2 + m_2\|\dot{\mathbf{q}}_2\|^2\right)$ and $V(\mathbf{q}_1, \mathbf{q}_2)$ is translationally invariant (see Definition 1.25). Using the change of coordinates $(\mathbf{r}_1, \mathbf{r}_2) := (\mathbf{q}_1 - \mathbf{q}_2, \mathbf{q}_1 + \mathbf{q}_2)$, show that the coordinates $r_2^1, r_2^2, r_2^3$ are all cyclic, and conclude that total linear momentum is conserved. Generalize this example to $N$-particle systems.

**Exercise 4.24**
If $L$ is invariant with respect to the tangent lift of a flow $\Phi$, show by direct calculation that $\left(T^*\Phi_s^{-1}\right) \circ \mathbb{F}L = \mathbb{F}L \circ T\Phi_s$ for every $s$.

**Exercise 4.25**
Let $E$ be the energy function corresponding to $L$, i.e. $E(v) = \langle \mathbb{F}L(v), v\rangle - L(v)$, for every $v \in TQ$. Show by direct calculation that $L$ is invariant with respect to the tangent lift of a flow $\Phi$ if and only if $E$ is.

## Solutions to selected exercises

**Solution to Exercise 4.2** $L$ is regular since

$$\frac{\partial^2 L}{\partial \dot{q}\,\partial \dot{q}} = m \neq 0.$$

The Euler–Lagrange equations are given in the region $q > 0$ by

$$\begin{aligned}\frac{\mathrm{d}}{\mathrm{d}t}\frac{\partial L}{\partial \dot{q}} - \frac{\partial L}{\partial q} &= m\ddot{q} + \frac{1}{q^2}\\ &= 0.\end{aligned}$$

Therefore, $q(t) \to 0$ as $t \to t_1 \in \mathbb{R}$. $q$ reaches 0 in finite time since it has a finite distance to travel and a monotonically increasing velocity in the direction of 0. The case with $q < 0$ is similar.

**Solution to Exercise 4.4** Let

$$L'(\boldsymbol{q}, \dot{\boldsymbol{q}}) = L(\boldsymbol{q}, \dot{\boldsymbol{q}}) + \frac{\mathrm{d}}{\mathrm{d}t}\gamma(\boldsymbol{q}).$$

and consider Hamilton's principle for the $\gamma$ term.

$$\begin{aligned}\delta \int_{t_0}^{t_1} \frac{\mathrm{d}}{\mathrm{d}t}\gamma \,\mathrm{d}t &= \int_{t_0}^{t_1} \frac{\mathrm{d}}{\mathrm{d}t}\left(\frac{\partial \gamma}{\partial \boldsymbol{q}} \cdot \delta \boldsymbol{q}\right) \mathrm{d}t \\ &= \left[\frac{\partial \gamma}{\partial \boldsymbol{q}} \cdot \delta \boldsymbol{q}\right]_{t_0}^{t_1} \\ &= 0,\end{aligned}$$

where the last step follows since the variations vanish at the end points. Therefore,

$$\delta \int_{t_0}^{t_1} L' \,\mathrm{d}t = \delta \int_{t_0}^{t_1} L \,\mathrm{d}t.$$

This is equivalent to the statement:

$$\begin{aligned}&\boldsymbol{q} \text{ satisfies the Euler–Lagrange equations for } L \\ &\quad \Longleftrightarrow \boldsymbol{q} \text{ satisfies the Euler–Lagrange equations for } L'.\end{aligned}$$

**Solution to Exercise 4.5** The Lagrangian vector field, $Z_L$, is given by

$$Z_L(\boldsymbol{q}, \dot{\boldsymbol{q}}) = \left(\boldsymbol{q}, \dot{\boldsymbol{q}}, \dot{\boldsymbol{q}}, \left(\frac{\partial^2 L}{\partial \dot{\boldsymbol{q}} \partial \dot{\boldsymbol{q}}}\right)^{-1} \left(\frac{\partial L}{\partial \boldsymbol{q}} - \frac{\partial^2 L}{\partial \boldsymbol{q} \partial \dot{\boldsymbol{q}}} \dot{\boldsymbol{q}}\right)\right).$$

The derivative of the fiber derivative is given by the matrix

$$\mathrm{D}\mathbb{F}L = \begin{pmatrix} I & 0 \\ \frac{\partial^2 L}{\partial \boldsymbol{q} \partial \dot{\boldsymbol{q}}} & \frac{\partial^2 L}{\partial \dot{\boldsymbol{q}} \partial \dot{\boldsymbol{q}}} \end{pmatrix}.$$

Therefore, the push-forward of $Z_L$ by $\mathbb{F}L$ is given by

$$\begin{aligned}(T\mathbb{F}L)\, Z_L(\boldsymbol{q}, \dot{\boldsymbol{q}}) &= (\mathbb{F}L, \mathrm{D}\mathbb{F}L) \left(\boldsymbol{q}, \dot{\boldsymbol{q}}, \dot{\boldsymbol{q}}, \left(\frac{\partial^2 L}{\partial \dot{\boldsymbol{q}} \partial \dot{\boldsymbol{q}}}\right)^{-1} \left(\frac{\partial L}{\partial \boldsymbol{q}} - \frac{\partial^2 L}{\partial \boldsymbol{q} \partial \dot{\boldsymbol{q}}} \dot{\boldsymbol{q}}\right)\right) \\ &= \left(\boldsymbol{q}, \frac{\partial L}{\partial \dot{\boldsymbol{q}}}, \dot{\boldsymbol{q}}, \frac{\partial L}{\partial \boldsymbol{q}}\right).\end{aligned}$$

This is exactly the result

$$\frac{\mathrm{d}\boldsymbol{p}}{\mathrm{d}t} - \frac{\partial L}{\partial \boldsymbol{q}} = 0, \qquad \frac{\mathrm{d}\boldsymbol{q}}{\mathrm{d}t} = \dot{\boldsymbol{q}}.$$

**Solution to Exercise 4.7** The Lagrangian for the spherical pendulum is given in spherical coordinates by

$$L = \frac{m}{2}\left(\dot{\theta}^2 + \sin^2\theta\dot{\phi}^2\right) + mg\cos\theta,$$

as in Example 1.42. To show that $L$ is hyperregular calculate the fiber derivative,

$$\begin{aligned}\mathbb{F}L(\theta,\phi,\dot{\theta},\dot{\phi}) &= \left(\boldsymbol{q}, \frac{\partial L}{\partial \dot{\boldsymbol{q}}}\right)\\ &= \left(\theta,\phi,m\dot{\theta},m\sin^2\theta\dot{\phi}\right).\end{aligned}$$

which is a diffeomorphism for $\theta \neq 0, \pi$ (which are the poles of the sphere). Therefore $L$ is hyperregular. Setting $p_\theta = m\dot{\theta}$ and $p_\phi = m\sin^2\theta\dot{\phi}$ and invoking the Legendre transform yields

$$\begin{aligned}H &= \left(p_\theta, p_\phi\right)\cdot\left(\dot{\theta},\dot{\phi}\right) - L\\ &= \frac{p_\theta^2}{2m} + \frac{p_\phi^2}{2m\sin^2\theta} - mg\cos\theta.\end{aligned}$$

Hamilton's equation for $p_\phi$ yields the conservation of angular momentum.

$$\dot{p}_\phi = -\frac{\partial H}{\partial \phi} = 0 \implies p_\phi = m\sin^2\theta\dot{\phi} = \text{const.}$$

**Solution to Exercise 4.9** The area form on the sphere is $\omega = \sin\theta(\mathrm{d}\theta\wedge\mathrm{d}\phi)$ as in Example 3.53. Given the Hamiltonian $H(x,y,z) = z = \cos\theta$, $X_H$ is defined by

$$X_H \lrcorner\, \omega = \mathrm{d}H = -\sin\theta\,\mathrm{d}\theta.$$

Therefore, $X_H = \frac{\partial}{\partial\phi}$.

**Solution to Exercise 4.10** The canonical symplectic form $\omega = \mathrm{d}q^i\wedge\mathrm{d}p_i$ gives Hamiltonian vector fields

$$X_H = \frac{\partial H}{\partial p_i}\frac{\partial}{\partial q^i} - \frac{\partial H}{\partial q^i}\frac{\partial}{\partial p_i}.$$

Therefore the Poisson bracket associated with $\omega$ is

$$\begin{aligned}\{F, H\} &= X_H \lrcorner (X_F \lrcorner \omega) \\ &= X_H \lrcorner \mathrm{d}F \\ &= \left(\frac{\partial F}{\partial q^i}\frac{\partial H}{\partial p_i} - \frac{\partial H}{\partial q^i}\frac{\partial F}{\partial p_i}\right).\end{aligned}$$

This is the canonical Poisson bracket.

**Solution to Exercise 4.11** Let

$$B = \begin{bmatrix} 0 & 1 & -2 \\ -1 & 0 & 0 \\ 2 & 0 & 0 \end{bmatrix},$$

$F = x \sin z$, and $G = y^2$. Then

$$\begin{aligned}\{F, G\} &= [\sin z, 0, x\cos z] \begin{bmatrix} 0 & 1 & -2 \\ -1 & 0 & 0 \\ 2 & 0 & 0 \end{bmatrix} \begin{bmatrix} 0 \\ 2y \\ 0 \end{bmatrix} \\ &= [\sin z, 0, x\cos z] \cdot \begin{bmatrix} 2y \\ 0 \\ 0 \end{bmatrix} \\ &= 2y \sin z,\end{aligned}$$

and by Remark 4.27,

$$X_F = \begin{bmatrix} 0 & 1 & -2 \\ -1 & 0 & 0 \\ 2 & 0 & 0 \end{bmatrix} \begin{bmatrix} \sin z \\ 0 \\ x\cos z \end{bmatrix} = \begin{bmatrix} -2x\cos z \\ -\sin z \\ 2\sin z \end{bmatrix}.$$

**Solution to Exercise 4.15** Consider the following calculation

$$\begin{aligned}\nabla(\alpha C + \beta H) \times \nabla(\gamma C + \delta H) &= (\alpha\delta - \beta\gamma)(\nabla C \times \nabla H) \\ &= \nabla C \times \nabla H \quad \Longleftrightarrow \quad \alpha\delta - \beta\gamma = 1.\end{aligned}$$

For a geometrical interpretation consider the map $(C, H) : \mathbb{R}^3 \to \mathbb{R}^2$. The above calculation shows that the equation

$$\dot{\boldsymbol{x}} = \nabla C \times \nabla H$$

is invariant under area-preserving transformations in $\mathbb{R}^2$, $A \in SL(2, \mathbb{R})$.

**Solution to Exercise 4.17** Let $(x_1, x_2)$ be coordinates of the level surface of $C$. The Poisson bracket is given by

$$\{F, G\} \mathrm{d}X_1 \wedge \mathrm{d}X_2 \wedge \mathrm{d}X_3 = \mathrm{d}C \wedge \mathrm{d}F \wedge \mathrm{d}G.$$

Using coordinates that align with a level set of $C$ gives

$$\begin{aligned}\{F, G\} \mathrm{d}X_1 \wedge \mathrm{d}X_2 \wedge \mathrm{d}X_3 &= \mathrm{d}C \wedge \mathrm{d}F \wedge \mathrm{d}G, \\ &= \mathrm{d}C \wedge \{F, G\}|_C \, \mathrm{d}x_1 \wedge \mathrm{d}x_2.\end{aligned}$$

Therefore,

$$\{F, G\}|_C = \left( \frac{\partial F}{\partial x^1} \frac{\partial G}{\partial x^2} - \frac{\partial G}{\partial x^1} \frac{\partial F}{\partial x^2} \right),$$

which is the canonical Poisson bracket in $\boldsymbol{x}$ coordinates.

**Solution to Exercise 4.18** Any function, $\Phi(C)$, is a Casimir function for $\{F, H\} = \nabla C \cdot (\nabla F \times \nabla H)$, since

$$\{\Phi(C), H\} = \Phi'(C) \nabla C \cdot (\nabla C \times \nabla H) = 0.$$

These are the only Casimirs since $\nabla F$ must be parallel to $\nabla C$ everywhere for $F$ to be a Casimir.

**Solution to Exercise 4.25** Suppose that $L \circ T\Phi_s = L$. Then

$$\begin{aligned}E \circ T\Phi_s(v) &= \langle \mathbb{F}L \circ T\Phi_s(v) \,,\, T\Phi_s(v) \rangle - L \circ T\Phi_s(v) \\ &= \left\langle T^* \Phi_s^{-1} \circ \mathbb{F}L(v) \,,\, T\Phi_s(v) \right\rangle - L(v) \\ &= \langle \mathbb{F}L(v) \,,\, v \rangle - L(v) \\ &= E(v).\end{aligned}$$

Therefore, $E$ is invariant under $T\Phi_s$. The converse is similar.

# 5 Lie groups and Lie algebras

In general, a system is symmetric when its state does not change under a certain transformation. For mechanical systems, a symmetry is a transformation of phase space, and possibly time, that leaves the Hamiltonian or Lagrangian invariant. Symmetries form groups, with the group operation being composition. For instance, a group consisting of a reflection and the identity transformation is isomorphic to $\mathbb{Z}_2$; while the group of rotations of $\mathbb{R}^3$ is the matrix Lie group $SO(3)$.

Lie groups are the mathematical concept appropriate for describing continuously varying groups of transformations. The present chapter introduces general Lie groups, Lie algebras and the exponential map, beginning with the special cases of matrix Lie groups (first introduced in Chapter 2) and matrix Lie algebras. Group actions, which associate group elements with transformations, will be defined in the next chapter.

## 5.1 Matrix Lie groups and Lie algebras

The following definition is repeated from Section 2.4.

**Definition 5.1** *A **matrix group** is a subset of $\mathcal{M}(n,\mathbb{R})$, or $\mathcal{M}(n,\mathbb{C})$, that is a group, with the group operation being matrix multiplication. A **matrix Lie group** is a matrix group that is also a submanifold of $\mathcal{M}(n,\mathbb{R})$ or $\mathcal{M}(n,\mathbb{C})$.*

Recall that $\mathcal{M}(n,\mathbb{R})$ is the space of $n \times n$ real matrices, which is a real vector space that can be identified with $\mathbb{R}^{n\times n} = \mathbb{R}^{n^2}$. Similarly $\mathcal{M}(n,\mathbb{C})$ is a complex vector space that can be identified with $\mathbb{C}^{n^2}$ or $\mathbb{R}^{(2n)^2}$. Note that, to satisfy the group axioms, the identity matrix $I$ must be contained in the group. Also, all matrices in the group must be invertible, so the group must be a subgroup of $GL(n,\mathbb{R})$ or $GL(n,\mathbb{C})$.

**Definition 5.2** *The **matrix commutator** of any pair of $n \times n$ matrices $A$ and $B$ is defined as $[A,B] := AB - BA$.*

**Proposition 5.3 (Properties of the matrix commutator)**
*The matrix commutator operation has the following properties for any $n \times n$ matrices $A$, $B$ and $C$:*

*(i)* $[B, A] = -[A, B]$ *(**skew-symmetry**); and*
*(ii)* $[[A, B], C] + [[B, C], A] + [[C, A], B] = 0$ *(the **Jacobi identity**).*

*Proof* These straightforward calculations are left as Exercise 5.1. ■

**Example 5.4 (The orthogonal group, $O(n)$)** Recall from Proposition 2.90 that the set $O(n)$ of real orthogonal $n \times n$ matrices is a matrix Lie group of dimension $n(n-1)/2$. The tangent space to $O(n)$ at the identity, $T_I O(n)$, is the vector subspace of skew-symmetric matrices, i.e. those matrices $A$ satisfying

$$A^T + A = 0.$$

For any pair of matrices $A, B \in T_I O(n)$, the matrix commutator $[A, B]$ is also in $T_I O(n)$. Indeed,

$$\begin{aligned}[A, B]^T + [A, B] &= (AB - BA)^T + (AB - BA) \\ &= B^T A^T - A^T B^T + AB - BA = 0,\end{aligned}$$

since $A^T = -A$ and $B^T = -B$.

The previous example shows that $T_I O(n)$ is closed under the matrix commutator operation. In fact, this is true for the tangent space at the identity $T_I G$ of any matrix Lie group $G$, as we will now show. The proof relies on the following lemma, which states that $T_I G$ is closed under conjugation[1] by any element of $G$.

**Lemma 5.5** *Let $R$ be an arbitrary element of a matrix Lie group $G$, and let $B \in T_I G$. Then $RBR^{-1} \in T_I G$.*

*Proof* Let $C_B(t)$ be a curve in $G$ such that $C_B(0) = I$ and $C'_B(0) = B$. Define $V(t) = RC_B(t)R^{-1} \in G$ for all $t$. Then $V(0) = I$ and $V'(0) = RBR^{-1}$. Hence, $RBR^{-1} = V'(0) \in T_I G$, which proves the lemma. ■

**Proposition 5.6** *Let $G$ be a matrix Lie group, and let $A, B \in T_I G$. Then $[A, B] \in T_I G$.*

---

[1] Conjugation by a group element $g$ is the map $h \mapsto ghg^{-1}$.

*Proof* Let $R_A(s)$ be a curve in $G$ such that $R_A(0) = I$ and $R'_A(0) = A$. Define $C(s) = R_A(s)B(R_A(s))^{-1}$. Then the previous lemma implies that $C(s) \in T_I G$ for every $s$. Hence, $C'(s) \in T_I G$, for every $s$, in particular for $s = 0$. Since $C'(0) = AB - BA = [A, B]$ (using a formula in Exercise 3.12), this shows that $[A, B] \in T_I G$ . ■

**Definition 5.7** *A* ***matrix Lie algebra*** *is a vector subspace of* $\mathcal{M}(n, \mathbb{R})$ *or* $\mathcal{M}(n, \mathbb{C})$ *for some* $n$*, with the usual operations of matrix addition and scalar multiplication, that is also closed under the matrix commutator* $[\cdot, \cdot]$.

**Proposition 5.8** *For any matrix Lie group* $G$*, the tangent space at the identity* $T_I G$ *is a matrix Lie algebra, called the* **Lie algebra of** $G$*. Conversely, any matrix Lie algebra is the Lie algebra of some matrix Lie group.*

*Sketch of proof* The first statement follows from Proposition 5.6 and the fact that $T_I G$ is a vector subspace of $\mathcal{M}(n, \mathbb{R})$ or $\mathcal{M}(n, \mathbb{C})$. For the converse, recall the matrix exponential operation exp, defined by

$$\exp(A) := I + A + \frac{1}{2!}A^2 + \frac{1}{3!}A^3 + \dots$$

Suppose we are given a matrix Lie algebra $V$. Let

$$G := \exp(V) := \{\exp(A) : A \in V\}.$$

It can be shown that $G$ is a matrix Lie group, and that $T_I G = V$. See [Bak02]. ■

Note that there may be more than one matrix Lie group with the same Lie algebra. For example, $O(n)$ and $SO(n)$ have the same Lie algebra: the set of skew-symmetric matrices, as shown in Proposition 2.90 and Example 5.12.

The remainder of this section consists of examples of matrix Lie groups and their Lie algebras. Parts of some examples have already appeared earlier in Section 2.4. All of the examples given here are groups of matrices with *real* entries; for example '$GL(n)$' denotes $GL(n, \mathbb{R})$ (some complex matrix groups appear in the exercises). We use the notational convention that the Lie algebra of a Lie group is denoted by the lower-case Gothic version of the letter(s) denoting the Lie group.

**Example 5.9 (General linear group $GL(n)$ and Lie algebra $\mathfrak{gl}(n)$)** The matrix Lie group $GL(n)$ is formed by all $n \times n$ matrices with non-zero determinant. It is an open subset of $\mathcal{M}(n, \mathbb{R})$, and hence an $n^2$-dimensional submanifold of $\mathcal{M}(n, \mathbb{R})$. Its Lie algebra is $\mathcal{M}(n, \mathbb{R})$, with the commutator operation, and is denoted in this context by $\mathfrak{gl}(n)$.

**Example 5.10 (Special linear group $SL(n)$ and Lie algebra $\mathfrak{sl}(n)$)**
The subgroup of $GL(n)$ formed by matrices with determinant 1 is called the special linear group and is denoted by $SL(n)$. It is a matrix Lie group of dimension $n^2 - 1$, as shown in Proposition 2.92. We shall now prove that its Lie algebra $\mathfrak{sl}(n)$ is the space of all traceless matrices. Since $SL(n) = \det^{-1}(1)$, its Lie algebra is the kernel of the derivative of the determinant map at the identity, that is,

$$\mathfrak{sl}(n) := T_I SL(n, \mathbb{R}) = \{A \in \mathfrak{gl}(n) : (T_I \det)(A) = 0\}.$$

To compute the derivative (or tangent map) $T_I \det$, note that every matrix $A$ can be expressed as $\left.\frac{\mathrm{d}}{\mathrm{d}t}\right|_{t=0} I + tA$. Using a standard formula for the determinant,

$$\begin{aligned} \big(T_I \det\big)(A) &= \left.\frac{d}{dt}\right|_{t=0} \det(I + tA) \\ &= \left.\frac{d}{dt}\right|_{t=0} I + t\big(\operatorname{tr} A\big) + \cdots + t^n \det(A) \\ &= \operatorname{tr} A. \end{aligned}$$

Therefore, the Lie algebra $\mathfrak{sl}(n)$ is the set of $n \times n$ matrices with trace zero.

**Example 5.11 (Orthogonal group $O(n)$ and Lie algebra $\mathfrak{o}(n)$)**
As shown in Example 5.4, the set of orthogonal $n \times n$ matrices forms a matrix Lie group $O(n)$ of dimension $n(n-1)/2$, with Lie algebra $\mathfrak{o}(n)$ equal to the space of skew-symmetric $n \times n$ matrices.

**Example 5.12 (Special orthogonal group $SO(n)$ and Lie algebra $\mathfrak{so}(n)$)**
The special orthogonal group $SO(n)$ consists of all orthogonal matrices with determinant 1,

$$SO(n) := O(n) \cap SL(n).$$

In Section 2.4, we proved that $O(n)$ has two connected components and $SO(n)$ is the connected component containing the identity. Its dimension is $n(n-1)/2$. For any path $C(t)$ in $O(n)$ such that $C(0) = I$, at least some portion of the path near $t = 0$, say for $t \in (-\varepsilon, \varepsilon)$, must be in $SO(n)$. Thus, the Lie algebra $\mathfrak{so}(n)$ of the $SO(n)$ coincides with $\mathfrak{o}(n)$. Note that $SO(3)$ is the group of rotations of $\mathbb{R}^3$.

**Example 5.13 (Symplectic group $Sp(2n)$ and Lie algebra $\mathfrak{sp}(2n)$)**
Consider the non-singular skew-symmetric $2n \times 2n$ matrix

$$J = \begin{bmatrix} 0 & I \\ -I & 0 \end{bmatrix}.$$

The real symplectic group $Sp(2n)$ is defined as

$$Sp(2n) = \{U \in GL(2n) : U^T J U = J\}.$$

We know from Exercise 2.31 that this is a matrix Lie group of dimension $2n^2 + n$, with Lie algebra

$$\mathfrak{sp}(2n) = \{B \in \mathcal{M}(2n, \mathbb{R}) : B^T J + J B = 0\}.$$

See also Exercise 5.3.

**Example 5.14 (Special Euclidean group $SE(3)$ and Lie algebra $\mathfrak{se}(3)$)** The special Euclidean group, denoted $SE(3)$, is the Lie group of $4 \times 4$ matrices of the form

$$E(\mathbf{R}, \mathbf{v}) = \begin{bmatrix} \mathbf{R} & \mathbf{v} \\ 0 & 1 \end{bmatrix},$$

where $\mathbf{R} \in SO(3)$ and $\mathbf{v} \in \mathbb{R}^3$. Note that, for any $\mathbf{w} \in \mathbb{R}^3$,

$$\begin{bmatrix} \mathbf{R} & \mathbf{v} \\ 0 & 1 \end{bmatrix} \begin{bmatrix} \mathbf{w} \\ 1 \end{bmatrix} = \begin{bmatrix} \mathbf{R}\mathbf{w} + \mathbf{v} \\ 1 \end{bmatrix},$$

so $E(\mathbf{R}, \mathbf{v})$ corresponds to rotation by $\mathbf{R}$ followed by translation by $\mathbf{v}$. Thus the special Euclidean group describes the set of rigid motions and coordinate transformations of three-dimensional space. From the example of $SO(3)$ above, it follows that the Lie algebra of $SE(3)$, written $\mathfrak{se}(3)$, is the set of all matrices of the form

$$\begin{bmatrix} \mathbf{A} & \mathbf{w} \\ 0 & 0 \end{bmatrix},$$

for $\mathbf{A}$ skew-symmetric and $\mathbf{w}$ arbitrary. See Exercise 5.2.

**Exercise 5.1**
Verify that the matrix commutator operation is skew-symmetric and satisfies the Jacobi identity, as stated in Proposition 5.3.

**Exercise 5.2**
A point $P$ in $\mathbb{R}^3$ undergoes a rigid motion associated with $E(R_1, v_1)$ followed by a rigid motion associated with $E(R_2, v_2)$. What matrix element of $SE(3)$ is associated with the composition of these motions in the given order? Compare this with the result of matrix multiplication of elements of $SE(3)$. Investigate the commutator operation in $\mathfrak{se}(3)$. Note: $SE(3)$ is an example of a *semidirect product* Lie group (see Section 9.6).

**Exercise 5.3**
Let $Y \in \mathfrak{sp}(2n, \mathbb{R})$ be partitioned into $n \times n$ blocks,

$$Y = \begin{bmatrix} A & B \\ C & D \end{bmatrix}.$$

Write down a complete set of equations involving $A$, $B$, $C$, and $D$ that must be satisfied for $Y$ to be in $\mathfrak{sp}(2n, \mathbb{R})$. Deduce that the dimension of $\mathfrak{sp}(2n, \mathbb{R})$, and consequently $Sp(2n, \mathbb{R})$, is $2n^2 + n$. This is an alternative to the dimension calculation method suggested in Exercise 2.31.

**Exercise 5.4**
Show that the set of $n \times n$ invertible complex matrices $GL(n, \mathbb{C})$ is a Lie group and describe its Lie algebra $\mathfrak{gl}(n, \mathbb{C})$. Note that the complex dimension of $GL(n, \mathbb{C})$ is $n^2$, while its real dimension is $2n^2$.

**Exercise 5.5**
Show that the ***complex special linear group***

$$SL(n, \mathbb{C}) := \{U \in GL(n, \mathbb{C}) : \det U = 1\}$$

is a Lie group of complex dimension $n^2 - 1$. Then show that its Lie algebra $\mathfrak{sl}(n, \mathbb{C})$ is given by all $n \times n$ complex matrices with trace zero.

**Exercise 5.6**
The **unitary group** is

$$U(n) := \{U \in GL(n, \mathbb{C}) : U^*U = I\}$$

where $U^*$ denotes the (complex) conjugate transpose of $U$. Note that the condition $U^*U = I$ is equivalent to the orthogonality condition $\langle Uz_1, Uz_2 \rangle = \langle z_1, z_2 \rangle$ for all $z_1, z_2 \in \mathbb{C}^n$ where $\langle z_1, z_2 \rangle := z_1^i \bar{z}_2^i$ is the (Hermitian) inner product on the $\mathbb{C}^n$. Show that $U(n)$ is a Lie group. Show that its Lie algebra $\mathfrak{u}(n)$ is given by the set of skew-symmetric Hermitian matrices, that is

$$\mathfrak{u}(n) = \{A \in \mathfrak{gl}(n, \mathbb{C}) : A = -A^*\}.$$

**Exercise 5.7**
Show that the **special unitary group** $SU(n, \mathbb{C}) := U(n) \cap SL(n, \mathbb{C})$ is a Lie group and describe its Lie algebra $\mathfrak{su}(n, \mathbb{C})$.

**Exercise 5.8**
Given $X$ a square matrix, recall the matrix exponential:

$$\exp(X) = I + X + \frac{1}{2!}X^2 + \frac{1}{3!}X^3 + \cdots.$$

Suppose the $n \times n$ matrices $A$ and $M$ satisfy

$$AM + MA^T = 0.$$

Show that $\exp(At)M\exp(A^Tt) = M$ for all $t$. This direct calculation shows that for $A \in \mathfrak{so}(n)$ or $A \in \mathfrak{sp}(2n)$, we have $\exp(At) \in SO(n)$ or $\exp(At) \in Sp(2n)$, respectively.

## 5.2 Abstract Lie groups and Lie algebras

In the previous section we defined matrix Lie groups and matrix Lie algebras. In this section we generalize these ideas.

**Definition 5.15** *A **Lie group** is a smooth manifold that is also a group, with the property that the operations of group multiplication, $(g,h) \mapsto gh$ and inversion, $g \mapsto g^{-1}$, are smooth.*

Every matrix Lie group is a Lie group. Indeed, it is a manifold by definition, and the operations of matrix multiplication and matrix inversion are smooth because they are rational operations on the matrix entries.

Here is an example of a Lie group that is not a matrix group:

**Example 5.16** In $\mathbb{R}^n$ consider the equivalence relation $x \sim y$ iff $x - y \in 2\pi\mathbb{Z}^n$. The set of equivalence classes $[x] = \{y \in \mathbb{R}^n \mid y - x = 2\pi k$ for some $k = (k_1, k_2, \ldots, k_n) \in \mathbb{Z}^n\}$, is denoted $\mathbb{R}^n/2\pi\mathbb{Z}^n$ or $\mathbb{T}^n$, and is called the $n$-torus. The $n$-torus is a smooth manifold, as can be shown by a direct generalization of Example 2.100. It is clearly an Abelian group with respect to the sum operation $([x],[y]) \to [x] + [y] := [x+y]$, with inverse $-[x] = [-x]$. By using local coordinate charts, one can check that the addition and inverse operations are smooth. The 1-torus, with this group operation, is called the circle group because it is diffeomorphic to the unit circle $S^1$.

One of the fundamental properties of a Lie group is that given a local chart, one may construct an entire atlas. This is achieved by using the group multiplication. For any given $g \in G$, the ***left translation*** by $g$ is the map

$$L_g : G \to G, \quad h \to L_g(h) := gh.$$

The definition of a Lie group ensures that this map and its inverse are smooth. Left translation by $g$ shifts any given chart $U$ that covers the identity to a chart $L_g(U)$ that covers $g$. The union of such charts, for all $g$, covers $G$, i.e., $\bigcup_{g\in G} L_g(U) = G$ and in fact they are compatible and hence form an atlas – see Exercise 5.11. Alternatively, one may use the ***right translation*** maps

$$R_g : G \to G, \quad h \to R_g(h) := hg.$$

We will now give an abstract definition of a Lie algebra and then show that the tangent space at the identity $T_eG$ of any Lie group $G$, with a particular bracket, is a Lie algebra.

**Definition 5.17** *A (real) **Lie algebra** is a (real) vector space $\mathcal{A}$ together with a bilinear operation $(v,w) \in \mathcal{A} \times \mathcal{A} \to [v,w] \in \mathcal{A}$, called the **bracket**, such that*

1. $[v, w] = -[w, v]$ *for all* $v, w \in V$ (**skew-symmetry**),
2. $[[v, w], u]+[[u, v], w]+[[w, u], v] = 0$ *for all* $v, w, u \in V$ (**Jacobi identity**).

**Example 5.18** The vector space $\mathbb{R}^3$ is a Lie algebra when endowed with a bracket given by the usual vector cross-product $[\mathbf{x}, \mathbf{y}] := \mathbf{x} \times \mathbf{y}$.

**Example 5.19** Every matrix Lie algebra, with bracket given by commutator $[A, B] := AB - BA$, is a Lie algebra, by Proposition 5.3.

**Example 5.20** The space of linear maps on a vector space $V$ is a Lie algebra, with bracket $[f, g] = f \circ g - g \circ f$.

**Example 5.21** The vector space $\mathfrak{X}(M)$ of all smooth vector fields on a smooth manifold $M$ is a Lie algebra, with the ***Jacobi-Lie bracket*** $[X, Y] := \mathcal{L}_X Y$. Recall from Chapter 3 that in local coordinates we have

$$\begin{aligned} [X, Y] &= (\mathrm{D}Y) \cdot X - (\mathrm{D}X) \cdot Y \\ &= X \cdot \nabla Y - Y \cdot \nabla X. \end{aligned}$$

**Definition 5.22** *A **Lie subalgebra** of a Lie algebra $\mathcal{A}$ is a subspace of $\mathcal{A}$ closed under the bracket. Clearly, any such subalgebra is itself a Lie algebra, with the same bracket as $\mathcal{A}$.*

Consider a Lie group $G$ and its tangent space at the identity $T_eG$. We are looking for a meaningful bracket $[\cdot, \cdot]_{T_eG}$ on $T_eG$. We have just learned from Example 5.21 that the vector space $\mathfrak{X}(G)$ of smooth vector fields on $G$ is a Lie algebra. Should we have an isomorphism between $T_eG$ and a subalgebra of $\mathfrak{X}(G)$, we could then define the bracket on $T_eG$ via this isomorphism. More precisely, given an isomorphism $\lambda : T_eG \to C \subseteq \mathfrak{X}(G)$, where $C$ is a subalgebra of $\mathfrak{X}(G)$, the bracket on $T_eG$ may be defined via:

$$[\xi, \eta]_{T_eG} := [\lambda(\xi), \lambda(\eta)]_{\mathfrak{X}(G)}.$$

The map $\lambda$ will be defined using the left translation maps $L_g$ introduced earlier. Recall that the tangent map $T_eL_g$ shifts vectors based at $e$ to vectors based at $g \in G$. By doing this operation for every $g \in G$, we define a vector field:

**Definition 5.23** *The **left extension** of any $\xi \in T_eG$ is the vector field $X_\xi^L$ given by*

$$X_\xi^L(g) := T_eL_g(\xi).$$

**Definition 5.24** *A vector field $X : G \to TG$, $h \to X(h)$, is called* ***left invariant*** *if*

$$(L_g)_*(X) = X \quad \text{for all} \ g \in G.$$

*The set of left invariant vector fields on G is denoted $\mathfrak{X}_L(G)$. By definition of the push-forward $(L_g)_*$, an equivalent definition of X being left invariant is*

$$T_h L_g(X(h)) = X(L_g(h)) = X(gh), \quad \text{for all } g, h \in G.$$

*Because of this second way of writing the definition, left invariant vector fields are also sometimes called* ***left equivariant*** *(see Definition 6.36).*

**Lemma 5.25** *A vector field on G is left invariant if and only if it equals $X_\xi^L$ for some $\xi \in T_e G$, i.e.*

$$\mathfrak{X}_L(G) = \left\{ X_\xi^L : \xi \in T_e G \right\}.$$

*The set $\mathfrak{X}_L(G)$ is a vector subspace of $\mathfrak{X}(G)$, and the following map is a vector space isomorphism,*

$$\begin{aligned} \lambda : T_e G &\longrightarrow \mathfrak{X}_L(G) \\ \xi &\longmapsto X_\xi^L. \end{aligned}$$

*Proof* We begin by showing that $X_\xi^L$ is left invariant, for every $\xi$:

$$T_h L_g(X_\xi^L(h)) = T_h L_g \circ T_e L_h(\xi) = T_e(L_g \circ L_h)(\xi) = T_e L_{gh}(\xi) = X_\xi^L(gh).$$

Next, for any left invariant vector field $X$ on $G$, if we define $\xi := X(e)$, then by left invariance

$$X(g) = T_e L_g(X(e)) = T_e L_g(\xi) = X_\xi^L(g), \quad \text{for any } g \in G,$$

so $X = X_\xi^L$. Thus, we have proven the first claim. The second claim is left as an exercise. ■

**Lemma 5.26** *The subspace $\mathfrak{X}_L(G)$ is a Lie subalgebra of $\mathfrak{X}(G)$.*

*Proof* Applying the left invariance property and Theorem 3.25, we see that, for any $\xi, \eta \in T_e G$,

$$[X_\xi^L, X_\eta^L] = [L_g^* X_\xi^L, L_g^* X_\eta^L] = L_g^*[X_\xi^L, X_\eta^L],$$

which shows that $[X_\xi^L, X_\eta^L]$ is left invariant. This shows that $\mathfrak{X}_L(G)$ is a Lie subalgebra of $\mathfrak{X}(G)$. ■

**Definition 5.27** *For any Lie group $G$, the* **Lie bracket** *on $T_eG$ is defined via the isomorphism $\lambda : T_eG \to \mathfrak{X}_L(G)$, $\xi \mapsto X_\xi^L$, as follows: for all $\xi, \eta \in T_eG$,*

$$[\xi, \eta] := [X_\xi^L, X_\eta^L](e).$$

*The Lie bracket is called the* **pull-back** *by $\lambda$ of the Jacobi–Lie bracket.*

Since $\mathfrak{X}_L(G)$, with the Jacobi-Lie bracket, is a Lie algebra, it follows directly that the space $T_eG$, together with the Lie bracket defined above, is a Lie algebra.

**Definition 5.28** *The tangent space at the identity $T_eG$ of a Lie group $G$, together with the Lie bracket defined above, is called the* **Lie algebra of** *$G$.*

It remains to show that this abstract definition is consistent with the corresponding definition for matrix Lie groups:

**Proposition 5.29** *The Lie algebra of a matrix Lie group is the space $T_IG$, together with the commutator operation.*

*Proof* This is left as Exercise 5.13. ■

The Lie algebra of a group is conventionally denoted by the same letter(s) as the group itself, but in lower-case gothic. For example, the Lie algebra of $G$ is denoted by $\mathfrak{g}$.

**Exercise 5.9** Show that $\mathbb{R}$, with addition, is a Lie group. Show that $\mathbb{R} \setminus \{0\}$, with multiplication, is also a Lie group. Describe the Lie algebra and Lie bracket for both Lie groups (see Example 5.38).

**Exercise 5.10 (Abelian Lie groups and Lie algebras)**
Let $G$ be an Abelian group and $\mathfrak{g}$ its Lie algebra. Show that $[\xi, \eta] = 0$ for all $\xi, \eta \in \mathfrak{g}$.

The converse is not true! For instance, take $G = \mathbb{R} \times S_3$ where $S_3$ is the permutation group of three elements. Show that $G$ is a *non-Abelian* Lie group when the multiplication operation is defined as $((x_1, \sigma_1), (x_2, \sigma_2)) \to (x_1 + x_2, \sigma_1 \circ \sigma_2)$. Then describe $\mathfrak{g}$, the Lie algebra of $G$, and show that $[\xi, \eta] = 0$ for all $\xi, \eta \in \mathfrak{g}$.

**Exercise 5.11 (A Lie group needs one chart to have an atlas)**
Let $G$ be a Lie group, and let $(U, \phi)$ be a chart around $e \in G$, i.e., such that $e \in U$. Given any element $g \in G$, the chart $(U_g, \phi_g)$ around $g \in G$ is defined by $U_g := L_g(U)$ and $\phi_g := \phi \circ L_{g^{-1}}$. Show that the set of charts $\{(U_g, \phi_g)\}_{g \in G}$ is an atlas for $G$.

**Exercise 5.12 (Lie algebra of right invariant vector fields)**
Define right extensions and $\mathfrak{X}_R(G)$, the set of right invariant vector fields, by analogy with Definitions 5.23 and 5.24. State and check the properties corresponding to Lemmas 5.25 and 5.26. Show that $[X_\xi^R, X_\eta^R] = -[X_\xi^L, X_\eta^L]$. Thus, though right invariant vector fields can be used to define a Lie bracket on $\mathfrak{g}$, the resulting bracket is *minus* the usual one. HINT: show that $X_\xi^R = -I_* X_\xi^L$, where $I_*$ is the push-forward of the inversion map $I(g) := g^{-1}$, and apply Theorem 3.25.

**Exercise 5.13 (Lie algebra of a matrix Lie group revisited)**
Describe the space $\mathfrak{X}_L(GL(n))$. Then using Definition 5.27, show that the Lie bracket on $\mathfrak{gl}(n)$ is indeed the commutator $[A, B] = AB - BA$. Generalize this argument to arbitrary matrix Lie groups.

**Exercise 5.14 (Structure constants)**
Let $\mathfrak{g}$ be a finite-dimensional Lie algebra and let $e_1, e_2, \ldots, e_n$ be a basis for $\mathfrak{g}$ as a vector space. Then for each $i, j$ there are unique **structure constants** $c_{ij}^k$, $k = 1, 2, \ldots, n$ such that

$$[e_i, e_j] = c_{ij}^k e_k.$$

The constants $c_{ij}^k$ depend on the chosen basis and determine the bracket on $\mathfrak{g}$. Show that the skew-symmetry of the bracket and the Jacobi identity translate into

$$c_{ij}^k + c_{ji}^k = 0$$

and

$$c_{ij}^k c_{km}^l + c_{jm}^k c_{ki}^l + c_{mi}^k c_{kj}^l = 0.$$

## 5.3 Isomorphisms of Lie groups and Lie algebras

**Definition 5.30** *Let $(G, \cdot)$ and $(H, \circ)$ be two Lie groups. A smooth map $\rho : G \to H$ is called a* ***Lie group homomorphism*** *if*

$$\rho(g \cdot h) = \rho(g) \circ \rho(h), \qquad \text{for all } g, h \in G.$$

*If in addition the map $\rho$ is bijective, then we call $\rho$ a* ***Lie group isomorphism*** *and we write $G \simeq H$.*

**Example 5.31** Consider the general linear group $GL(n)$ and the multiplication group of real numbers $\mathbb{R} \setminus \{0\}$. Then, the map $\det : GL(n) \to \mathbb{R} \setminus \{0\}$ is a Lie group homomorphism.

**Definition 5.32** *Let $\mathfrak{g}$ and $\mathfrak{h}$ be Lie algebras. A linear map $\rho : \mathfrak{g} \to \mathfrak{h}$ is called a* ***Lie algebra homomorphism*** *if*

$$\rho([\xi, \eta]) = [\rho(\xi), \rho(\eta)], \qquad \text{for all } \xi, \eta \in \mathfrak{g}.$$

*If in addition $\rho$ is bijective, then we call $\rho$ a* ***Lie algebra isomorphism*** *and we write $\mathfrak{g} \simeq \mathfrak{h}$.*

**Example 5.33** Let $\mathfrak{g}$ be the Lie algebra of a Lie group $G$. The map $\mathfrak{g} \to \mathfrak{X}_L(G)$, $\xi \mapsto X_\xi^L$, is a Lie algebra isomorphism, by Lemma 5.25 and Definition 5.27.

**Example 5.34 (Vector representation of $\mathfrak{so}(2)$)** The Lie algebra $\mathfrak{so}(2)$ is isomorphic with the real line $\mathbb{R}$, via

$$\xi \to \begin{bmatrix} 0 & -\xi \\ \xi & 0 \end{bmatrix}.$$

This is clearly a linear map. The commutator of any two elements of $\mathfrak{so}(2)$ is zero, so the Lie bracket is trivial, and the usual Lie bracket on $\mathbb{R}$ is trivial (see Exercise 5.9), so this map is (trivially) a Lie algebra isomorphism.

**Example 5.35 (Vector representation of $\mathfrak{so}(3)$ (the 'hat' map))** The Lie algebra $\mathfrak{so}(3)$ can be identified with the vector space $\mathbb{R}^3$ via the so-called 'hat' map, introduced in Section 1.5 and defined by

$$(\widehat{\ }) : \mathbb{R}^3 \to \mathfrak{so}(3), \quad \mathbf{x} = (x_1, x_2, x_3) \to \widehat{\mathbf{x}} = \begin{bmatrix} 0 & -x_3 & x_2 \\ x_3 & 0 & -x_1 \\ -x_2 & x_1 & 0 \end{bmatrix}.$$

It is easily verified from this definition that, for any $\mathbf{x}, \mathbf{y} \in \mathbb{R}^3$,

$$\widehat{\mathbf{x}}\mathbf{y} = \mathbf{x} \times \mathbf{y}.$$

An equivalent way of writing the definition of the hat map is

$$\widehat{\mathbf{x}}_{ij} = -\epsilon_{ijk}\, x^k, \quad \text{where} \quad (x^1, x^2, x^3) = (x, y, z).$$

Here, $\epsilon_{123} = 1$ and $\epsilon_{321} = -1$, with cyclic permutations, i.e. $\epsilon_{ijk} = \epsilon_{jki} = \epsilon_{kij}$. Note that the tensor $\epsilon_{ijk}$ is also commonly used to define the cross-product of vectors in $\mathbb{R}^3$, as follows:

$$(\mathbf{x} \times \mathbf{y})^i = \epsilon_{ijk}\, x^j\, y^k,$$

which provides a second way to verify the identity $\widehat{\mathbf{x}}\mathbf{y} = \mathbf{x} \times \mathbf{y}$.

A straightforward calculation, left to Exercise 5.19, shows that

$$[\widehat{\mathbf{x}}, \widehat{\mathbf{y}}] = \widehat{\mathbf{x} \times \mathbf{y}}.$$

It follows that the hat map is a Lie algebra homomorphism from $\mathfrak{so}(3)$ to $\mathbb{R}^3$, where the Lie bracket on $\mathbb{R}^3$ is the cross-product.

**Example 5.36 (Vector representation of $\mathfrak{so}(3)^*$ (the 'breve' map))** The dual space[2] $\mathfrak{so}(3)^*$ may be identified with $\mathbb{R}^3$ via the map

$$(\breve{\ }) : \mathbb{R}^3 \to \mathfrak{so}(3)^*, \qquad \mathbf{\Pi} = (\Pi^1, \Pi^2, \Pi^3) \mapsto \breve{\mathbf{\Pi}},$$

where

$$\left\langle \breve{\mathbf{\Pi}}, \widehat{\mathbf{x}} \right\rangle := \mathbf{\Pi} \cdot \mathbf{x} = \Pi^1 x^1 + \Pi^2 x^2 + \Pi^3 x^3,$$

for all $\mathbf{x} \in \mathbb{R}^3$. It can be checked (see Exercise 5.15) that

$$\left\langle \breve{\mathbf{\Pi}}, \widehat{\mathbf{x}} \right\rangle = \frac{1}{2}\operatorname{tr}(\widehat{\mathbf{\Pi}}\, \widehat{\mathbf{x}}^T).$$

Thus, $\breve{\mathbf{\Pi}}$ corresponds to $\frac{1}{2}\widehat{\mathbf{\Pi}}$ under the isomorphism $\mathfrak{so}(3) \to \mathfrak{so}(3)^*$ given by the trace pairing. If $\breve{\mathbf{\Pi}}$ is identified with the matrix $\frac{1}{2}\widehat{\mathbf{\Pi}}$, then

$$\breve{\mathbf{\Pi}}_{ij} = \frac{1}{2}\widehat{\mathbf{\Pi}}_{ij} = -\frac{1}{2}\epsilon_{ijk}\Pi^k$$

[2] Recall that the dual of a vector space $V$ is the set $V^*$ of all linear maps from $V$ to $\mathbb{R}$, and that $V^*$ is itself a vector space. As vector spaces, Lie algebras have duals. As usual, these are denoted with the asterisk notation.

and the corresponding inverse formula is

$$\Pi^1 = \breve{\Pi}_{32} - \breve{\Pi}_{23}$$

and cyclic permutations.

**Exercise 5.15**
Show that

$$\mathbf{u} \cdot \mathbf{v} = -\frac{1}{2}\operatorname{tr}(\widehat{\mathbf{u}}\widehat{\mathbf{v}}) = \frac{1}{2}\operatorname{tr}(\widehat{\mathbf{u}}\widehat{\mathbf{v}}^T).$$

Also verify that $\mathbf{u} \cdot \mathbf{v} = \operatorname{tr}(\mathbf{u}\mathbf{v}^T)$.

**Exercise 5.16**
For every $\xi \in \mathfrak{so}(3)$, and every $t \in \mathbb{R}$, show that $\exp(t\xi)$ is a rotation around the axis parallel to $\xi$, with the direction of the rotation given by the right-hand rule. What is the angle of rotation?

**Exercise 5.17**
If we denote by $(\mathbf{e}_1, \mathbf{e}_2, \mathbf{e}_3)$ the standard basis of $\mathbb{R}^3$, the corresponding basis $(\hat{\mathbf{e}}_1, \hat{\mathbf{e}}_2, \hat{\mathbf{e}}_3)$ for $\mathfrak{so}(3)$ reads

$$\hat{\mathbf{e}}_1 = \begin{bmatrix} 0 & 0 & 0 \\ 0 & 0 & -1 \\ 0 & 1 & 0 \end{bmatrix}, \quad \hat{\mathbf{e}}_2 = \begin{bmatrix} 0 & 0 & 1 \\ 0 & 0 & 0, \\ -1 & 0 & 0 \end{bmatrix}, \quad \hat{\mathbf{e}}_3 = \begin{bmatrix} 0 & -1 & 0 \\ 1 & 0 & 0 \\ 0 & 0 & 0 \end{bmatrix}.$$

Show that $[\hat{\mathbf{e}}_1, \hat{\mathbf{e}}_2] = \hat{\mathbf{e}}_3$ and cyclic permutations, while all other matrix commutators among the basis elements vanish. Note that this can either be checked directly or deduced from the correspondence of the bracket on $\mathfrak{so}(3)$ with the cross product.

**Exercise 5.18**
Let $S^1$ be the circle group, defined as in Example 5.16. Show that the following map $R : S^1 \to SO(2)$ is a Lie group isomorphism,

$$[\theta] \to R_\theta := \begin{bmatrix} \cos\theta & -\sin\theta \\ \sin\theta & \cos\theta \end{bmatrix}.$$

**Exercise 5.19**
Show that $[\widehat{\mathbf{x}}, \widehat{\mathbf{y}}] = \widehat{\mathbf{x} \times \mathbf{y}}$.

**Exercise 5.20 (The Lie algebra $\mathfrak{su}(2)$)**
Consider the Lie algebra $\mathfrak{su}(2)$ of $SU(2)$, that is, the set of $2 \times 2$ skew Hermitian matrices of trace zero (see Exercise 5.7). Show that this Lie algebra is isomorphic to $(\mathbb{R}^3, \times)$, and therefore to $\mathfrak{so}(3)$, by the isomorphism given by the ***tilde map***$(\tilde{\ })$ : $\mathbb{R}^3 \to \mathfrak{su}(2)$

$$\mathbf{x} = (x^1, x^2, x^3) \in \mathbb{R}^3 \to \widetilde{\mathbf{x}}$$
$$:= \frac{1}{2}\begin{bmatrix} -ix^3 & -ix^1 - x^2 \\ -ix^1 + x^2 & ix^3 \end{bmatrix} \in \mathfrak{su}(2). \qquad (5.1)$$

That is, show that $[\widetilde{\mathbf{x}}, \widetilde{\mathbf{y}}] = \widetilde{\mathbf{x} \times \mathbf{y}}$. Also show that $\det(2\widetilde{\mathbf{x}}) = \|\mathbf{x}\|^2$ and $\operatorname{tr}(\widetilde{\mathbf{x}}\widetilde{\mathbf{y}}) = -\frac{1}{2}\mathbf{x} \cdot \mathbf{y}$. Note that $\mathfrak{su}(2)^*$ may be identified with $\mathbb{R}^3$ via the ***'check' map*** $\mu \in \mathfrak{su}(2)^* \to \check{\mu} \in \mathbb{R}^3$ defined by

$$\check{\mu} \cdot \mathbf{x} := -2\langle \mu, \widetilde{\mathbf{x}} \rangle$$

for any $\mathbf{x} \in \mathbb{R}^3$.

**Exercise 5.21**
Find an analogue of the 'hat map' for the six-dimensional Lie algebra $\mathfrak{so}(4)$.

**Exercise 5.22**
Find an analog of the "hat map" for the three dimensional Lie algebra $\mathfrak{sp}(2)$.

**Exercise 5.23**
Show that two isomorphic Lie groups have isomorphic Lie algebras. Note: the converse is not true! For example, $\mathfrak{su}(2) \simeq \mathfrak{so}(3)$ but $SU(2)$ is not isomorphic to $SO(3)$; see [MR02] for details.

## 5.4 The exponential map

Let $\xi \in \mathfrak{g}$ and consider the left extension $X_\xi$, i.e., the left-invariant vector field on $G$ defined by

$$X_\xi(g) := T_e L_g(\xi)$$

(for convenience we drop the superscript $L$ in the notation for the left extension). The ***one-parameter subgroup*** corresponding to $\xi$, denoted $\gamma_\xi : \mathbb{R} \to G$, is the unique solution curve of the initial-value problem

$$\frac{\mathrm{d}g}{\mathrm{d}t} = X_\xi(g),$$
$$g(0) = e.$$

(The fact that the image of this curve is a subgroup of $G$ is shown in part 2 of Proposition 5.40.)

**Definition 5.37** *The Lie **exponential map** is the map*

$$\exp : \mathfrak{g} \to G, \qquad \xi \to \exp(\xi) := \gamma_\xi(1).$$

**Example 5.38 (The exponential map for $GL(1)$)** Consider the set of non-zero real numbers, with the multiplication operation. This is a Lie group. In fact, it is the matrix Lie group $GL(1)$, if we identify the space of $1 \times 1$ matrices $M(1, \mathbb{R})$ with $\mathbb{R}$. The group identity is 1. The Lie algebra of $GL(1)$ is $\mathfrak{gl}(1) \simeq \mathbb{R}$. For any given $x \in GL(1)$, the left translation map $L_x : GL(1) \to GL(1)$ is given by $L_x(y) = xy$. The left extension of a vector $\xi \in \mathfrak{gl}(1)$ is the vector field $X_\xi$ given by $X_\xi(x) := T_1 L_x(\xi) = x\xi$. This implies that the Lie bracket is the trivial one, i.e., $[\xi, \eta] = 0$ for all $\xi, \eta \in \mathbb{R}$; which can be verified directly (check this!), and also follows from a general property of Abelian Lie groups – see Exercise 5.10. The integral curve $t \to \gamma_\xi(t)$ is the solution of corresponding ODE

$$\frac{\mathrm{d}x}{\mathrm{d}t} = \xi x,$$
$$x(0) = 1.$$

By direct integration, we have $\gamma_\xi(t) = e^{t\xi}$. The Lie exponential map here is the usual exponential of real numbers $\exp(\xi) = \gamma_\xi(1) = e^\xi$.

The next proposition generalizes the previous example.

**Proposition 5.39** *The Lie exponential map for matrix groups is the usual exponential of matrices,*

$$\exp(A) = \gamma_A(1) = e^A = I + \frac{A}{1!} + \frac{A^2}{2!} + \cdots .$$

*Proof* Let $A \in \mathfrak{gl}(n) = \mathcal{M}(n, \mathbb{R})$. The convergence of the series for $e^A$ is treated in standard ODE texts, such as [HS74], where it is also shown that $e^A$ is invertible, with inverse $e^{-A}$. It is straightforward to show that

$$\frac{\mathrm{d}}{\mathrm{d}t} e^{tA} = e^{tA} A. \tag{5.2}$$

(This also equals $Ae^{tA}$, which we don't need here.) The left extension of any $A \in \mathfrak{gl}(n)$ is given by $X_A(B) = T_I L_B(A) = BA$. Thus

$$\frac{\mathrm{d}}{\mathrm{d}t} e^{tA} = e^{tA} A = X_A(e^{tA}).$$

Since $\gamma_A(0) = I$ it follows that the curve in $GL(n)$ defined by $\gamma_A(t) := e^{At}$ is the one-parameter subgroup generated by $A$. It follows that the exponential map on $GL(n)$ is $\exp(A) = \gamma_A(1) = e^A$.

For a matrix Lie group $G \subset GL(n)$, one also needs to show that $\exp(A) \in G$ for every $A \in \mathfrak{g}$. But this is a consequence of the existence and uniqueness of solutions of the vector field $X_A$. Indeed, the vector field $X_A$ on $G$ is the restriction to $G$ of a vector field of the same name on $GL(n)$. The restricted $X_A$ has a unique solution on $G$ with $\gamma_A(0) = I$. This curve must also be the unique solution to the whole vector field $X_A$ on $GL(n)$ with $\gamma_A(0) = I$. But we have just shown that that solution is $\gamma_A(t) := e^{At}$, so this curve must remain in $G$. ■

**Proposition 5.40 (Lie exponential map and one-parameter subgroups)** *The following statements hold:*

1. $$\exp(t\xi) = \gamma_\xi(t) \qquad \textit{for all } t \in \mathbb{R} \tag{5.3}$$

   *and therefore the Lie exponential maps the line* $\{t\xi : t \in \mathbb{R}\} \subseteq \mathfrak{g}$ *onto the set* $\{\gamma_\xi(t) : t \in \mathbb{R}\} \subseteq G$.

2. $\{\gamma_\xi(t) : t \in \mathbb{R}\}$ *is a smooth one-parameter subgroup of* $G$, *that is*

   $$\gamma_\xi(s + t) = \gamma_\xi(s)\gamma_\xi(t) \quad \textit{for all } s, t \in \mathbb{R}.$$

3. *All smooth one-parameter subgroups of* $G$ *are of the form* $\{\exp(t\xi) : t \in \mathbb{R}\}$ *for some* $\xi \in \mathfrak{g}$.

4. *The Lie exponential is a local $C^\infty$ diffeomorphism from a neighbourhood of $0 \in \mathfrak{g}$ onto a neighbourhood of $e \in G$.*

*Proof*

1. By definition $\exp t\xi = \gamma_{t\xi}(1)$ and therefore we have to show that $\gamma_{t\xi}(1) = \gamma_\xi(t)$. We will prove that

$$\gamma_{t\xi}(\tau) = \gamma_\xi(\tau t) \quad \text{for all } \tau \in \mathbb{R},$$

from where choosing $\tau = 1$ the conclusion is immediate. To prove the above equality it is sufficient to show that $\gamma_{t\xi}(\tau)$ and $\gamma_\xi(t\tau)$ satisfy the same differential equation and have the same initial conditions.

For a fixed $t \in \mathbb{R}$, the integral curve $\tau \to \gamma_{t\xi}(\tau)$ satisfies $\gamma_{t\xi}(0) = e$ and

$$\frac{\mathrm{d}\gamma_{t\xi}(\tau)}{\mathrm{d}\tau} = X_{t\xi}(\gamma_{t\xi}(\tau)) = T_e L_{\gamma_{t\xi}(\tau)}(t\xi) = t T_e L_{\gamma_{t\xi}(\tau)}(\xi) = t X_\xi(\gamma_{t\xi}(\tau)),$$

where we used that the tangent map is linear. In other words, the curve $\tau \to \gamma_{t\xi}(\tau)$ is the solution of the initial value problem

$$\frac{\mathrm{d}g}{\mathrm{d}\tau} = t X_\xi(g), \quad g(0) = e. \tag{5.4}$$

Looking now at $\tau \to \gamma_\xi(\tau t)$, we have $\gamma_\xi(0) = e$ and

$$\frac{\mathrm{d}\gamma_\xi(\tau t)}{\mathrm{d}\tau} = t\frac{\mathrm{d}\gamma_\xi(\tau t)}{\mathrm{d}(\tau t)} = t X_\xi(\gamma_\xi(\tau t)),$$

that is, $\tau \to \gamma_\xi(\tau t)$ verifies the same initial value problem (5.4).

2. At $t = 0$ we have both sides equal to $\gamma_\xi(s)$. Also, as functions of $t$, both sides verify

$$\frac{\mathrm{d}g}{\mathrm{d}t} = X_\xi(g)$$

and therefore they are solutions of the same initial value problem.

3. From parts 1 and 2 it is clear that for each $\xi \in \mathfrak{g}$ the set $\{\exp(t\xi) \mid t \in \mathbb{R}\} = \{\gamma_\xi(t) \mid t \in \mathbb{R}\}$ is a one-parameter subgroup of $G$. It remains to show that to any one-parameter $\gamma(t)$ subgroup of $G$ there corresponds a $\xi \in \mathfrak{g}$ such

that $\gamma(t) = \gamma_\xi(t)$ for all $t$. For any such subgroup $\gamma$, let $\xi = \gamma'(0)$. Then,

$$\begin{aligned}\frac{\mathrm{d}\gamma}{\mathrm{d}t} &= \left.\frac{\mathrm{d}}{\mathrm{d}s}\right|_{s=0} \gamma(t+s) = \left.\frac{\mathrm{d}}{\mathrm{d}s}\right|_{s=0} \gamma(t)\gamma(s) \\ &= (T_e L_{\gamma(t)})(\gamma'(0)) \\ &= X_\xi(\gamma(t)).\end{aligned}$$

Also, it follows from the definition of a one-parameter subgroup that $\gamma(0) = e$. Thus $\gamma$ satisfies the initial value problem

$$\frac{\mathrm{d}g}{\mathrm{d}t} = X_\xi(g(t)), \qquad g(0) = e,$$

which is the one that defines $\gamma_\xi$. Therefore, $\gamma = \gamma_\xi$.

4. We refer to [War83] for the proof that exp is a $C^\infty$ map. Differentiating eqn (5.3) at $t = 0$ we have $T_{0_\mathfrak{g}} \exp = \mathrm{Id}_\mathfrak{g}$ (the identity map on $\mathfrak{g}$. By the Inverse Function Theorem, the map $\exp : \mathfrak{g} \to G$ is has a local $C^\infty$ inverse near $\exp(0_\mathfrak{g}) = e$. ■

**Corollary 5.41** *The exponential map induces a coordinate chart in a neighborhood of e. The coordinates associated to this chart are called* ***canonical coordinates*** *of the Lie group G.*

**Exercise 5.24**
Show that $\exp(0_\mathfrak{g}) = e$ and that if $[\xi, \eta] = 0$ then $\exp(\xi)\exp(\eta) = \exp(\xi + \eta)$.

## Solutions to selected exercises

**Solution to Exercise 5.6** Let $U(t)$ be a path through $U(n)$ such that $U(0) = I$ and $\dot{U}(0) = A \in \mathfrak{u}(n)$. The defining relationship for the Lie algebra is found by differentiating the relationship, $U^* U = I$, and evaluate at $t = 0$.

$$\left.\frac{\mathrm{d}}{\mathrm{d}t}\right|_{t=0} U^* U = \dot{U}^* U + U^* \dot{U}\Big|_{t=0} = A^* + A = 0.$$

Therefore, elements $A \in \mathfrak{u}(n)$ satisfy $A^* = -A$.

**Solution to Exercise 5.11** Since group multiplication in a Lie group is smooth, $U_g = L_g(U) \subset G$ is open for all $g$ and for all open sets $U \subset G$.

Since $e \in U$, $g \in U_g$ for all $g \in G$. Also $\phi_g : U_g \to \phi(U)$ is an isomorphism since $L_g$ is a diffeomorphism. Finally, the transition maps, $\phi_g \circ \phi_h^{-1} = \phi \circ L_{g^{-1}h} \circ \phi^{-1}$, are smooth since they are the composition of smooth maps.

**Solution to Exercise 5.13** The space of left invariant vector fields in $GL(n, \mathbb{R})$ are vector fields $X(G)$ such that $X(HG) = HX(G)$. These are characterised by the equality $X(G) = GA$ for $A = X(I) \in \mathfrak{gl}(n, \mathbb{R})$. The Lie derivative of a vector field is given by

$$\mathcal{L}_X Y = \mathrm{D}Y \cdot X - \mathrm{D}X \cdot Y,$$

where, if $X(G) = GA$ and $Y(G) = GB$,

$$\mathrm{D}Y \cdot X(G) = \left.\frac{\mathrm{d}}{\mathrm{d}t}\right|_{t=0} Y(G + tGA) = \left.\frac{\mathrm{d}}{\mathrm{d}t}\right|_{t=0} (G + tGA)B = GAB.$$

Therefore,

$$\mathcal{L}_X Y(G) = GAB - GBA = G[A, B].$$

Therefore, the Jacobi–Lie bracket is the left invariant vector field associated with the matrix commutator.

**Solution to Exercise 5.20** This calculation is long but is best performed directly. It is possible to simplify the calculation immeasurably by the use of quaternions.

$$\begin{aligned}
&\left[\frac{1}{2}\begin{bmatrix} -ix^3 & -ix^1 - x^2 \\ -ix^1 + x^2 & ix^3 \end{bmatrix}, \frac{1}{2}\begin{bmatrix} -iy^3 & -iy^1 - y^2 \\ -iy^1 + y^2 & iy^3 \end{bmatrix}\right] \\
&= \frac{1}{2}\begin{bmatrix} -i(x^1y^2 - x^2y^1) & -i(x^2y^3 - x^3y^2) - (x^3y^1 - y^3x^1) \\ -i(x^2y^3 - x^3y^2) + (x^3y^1 - y^3x^1) & i(x^1y^2 - x^2y^1) \end{bmatrix} \\
&= (\mathbf{x} \times \mathbf{y})\tilde{}.
\end{aligned}$$

The verifications of $\det(2\tilde{\boldsymbol{x}}) = \|\boldsymbol{x}\|^2$ and $\mathrm{tr}(\tilde{\boldsymbol{x}}\tilde{\boldsymbol{y}}) = -\frac{1}{2}\boldsymbol{x} \cdot \boldsymbol{y}$ are similar.

**Solution to Exercise 5.21** Consider the basis

$$\widehat{J}_1 = \begin{bmatrix} 0 & 0 & 0 & 0 \\ 0 & 0 & -1 & 0 \\ 0 & 1 & 0 & 0 \\ 0 & 0 & 0 & 0 \end{bmatrix}, \quad \widehat{J}_2 = \begin{bmatrix} 0 & 0 & -1 & 0 \\ 0 & 0 & 0 & 0 \\ 1 & 0 & 0 & 0 \\ 0 & 0 & 0 & 0 \end{bmatrix},$$

$$\widehat{J}_3 = \begin{bmatrix} 0 & -1 & 0 & 0 \\ 1 & 0 & 0 & 0 \\ 0 & 0 & 0 & 0 \\ 0 & 0 & 0 & 0 \end{bmatrix}, \quad \widehat{K}_1 = \begin{bmatrix} 0 & 0 & 0 & -1 \\ 0 & 0 & 0 & 0 \\ 0 & 0 & 0 & 0 \\ 1 & 0 & 0 & 0 \end{bmatrix},$$

$$\widehat{K}_2 = \begin{bmatrix} 0 & 0 & 0 & 0 \\ 0 & 0 & 0 & -1 \\ 0 & 0 & 0 & 0 \\ 0 & 1 & 0 & 0 \end{bmatrix}, \quad \widehat{K}_3 = \begin{bmatrix} 0 & 0 & 0 & 0 \\ 0 & 0 & 0 & 0 \\ 0 & 0 & 0 & -1 \\ 0 & 0 & 1 & 0 \end{bmatrix}.$$

The hat map $\widehat{\ } : \mathbb{R}^6 \to \mathfrak{so}(4)$ is given by $(\boldsymbol{a}, \boldsymbol{b}) \to \boldsymbol{a} \cdot \widehat{\boldsymbol{J}} + \boldsymbol{b} \cdot \widehat{\boldsymbol{K}}$. For more on this map see [Hol08].

**Solution to Exercise 5.22** Consider the basis

$$I = \begin{bmatrix} 0 & 1 \\ 0 & 0 \end{bmatrix}, \quad J = \begin{bmatrix} 0 & 0 \\ 1 & 0 \end{bmatrix}, \quad K = \begin{bmatrix} 1 & 0 \\ 0 & -1 \end{bmatrix}.$$

In this basis the bracket relations are given by

$$[I, J] = K, \quad [J, K] = 2J, \quad [K, I] = 2I.$$

Therefore, setting $\widehat{\boldsymbol{a}} = \boldsymbol{a} \cdot (I, J, K) = a_1 I + a_2 J + a_3 K$ gives

$$\left[\widehat{\boldsymbol{a}}, \widehat{\boldsymbol{b}}\right] = \left( \begin{bmatrix} 0 & 2 & 0 \\ 2 & 0 & 0 \\ 0 & 0 & 1 \end{bmatrix} (\boldsymbol{a} \times \boldsymbol{b}) \right) \cdot (I, J, K).$$

# 6 Group actions, symmetries and reduction

An *action* of a group on a set is a map that associates to each element of the group an invertible transformation of the given set, in such a way that the group operation corresponds to composition of transformations. Thus, the group may be thought of as a group of transformations. For instance, every matrix group is naturally thought of as a group of invertible linear transformations of a Euclidean space. A smooth action of a Lie group on a manifold associates to each group element a diffeomorphism from the manifold to itself, so the Lie group may be identified with a group of diffeomorphisms.

If the diffeomorphisms corresponding to the group elements all leave a certain function invariant, then the group (with the specified action) is a *symmetry group* of that function. For example, $SO(3)$, with its standard action on $\mathbb{R}^3$, is a symmetry group of any rotationally invariant function on $\mathbb{R}^3$.

In mechanics, symmetry can be used to *reduce* the dynamics, that is, to transform the equations of motion into a set of equations easier to deal with. A first illustration of this is 'naive reduction' using cyclic variables, introduced in Section 4.4. There are several general kinds of reduction, all based on Lie group actions, and all with the property that the reduced system inherits the mechanical structure (Lagrangian or Hamiltonian) of the original system. In the last section of this chapter, we introduce Poisson reduction, which applies to all symmetrical Hamiltonian systems.

## 6.1 Lie group actions

**Definition 6.1** *A (smooth)* ***left action*** *of a Lie group $G$ on manifold $M$ is a smooth mapping $\Phi : G \times M \to M$ such that*

*(i) $\Phi(e, x) = x$ for all $x \in M$,*

*(ii) $\Phi(g, \Phi(h, x)) = \Phi(gh, x)$ for all $g, h \in G$ and $x \in M$, and*

*(iii) For every $g \in G$, the map $\Phi_g : M \to M$, defined by*

$$\Phi_g(x) := \Phi(g, x),$$

*is a diffeomorphism.*

Note that, the condition (ii) may also be written as $\Phi_g \circ \Phi_h = \Phi_{gh}$. We often use the notation

$$gx := \Phi_g(x) = \Phi(g, x)$$

and say that $g$ acts on $x$. With this notation, condition (ii) above then simply reads

$$g(hx) = (gh)x.$$

All actions of Lie groups will be assumed to be smooth.

**Example 6.2** The ***standard action*** of a matrix Lie group $G \subseteq GL(n, \mathbb{R})$ on $\mathbb{R}^n$ is given by $\Phi(A, \mathbf{v}) = A\mathbf{v}$ (matrix multiplication). It is easily verified that this is a left action. In particular, note that condition (ii) in the definition follows from associativity of matrix multiplication:

$$\Phi(A, \Phi(B, \mathbf{v})) = A(B\mathbf{v}) = (AB)\mathbf{v} = \Phi(AB, \mathbf{v}).$$

**Definition 6.3** *A **right action** of a Lie group $G$ on manifold $M$ is a smooth mapping $\Phi : G \times M \to M$ satisfying the same conditions as for a left action given in Definition 6.1, except that condition (ii) is replaced by:*

*(ii)′* $\Phi(g, \Phi(h, x)) = \Phi(hg, x)$ *for all* $g, h \in G$ *and* $x \in M$.

The convenient notation for a right action is $xg := \Phi_g(x) = \Phi(g, x)$, because with this notation, condition (ii)′ above then simply reads

$$(xh)g = x(hg).$$

A left (or right) action by a group $G$ is called a left (or right) $G$ action. Note that if $G$ is abelian, then every left $G$ action is a right $G$ action and vice versa.

**Example 6.4 (Group actions generalise flows)** The flow of any complete vector field is both a left and a right $\mathbb{R}$ action, as can be seen by comparing the previous definitions with Theorem 3.9. (Recall that $\mathbb{R}$ is an abelian Lie group, with the group operation being addition.) The "flow property" of the flow of a vector field is a special case of both conditions (ii) and (ii)′ above.

**Remark 6.5** *Any left action* $(g, x) \to gx$ *gives rise to a right action by* $(g, x) \to g^{-1}x$. *See Exercise 6.1.*

The next definitions will use the notation of left actions, but they apply equally well to right actions.

**Definition 6.6 (Isotropy)** *Let $G$ be a group acting on $M$. The **isotropy subgroup** of any $x \in M$ is*

$$G_x := \{g \in G : gx = x\}.$$

*If $G_x$ is nontrivial, i.e., contains elements other than the group identity, then $x$ is called an **isotropic** (or **symmetric**) point.*

**Example 6.7** For the standard action of any nontrivial matrix Lie group, the origin is always an isotropic point.

**Definition 6.8 (Orbits)** *Let $G$ act on $M$. For a given point $x \in M$, the subset*

$$\mathrm{Orb}\,(x) := \{gx : g \in G\} \subseteq M,$$

*is called the **group orbit** through $x$.*

**Example 6.9** For the flow of a complete field, considered as an $\mathbb{R}$ action, this definition of an orbit agrees with the familiar notion of orbit in dynamical systems. Also, a point is isotropic if and only if it is either fixed or periodic.

**Definition 6.10 (Properties of group actions)** *The action $\Phi : G \times M \to M$ of a group $G$ on a manifold $M$ is said to be:*

1. ***transitive** if for every $x, y \in M$, there exists a $g \in G$ such that $gx = y$;*
2. ***free** if it has no isotropic points, i.e. for all $x$, if $gx = x$ then $g = e$;*
3. ***faithful** (or **effective**) if for all $g \in G$ such that $g \neq e$, there exists $x \in M$ such that $gx \neq x$; and*
4. ***proper** if, whenever the sequences $\{x_n\}$ and $\{g_n x_n\}$ converge in $M$, the sequence $\{g_n\}$ has a convergent subsequence in $G$.*

**Remark 6.11** *A transitive action has only one group orbit.*

**Remark 6.12** *An equivalent definition of a faithful action is: if $\Phi_g$ is the identity transformation on $M$, then $g = e$.*

**Proposition 6.13** *Every action of a compact group is proper.*

*Proof* In a compact group, every sequence $\{g_n\}$ has a convergent subsequence. ■

**Example 6.14 ($SO(2)$ action on $\mathbb{R}^2$)** Consider the standard action of $SO(2)$ on $\mathbb{R}^2$,

$$\left(\begin{bmatrix}\cos\theta & -\sin\theta\\ \sin\theta & \cos\theta\end{bmatrix}, \begin{bmatrix}x\\ y\end{bmatrix}\right) \to \begin{bmatrix}\cos\theta & -\sin\theta\\ \sin\theta & \cos\theta\end{bmatrix}\cdot\begin{bmatrix}x\\ y\end{bmatrix} = \begin{bmatrix}x\cos\theta - y\sin\theta\\ x\sin\theta + y\cos\theta\end{bmatrix}.$$

The group orbits are circles around the origin, except for the one-point orbit $\{(0,0)\}$. The isotropy group of the origin is $SO(2)$, and the isotropy group of every other point $(x,y) \in \mathbb{R}^2$ is trivial, i.e, $G_{(x,y)} = \{\mathbf{I}\}$, the set containing only the group identity. This action is non-transitive, non-free (since $(0,0)$ is isotropic), faithful and proper (since $SO(2)$ is compact).

The following property of proper actions is proven in [AM78].

**Proposition 6.15** *Orbits of proper group actions are embedded submanifolds.*

**Example 6.16 (A non-proper group action)** Consider the left action of $\mathbb{R}$ on $\mathbb{T}^2$ given by

$$\big(\theta, (\phi_1, \phi_2)\big) \to (\phi_1 + \theta, \phi_2 + \sqrt{2}\,\theta).$$

The orbit of any point is dense in $\mathbb{T}^2$ (see Exercise 6.4). This action is non-transitive, free, faithful and non-proper. If the irrational $\sqrt{2}$ is replaced by a rational number, then the action becomes proper and the orbit of any point is closed.

**Definition 6.17** *A $G$ action on a vector space $V$ is called **linear** if $\Phi_g$ is a linear map, for every $g \in G$.*

Clearly, the standard action of any matrix Lie group is linear.

**Remark 6.18** *A linear action of a group $G$ on $\mathbb{R}^n$ is called a **representation** of $G$. Note that if $\Phi$ is such an action, then each $\Phi_g$ corresponds to an element of $GL(n)$.*

**Definition 6.19** *Let $G$ act on $Q$, and let $\omega$ be a symplectic form on $Q$. The action is **symplectic** with respect to $\omega$ if, for every $g \in G$, the map $\Phi_g$ is symplectic, that is (see Definition 3.109), $\Phi_g^*\omega = \omega$. Similarly, if $Q$ has a Poisson bracket $\{\cdot,\cdot\}$, then the $G$ action is **Poisson** with respect to* *that bracket if, for every $g \in G$, the map $\Phi_g$ is a Poisson map, that is (see Definition 4.32), for all scalar fields $F, G$ on $Q$,*

$$\{F \circ \Phi_g, G \circ \Phi_g\} = \{F, G\} \circ \Phi_g.$$

*Both symplectic and Poisson actions are also called* ***canonical*** *actions.*

**Example 6.20** The standard action of $SO(2)$ on $\mathbb{R}^2$ (see Example 6.14) is symplectic with respect to the canonical symplectic form on $\mathbb{R}^2$, which is $\omega = \mathrm{d}x \wedge \mathrm{d}y$. Indeed, for any $\mathbf{R} \in SO(2)$, the map $\Phi_\mathbf{R}$ is symplectic if and only if $\Phi_\mathbf{R}^* \omega = \omega$, that is,

$$\omega(\Phi_\mathbf{R}(\mathbf{x}))\,(\mathrm{D}\Phi_\mathbf{R}(\mathbf{x})\mathbf{v}, \mathrm{D}\Phi_\mathbf{R}(\mathbf{x})\mathbf{w}) = \omega(\mathbf{v}, \mathbf{w}),$$

for all $\mathbf{x}, \mathbf{v}, \mathbf{w} \in \mathbb{R}^2$. The map $\Phi_\mathbf{R}$ is $\mathbf{R}$ itself, viewed as a linear transformation. Since it is linear its derivative, at any point, is itself: $\mathrm{D}\Phi_\mathbf{R}(\mathbf{x})\mathbf{v} = \Phi_\mathbf{R}(\mathbf{v}) = \mathbf{R}\mathbf{v}$, for all $\mathbf{x}, \mathbf{v}$. Also, recall from Example 3.108 that the matrix of $\omega(\mathbf{x})$, for any $\mathbf{x} \in \mathbb{R}^2$, is

$$\mathbf{J} = \begin{bmatrix} 0 & 1 \\ -1 & 0 \end{bmatrix}.$$

Thus, $\Phi_\mathbf{R}$ is symplectic if and only if

$$(\mathbf{R}\mathbf{v})^T\,\mathbf{J}\,(\mathbf{R}\mathbf{w}) = \mathbf{v}^T\,\mathbf{J}\,\mathbf{w},$$

for all $\mathbf{v}, \mathbf{w} \in \mathbb{R}^2$, which is equivalent to

$$\mathbf{R}^T\,\mathbf{J}\,\mathbf{R} = \mathbf{J}.$$

This can be verified by direct computation, for any $\mathbf{R} \in SO(2)$, which shows that the action is symplectic. Note that the previous equation is the condition for $\mathbf{R}$ to be an element of the symplectic group $Sp(2, \mathbb{R})$, as defined in Example 5.13. So we have also shown that $SO(2) \subset Sp(2, \mathbb{R})$.

The next proposition generalizes the previous example.

**Proposition 6.21** *The standard action of a matrix Lie group $G$ on $\mathbb{R}^{2n}$ is symplectic with respect to the canonical symplectic form on $\mathbb{R}^n$ if and only if $G \subseteq Sp(2n, \mathbb{R})$.*

*Proof* Example 3.110 showed that an action $\Phi$ of $G$ on $\mathbb{R}^{2n}$ is symplectic if and only if $\mathrm{D}\Phi_A(z) \in Sp(2n)$ for all $A \in G$. For the standard action of $G \subseteq GL(n)$, the map $\Phi_A$ is the linear transformation $A$, and the derivative of $A$, at any point, is $A$ itself. Thus, the standard action of $G$ is symplectic if and only if $A \in Sp(2n, \mathbb{R})$, for all $A \in G$. ■

**Definition 6.22 (Infinitesimal generator)** *Consider the left action of a Lie group $G$ on the manifold $M$, $(g, x) \to gx$. Let $\xi \in \mathfrak{g}$ be a vector in the Lie algebra of $G$ and consider the one-parameter subgroup $\{\exp(t\xi) : t \in \mathbb{R}\} \subseteq G$. The orbit of an element $x$ with respect to this subgroup is a smooth path $t \to (\exp(t\xi))\, x$ in $M$. The* ***infinitesimal generator*** *associated to $\xi$ at $x \in M$, denoted $\xi_M(x)$, is the tangent (or velocity) vector to this curve at $x$, that is:*

$$\xi_M(x) := \left.\frac{\mathrm{d}}{\mathrm{d}t}\right|_{t=0} (\exp(t\xi)\, x) \in T_x M. \tag{6.1}$$

*The smooth vector field $\xi_M : M \to TM$, $x \to \xi_M(x)$, is called the* ***infinitesimal generator vector field associated to*** *$\xi$.*

**Remark 6.23** *The infinitesimal generator map $\mathfrak{g} \times M \to TM$, $(\xi, x) \to \xi_M(x)$ can be thought of as the 'infinitesimal action' of $\mathfrak{g}$ on $M$. It is the expression at the tangent level of the action of $G$ on $M$.*

**Remark 6.24** *For the standard action of a matrix Lie group on $\mathbb{R}^n$, the expression $\exp(t\boldsymbol{\xi})\mathbf{x}$ is the matrix product of $\exp(t\boldsymbol{\xi})$ and $\mathbf{x}$. Hence,*

$$\boldsymbol{\xi}_M(\mathbf{x}) = \left.\frac{\mathrm{d}}{\mathrm{d}t}\right|_{t=0} (\exp(t\boldsymbol{\xi})\mathbf{x}) = \left(\left.\frac{\mathrm{d}}{\mathrm{d}t}\right|_{t=0} \exp(t\boldsymbol{\xi})\right)\mathbf{x} = \boldsymbol{\xi}\mathbf{x} \quad \textit{(matrix product).}$$

**Example 6.25 (The infinitesimal generator for the $SO(2)$ action on $\mathbb{R}^2$)** Let

$$\boldsymbol{\xi} := \begin{bmatrix} 0 & -\xi \\ \xi & 0 \end{bmatrix} \in \mathfrak{so}(2) \quad \text{and} \quad \mathbf{x} := \begin{bmatrix} x_1 \\ x_2 \end{bmatrix} \in \mathbb{R}^2.$$

Then

$$\boldsymbol{\xi}_{\mathbb{R}^2}(\mathbf{x}) = \boldsymbol{\xi} \begin{bmatrix} x_1 \\ x_2 \end{bmatrix} = \begin{bmatrix} 0 & -\xi \\ \xi & 0 \end{bmatrix} \begin{bmatrix} x_1 \\ x_2 \end{bmatrix} = \begin{bmatrix} -\xi x_2 \\ \xi x_1 \end{bmatrix}.$$

**Proposition 6.26** *The map $\mathfrak{g} \to \mathfrak{X}(M)$, $\xi \to \xi_M$, is linear and satisfies*

$$[\xi, \eta] = -[\xi_M, \eta_M],$$

*for all $\xi, \eta \in \mathfrak{g}$ (such a map is called a Lie algebra anti-homomorphism).*

*Proof* See for example [MR02]. ∎

**Corollary 6.27** *Given a $G$ action on $M$, the set of infinitesimal generator vector fields forms a subalgebra of $\mathfrak{X}(M)$.*

Infinitesimal generator vector fields are not necessarily left invariant, as shown by Exercise 6.6.

Any action of $G$ on a manifold $Q$ induces corresponding 'lifted' actions on $TQ$ and $T^*Q$:

**Definition 6.28** *Let $\Phi : G \times Q \to Q$ be a (left or right) action, so $\Phi_g : Q \to Q$ for every $g \in G$. The **tangent lift** of $\Phi$ is the action*

$$\begin{aligned} G \times TQ &\to TQ \\ (g,(q,v)) &\to T\Phi_g(q,v) = \left(\Phi_g(q), T_q\Phi_g(v)\right). \end{aligned}$$

*The **cotangent lift** of $\Phi$ is the action*

$$\begin{aligned} G \times T^*Q &\to T^*Q \\ (g,(q,\alpha)) &\to T^*\Phi_{g^{-1}}(q,\alpha) = \left(\Phi_g(q), T^*_{\Phi_g(q)}\Phi_{g^{-1}}(\alpha)\right). \end{aligned}$$

Thus, for the tangent lift, the maps $\Phi_g : Q \to Q$, for every $g$, are each tangent lifted to $T\Phi_g : TQ \to TQ$, as defined in Section 2.2. If $\Phi$ is a left action, then its tangent lift is a left action, because

$$T\Phi_g\left(T\Phi_h(q,v)\right) = T\left(\Phi_g \circ \Phi_h\right)(q,v) = T\Phi_{gh}(q,v)$$

(see Exercise 6.5). Similarly, if $\Phi$ is a right action, then its tangent lift is a right action.

For the cotangent lift, the action of each $g \in G$ on $T^*Q$ is the cotangent lift of $\Phi_g : Q \to Q$, that is, $T^*\Phi_{g^{-1}}$. Recall from Section 2.3,

$$\left\langle T^*_{\Phi_g(q)}\Phi_{g^{-1}}(\alpha)), w\right\rangle := \left\langle \alpha, T_{\Phi_g(q)}\Phi_{g^{-1}}w\right\rangle, \quad \text{for all} \quad w \in T_{\Phi_g(q)}Q.$$

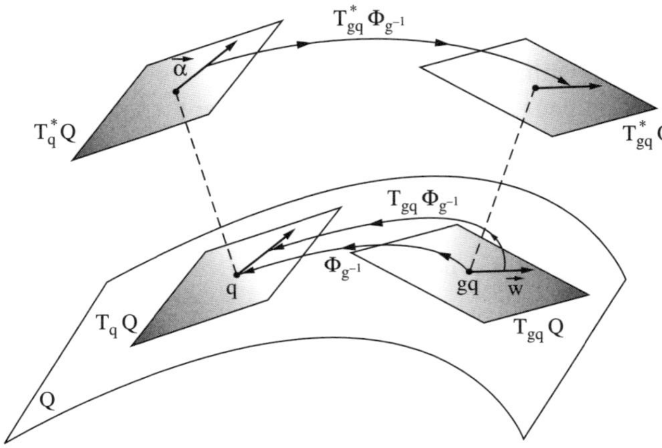


**Figure 6.1** The cotangent lift of an action $\Phi$ on $Q$ takes any cotangent vector $\alpha \in T^*_qQ$ to the cotangent vector $T^*_{gq}\Phi_{g^{-1}}(\alpha) \in T^*_{gq}Q$.

It is straightforward to verify that

$$T^*\Phi_{g^{-1}}\left(T^*\Phi_{h^{-1}}(q,\alpha)\right) = T^*\left(\Phi_{h^{-1}} \circ \Phi_{g^{-1}}\right)(q,v) = T^*\Phi_{(gh)^{-1}}(q,v).$$

This implies that the cotangent lift of a left (resp. right) action is a left (resp. right) action – see Exercise 6.5. This is the reason for the inverse in the definition.

**Example 6.29** It follows from Example 2.79 that the tangent lift of the standard action of a matrix group is given, in tangent-lifted coordinates, by

$$T\Phi_A(\mathbf{q},\dot{\mathbf{q}}) = (A\mathbf{q}, A\dot{\mathbf{q}}).$$

The cotangent lifted action is given, in cotangent-lifted coordinates, by

$$T^*\Phi_{A^{-1}}(\mathbf{q},\mathbf{p}) = (A\mathbf{q}, \mathbf{p}A^{-1})$$

(if we represent $\mathbf{p}$ as a row vector), or

$$T^*\Phi_{A^{-1}}(\mathbf{q},\mathbf{p}) = (A\mathbf{q}, A^{-T}\mathbf{p}) = (A\mathbf{q}, \left(A^{-1}\right)^T \mathbf{p})$$

(if we represent $\mathbf{p}$ as a column vector).

**Remark 6.30** *It is straightforward to verify that*

$$\left\langle T^*\Phi_{g^{-1}}(q,\alpha), T\Phi_g(q,v)\right\rangle = \langle (q,\alpha),(q,v)\rangle .$$

*Equivalently, using concatenation notation,*

$$\langle g(q,\alpha), g(q,v)\rangle = \langle (q,\alpha),(q,v)\rangle . \tag{6.2}$$

**Proposition 6.31 (Infinitesimal lifted actions)** *Let $G$ act on $Q$, and by tangent and cotangent lifts on $TQ$ and $T^*Q$. Let $\xi \in \mathfrak{g}$. Note that, in local coordinates, the infinitesimal generator vector field on $Q$ reads,*

$$\xi_Q : \mathbb{R}^n \to \mathbb{R}^n, \quad \mathbf{q} \to \xi_Q(\mathbf{q}).$$

*The infinitesimal generator vector fields on $TQ$ and $T^*Q$ are given by*

$$\xi_{TQ}(\mathbf{q},\dot{\mathbf{q}}) = \left(\xi_Q(\mathbf{q}), D\xi_Q(\mathbf{q})\cdot\dot{\mathbf{q}}\right),$$
$$\xi_{T^*Q}(\mathbf{q},\mathbf{p}) = \left(\xi_Q(\mathbf{q}), -\left(D\xi_Q(\mathbf{q})\right)^T\cdot\mathbf{p}\right).$$

*Proof* Let $\Phi$ be the action of $G$ on $Q$, so that

$$\xi_Q(\mathbf{q}) = \left.\frac{\mathrm{d}}{\mathrm{d}t}\right|_{t=0} \Phi_{\exp t\xi}(\mathbf{q}).$$

By the equality of mixed partials,

$$\left.\frac{\mathrm{d}}{\mathrm{d}t}\right|_{t=0} \mathrm{D}\Phi_{\exp t\xi}(\mathbf{q}) \cdot \dot{\mathbf{q}} = \mathrm{D}\left( \left.\frac{\mathrm{d}}{\mathrm{d}t}\right|_{t=0} \Phi_{\exp t\xi}(\mathbf{q}) \right) \cdot \dot{\mathbf{q}} = \mathrm{D}\xi_Q(\mathbf{q}) \cdot \dot{\mathbf{q}}.$$

Thus,

$$\begin{aligned} \xi_{TQ}(\mathbf{q}, \dot{\mathbf{q}}) &= \left.\frac{d}{dt}\right|_{t=0} (\exp t\xi)(\mathbf{q}, \dot{\mathbf{q}}) \\ &= \left.\frac{d}{dt}\right|_{t=0} \left(\Phi_{\exp t\xi}(\mathbf{q}), \mathrm{D}\Phi_{\exp t\xi}(\mathbf{q}) \cdot \dot{\mathbf{q}}\right) \\ &= \left(\xi_Q(\mathbf{q}), \mathrm{D}\xi_Q(\mathbf{q}) \cdot \dot{\mathbf{q}}\right). \end{aligned}$$

This proves the first claim. Recall from Section 2.3 that if cotangent vectors are are considered as columns, then

$$\left(T^* f\right)^{-1}(\mathbf{q}, \mathbf{p}) = \left( f(\mathbf{q}), (\mathrm{D} f(\mathbf{q}))^{-T} \cdot \mathbf{p} \right).$$

It follows that

$$\begin{aligned} \xi_{T^*Q}(\mathbf{q}, \mathbf{p}) &= \left.\frac{\mathrm{d}}{\mathrm{d}t}\right|_{t=0} \left(T^*\Phi_{\exp -t\xi}\right)(\mathbf{q}, \mathbf{p}) \\ &= \left.\frac{\mathrm{d}}{\mathrm{d}t}\right|_{t=0} \left( \Phi_{\exp t\xi}(\mathbf{q}), \left(\mathrm{D}\Phi_{\exp -t\xi}(\mathbf{q})\right)^{-T} \cdot \mathbf{p} \right) \\ &= \left( \xi_Q(\mathbf{q}), \left.\frac{\mathrm{d}}{\mathrm{d}t}\right|_{t=0} \left(\mathrm{D}\Phi_{\exp -t\xi}(\mathbf{q})\right)^{-T} \cdot \mathbf{p} \right). \end{aligned}$$

Since

$$\left\langle \left(\mathrm{D}\Phi_{\exp -t\xi}(\mathbf{q})\right)^{-T} \cdot \mathbf{p}, \mathrm{D}\Phi_{\exp -t\xi}(\mathbf{q}) \cdot \dot{\mathbf{q}} \right\rangle = \langle \dot{\mathbf{q}}, \mathbf{p} \rangle,$$

for all $t$, differentiation with respect to $t$ (in local coordinates) gives,

$$\begin{aligned} \left\langle \left.\frac{\mathrm{d}}{\mathrm{d}t}\right|_{t=0} \left(\mathrm{D}\Phi_{\exp -t\xi}(\mathbf{q})\right)^{-T} \cdot \mathbf{p}, \dot{\mathbf{q}} \right\rangle &= -\left\langle \mathbf{p}, \left.\frac{\mathrm{d}}{\mathrm{d}t}\right|_{t=0} \mathrm{D}\Phi_{\exp -t\xi}(\mathbf{q}) \cdot \dot{\mathbf{q}} \right\rangle \\ &= -\left\langle \mathbf{p}, \mathrm{D}\xi_Q(\mathbf{q}) \cdot \dot{\mathbf{q}} \right\rangle \\ &= -\left\langle \left(\mathrm{D}\xi_Q(\mathbf{q})\right)^T \cdot \mathbf{p}, \dot{\mathbf{q}} \right\rangle. \end{aligned}$$

This completes the proof, since $\dot{\mathbf{q}}$ is arbitrary. ■

**Proposition 6.32** *The tangent and cotangent lifts of any proper action are proper.*

*Proof* Let $G$ act properly on $Q$, and consider the tangent-lifted action on $TQ$. Suppose that $\{(q_i, v_i)\}$ and $\{g_i(q_i, v_i)\}$ are convergent sequences in $TQ$. Considering the base points of each vector in each of these sequences, it follows that $\{q_i\}$ and $\{g_i q_i\}$ are convergent too. Since the action on $Q$ is assumed to be proper, it follows that $\{g_i\}$ has a convergent subsequence. The argument for cotangent lifts is analogous. ■

**Proposition 6.33** *Every cotangent-lifted action is symplectic and Poisson, with respect to the canonical symplectic form and the canonical Poisson bracket.*

*Proof* This follows directly from Propositions 3.117 and 4.34. ■

The following definition generalizes Definition 4.44.

**Definition 6.34** *A function $F$ is* **invariant** *(or* **symmetric***) with respect to an action $\Phi$ of a Lie group $G$ if, for every $g \in G$, the map $\Phi_g$ is a symmetry of $F$, that is, $F \circ \Phi_g = F$. The group $G$ is called a* **Lie group symmetry** *or* **symmetry group** *of $F$.*

The following proposition generalizes Proposition 4.52. Its proof is left as Exercise 6.10.

**Proposition 6.35** *Let $\Phi : G \times Q \to Q$ be an action. Let $L : TQ \to \mathbb{R}$ be a hyperregular Lagrangian, with corresponding Hamiltonian $H : T^*Q \to \mathbb{R}$. Then $L$ is invariant with respect to the tangent lift of $\Phi$ if and only if $H$ is invariant with respect to the cotangent lift of $\Phi$.*

**Definition 6.36** *Let $G$ act on two manifolds $M$ and $N$. A map $f : M \to N$ is* **equivariant** *if*

$$f(gz) = gf(z),$$

*for all $g \in G, z \in M$.*

**Exercise 6.1**
If $(g, x) \to gx$ is a left action, verify that $(g, x) \to g^{-1}x$ is a right action. Similary, if $(g, x) \to xg$ is a right action, verify that $(g, x) \to xg^{-1}$ is a left action.

**Exercise 6.2**
Show that an action is faithful if and only if the map $g \to \Phi_g$ is injective.

**Exercise 6.3 ($SO(3)$ action on $\mathbb{R}^3$)**
Show that the standard action of $SO(3)$ is faithful, non-transitive, non-free and proper. Find all of the orbits.

**Exercise 6.4**
Show that, for the action in Example 6.16, the orbit of any point is dense in $\mathbb{T}^2$, but does not equal $\mathbb{T}^2$. Conclude that the orbit is not an embedded submanifold of $\mathbb{T}^2$.

**Exercise 6.5**
Show that the tangent and cotangent lifts of a left (resp., right) action are left (resp., right) actions.

**Exercise 6.6**
Compute the infinitesimal generator for the $SO(3)$ action on $\mathbb{R}^3$. Use the hat map to show that for $\hat{\boldsymbol{\xi}} \in \mathfrak{so}(3)$, $\hat{\boldsymbol{\xi}}_{SO(3)}(\mathbf{x}) = \boldsymbol{\xi} \times \mathbf{x}$. Show that, for all non-zero $\hat{\boldsymbol{\xi}} \in \mathfrak{so}(3)$, the infinitesimal generator vector field corresponding to $\hat{\boldsymbol{\xi}}$ is not left invariant.

**Exercise 6.7**
Given $\xi \in \mathfrak{g}$, consider $\xi_M$, the corresponding (left) invariant vector field. Show that the integral curve of $\xi_M$ through a point $x_0 \in M$ is $t \to \exp(t\xi)x_0$.

**Exercise 6.8**
Consider the left action of a Lie group $G$ on the manifold $M$. Assume $x \in M$ a point with a non-trivial isotropy subgroup $G_x$. Show that for any $\xi$ in the Lie subalgebra $\mathfrak{g}_x$ we have $\xi_M(x) = 0$.

**Exercise 6.9 (preview of Adjoint and adjoint actions on $\mathfrak{so}(3)$)**
The Adjoint action of $SO(3)$ and $\mathfrak{so}(3)$ is $(\mathbf{R}, \hat{\mathbf{w}}) \mapsto \mathbf{R}\hat{\mathbf{w}}\mathbf{R}^{-1}$. Show that the infinitesimal generator of $\hat{\mathbf{v}} \in \mathfrak{so}(3)$, which is denoted by $\mathrm{ad}_{\hat{\mathbf{v}}}$, is given by

$$\mathrm{ad}_{\hat{\mathbf{v}}}\hat{\mathbf{w}} := \hat{\mathbf{v}}_{\mathfrak{so}(3)}(\hat{\mathbf{w}}) = [\hat{\mathbf{v}}, \hat{\mathbf{w}}] = \widehat{\mathbf{v} \times \mathbf{w}}.$$

**Exercise 6.10**
Check that Proposition 6.35 is a generalization of Proposition 4.52, and prove the former by generalizing the proof of the latter.

## 6.2 Actions of a Lie group on itself

Let $G$ be a Lie group. There are four different translation actions of $G$ on itself, defined as follows:

| | Left action | Right action |
|---|---|---|
| Left multiplication | $(g,h) \to L_g(h) := gh$ | $(g,h) \to L_{g^{-1}}(h) := g^{-1}h$ |
| Right multiplication | $(g,h) \to R_{g^{-1}}h := hg^{-1}$ | $(g,h) \to R_g(h) := hg$ |

From now on, unless otherwise specified, we use the following terminology:

**Definition 6.37** *The action of $G$ on itself by **left translation** (or **left multiplication**) is the left action defined by*

$$(g,h) \to L_g(h) := gh.$$

*Analogously, the **right translation** (or **right multiplication**) action is the right action defined by*

$$(g,h) \to R_g(h) := hg.$$

All actions above can be lifted to actions on the tangent bundle $TG$ and the cotangent bundle $T^*G$. For instance, for the left multiplicative action we have:

$$G \times TG \to TG \qquad \text{(tangent-lifted left translation)}$$

$$(g,(h,v)) \to (gh, gv) := (gh, T_h L_g(v)) = \left( gh, \frac{\mathrm{d}}{\mathrm{d}t} (gc(t)) \Big|_{t=0} \right),$$

where $c(t)$ is any path in $G$ with $c(0) = h$ and $c'(0) = v$; and

$$G \times T^*G \to T^*G \quad \text{(cotangent-lifted left translation)}$$

$$(g,(h,\alpha)) \to (gh, g\alpha) := (gh, T^*_{gh} L_{g^{-1}}(\alpha)),$$

where

$$\left\langle T^*_{gh} L_{g^{-1}}(\alpha), w \right\rangle = \left\langle \alpha, T_{gh} L_{g^{-1}}(w) \right\rangle \quad \text{for all} \quad w \in T_{gh}G.$$

**Example 6.38 (Left translation action of a matrix Lie group on itself)** Let $G$ be a matrix Lie group and consider the left translation action of $G$ on itself. For any $R \in G$ and any $(A, \dot{A}) \in TG$, if $C(t)$ is a path in $G$ with $C(0) = A$ and $C'(0) = \dot{A}$, then

$$T_A L_R(\dot{A}) = \frac{\mathrm{d}}{\mathrm{d}t} (R\, C(t)) \Big|_{t=0} = R \frac{\mathrm{d}}{\mathrm{d}t} C(t) \Big|_{t=0} = R\dot{A}.$$

Thus the tangent lifted left translation action is

$$(R,(A,\dot{A})) \to (RA, T_A L_R(\dot{A})) = (RA, R\dot{A}).$$

The cotangent-lifted left translation action is:

$$(R,(A,P)) \to (RA, T^*_{RA} L_{R^{-1}}(P)),$$

for every $P \in T^*_A G$, where

$$\langle T^*_{RA} L_{R^{-1}}(P), \dot{R} \rangle = \langle P, T_{RA} L_{R^{-1}}(\dot{R}) \rangle = \left\langle P, R^{-1}\dot{R} \right\rangle \quad \text{for all} \quad \dot{R} \in T_{RA}G.$$

Identifying $T^*_A G$ with $T_A G$ via the trace pairing (see Example 3.74),

$$\left\langle P, R^{-1}\dot{R} \right\rangle = \operatorname{tr}\left( P(R^{-1}\dot{R})^T \right) = \operatorname{tr}\left( R^{-T} P \dot{R}^T \right) = \left\langle R^{-T} P, \dot{R} \right\rangle,$$

so

$$T^*_{RA} L_{R^{-1}}(P) = R^{-T} P.$$

Thus, using the identification above, the cotangent-lifted left translation action is

$$(R,(A,P)) \to (RA, T^*_{RA}L_{R^{-1}}P) = (RA, R^{-T}P).$$

**Definition 6.39** *The action of $G$ on itself by* ***conjugation,*** *or* ***inner automorphism,*** *is*

$$G \times G \to G, \quad (g,h) \to I_g(h) := (L_g \circ R_{g^{-1}})(h) = ghg^{-1}.$$

*Orbits of this action are called* ***conjugacy classes.***

**Definition 6.40** *The* ***Adjoint action*** *of $G$ on $\mathfrak{g}$ is*

$$G \times \mathfrak{g} \to \mathfrak{g}, \quad (g,\xi) \to \mathrm{Ad}_g\xi := T_e I_g(\xi) = T_e(L_g \circ R_{g^{-1}})\xi.$$

**Example 6.41 (Adjoint action for matrix Lie groups)** For any matrix Lie group $G$, the inner automorphisms are $I_R(B) = RBR^{-1}$, $R \in G$. Taking a path $B(t)$ with $B(0) = I$ and $B'(0) = \xi \in \mathfrak{g}$, we have

$$\mathrm{Ad}_R\xi = T_I I_R\xi = \left.\frac{\mathrm{d}}{\mathrm{d}t}\right|_{t=0} I_R(B(t)) = \left.\frac{\mathrm{d}}{\mathrm{d}t}\right|_{t=0} R\,B(t)\,R^{-1} = R\,\xi\,R^{-1}.$$

Note that we have already seen an example of the Adjoint action for matrix Lie groups acting on matrix Lie algebras, when we defined

$$C(s) = R_A(s)\,B\,R_A(s)^{-1},$$

in the proof of Proposition 5.6.

**Definition 6.42** *The* ***Co-Adjoint action*** *of $G$ on $\mathfrak{g}^*$ is the inverse dual of the Adjoint action:*

$$G \times \mathfrak{g}^* \to \mathfrak{g}^*, \quad (g,\mu) \to \mathrm{Ad}^*_{g^{-1}}\mu,$$

*where*

$$\langle \mathrm{Ad}^*_{g^{-1}}\mu, \xi\rangle = \langle \mu, \mathrm{Ad}_{g^{-1}}\xi\rangle$$

*for all $\mu \in \mathfrak{g}^*$, $\xi \in \mathfrak{g}$ and $\langle \cdot, \cdot \rangle : \mathfrak{g}^* \times \mathfrak{g} \to \mathbb{R}$ is the natural pairing.*

Note that for a fixed $g \in G$, the maps $\mathrm{Ad}_g$ and $\mathrm{Ad}^*_{g^{-1}}$ are the tangent and cotangent lifts, respectively, at the identity, of the inner automorphism $I_g$.

When there is no confusion, it is common to use either concatenation notation or 'dot' notation for both Adjoint and Co-Adjoint action. Thus:

$$g\xi = g \cdot \xi := \mathrm{Ad}_g \xi \quad \text{and} \quad g\mu = g \cdot \mu := \mathrm{Ad}^*_{g^{-1}}\mu. \tag{6.3}$$

For the inverse Adjoint and Co-Adjoint actions, it is also common to denote

$$\xi g = \xi \cdot g := \mathrm{Ad}_{g^{-1}}\xi \quad \text{and} \quad \mu g = \mu \cdot g := \mathrm{Ad}^*_g \mu. \tag{6.4}$$

**Example 6.43 (Co-Adjoint action for matrix Lie groups)** For any matrix Lie group $G$, we have seen in Example 6.41 that $\mathrm{Ad}_R \xi = R\,\xi\,R^{-1}$. It follows that, for any $\mu \in \mathfrak{g}^*, \xi \in \mathfrak{g}$,

$$\langle \mathrm{Ad}^*_{R^{-1}}\,\mu, \xi \rangle = \langle \mu, \mathrm{Ad}_{R^{-1}}\,\xi \rangle = \left\langle \mu, R^{-1}\xi R \right\rangle.$$

Now, if we identify $\mu \in \mathfrak{g}^*$ with a matrix $\mu \in \mathfrak{g}$ via the trace pairing, as explained in Remark 3.75, then

$$\begin{aligned} \left\langle \mu, R^{-1}\xi R \right\rangle &= \mathrm{tr}\left( \mu\, R^T \xi^T R^{-T} \right) \\ &= \mathrm{tr}\left( R^{-T} \mu R^T \xi^T \right) \\ &= \left\langle R^{-T}\mu R^T, \xi \right\rangle. \end{aligned}$$

Therefore,

$$\mathrm{Ad}^*_{R^{-1}}\,\mu = R^{-T}\mu R^T.$$

**Example 6.44 (Adjoint and Co-Adjoint actions for $SO(3)$)** We use the hat map $(\widehat{\ }) : \mathbb{R}^3 \to \mathfrak{so}(3)$ as defined in Example 5.35 and the formula for the Adjoint action of matrix groups as given in Example 6.41. Let $\boldsymbol{R} \in SO(3)$ and $\widehat{\boldsymbol{\Omega}} \in \mathfrak{so}(3)$. We have

$$\mathrm{Ad}_{\boldsymbol{R}}\widehat{\boldsymbol{\Omega}} = \boldsymbol{R}\widehat{\boldsymbol{\Omega}}\boldsymbol{R}^{-1}.$$

For any $\mathbf{w} \in \mathbb{R}^3$,

$$\begin{aligned} \left(\mathrm{Ad}_{\boldsymbol{R}}\widehat{\boldsymbol{\Omega}}\right)\mathbf{w} &= \left(\boldsymbol{R}\widehat{\boldsymbol{\Omega}}\boldsymbol{R}^{-1}\right)\mathbf{w} = \boldsymbol{R}\left(\widehat{\boldsymbol{\Omega}}(\boldsymbol{R}^{-1}\mathbf{w})\right) \\ &= \boldsymbol{R}(\boldsymbol{\Omega} \times \boldsymbol{R}^{-1}\mathbf{w}) = (\boldsymbol{R}\boldsymbol{\Omega}) \times \mathbf{w} = \widehat{\boldsymbol{R}\boldsymbol{\Omega}}\,\mathbf{w}, \end{aligned}$$

where we have used the relation $\boldsymbol{R}(\mathbf{u} \times \mathbf{v}) = \boldsymbol{R}\mathbf{u} \times \boldsymbol{R}\mathbf{v}$, which holds for any $\mathbf{u}, \mathbf{v} \in \mathbb{R}^3$ and $\boldsymbol{R} \in SO(3)$. Consequently,

$$\mathrm{Ad}_{\boldsymbol{R}} \widehat{\boldsymbol{\Omega}} = \widehat{\boldsymbol{R}\boldsymbol{\Omega}}.$$

Identifying $\mathfrak{so}(3) \simeq \mathbb{R}^3$ then gives

$$\mathrm{Ad}_{\boldsymbol{R}} \boldsymbol{\Omega} = \boldsymbol{R}\boldsymbol{\Omega}.$$

Thus, the Adjoint action of $SO(3)$ on $\mathfrak{so}(3)$ may be identified with the left multiplicative action of $SO(3)$ on $\mathbb{R}^3$.

To compute the Co-Adjoint action of $SO(3)$ on $\mathfrak{so}(3)^*$, first recall that an element of $\breve{\boldsymbol{\Pi}} \in \mathfrak{so}(3)^*$ may be identified with a vector $\boldsymbol{\Pi} \in \mathbb{R}^3$ by $\langle \breve{\boldsymbol{\Pi}}, \widehat{\boldsymbol{\Omega}} \rangle = \boldsymbol{\Pi} \cdot \boldsymbol{\Omega}$ for all $\boldsymbol{\Omega} \in \mathbb{R}^3$ (see Example 5.36). We have

$$\begin{aligned}\left\langle \mathrm{Ad}^*_{\boldsymbol{R}^{-1}} \breve{\boldsymbol{\Pi}}, \widehat{\boldsymbol{\Omega}} \right\rangle &= \left\langle \breve{\boldsymbol{\Pi}}, \mathrm{Ad}_{\boldsymbol{R}^{-1}} \widehat{\boldsymbol{\Omega}} \right\rangle = \left\langle \breve{\boldsymbol{\Pi}}, \widehat{\boldsymbol{R}^{-1}\boldsymbol{\Omega}} \right\rangle \\ &= \boldsymbol{\Pi} \cdot \boldsymbol{R}^{-1}\boldsymbol{\Omega} = (\boldsymbol{R}\boldsymbol{\Pi}) \cdot \boldsymbol{\Omega} = \left\langle (\boldsymbol{R}\boldsymbol{\Pi})\breve{}, \widehat{\boldsymbol{\Omega}} \right\rangle.\end{aligned}$$

Hence, the Co-Adjoint action of $SO(3)$ on $\mathfrak{so}(3)^*$ has the expression

$$\mathrm{Ad}^*_{\boldsymbol{R}^{-1}} \breve{\boldsymbol{\Pi}} = (\boldsymbol{R}\boldsymbol{\Pi})\breve{}.$$

Identifying $\mathfrak{so}(3)^* \simeq \mathbb{R}^3$ then gives

$$\mathrm{Ad}^*_{\boldsymbol{R}^{-1}} \Pi = \boldsymbol{R}\Pi.$$

Thus, the Co-Adjoint action of $SO(3)$ on $\mathfrak{so}(3)^*$ may *also* be identified with the left multiplicative action of $SO(3)$ on $\mathbb{R}^3$.

**Definition 6.45** *For any Lie group $G$, and any $\mu \in \mathfrak{g}^*$, the* ***Co-Adjoint orbit*** *of $\mu$ is the orbit of $\mu$ under the Co-Adjoint action of $G$ on $\mathfrak{g}^*$,*

$$\mathcal{O}_\mu := \{g \cdot \mu : g \in G\} = \{\mathrm{Ad}^*_{g^{-1}} \mu : g \in G\}.$$

**Example 6.46 (Co-Adjoint orbit for $SO(3)$)** Example 6.44 shows that the Co-Adjoint action of $SO(3)$ on $\mathfrak{so}(3)^*$ may be identified with the left multiplicative action of $SO(3)$ on $\mathbb{R}^3$. Thus, the Co-Adjoint orbit of $\boldsymbol{\Pi} \in \mathfrak{so}(3)^* \simeq \mathbb{R}^3$ is

$$\mathcal{O}_{\boldsymbol{\Pi}} = \{\boldsymbol{R}\boldsymbol{\Pi} : \boldsymbol{R} \in SO(3)\} \subset \mathbb{R}^3,$$

which is a 2-sphere of radius $\|\boldsymbol{\Pi}\|$.

The infinitesimal generator (see Definition 6.22) of the Adjoint action of $G$ is the vector field $\xi_{\mathfrak{g}}$ defined by

$$\xi_{\mathfrak{g}}(\eta) = \left.\frac{\mathrm{d}}{\mathrm{d}t}\right|_{t=0} \mathrm{Ad}_{\exp t\xi}\,\eta = T_e(\mathrm{Ad}_{(\cdot)}\,\eta)\,\xi,$$

where $\mathrm{Ad}_{(\cdot)}\eta$ is the map from $G$ to $\mathfrak{g}$ that takes $g$ to $\mathrm{Ad}_g\eta$. Since $\mathfrak{g}$ is a vector space, $T\mathfrak{g} \simeq \mathfrak{g} \times \mathfrak{g}$, a vector field on $\mathfrak{g}$ can be considered as a map from $\mathfrak{g}$ to $\mathfrak{g}$. Considered in this way, the vector field $\xi_{\mathfrak{g}}$ is denoted by $\mathrm{ad}_\xi$.

**Definition 6.47** *The infinitesimal generator map,*

$$\mathfrak{g} \times \mathfrak{g} \to \mathfrak{g}, \quad (\xi, \eta) \mapsto \mathrm{ad}_\xi(\eta) = \xi_{\mathfrak{g}}(\eta) = \left.\frac{\mathrm{d}}{\mathrm{d}t}\right|_{t=0} \mathrm{Ad}_{\exp t\xi}\,\eta,$$

*is called the* ***adjoint action*** *of* $\mathfrak{g}$ *on itself, even though it is not a group action. The* ***adjoint operator*** *on* $\mathfrak{g}$, *denoted by* ad, *is defined by*

$$\mathrm{ad}_\xi\eta = \xi_{\mathfrak{g}}(\eta) \quad \textit{for all} \quad \eta \in \mathfrak{g}.$$

**Example 6.48 (adjoint action for matrix Lie algebras)** Recall from Example 6.41 that for any matrix group,

$$\mathrm{Ad}_R\,\eta = R\eta R^{-1}.$$

It follows that

$$\begin{aligned}\mathrm{ad}_\xi\eta &= \left.\frac{\mathrm{d}}{\mathrm{d}t}\right|_{t=0} \mathrm{Ad}_{\exp t\xi}\,\eta \\ &= \left.\frac{\mathrm{d}}{\mathrm{d}t}\right|_{t=0} (\exp t\xi)\,\eta\,\exp(-t\xi) \\ &= \xi\eta - \eta\xi = [\xi, \eta].\end{aligned}$$

In fact, this result holds for all Lie algebras:

**Proposition 6.49**

$$\mathrm{ad}_\xi\,\eta = [\xi, \eta], \tag{6.5}$$

*where* $[\xi, \eta] := \left[X^L_\xi, X^L_\eta\right]$ *is the Lie bracket defined in Definition 5.27.*

*Proof*

$$\begin{aligned}
\mathrm{ad}_\xi \eta = \left.\frac{\mathrm{d}}{\mathrm{d}t}\right|_{t=0} \mathrm{Ad}_{\exp t\xi}\eta &= \left.\frac{\mathrm{d}}{\mathrm{d}t}\right|_{t=0} T_e\left(L_{\exp t\xi} \circ R_{-\exp t\xi}\right)\eta \\
&= \left.\frac{\mathrm{d}}{\mathrm{d}t}\right|_{t=0} T_{\exp t\xi} R_{-\exp t\xi}\left(T_e L_{\exp t\xi}(\eta)\right) \\
&= \left.\frac{\mathrm{d}}{\mathrm{d}t}\right|_{t=0} T_{\exp t\xi} R_{-\exp t\xi}\left(X_\eta^L(\exp t\xi)\right).
\end{aligned}$$

The time-$t$ flow of $X_\xi^L$ is $\Phi_t(g) := g\exp(t\xi) = R_{\exp(t\xi)}g$. Hence

$$\begin{aligned}
\mathrm{ad}_\xi \eta &= \left.\frac{\mathrm{d}}{\mathrm{d}t}\right|_{t=0} T_{\Phi_t(e)}\Phi_t^{-1}\left(X_\eta^L(\Phi_t(e))\right) \\
&= \left.\frac{\mathrm{d}}{\mathrm{d}t}\right|_{t=0} (\Phi_t^* X_\eta^L)(e) \\
&= \mathcal{L}_{X_\xi^L} X_\eta^L(e) \\
&= \left[X_\xi^L, X_\eta^L\right](e) = [\xi, \eta].
\end{aligned}$$

■

**Definition 6.50** *The* ***coadjoint operator*** *is the map*

$$\begin{aligned}
\mathrm{ad}^* : \mathfrak{g} \times \mathfrak{g}^* &\to \mathfrak{g}^*, \\
(\xi, \mu) &\mapsto \mathrm{ad}_\xi^*(\mu),
\end{aligned}$$

*such that, for every $\xi \in \mathfrak{g}$, the map $\mathrm{ad}_\xi^* : \mathfrak{g}^* \to \mathfrak{g}^*$ is the dual of the adjoint operator, that is,*

$$\langle \mathrm{ad}_\xi^* \mu, \eta\rangle = \langle \mu, \mathrm{ad}_\xi \eta\rangle \quad \textit{for all} \quad \eta \in \mathfrak{g}.$$

**Remark 6.51** *The infinitesimal generator of the Co-Adjoint action is*

$$\xi_{\mathfrak{g}^*}(\mu) = -\mathrm{ad}_\xi^*(\mu)$$

*(see Exercise 6.14). The minus sign is because of the inverse in the definition of the Co-Adjoint action.*

**Example 6.52 (coadjoint operator for matrix Lie algebras)** Let $\mathfrak{g}$ be a matrix Lie algebra. For every $\xi, \eta \in \mathfrak{g}$ and every $\mu \in \mathfrak{g}^*$,

$$\begin{aligned}\langle \mathrm{ad}^*_\xi \mu, \eta \rangle &= \langle \mu, \mathrm{ad}_\xi\, \eta \rangle \\ &= \langle \mu, [\xi\ \eta] \rangle = \langle \mu, \xi\eta - \eta\xi \rangle = \langle \mu, \xi\eta \rangle - \langle \mu, \eta\xi \rangle.\end{aligned}$$

Identifying $\mu \in \mathfrak{g}^*$ with $\boldsymbol{\mu} \in \mathfrak{g}$ via the trace pairing, as in Remark 3.75, we have

$$\begin{aligned}\langle \mu, \xi\eta \rangle - \langle \mu, \eta\xi \rangle &= \mathrm{tr}\left(\mu\,(\xi\eta)^T\right) - \mathrm{tr}\left(\mu\,(\eta\xi)^T\right) \\ &= \mathrm{tr}(\mu\,\eta^T\xi^T) - \mathrm{tr}(\mu\,\xi^T\eta^T) \\ &= \mathrm{tr}(\xi^T\mu\,\eta^T) - \mathrm{tr}(\mu\,\xi^T\eta^T) \\ &= \mathrm{tr}\left((\xi^T\mu - \mu\,\xi^T)\eta^T\right) = \langle -[\mu, \xi^T], \eta \rangle,\end{aligned}$$

so

$$\mathrm{ad}^*_\xi \mu = -[\mu, \xi^T] \quad \text{(for matrix Lie algebras)}.$$

Note that in $\mathfrak{so}(n)$, since $\xi = -\xi^T$, we have

$$\mathrm{ad}^*_\xi \mu = [\mu, \xi] \quad (\text{for } \mathfrak{so}(n)) \,. \tag{6.6}$$

**Example 6.53 (adjoint and coadjoint operators for $\mathfrak{so}(3)$ and $\mathfrak{so}(3)^*$)** Given $\hat{\boldsymbol{\Omega}}, \hat{\mathbf{w}} \in \mathfrak{so}(3)$, we have $\mathrm{ad}_{\hat{\boldsymbol{\Omega}}}\hat{\mathbf{w}} = [\hat{\boldsymbol{\Omega}}, \hat{\mathbf{w}}] = \widehat{\boldsymbol{\Omega} \times \mathbf{w}}$ (see Exercise 6.9). Further, if $\breve{\boldsymbol{\Pi}} \in \mathfrak{so}(3)^*$, then

$$\langle \mathrm{ad}^*_{\hat{\boldsymbol{\Omega}}}\, \breve{\boldsymbol{\Pi}}, \hat{\mathbf{w}} \rangle = \langle \breve{\boldsymbol{\Pi}}, \mathrm{ad}_{\hat{\boldsymbol{\Omega}}}\hat{\mathbf{w}} \rangle = \langle \breve{\boldsymbol{\Pi}}, [\hat{\boldsymbol{\Omega}}, \hat{\mathbf{w}}] \rangle = \boldsymbol{\Pi} \cdot (\boldsymbol{\Omega} \times \mathbf{w}) = (\boldsymbol{\Pi} \times \boldsymbol{\Omega}) \cdot \mathbf{w}.$$

Thus,

$$\mathrm{ad}^*_{\hat{\boldsymbol{\Omega}}} \breve{\boldsymbol{\Pi}} = (\boldsymbol{\Pi} \times \boldsymbol{\Omega})\breve{}.$$

**Proposition 6.54 (Derivation along a Co-Adjoint orbit)** *If $g(t)$ is a path in a Lie group $G$, and $\mu(t)$ is a path in $\mathfrak{g}^*$, then*

$$\frac{\mathrm{d}}{\mathrm{d}t}\mathrm{Ad}^*_{g(t)^{-1}}\mu(t) = \mathrm{Ad}^*_{g(t)^{-1}}\left[\frac{\mathrm{d}\mu}{\mathrm{d}t} - \mathrm{ad}^*_{\xi(t)}\mu(t)\right], \tag{6.7}$$

*where $\xi(t) = g(t)^{-1}\dot{g}(t)$.*

*Proof* For any $\eta \in \mathfrak{g}$,

$$\left.\frac{\mathrm{d}}{\mathrm{d}t}\right|_{t=t_0} \mathrm{Ad}_{g(t)^{-1}}\eta = \left.\frac{\mathrm{d}}{\mathrm{d}t}\right|_{t=t_0} \mathrm{Ad}_{g(t)^{-1}g(t_0)}\left(\mathrm{Ad}_{g(t_0)^{-1}}\eta\right)$$
$$= -\mathrm{ad}_{\xi(t_0)}\left(\mathrm{Ad}_{g(t_0)^{-1}}\eta\right)$$

since

$$\left.\frac{\mathrm{d}}{\mathrm{d}t}\right|_{t=t_0} g(t)^{-1}g(t_0) = \left(-g(t_0)^{-1}\dot{g}(t_0)g(t_0)^{-1}\right)g(t_0) = -\xi(t_0).$$

Thus

$$\frac{\mathrm{d}}{\mathrm{d}t}\mathrm{Ad}_{g(t)^{-1}}\eta = -\mathrm{ad}_{\xi(t)}\left(\mathrm{Ad}_{g(t)^{-1}}\eta\right)$$

So

$$\left\langle\frac{\mathrm{d}}{\mathrm{d}t}\mathrm{Ad}^*_{g(t)^{-1}}\mu(t), \eta\right\rangle = \frac{\mathrm{d}}{\mathrm{d}t}\left\langle\mathrm{Ad}^*_{g(t)^{-1}}\mu(t), \eta\right\rangle = \frac{\mathrm{d}}{\mathrm{d}t}\left\langle\mu(t), \mathrm{Ad}_{g(t)^{-1}}\eta\right\rangle$$
$$= \left\langle\frac{\mathrm{d}\mu}{\mathrm{d}t}, \mathrm{Ad}_{g(t)^{-1}}\eta\right\rangle + \left\langle\mu(t), \frac{\mathrm{d}}{\mathrm{d}t}\mathrm{Ad}_{g(t)^{-1}}\eta\right\rangle$$
$$= \left\langle\frac{\mathrm{d}\mu}{\mathrm{d}t}, \mathrm{Ad}_{g(t)^{-1}}\eta\right\rangle + \left\langle\mu(t), -\mathrm{ad}_{\xi(t)}\left(\mathrm{Ad}_{g(t)^{-1}}\eta\right)\right\rangle$$
$$= \left\langle\frac{\mathrm{d}\mu}{\mathrm{d}t}, \mathrm{Ad}_{g(t)^{-1}}\eta\right\rangle - \left\langle\mathrm{ad}^*_{\xi(t)}\mu(t), \mathrm{Ad}_{g(t)^{-1}}\eta\right\rangle$$
$$= \left\langle\mathrm{Ad}^*_{g(t)^{-1}}\frac{\mathrm{d}\mu}{\mathrm{d}t}, \eta\right\rangle - \left\langle\mathrm{Ad}^*_{g(t)^{-1}}\mathrm{ad}^*_{\xi(t)}\mu(t), \eta\right\rangle$$
$$= \left\langle\mathrm{Ad}^*_{g(t)^{-1}}\left[\frac{\mathrm{d}\mu}{\mathrm{d}t} - \mathrm{ad}^*_{\xi(t)}\mu(t)\right], \eta\right\rangle.$$

■

In the 'dot' notation (6.3), the above reads:

$$\frac{\mathrm{d}}{\mathrm{d}t}\left(g(t)\cdot\mu(t)\right) = g(t)\cdot\left[\frac{\mathrm{d}\mu}{\mathrm{d}t} - \mathrm{ad}^*_{\xi(t)}\mu(t)\right]. \tag{6.8}$$

**Remark 6.55** *Analogously, the following holds:*

$$\frac{\mathrm{d}}{\mathrm{d}t}\mathrm{Ad}^*_{g(t)}\mu(t) = \mathrm{Ad}^*_{g(t)}\left[\frac{\mathrm{d}\mu}{\mathrm{d}t} + \mathrm{ad}^*_{\eta(t)}\mu(t)\right], \tag{6.9}$$

*or, in the dot notation,*

$$\frac{\mathrm{d}}{\mathrm{d}t}\left(\mu(t)\cdot g(t)\right) = \left[\frac{\mathrm{d}\mu}{\mathrm{d}t} + \mathrm{ad}^*_{\eta(t)}\mu(t)\right]\cdot g(t) \tag{6.10}$$

*where* $\eta(t) = \dot{g}(t)\, g(t)^{-1}$.

**Exercise 6.11**
Show that the left and right translation actions defined at the beginning of the section are indeed actions. Show that they are transitive, free and proper.

**Exercise 6.12**
Show that the Adjoint and Co-Adjoint actions are left actions.

**Exercise 6.13**
Show that the Adjoint and Co-Adjoint actions of an Abelian group are the trivial, i.e., for every $g \in G$, the map $\mathrm{Ad}_g$ is the identity map on $\mathfrak{g}$ and $\mathrm{Ad}^*_g$ is the identity map on $\mathfrak{g}^*$. It follows that the Co-Adjoint orbits are trivial, i.e. $\mathcal{O}_\mu = \{\mu\}$ for every $\mu \in \mathfrak{g}^*$.

**Exercise 6.14**
Show that the infinitesimal generator of the Co-Adjoint action is

$$\xi_{\mathfrak{g}^*}(\mu) = -\mathrm{ad}^*_\xi(\mu).$$

**Exercise 6.15**
Compute the adjoint and coadjoint actions for $\mathfrak{se}(2)$ and $\mathfrak{se}(3)$, and $\mathfrak{se}^*(2)$ and $\mathfrak{se}^*(3)$, respectively.

**Exercise 6.16**
Consider a left action of a Lie group $G$ on a manifold $M$, and its tangent-lifted action on TM. Prove that for any $g \in G$, $\xi \in \mathfrak{g}$ and $z \in M$,

$$g\left(\xi_M(g^{-1}z)\right) = (\mathrm{Ad}_g\xi)_M(z).$$

## 6.3 Quotient spaces

Consider a smooth action of a Lie group $G$ on a manifold $M$. It can be easily verified that the relation

$$x \sim y \quad \text{if and only if there exists } g \in G \quad \text{such that} \quad gx = y$$

is an equivalence relation on $M$. The equivalence class containing $x$ is precisely the orbit of $x$,

$$[x] := \{y : y \sim x\} = \{y : y \in Orb(x)\} = Orb(x).$$

The set of all orbits is called the ***orbit space***, or ***quotient space***, and denoted $M/G$. Let

$$\pi : M \to M/G$$

be the map assigning to each $x \in M$ its orbit $Orb(x) \in M/G$, that is $\pi(x) = [x] = Orb(x)$. The orbit space is endowed with a topological structure, called the quotient topology, defined by

$$U \subseteq M/G \text{ is open} \quad \text{if and only if } \pi^{-1}(U) \text{ is open in } M.$$

It is natural to ask whether $M/G$ has a smooth manifold structure. A widely applicable, though partial, answer is given in the following proposition:

**Proposition 6.56** *If the action of a Lie group $G$ on a manifold $M$ is free, proper and smooth, then the quotient space $M/G$ is has a unique smooth manifold structure such that $\pi : M \to M/G$, defined by $\pi(x) = Orb(x)$, is a submersion. This manifold structure is compatible with the quotient topology defined above, and has the following properties:*

1. *A map $f : M/G \to N$, for any manifold $N$, is smooth if and only if the composition $f \circ \pi$ is smooth.*

2. *If* $f : M/G \to N$ *is a bijection and* $f \circ \pi : M \to N$ *is a submersion, then* $f$ *is a diffeomorphism.*
3. *Suppose* $\varphi : M \to P$ *is a G-equivariant diffeomorphism. Let* $\pi_M : M \to M/G$ *and* $\pi_P : P \to P/G$ *be the projections* $z \to [z]$, *and define* $\overline{\varphi}$ *by requiring that the following diagram commute,*

$$\begin{array}{ccc} M & \stackrel{\varphi}{\longrightarrow} & P \\ \pi_M \downarrow & & \pi_P \downarrow \\ M/G & \stackrel{\overline{\varphi}}{\longrightarrow} & P/G, \end{array}$$

*i.e.* $\overline{\varphi} \circ \pi_M = \pi_P \circ \varphi$, *or in other notation,* $\overline{\varphi}[z] = [\varphi(z)]$ *for all* $z \in M$. *Then* $\overline{\varphi}$ *is a diffeomorphism.*

Note that the proposition applies to all smooth free actions of compact groups, since actions of compact groups are automatically proper (Theorem 6.13).

*Proof* Most of the proof goes beyond our present scope, but the interested reader may consult [AM78, 4.1.20 and 4.1.23]; and for uniqueness [AMR88, 3.5.21] (and see also [DK04]).

For part 2, let $(M/G)_1$ be the manifold structure given earlier in the proposition. Note that the manifold structure on $N$ can be pulled back by $f$ to give another manifold structure $(M/G)_2$ on $M/G$, as in Exercise 2.40, which automatically makes $f : (M/G)_2 \to N$ a diffeomorphism. Since $\pi = f^{-1} \circ (f \circ \pi)$ and $f \circ \pi$ is a submersion, it follows that $\pi : M \to (M/G)_2$ is a submersion. By the uniqueness property stated earlier in the proposition, $(M/G)_1 = (M/G)_2$. Hence, $f$ is a diffeomorphism with respect to the manifold structure on $M/G$ defined earlier in the proposition.

The proof of part 3 is left as Exercise 6.18. ■

**Example 6.57** Let $G = SO(2)$ and consider the usual action of $SO(2)$ on $\mathbb{R}^2$. This action is not free, since the isotropy subgroup of $(0,0)$ equals $SO(2)$. If we remove this 'singular point', and restrict the action to $M := \mathbb{R}^2 \setminus \{(0,0)\}$, then the restricted action is free. Let $f : M/G \to (0,\infty)$ be given by $f[x,y] = \sqrt{x^2+y^2}$. Then $f$ is a bijection, and $f \circ \pi : M \to (0,\infty)$ is given by $f \circ \pi(x,y) = \sqrt{x^2+y^2}$, which is a submersion. By part 2 of Proposition 6.56, $f$ is a diffeomorphism.

Note that the quotient space $\mathbb{R}^2/SO(2)$ is homeomorphic to $[0,\infty)$, which cannot be given a manifold structure (see Exercise 2.39).

An important special case occurs when $M = G \times N$, for some manifold $N$, and $G$ acts on $M$ by $g \cdot (h,n) = (gh,n)$. The equivalence class of any pair $(h,n)$ is $[(h,n)] = Orb(h,n) = G \times \{n\}$.

**Remark 6.58** *We use the notation* $[h,n] := [(h,n)]$.

There is a bijection

$$M/G \simeq (G \times N)/G \xrightarrow{f} N$$
$$[h,n] \longmapsto n.$$

Clearly $f \circ \pi : M \to N$, given by $(h,n) \to n$, is a submersion. Thus, by part 2 of the previous proposition, $M/G$ is diffeomorphic to $N$. In summary,

**Proposition 6.59** *If* $M = G \times N$ *and* $G$ *acts on* $M$ *by* $g(h,n) = (gh,n)$, *then* $M/G$ *is diffeomorphic to* $N$ *via the map* $[h,n] \longmapsto n$.

Now, we consider the cases where the manifold $M$ is either the tangent or the cotangent bundle of a Lie group $G$, with $G$ acting on $M$ by either tangent or cotangent lifts (respectively) of the left multiplicative action of $G$ on itself. We begin with the case $M = TG$. Consider the *left trivialization* map

$$\lambda : TG \to G \times \mathfrak{g}, \quad (h,\dot{h}) \to \lambda(h,\dot{h}) := (h, h^{-1}\dot{h}) = (h, T_h L_{h^{-1}}\dot{h}), \quad (6.11)$$

which has inverse $\lambda^{-1}(h,\xi) = (h, h\xi)$. Note that, since the left multiplicative operation and the projection $(h,\dot{h}) \to h$ are both smooth, it follows that $\lambda$ is a diffeomorphism. The left multiplicative action of $G$ on $TG$ induces, via $\lambda$, a corresponding action of $G$ on $G \times \mathfrak{g}$ that makes $\lambda$ equivariant, defined by $g(\lambda(h,\dot{h})) = \lambda(g(h,\dot{h}))$ (this is called 'the push-forward by $\lambda$' of the original action). It is straightforward to compute this action:

$$g(h,\xi) = g(\lambda(h,h\xi)) = \lambda(g(h,h\xi)) = \lambda(gh,gh\xi) = (gh,\xi).$$

Note that this action on $G \times \mathfrak{g}$ is a special case of the action in Proposition 6.59, with $N = \mathfrak{g}$. Applying that proposition, we see that $(G \times \mathfrak{g})/G$ is diffeomorphic to $\mathfrak{g}$ via the map $[g,\xi] \to \xi$. Now, since $\lambda$ is a $G$-equivariant diffeomorphism, part 3 of Proposition 6.56 shows that $(TG)/G$ is diffeomorphic to $(G \times \mathfrak{g})/G$, via the diffeomorphism $\overline{\lambda}$ defined by $\overline{\lambda}([h,\dot{h}]) = [\lambda(h,\dot{h})] = [h, h^{-1}\dot{h}]$. We have shown the following.:

**Proposition 6.60** *Let* $G$ *act on* $TG$ *by tangent-lifted left multiplication. Then the following are diffeomorphisms,*

$$\begin{array}{ccccc} (TG)/G & \simeq & (G \times \mathfrak{g})/G & \simeq & \mathfrak{g}, \\ [h,\dot{h}] & \longmapsto & [h, h^{-1}\dot{h}] & \longmapsto & h^{-1}\dot{h}. \end{array}$$

Analogously one can show that under the *cotangent left trivialization*

$$\lambda^* : T^*G \to G \times \mathfrak{g}^*, \quad (h, \alpha) \to (h, h^{-1}\alpha) := (h, T_e^* L_h \alpha), \tag{6.12}$$

the induced left action of $G$ on $G \times \mathfrak{g}^*$ is given by

$$g(h, \mu) = (gh, \mu).$$

Again, Proposition 6.59 applies, this time with $N = \mathfrak{g}^*$, so $(G \times \mathfrak{g}^*)/G$ is diffeomorphic to $\mathfrak{g}^*$ via the map $[g, \mu] \to \mu$. Thus, we have,

**Proposition 6.61** *Let $G$ act on $T^*G$ by cotangent-lifted left multiplication. Then the following are diffeomorphisms,*

$$\begin{array}{ccccc} (T^*G)/G & \simeq & (G \times \mathfrak{g}^*)/G & \simeq & \mathfrak{g}^*, \\ [h, \alpha] & \longmapsto & [h, h^{-1}\alpha] & \longmapsto & h^{-1}\alpha. \end{array}$$

**Remark 6.62** *The identification of $(G \times \mathfrak{g})/G$ with $\mathfrak{g}$, and $(G \times \mathfrak{g}^*)/G$ with $\mathfrak{g}^*$, may equally well be achieved using the right trivializations*

$$\begin{aligned} \mu : TG \to G \times \mathfrak{g}, \qquad & (h, \dot{h}) \to (h, \dot{h}h^{-1}) := (h, T_h R_{h^{-1}} \dot{h}); \\ \mu^* : T^*G \to G \times \mathfrak{g}^*, \qquad & (h, \alpha) \to (h, \alpha h^{-1}) := (h, T_e^* R_h \alpha). \end{aligned}$$

**Exercise 6.17**
Let $\pi : M \to M/G$ be as defined in the text. Show that a function $f : M \to P$, where $P$ is any manifold, is $G$-invariant if and only if $f$ can be written as $\varphi \circ \pi$ for some $\varphi : M/G \to P$. Note that, by Proposition 6.56, $\varphi$ is smooth if and only if $f$ is.

**Exercise 6.18**
Prove part 3 Proposition 6.56, using part 2 of the same proposition.

## 6.4 Poisson reduction

We start with a general theorem concerning reduction on a Poisson manifold $P$. Recall from Definition 6.19 that an action $\Phi$ of some Lie group $G$ on $P$

is called canonical if $\Phi_g$ is a Poisson map for all $g \in G$, that is

$$\{F, K\} \circ \Phi_g = \{F \circ \Phi_g, K \circ \Phi_g\}, \quad \text{for all} \quad F, K \in \mathcal{F}(P) \quad \text{and} \quad g \in G. \tag{6.13}$$

For a canonical action of $G$ on $P$, the Poisson Reduction Theorem states that the Poisson structure on $P$ induces a Poisson structure on the quotient space $P/G$, and solutions of an invariant Hamiltonian vector field on $P$ project onto solutions of a corresponding 'reduced' Hamiltonian vector field on $P/G$. In the next section, we will consider the special case where $P$ is a cotangent bundle of a Lie group $T^*G$ and $G$ acts on $P$ by cotangent lifts of left (or right) translations.

**Theorem 6.63 (Poisson Reduction Theorem)** *Let $G$ be a Lie group acting canonically on a Poisson manifold. Suppose that $P/G$ is a smooth manifold and the projection $\pi : P \to P/G$ is a submersion. Then there is an unique Poisson bracket $\{\cdot,\cdot\}_{red}$ on $P/G$, called the* ***reduced Poisson bracket****, such that $\pi$ is a Poisson map, i.e.,*

$$\{F, K\}_{red} \circ \pi := \{F \circ \pi, K \circ \pi\}, \qquad \textit{for any } F, K \in \mathcal{F}(P/G). \tag{6.14}$$

*Further, suppose that $H : P \to \mathbb{R}$ is $G$-invariant, and define $h^{red} : P/G \to \mathbb{R}$ by $H = h^{red} \circ \pi$. If $\phi$ and $\phi^{red}$ are the Hamiltonian flows corresponding to $H$ and $h^{red}$, respectively, then*

$$\phi_t^{red} \circ \pi = \pi \circ \phi_t, \qquad \textit{for all } t.$$

*If $z(t)$ is a solution of $X_H$, then $\pi \circ z(t)$ is a solution of $X_h^{red}$.*

**Remark 6.64** *As stated earlier in Proposition 6.56 (without proof), a sufficient condition for $P/G$ to be a smooth manifold and the projection $\pi : P \to P/G$ to be a submersion is that the action be free and proper.*

*Proof* For every $F, K \in \mathcal{F}(P/G)$, since $G$ acts canonically and $\pi$ is $G$-invariant, it follows that,

$$\{F \circ \pi, K \circ \pi\} \circ \Phi_g = \{F \circ \pi \circ \Phi_g, K \circ \pi \circ \Phi_g\} = \{F \circ \pi, K \circ \pi\},$$

for every $g \in G$, and thus $\{F \circ \pi, K \circ \pi\}$ is $G$-invariant. Hence the function $\{F \circ \pi, K \circ \pi\}$ can be expressed as $\varphi \circ \pi$, for some $\varphi \in \mathcal{F}(P/G)$ (see Exercise 6.17). We denote this $\varphi$ by $\{F, K\}_{red}$. This defines a bracket operation $\{\cdot,\cdot\}_{red}$ on $P/G$, given by eqn (6.14). It is straightforward to verify that this bracket satisfies the definition of a Poisson bracket – we leave this as

Exercise 6.19. Note that eqn (6.14) is also the definition of $\pi$ being a Poisson map, so $\{\cdot,\cdot\}_{red}$ is the only bracket on $P/G$ that makes $\pi$ a Poisson map.

Now consider the Hamiltonian flows $\phi$ and $\phi^{red}$ corresponding to $H$ and $h^{red}$. For every $F \in \mathcal{F}(P/G)$, since $F \circ \pi \in \mathcal{F}(P)$, the definition of $\phi$ implies that

$$\frac{\mathrm{d}}{\mathrm{d}t}((F \circ \pi) \circ \phi_t(z)) = \{F \circ \pi, H\} \circ \phi_t(z), \quad \text{for all } z \in P.$$

Since $H = h^{red} \circ \pi$, this is equivalent to

$$\frac{\mathrm{d}}{\mathrm{d}t}(F \circ \pi \circ \phi_t(z)) = \{F, h^{red}\}_{red} \circ \pi \circ \phi_t(z). \tag{6.15}$$

Similarly, the definition of $\phi^{red}$ is that, for every $F \in \mathcal{F}(P/G)$,

$$\frac{\mathrm{d}}{\mathrm{d}t}(F \circ \phi_t^{red}([z])) = \{F, h^{red}\}_{red} \circ \phi_t^{red}([z]), \quad \text{for all } [z] \in P/G.$$

Since $[z] = \pi(z)$, this is equivalent to

$$\frac{\mathrm{d}}{\mathrm{d}t}(F \circ \phi_t^{red} \circ \pi(z)) = \{F, h^{red}\}_{red} \circ \phi_t^{red} \circ \pi(z), \quad \text{for all } z \in P.$$

This equation uniquely defines the flow $\phi^{red}$. Comparison of this definition with eqn (6.15) shows that $\phi_t^{red} \circ \pi = \pi \circ \phi_t$.

Finally, a path $z(t)$ in $P$ is a solution of $X_H$ if and only if $z(t) = \phi_t(z(0))$ for all $t$, which implies that $\pi \circ z(t) = \pi \circ \phi_t(z(0))$. As we have just shown, this is equivalent to $\pi \circ z(t) = \phi_t^{red} \circ \pi(z(0))$, which is equivalent to $\pi \circ z(t)$ being a solution of $X_{h^{red}}$. ■

We leave the proofs of the next two propositions as Exercises 6.21 and 6.22.

**Proposition 6.65** *Let $P_1$ and $P_2$ be two Poisson manifolds with brackets given by $\{\cdot,\cdot\}_{P_1}$, and $\{\cdot,\cdot\}_{P_2}$, respectively. Then the product space $P := P_1 \times P_2$ is a Poisson manifold with respect to the bracket*

$$\{\cdot,\cdot\}_P = \{\cdot,\cdot\}_{P_1} + \{\cdot,\cdot\}_{P_2}. \tag{6.16}$$

**Proposition 6.66** **(Poisson reduction of product spaces)** *Let $G$ and $H$ be Lie groups acting smoothly on $P_1$ and $P_2$ respectively, and suppose that $P_1/G$ and $P_2/H$ are smooth manifolds and $\pi_1 : P_1 \to P_1/G$ and $\pi_2 : P_2 \to P_2/H$ are submersions. Let $G \times H$ have the usual product action on $P_1 \times P_2$, i.e., $(g,h)(z_1,z_2) = (gz_1,hz_2)$. Then, $(P_1 \times P_2)/(G \times H)$ has a smooth manifold structure such that the projection $\pi : (P_1 \times P_2) \to$*

*$(P_1 \times P_2)/(G \times H)$ defined by $\pi(z_1, z_2) = ([z_1], [z_2]) = (\pi_1(z_1), \pi_2(z_2))$, is a submersion and the following map is a diffeomorphism:*

$$\varphi : (P_1 \times P_2)/(G \times H) \longrightarrow (P_1/G) \times (P_2/H),$$
$$[z_1, z_2] \longmapsto ([z_1], [z_2]).$$

*$(P_1 \times P_2)/(G \times H)$ is a Poisson manifold with respect to the bracket*

$$\{\cdot,\cdot\}_{(P_1\times P_2)/(G\times H)} = \{\cdot,\cdot\}_{P_1/G} + \{\cdot,\cdot\}_{P_2/H}. \tag{6.17}$$

**Corollary 6.67** *Consider the action of a Lie group $G$ on a Poisson manifold $P_1$ as in the statement of Theorem 6.63 and let $P_2$ be a Poisson manifold on which $G$ does not act. Then $P_1/G \times P_2$ is a Poisson manifold with respect to the bracket*

$$\{\cdot,\cdot\}_{(P_1\times P_2)/G} = \{\cdot,\cdot\}_{P_1/G} + \{\cdot,\cdot\}_{P_2}. \tag{6.18}$$

**Exercise 6.19**
Show that the bracket on $P/G$ defined in Proposition 6.63 satisfies the definition of a Poisson bracket.

**Exercise 6.20**
The last claim of Proposition 6.63 is: 'If $z(t)$ is a solution of $X_H$, then $\pi \circ z(t)$ is a solution of $X_h^{red}$.' Is the converse true? HINT: see Proposition 9.18.

**Exercise 6.21**
Prove Proposition 6.65, which concerns the Poisson bracket on a product space.

**Exercise 6.22**
Prove Proposition 6.66, which concerns Poisson reduction by a product action.

## Solutions to selected exercises

**Solution to Exercise 6.3** The following proofs are that the action of $SO(3)$ on $\mathbb{R}^3$ is faithful, non-transitive, non-free and proper, in that order.

1. Let $\boldsymbol{R} \in SO(3)$ such that $\boldsymbol{R} \neq \boldsymbol{I}$, then $\boldsymbol{R}$ has an axis or rotation when it acts on $\mathbb{R}^3$ (this was first proven by Euler). If $\boldsymbol{x}$ is not on the axis of rotation of $\boldsymbol{R}$ then $\boldsymbol{R}\boldsymbol{x} \neq \boldsymbol{x}$.
2. Note that $\|\boldsymbol{R}\boldsymbol{x}\|^2 = \|\boldsymbol{x}\|^2$ for all $g \in SO(3)$ and $\boldsymbol{x} \in \mathbb{R}^3$. Take $\boldsymbol{y} \in \mathbb{R}^3$ such that $\|\boldsymbol{y}\|^2 \neq \|\boldsymbol{x}\|^2$. Then $\boldsymbol{R} \cdot \boldsymbol{x} \neq \boldsymbol{y}$ for all $\boldsymbol{R} \in SO(3)$.
3. Suppose that $\boldsymbol{x} \in \mathbb{R}^3$ is on the axis of rotation of $\boldsymbol{R} \in SO(3)$. Then $\boldsymbol{R} \cdot \boldsymbol{x} = \boldsymbol{x}$, but $\boldsymbol{R} \neq \boldsymbol{I}$.
4. Every action by a compact group is proper and $SO(3)$ is compact, as in 6.13. Therefore the action of $SO(3)$ on $\mathbb{R}^3$ is proper.

The $SO(3)$-orbits in $\mathbb{R}^3$ are spheres and the origin.

**Solution to Exercise 6.5** Let $\Phi : G \times Q \to Q$ be a left action, then $\Phi_h \circ \Phi_g = \Phi_{hg}$. Differentiating this relation gives the tangent lift of $\Phi$,

$$T\Phi_h \circ T\Phi_g = T\Phi_{hg}.$$

Therefore the tangent lift of $\Phi$ is also a left action. Consider the following calculation for $\boldsymbol{p} \in T^*Q$ and $\dot{\boldsymbol{q}} \in TQ$:

$$\begin{aligned}\langle T^*\Phi_{hg}\boldsymbol{p}\,,\,\dot{\boldsymbol{q}}\rangle &= \langle \boldsymbol{p}\,,\,T\Phi_{hg}\dot{\boldsymbol{q}}\rangle \\ &= \langle \boldsymbol{p}\,,\,T\Phi_h \circ T\Phi_g\dot{\boldsymbol{q}}\rangle \\ &= \langle T^*\Phi_h\boldsymbol{p}\,,\,T\Phi_g\dot{\boldsymbol{q}}\rangle \\ &= \langle T^*\Phi_g \circ T^*\Phi_h\boldsymbol{p}\,,\,\dot{\boldsymbol{q}}\rangle.\end{aligned}$$

This yields the relation

$$T^*\Phi_g \circ T^*\Phi_h = T^*\Phi_{hg}.$$

The cotangent action defined by $T^*\Phi_{g^{-1}}$ is a left action since

$$T^*\Phi_{g^{-1}} \circ T^*\Phi_{h^{-1}} = T^*\Phi_{h^{-1}g^{-1}} = T^*\Phi_{(gh)^{-1}}.$$

The proofs for the right action follow in the same way.

**Solution to Exercise 6.6** The infinitesimal generator is given by

$$\left.\frac{\mathrm{d}}{\mathrm{d}t}\right|_{t=0}\left(\exp\left(t\hat{\boldsymbol{\xi}}\right)\boldsymbol{x}\right) = \hat{\boldsymbol{\xi}}\boldsymbol{x} = \boldsymbol{\xi} \times \boldsymbol{x}.$$

Therefore, $\hat{\boldsymbol{\xi}}_{\mathbb{R}^3}(\boldsymbol{x}) = \boldsymbol{\xi} \times \boldsymbol{x}$. Note that

$$\begin{aligned}\left((L_R)_*\hat{\boldsymbol{\xi}}_{\mathbb{R}^3}\right)(\boldsymbol{x}) &= \boldsymbol{R}\hat{\boldsymbol{\xi}}_{\mathbb{R}^3}(\boldsymbol{R}^{-1}\boldsymbol{x}) \\ &= \boldsymbol{R}(\boldsymbol{\xi} \times \boldsymbol{R}^{-1}\boldsymbol{x}) \\ &= (\boldsymbol{R}\boldsymbol{\xi}) \times \boldsymbol{x} \\ &\neq \boldsymbol{\xi} \times \boldsymbol{x} \\ &= \hat{\boldsymbol{\xi}}_{\mathbb{R}^3}(\boldsymbol{x}).\end{aligned}$$

Therefore, $\hat{\boldsymbol{\xi}}_{\mathbb{R}^3}$ is not left invariant.

**Solution to Exercise 6.13** Suppose that $G$ is Abelian, then $I_g h = ghg^{-1} = gg^{-1}h = h$. Therefore the Ad action is given by

$$\mathrm{Ad}_g\,\xi = \left.\frac{\mathrm{d}}{\mathrm{d}t}\right|_{t=0} I_g \exp(t\xi) = \left.\frac{\mathrm{d}}{\mathrm{d}t}\right|_{t=0} \exp(t\xi) = \xi.$$

The coadjoint action is also trivial since for all $\xi \in \mathfrak{g}$

$$\left\langle \mathrm{Ad}_g^*\,\mu\,,\,\xi\right\rangle = \langle \mu\,,\,\mathrm{Ad}_g\,\xi\rangle = \langle \mu\,,\,\xi\rangle.$$

Since the pairing is non-degenerate, $\mathrm{Ad}_g^*\,\mu = \mu$.

**Solution to Exercise 6.14** Let $g(t)$ be a smooth curve in $G$ such that $g(0) = e$ and $\dot{g}(0) = \xi \in \mathfrak{g}$. Let $\mu \in \mathfrak{g}^*$ and $\eta \in \mathfrak{g}$ and consider the following calculation of the infinitesimal generator of the Co-Adjoint action of $G$ on $\mathfrak{g}^*$.

$$\begin{aligned}\langle \xi_{\mathfrak{g}^*}(\mu)\,,\,\eta\rangle &= \left\langle \left.\frac{\mathrm{d}}{\mathrm{d}t}\right|_{t=0} \mathrm{Ad}^*_{g^{-1}(t)}\,\mu\,,\,\eta\right\rangle \\ &= \left.\frac{\mathrm{d}}{\mathrm{d}t}\right|_{t=0} \left\langle \mathrm{Ad}^*_{g^{-1}(t)}\,\mu\,,\,\eta\right\rangle\end{aligned}$$

$$
\begin{aligned}
&= \frac{d}{dt}\bigg|_{t=0} \left\langle \mu\,,\, \mathrm{Ad}_{g^{-1}(t)}\,\eta \right\rangle \\
&= \left\langle \mu\,,\, \frac{d}{dt}\bigg|_{t=0} \mathrm{Ad}_{g^{-1}(t)}\,\eta \right\rangle \\
&= \left\langle \mu\,,\, -\,\mathrm{ad}_\xi\,\eta \right\rangle \\
&= \left\langle -\,\mathrm{ad}^*_\xi\,\mu\,,\, \eta \right\rangle.
\end{aligned}
$$

Therefore, since the pairing is non-degenerate, the infinitesimal generator is given by

$$
\xi_{\mathfrak{g}^*}(\mu) = -\,\mathrm{ad}^*_\xi\,\mu.
$$

**Solution to Exercise 6.15** Both $\mathfrak{se}(2)$ and $\mathfrak{se}(3)$ have the following formulation of the Adjoint action.

$$
\begin{aligned}
\mathrm{Ad}_{(\boldsymbol{R},\boldsymbol{r})}(\xi, \boldsymbol{v}) &= (\boldsymbol{R}, \boldsymbol{r})(\xi, \boldsymbol{v})(\boldsymbol{R}^{-1}, -\boldsymbol{R}^{-1}\boldsymbol{r}) \\
&= (\boldsymbol{R}\xi, \boldsymbol{R}\boldsymbol{v})(\boldsymbol{R}^{-1}, -\boldsymbol{R}^{-1}\boldsymbol{r}) \\
&= (\mathrm{Ad}_{\boldsymbol{R}}\,\xi, \boldsymbol{R}\boldsymbol{v} - \left(\mathrm{Ad}_{\boldsymbol{R}}\,\xi\right)\boldsymbol{r}).
\end{aligned}
$$

Since $\mathfrak{so}(2)$ is Abelian, $\mathfrak{se}(2)$ also has $\mathrm{Ad}_{\boldsymbol{R}} = \boldsymbol{I}$. Therefore, for $\mathfrak{se}(2)$,

$$
\mathrm{Ad}_{(\boldsymbol{R},\boldsymbol{r})}(\xi, \boldsymbol{v}) = (\xi, \boldsymbol{R}\boldsymbol{v} - \xi\boldsymbol{r}).
$$

Differentiating the relationships above we find that in $\mathfrak{se}(3)$ the adjoint action is given by

$$
[(\eta, \boldsymbol{u})\,, (\xi, \boldsymbol{v})] = ([\eta, \xi]\,, \eta\boldsymbol{v} - \xi\boldsymbol{u})\,.
$$

Using the additional relationship in $\mathfrak{se}(2)$ gives

$$
[(\eta, \boldsymbol{u})\,, (\xi, \boldsymbol{v})] = (0, \eta\boldsymbol{v} - \xi\boldsymbol{u}).
$$

For the coadjoint actions it is convenient to use the hat map. The adjoint action is given by

$$
\mathrm{ad}_{(\hat{\eta},\boldsymbol{u})}(\hat{\xi}, \boldsymbol{v}) = (\eta \times \xi, \eta \times \boldsymbol{v} - \xi \times \boldsymbol{u})\,.
$$

The following calculation yields the coadjoint action,

$$\begin{aligned}
\left\langle \mathrm{ad}^*_{(\widehat{\boldsymbol{\eta}},\boldsymbol{u})}\,(\widehat{\boldsymbol{\mu}},\boldsymbol{m})\,,\,\left(\widehat{\boldsymbol{\xi}},\boldsymbol{v}\right)\right\rangle &= \left\langle (\widehat{\boldsymbol{\mu}},\boldsymbol{m})\,,\,\mathrm{ad}_{(\widehat{\boldsymbol{\eta}},\boldsymbol{u})}\left(\widehat{\boldsymbol{\xi}},\boldsymbol{v}\right)\right\rangle \\
&= \langle (\widehat{\boldsymbol{\mu}},\boldsymbol{m})\,,\,(\boldsymbol{\eta}\times\boldsymbol{\xi},\boldsymbol{\eta}\times\boldsymbol{v}-\boldsymbol{\xi}\times\boldsymbol{u})\rangle \\
&= \boldsymbol{\mu}\cdot(\boldsymbol{\eta}\times\boldsymbol{\xi})+\boldsymbol{m}\cdot(\boldsymbol{\eta}\times\boldsymbol{v}-\boldsymbol{\xi}\times\boldsymbol{u}) \\
&= -\,(\boldsymbol{\eta}\times\boldsymbol{\mu}+\boldsymbol{u}\times\boldsymbol{v})\cdot\boldsymbol{\xi}-\boldsymbol{\eta}\times\boldsymbol{m}\cdot\boldsymbol{v} \\
&= \left\langle\left((-\boldsymbol{\eta}\times\boldsymbol{\mu}-\boldsymbol{u}\times\boldsymbol{v})\widehat{\ },-\boldsymbol{\eta}\times\boldsymbol{m}\right),\,\left(\widehat{\boldsymbol{\xi}},\boldsymbol{v}\right)\right\rangle.
\end{aligned}$$

Therefore, we obtain the expression for the coadjoint action on $\mathfrak{se}^*(3)$ as

$$\mathrm{ad}^*_{(\widehat{\boldsymbol{\eta}},\boldsymbol{u})}\,(\widehat{\boldsymbol{\mu}},\boldsymbol{m}) = \left((-\boldsymbol{\eta}\times\boldsymbol{\mu}-\boldsymbol{u}\times\boldsymbol{v})\widehat{\ },-\boldsymbol{\eta}\times\boldsymbol{m}\right).$$

For $\mathfrak{se}^*(2)$ it is convenient to use the identification with $S^1$ parameterized in $\mathbb{C}$ by $e^{i\eta}$ for $\eta\in\mathbb{R}$. In this parametrization

$$\begin{aligned}
\left\langle \mathrm{ad}^*_{(i\eta,u)}\,(i\mu,m)\,,\,(i\xi,v)\right\rangle &= \langle (i\mu,m)\,,\,\mathrm{ad}_{(i\eta,u)}\,(i\xi,v)\rangle \\
&= \langle (i\mu,m)\,,\,(0,i\eta v-i\xi u)\rangle \\
&= \Re\left(i\overline{m}\,(\eta v-\xi u)\right) \\
&= \Im\left(\overline{m}u\right)\xi+\Re\left(i\overline{m}\eta v\right) \\
&= \langle (i\Im\left(\overline{m}u\right),-i\,(m\eta))\,,\,(i\xi,v)\rangle.
\end{aligned}$$

In $\mathbb{R}^3$ notation this reads

$$\mathrm{ad}^*_{((0,0,\eta)\widehat{\ },\boldsymbol{u})}\left((0,0,\mu)\widehat{\ },\boldsymbol{m}\right) = (-(\boldsymbol{u}\times\boldsymbol{m}),-(0,0,\xi)\times\boldsymbol{m}).$$

This is what one would expect since it is the restriction of the ad$^*$ action of $\mathfrak{se}^*(3)$ given by taking $\boldsymbol{m}$ and $\boldsymbol{u}$ to be in the $xy$-plane and the angular velocity to be on the $z$-axis.

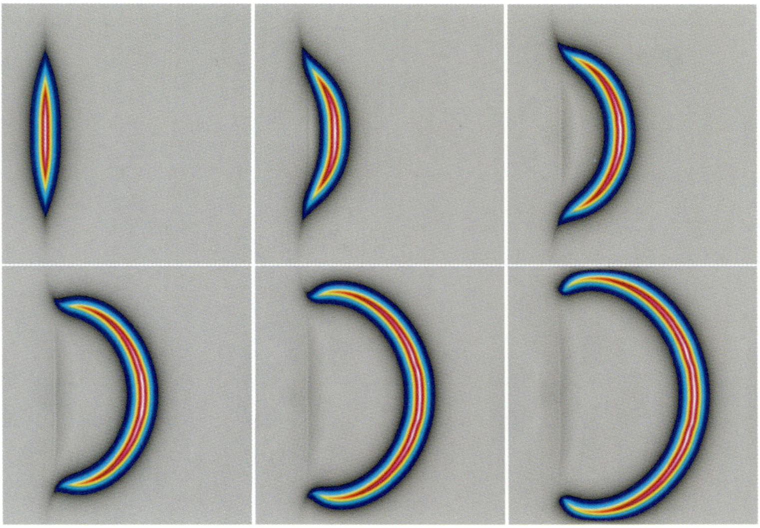

**Plate 1** A peakon segment of finite length is initially moving rightward (East). Because its speed vanishes at its ends and it has fully two-dimensional spatial dependence, it expands into a peakon 'bubble' as it propagates. (The colors indicate speed: red is highest, yellow is less, blue low, grey zero.) See page 410.

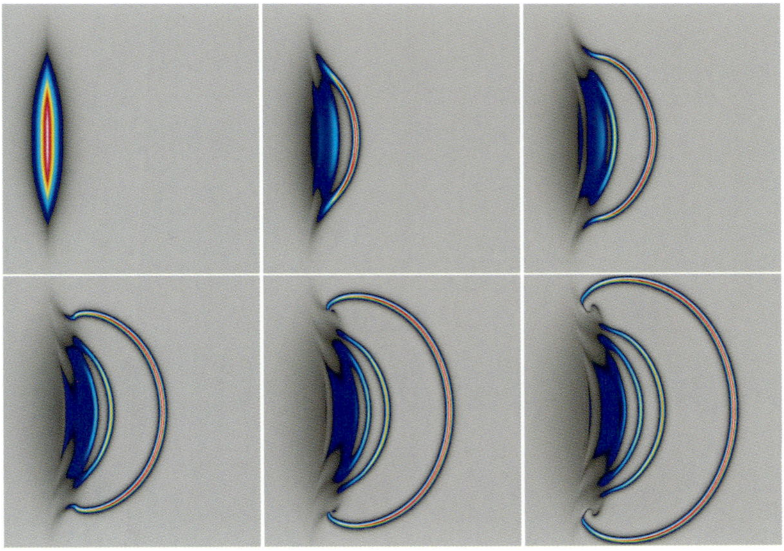

**Plate 2** An initially straight segment of velocity distribution whose exponential profile is wider than the width of the peakon solution will break up into a ***train*** of curved peakon 'bubbles'. This example illustrates the emergent property of the peakon solutions in two dimensions. See page 411.

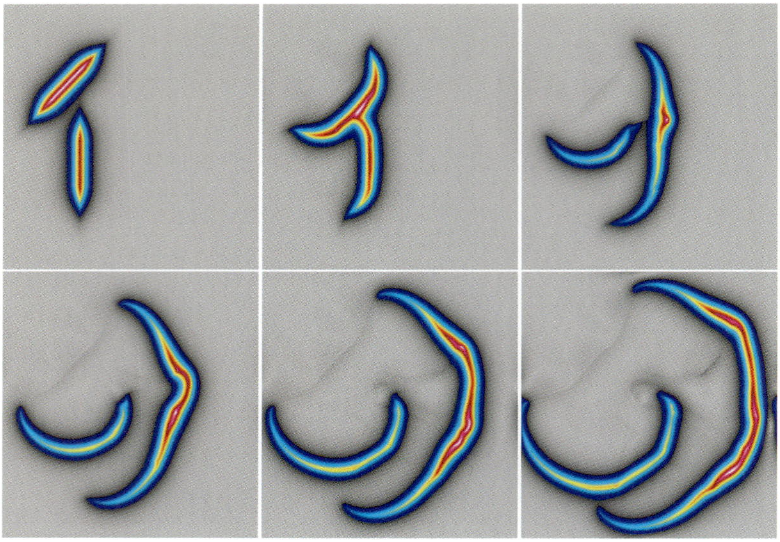

**Plate 3** A single collision is shown involving reconnection as the faster peakon segment initially moving Southeast along the diagonal expands, curves and obliquely overtakes the slower peakon segment initially moving rightward (East). This reconnection illustrates one of the collision rules for the strongly two-dimensional EPDiff flow. See page 412.

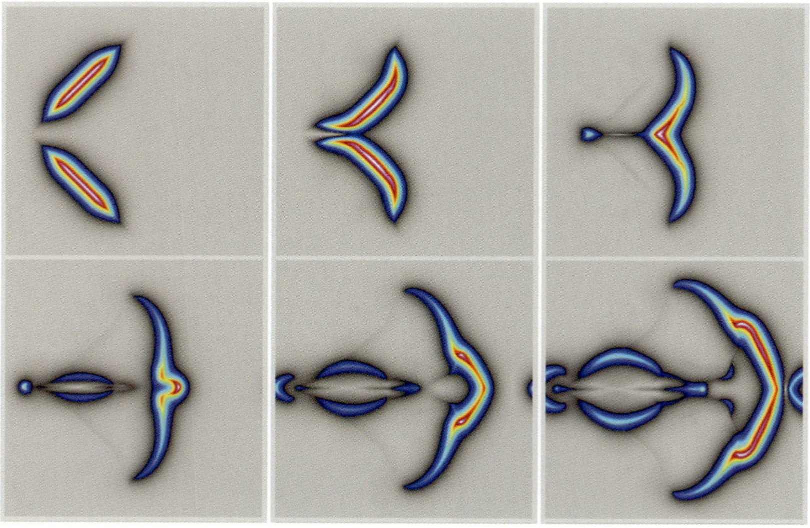

**Plate 4** The convergence of two peakon segments moving with reflection symmetry generates considerable acceleration along the midline, which continues to build up after the initial collision. See page 413.

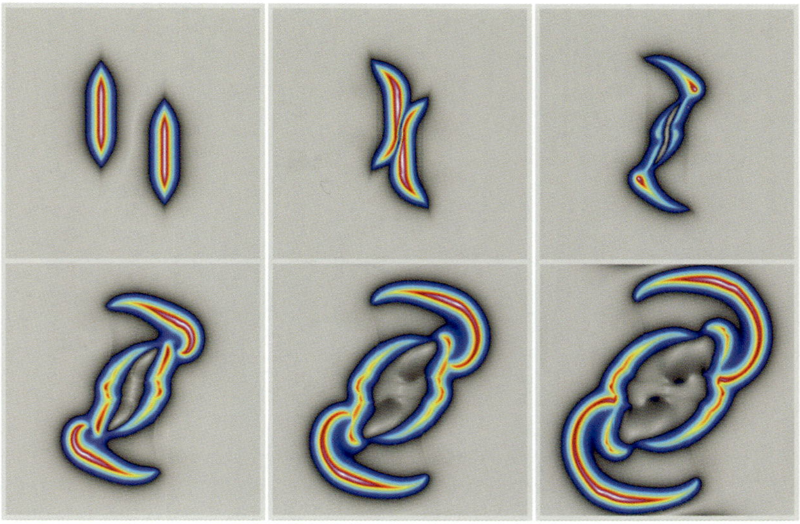

**Plate 5** The head-on collision of two offset peakon segments generates considerable complexity. Some of this complexity is due to the process of annihilation and recreation that occurs in the 1D antisymmetric head-on collisions of a peakon with its reflection, the antipeakon, as shown in Figure 12.3. Other aspects of it involve flow along the crests of the peakon segments as they stretch. See page 414.

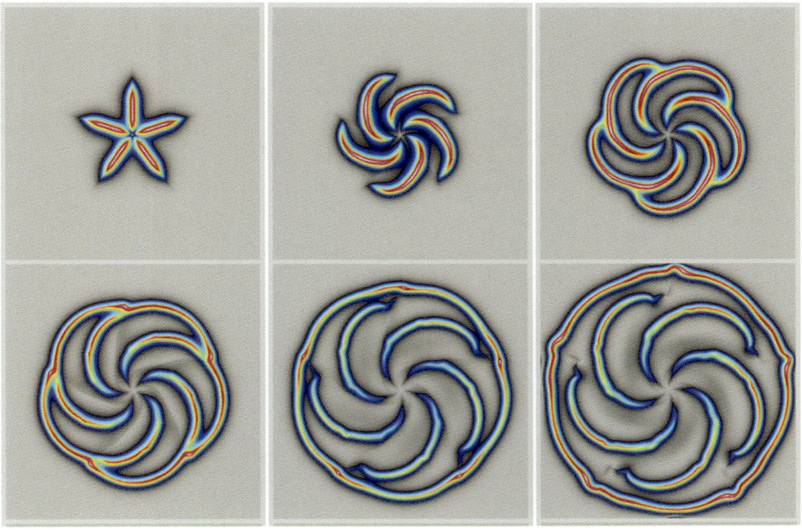

**Plate 6** The overtaking collisions of these rotating peakon segments with five-fold symmetry produces many reconnections (mergers), until eventually one peakon ring surrounds five curved peakon segments. If the evolution were allowed to proceed further, reconnections would tend to produce additional concentric peakon rings. See page 414.

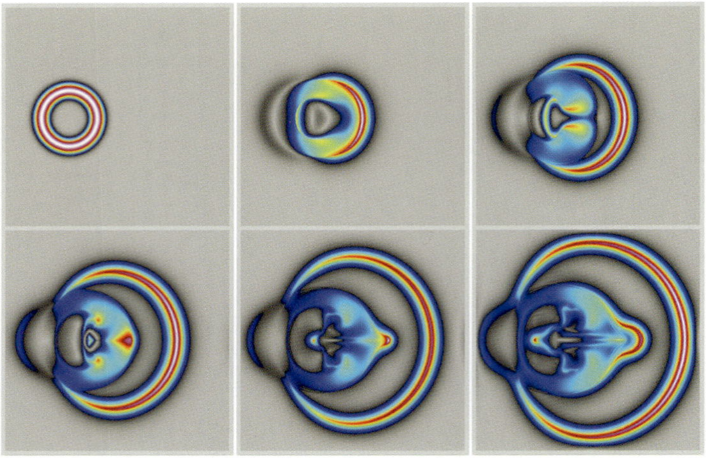

**Plate 7** A circular peakon ring initially undergoes uniform rightward translational motion along the $x$ axis. The right outer side of the ring produces diverging peakon curves, which slow as they propagate outward. The left inner side of the ring, however, produces converging peakon segments, which accelerate as they converge. They collide along the midline, then develop into divergent peakon curves still moving rightward that overtake the previous ones and collide with them from behind. These overtaking collisions impart momentum, but they apparently do not produce reconnections. See page 415.

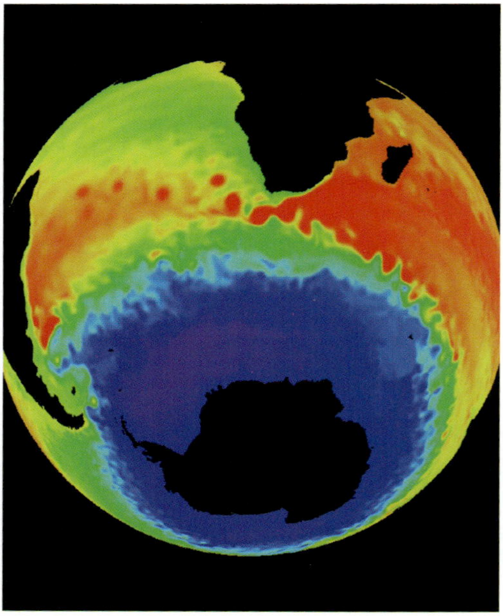

**Plate 8** Numerically computed isobars of the sea surface elevation. In the colour scheme of the figure, red is higher sea surface elevation and blue is lower. Geostrophically balanced flow takes place along these isobars. This view of the Southern Ocean shows the collision of the wide eastward Antarctic Circumpolar Current (ACC, blue) with the narrow westward Agulhas Current (red) off the coast of South Africa. The collision with the ACC turns the Agulhas back on itself and makes it flow eastward in an undulating path. Where the Agulhas turns back on itself a swirling loop pinches off and periodically releases eddies into the South Atlantic Ocean. See page 470.

# 7 Euler–Poincaré reduction: Rigid body and heavy top

In the absence of external torque, Euler's equation for a rigid body with a fixed point is

$$\mathbb{I}\dot{\boldsymbol{\Omega}} = \mathbb{I}\boldsymbol{\Omega} \times \boldsymbol{\Omega}, \tag{7.1}$$

where $\boldsymbol{\Omega}$ is the angular velocity of the body and $\mathbb{I}$ is the moment of inertia, both expressed in body coordinates. We derived this equation in Chapter 1, by showing that it is equivalent to the conservation of spatial angular momentum. How does this equation fit into the framework of Lagrangian mechanics, and how can it be generalized to apply to other mechanical systems?

The Lie group $SO(3)$ plays a double role in the mechanics of the rigid body: it is the configuration space, and it is also the symmetry group of the Lagrangian. This allows us to introduce a specific procedure for obtaining the reduced dynamics on the quotient space $TSO(3)/SO(3)$. This procedure, called Euler–Poincaré reduction, applies to arbitrary Lie groups, which makes it powerful in a variety of applications. For instance, we will apply it in later chapters in the study of pseudo-rigid bodies and fluids. Further, an adapted version of the Euler–Poincaré reduction extends to systems with broken symmetry such as the rigid body in a gravitational field, i.e. a heavy top. The Hamiltonian counterpart of this reduction theory will be the subject of Chapter 9.

## 7.1 Rigid body dynamics

Consider a rigid body with a fixed point. It is often assumed that this fixed point is the centre of mass of the body, but this is not necessary, and will not be true when we later study heavy tops.

Recall from Chapter 1 that given a reference configuration of the body, two systems of coordinates are introduced: a fixed inertial ***spatial coordinate system***, and a moving ***body coordinate system***, both with origin at the fixed point of the body. When the body is in its reference position, the two systems coincide. The body coordinate system moves with the body, which means that the position of any given particle, in the body coordinate system, remains fixed. The position in body coordinates is called the particle's ***label***. The configuration of the body at time $t$ is determined by a rotation matrix $\mathbf{R}(t)$ that takes the label $\mathbf{X}$ of any particle in the body to its spatial position $\mathbf{x}(t)$. Thus, the configuration space of the rigid body is $SO(3)$, and $\mathbf{R}(t)$ is a path in $SO(3)$. The position and velocity at time $t$ of the particle with label $\mathbf{X}$ are given as in eqn (1.35)

$$\mathbf{x}(t) = \mathbf{R}(t)\mathbf{X}, \quad \dot{\mathbf{x}}(t) = \dot{\mathbf{R}}(t)\mathbf{X} = \dot{\mathbf{R}}(t)\mathbf{R}^{-1}(t)\mathbf{x}(t). \tag{7.2}$$

Since $\mathbf{R}(t)$ is a curve in the space of orthogonal matrices, the quantities $\mathbf{R}^{-1}\dot{\mathbf{R}}$ and $\dot{\mathbf{R}}\mathbf{R}^{-1}$ are both skew-symmetric matrices. This can be checked directly, as in Chapter 1, but we now have another way of seeing this. Any tangent vector $(\mathbf{R}, \dot{\mathbf{R}}) \in TSO(3)$ can be translated to $\mathfrak{so}(3)$ (which equals $T_{\mathbf{I}}SO(3)$), by the tangent lift of either left multiplication or right multiplication by $\mathbf{R}^{-1}$:

$$TL_{\mathbf{R}^{-1}}(\mathbf{R}, \dot{\mathbf{R}}) = (\mathbf{I}, \mathbf{R}^{-1}\dot{\mathbf{R}}),$$

$$TR_{\mathbf{R}^{-1}}(\mathbf{R}, \dot{\mathbf{R}}) = (\mathbf{I}, \dot{\mathbf{R}}\mathbf{R}^{-1}),$$

where $\mathbf{I}$ is the identity matrix. Dropping the $\mathbf{I}$'s, this shows that $\mathbf{R}^{-1}\dot{\mathbf{R}}$ and $\dot{\mathbf{R}}\mathbf{R}^{-1}$ are elements of $\mathfrak{so}(3)$, which is the set of $3 \times 3$ antisymmetric matrices.

The ***spatial angular velocity vector***, $\boldsymbol{\omega}$, is defined by

$$\widehat{\boldsymbol{\omega}} = \dot{\mathbf{R}}\mathbf{R}^{-1},$$

using the 'hat map' defined in eqn (1.39). Note that

$$\dot{\mathbf{x}} = \dot{\mathbf{R}}\mathbf{R}^{-1}\mathbf{x} = \widehat{\boldsymbol{\omega}}\mathbf{x} = \boldsymbol{\omega} \times \mathbf{x}.$$

The ***body angular velocity***, $\boldsymbol{\Omega}$, is the spatial angular velocity vector expressed with respect to the body coordinate system,

$$\boldsymbol{\Omega} = \mathbf{R}^{-1}\boldsymbol{\omega}. \tag{7.3}$$

By a straightforward calculation, given in eqn (1.43), it follows that

$$\widehat{\boldsymbol{\Omega}} = \mathbf{R}^{-1}\dot{\mathbf{R}} = \mathrm{Ad}_{\mathbf{R}^{-1}}\,\widehat{\boldsymbol{\omega}}. \tag{7.4}$$

Thus, the matrix for *spatial* angular velocity $\boldsymbol{\omega}$ is given by *right* translation of $\dot{\mathbf{R}}$ to the identity, while the matrix for *body* angular velocity $\boldsymbol{\Omega}$ is given by *left* translation of $\dot{\mathbf{R}}$ to the identity. The two angular velocities are related by the Adjoint operation of the rotation group $SO(3)$ on its Lie algebra $\mathfrak{so}(3)$.

Let $\rho(\mathbf{X})$ be the density of the body at position $\mathbf{X}$ in body coordinates, which we assume to be constant. Let $\mathcal{B}$ be the region occupied by the body in its reference configuration. Then the mass of the body is

$$m = \int_{\mathcal{B}} \rho(\mathbf{X})\, \mathrm{d}^3\mathbf{X}.$$

By analogy with the kinetic energy $\frac{1}{2}\sum_i m_i \|\dot{\mathbf{x}}_i\|^2$ of a system of particles, we define the ***kinetic energy*** of the rigid body as

$$K = \frac{1}{2}\int_{\mathcal{B}} \rho(\mathbf{X}) \|\dot{\mathbf{x}}\|^2\, \mathrm{d}^3\mathbf{X}.$$

We can rewrite this definition as follows:

$$\begin{aligned} K &= \frac{1}{2}\int_{\mathcal{B}} \rho(\mathbf{X}) \|\dot{\mathbf{R}}\mathbf{X}\|^2\, \mathrm{d}^3\mathbf{X} = \frac{1}{2}\int_{\mathcal{B}} \rho(\mathbf{X}) \|\mathbf{R}^{-1}\dot{\mathbf{R}}\mathbf{X}\|^2\, \mathrm{d}^3\mathbf{X} \\ &= \frac{1}{2}\int_{\mathcal{B}} \rho(\mathbf{X}) \|\widehat{\boldsymbol{\Omega}}\mathbf{X}\|^2\, \mathrm{d}^3\mathbf{X} = \frac{1}{2}\int_{\mathcal{B}} \rho(\mathbf{X})\, \mathrm{tr}\left((\widehat{\boldsymbol{\Omega}}\mathbf{X})(\widehat{\boldsymbol{\Omega}}\mathbf{X})^T\right) \mathrm{d}^3\mathbf{X}, \end{aligned} \tag{7.5}$$

where in the last step we have used the identity $\mathbf{v}^T\mathbf{w} = \mathrm{tr}\left(\mathbf{v}\mathbf{w}^T\right)$, which appeared in Exercise 5.15. We now move the integration inside the trace. Since the body angular velocity, $\widehat{\boldsymbol{\Omega}}$, is independent of the label $\mathbf{X}$, we can see that

$$K = \frac{1}{2}\,\mathrm{tr}\left(\widehat{\boldsymbol{\Omega}}\left(\int_{\mathcal{B}} \rho(\mathbf{X})\mathbf{X}\mathbf{X}^T\, \mathrm{d}^3\mathbf{X}\right)\widehat{\boldsymbol{\Omega}}^T\right).$$

Let the ***coefficient of inertia matrix*** of the body, with respect to the origin, be

$$\mathbb{J} = \int_{\mathcal{B}} \rho(\mathbf{X})\mathbf{X}\mathbf{X}^T\, \mathrm{d}^3\mathbf{X}.$$

Note that $\mathbb{J}$ is symmetric and constant.

The kinetic energy may be written as

$$K = \frac{1}{2}\,\mathrm{tr}\left(\widehat{\boldsymbol{\Omega}}\,\mathbb{J}\,\widehat{\boldsymbol{\Omega}}^T\right). \tag{7.6}$$

Since $\widehat{\boldsymbol{\Omega}} = \mathbf{R}^{-1}\dot{\mathbf{R}}$, we also have

$$K = \frac{1}{2}\,\mathrm{tr}\left(\mathbf{R}^{-1}\dot{\mathbf{R}}\,\mathbb{J}\left(\mathbf{R}^{-1}\dot{\mathbf{R}}\right)^T\right) = \frac{1}{2}\,\mathrm{tr}\left(\dot{\mathbf{R}}\,\mathbb{J}\,\dot{\mathbf{R}}^T\right), \tag{7.7}$$

where we have used the identity $\mathrm{tr}(\mathbf{A}\mathbf{B}) = \mathrm{tr}(\mathbf{B}\mathbf{A})$ and the orthogonality of $\mathbf{R}$.

We can also express $K$ as

$$K = \frac{1}{2}\|\dot{\mathbf{R}}\|^2,$$

where the norm corresponds to the following Riemannian metric on $SO(3)$, called the ***kinetic energy metric***,

$$\langle\langle \dot{\mathbf{R}}_1, \dot{\mathbf{R}}_2 \rangle\rangle_{\mathbf{R}} := \operatorname{tr}\left(\dot{\mathbf{R}}_1\, \mathbb{J}\, \dot{\mathbf{R}}_2^T\right). \tag{7.8}$$

An important property of this metric is that it is invariant with respect to tangent-lifted *left* translation:

$$\begin{aligned}\langle\langle \mathbf{Q}\dot{\mathbf{R}}_1, \mathbf{Q}\dot{\mathbf{R}}_2 \rangle\rangle_{\mathbf{QR}} &= \operatorname{tr}\left(\left(\mathbf{Q}\dot{\mathbf{R}}_1\right) \mathbb{J} \left(\mathbf{Q}\dot{\mathbf{R}}_2\right)^T\right)\\ &= \operatorname{tr}\left(\dot{\mathbf{R}}_1\, \mathbb{J}\, \dot{\mathbf{R}}_2^T\right) = \langle\langle \dot{\mathbf{R}}_1, \dot{\mathbf{R}}_2 \rangle\rangle_{\mathbf{R}}.\end{aligned}$$

The left invariance of this metric implies that the kinetic energy $K$ is also left-invariant. In fact, this left invariance is why it is possible to express $K$ in terms of $\boldsymbol{\Omega}$ only: note that the first step in eqn (7.5) is actually an alternative proof of the left invariance of $K$.

The left invariance of $K$ is called its ***spatial symmetry***. To understand this terminology, consider a configuration $\mathbf{R}$ of the rigid body, which maps every label $\mathbf{X}$ to the corresponding spatial position $\mathbf{x} = \mathbf{RX}$. Left multiplying $\mathbf{R}$ by $\mathbf{Q}$ gives a configuration $\mathbf{QR}$, which maps the label $\mathbf{X}$ to spatial position $\mathbf{QRX} = \mathbf{Qx}$. In other words, a particle that has spatial position $\mathbf{x}$ when the body is in configuration $\mathbf{R}$ has spatial position $\mathbf{Qx}$ when the body is in configuration $\mathbf{QR}$. So to get from configuration $\mathbf{R}$ to configuration $\mathbf{QR}$, the body must undergo a rotation that has matrix $\mathbf{Q}$ in ***spatial coordinates***.[1]

In contrast, *right* multiplication by $\mathbf{Q}$ causes points with spatial position $\mathbf{RX}$ to move to spatial position $\mathbf{RQX}$, which we can think of as moving a point with label $\mathbf{X}$ to a point with label $\mathbf{QX}$. In other words, $\mathbf{Q}$ is acting as a rotation in ***body coordinates***.

There is another useful way to express the kinetic energy, in terms of the angular velocity vector $\boldsymbol{\Omega}$, instead of the corresponding matrix $\widehat{\boldsymbol{\Omega}}$. By definition of the hat map, $\widehat{\boldsymbol{\Omega}}\mathbf{X} = \boldsymbol{\Omega} \times \mathbf{X} = -\mathbf{X} \times \boldsymbol{\Omega} = -\widehat{\mathbf{X}}\boldsymbol{\Omega}$, so from

[1] Note that the correspondence of spatial rotations with *left* multiplication depends on the convention that linear transformations correspond to *left* multiplying by the appropriate matrix.

eqn (7.5),

$$\begin{aligned}K &= \frac{1}{2}\int_{\mathcal{B}} \rho(\mathbf{X})\|\widehat{\boldsymbol{\Omega}}\mathbf{X}\|^2\,\mathrm{d}^3\mathbf{X} = \frac{1}{2}\int_{\mathcal{B}} \rho(\mathbf{X})\|\widehat{\mathbf{X}}\boldsymbol{\Omega}\|^2\,\mathrm{d}^3\mathbf{X}\\ &= \frac{1}{2}\int_{\mathcal{B}} \rho(\mathbf{X})\boldsymbol{\Omega}^T\widehat{\mathbf{X}}^T\widehat{\mathbf{X}}\boldsymbol{\Omega}\,\mathrm{d}^3\mathbf{X}\\ &= \frac{1}{2}\boldsymbol{\Omega}^T\left(\int_{\mathcal{B}} \rho(\mathbf{X})\widehat{\mathbf{X}}^T\widehat{\mathbf{X}}\,\mathrm{d}^3\mathbf{X}\right)\boldsymbol{\Omega}.\end{aligned} \tag{7.9}$$

We now need one further identity concerning the hat map (see Exercise 7.4):

$$\widehat{\mathbf{X}}^T\widehat{\mathbf{X}} = \|\mathbf{X}\|^2\mathbf{I} - \mathbf{X}\mathbf{X}^T.$$

The right-hand side of this identity looks remarkably like the ***moment of inertia tensor***, defined in eqn (1.49):

$$\mathbb{I} := \int_{\mathcal{B}} \rho(\mathbf{X})\left(\|\mathbf{X}\|^2\,\mathbf{I} - \mathbf{X}\mathbf{X}^T\right)\mathrm{d}^3\mathbf{X}.$$

Indeed, it now follows from the last line of eqn (7.9) that

$$K = \frac{1}{2}\boldsymbol{\Omega}^T\mathbb{I}\,\boldsymbol{\Omega}, \tag{7.10}$$

which we can also express using the Euclidean inner product as $K = \frac{1}{2}\langle\boldsymbol{\Omega}, \mathbb{I}\,\boldsymbol{\Omega}\rangle$.

**Remark 7.1** *By a slight generalization of the arguments above, one may show that*

$$\mathrm{tr}\left(\hat{\boldsymbol{\xi}}\,\mathbb{J}\,\hat{\boldsymbol{\eta}}^T\right) = \boldsymbol{\xi}^T\,\mathbb{I}\,\boldsymbol{\eta},$$

*for any* $\boldsymbol{\xi}, \boldsymbol{\eta} \in \mathfrak{so}(3)$.

The motion of the body can be determined by Hamilton's principle, for a Lagrangian $L : TSO(3) \to \mathbb{R}$. If there are no external forces on the body, then the Lagrangian is equal to the kinetic energy, and the system is called the ***free rigid body***. More generally, in the absence of dissipative or magnetic forces, the system is often 'simple mechanical', i.e. the Lagrangian function is of the form 'kinetic minus potential',

$$L(\mathbf{R}, \dot{\mathbf{R}}) = \frac{1}{2}\,\mathrm{tr}(\dot{\mathbf{R}}\,\mathbb{J}\dot{\mathbf{R}}^T) - V(\mathbf{R}),$$

for some function $V : SO(3) \to \mathbb{R}$ called the potential energy. Once we have the Lagrangian, we can formulate the equations of motion, which are the Euler–Lagrange equations on $TSO(3)$:

$$\frac{\mathrm{d}}{\mathrm{d}t}\left(\frac{\partial L}{\partial \dot{\mathrm{R}}}\right) = \frac{\partial L}{\partial \mathrm{R}}. \tag{7.11}$$

The interpretation of this equation is tricky. In this form, it is only valid in local coordinates on the manifold $SO(3)$, for example Euler angles (see Exercise 4.20). Further, even though $L$ is left $G$-invariant, $\partial L/\partial \mathrm{R}$ won't necessarily be zero in local coordinates. If we wish instead to proceed using only matrix calculations, we have to consider $SO(3)$ as a submanifold of $\mathcal{M}(3,\mathbb{R}) \cong \mathbb{R}^9$, and use Lagrange multipliers to describe the motion as a system on $\mathbb{R}^9$ with holonomic constraints (see Exercise 7.7). Thus, while correct and very helpful in understanding the problem as a mechanical system, the Euler–Lagrange equations are unwieldy to work with directly. However, for the free rigid body, and many other systems, symmetry provides a great simplification, as we will see in the next section.

**Exercise 7.1**
Consider an ellipsoidal body with uniform density, with axes aligned with the coordinate axes when the body is in the reference configuration. Calculate $\mathbb{J}$ and show that, in this example, it equals one half of the moment of inertia tensor $\mathbb{I}$, given in Exercise 1.13. (This is not a generally valid relationship.)

**Exercise 7.2**
Show that the eqn (7.8) defines a Riemannian metric.

**Exercise 7.3**
Consider an ellipsoidal body with uniform density, with axes aligned with the coordinate axes when the body is in the reference configuration. Show that the kinetic energy metric need *not* be invariant with respect to the tangent-lifted *right* multiplication action of $SO(3)$ on $TSO(3)$, and in fact it is only right invariant if the ellipsoid is a sphere.

**Exercise 7.4**
Verify the following by direct computation:

$$\widehat{\mathbf{X}}^T\widehat{\mathbf{X}} = \|\mathbf{X}\|^2\mathbf{I} - \mathbf{X}\mathbf{X}^T$$

for all $\mathbf{X} \in \mathbb{R}^n$.

**Exercise 7.5**
Show that $\mathbb{I} = \mathrm{tr}\,(\mathbb{J})\,\mathbf{I} - \mathbb{J}$.

**Exercise 7.6**
Suppose that $\mathbb{I}$ is diagonal, with $\mathbb{I} = diag(I_1, I_2, I_3)$. Show that $\mathbb{J}\widehat{\boldsymbol{\Omega}} := D\widehat{\boldsymbol{\Omega}} + \widehat{\boldsymbol{\Omega}}D$, with $D = diag(d_1, d_2, d_3)$ with $d_1 = I_2^2 + I_3^2$, $d_2 = I_3^2 + I_1^2$, and $d_3 = I_1^2 + I_2^2$.

**Exercise 7.7 (Rigid body modelled using holonomic constraints)**
Consider the motion of a rigid body as modelled by a given Lagrangian

$$L : TSO(3) \to \mathbb{R}, \quad L(\mathbf{R}, \dot{\mathbf{R}}) = \frac{1}{2}\mathrm{tr}(\dot{\mathbf{R}}\mathbb{J}\dot{\mathbf{R}}^T) - V(\mathbf{R}).$$

The system may be seen as defined on the space of $3 \times 3$ matrices subject to the holonomic constraints $\mathbf{R}\mathbf{R}^T = \mathbf{R}^T\mathbf{R} = \mathbf{I}$, $\mathbf{R} \in \mathcal{M}_{3\times3}$. The constrained Lagrangian (see Section 1.3) is given by

$$L_{gen}(\mathbf{R}, \dot{\mathbf{R}}) = \frac{1}{2}\mathrm{tr}(\dot{\mathbf{R}}\mathbb{J}\dot{\mathbf{R}}^T) - V(\mathbf{R}) + \frac{1}{2}\mathrm{tr}\left((\mathbf{R}\mathbf{R}^T - \mathbf{I})\Lambda^T\right) + \frac{1}{2}\mathrm{tr}\left((\mathbf{R}^T\mathbf{R} - \mathbf{I})\Lambda^T\right),$$

where the matrix $\Lambda$ is formed by the Lagrange multipliers $(\Lambda_{ij})$ with $i, j = 1, 2, 3$.

a) Show that since $\mathbf{R}\mathbf{R}^T = \mathbf{R}^T\mathbf{R}$, the matrix $\Lambda$ must be symmetric, i.e. $\Lambda = \Lambda^T$ and so

$$L_{gen}(\mathbf{R}, \dot{\mathbf{R}}) = \frac{1}{2}\mathrm{tr}(\dot{\mathbf{R}}\mathbb{J}\dot{\mathbf{R}}^T) - V(\mathbf{R}) + \mathrm{tr}\left(\Lambda(\mathbf{R}\mathbf{R}^T - \mathbf{I})\right),$$

b) Show that the Euler–Lagrange equation for $L_{gen}$ is

$$\ddot{\mathrm{R}}\mathbb{J} + \frac{\partial V}{\partial \mathrm{R}} = 2\mathrm{R}\Lambda.$$

Here $\partial V/\partial \mathrm{R}$ is understood as the matrix whose entries are $(\partial V/\partial \mathrm{R}_{ij})$.

c) Use the symmetry of $\Lambda$ to show that the equations above are equivalent to

$$\mathrm{R}^T\ddot{\mathrm{R}}\mathbb{J} - (\mathrm{R}^T\ddot{\mathrm{R}}\mathbb{J})^T = \frac{\partial V}{\partial \mathrm{R}^T}\mathrm{R} - \mathrm{R}^T\frac{\partial V}{\partial \mathrm{R}} =: \breve{\Xi}.$$

d) Use that $\mathrm{R}^T\dot{\mathrm{R}} = \widehat{\boldsymbol{\Omega}}$ and that $\boldsymbol{\Omega} + \boldsymbol{\Omega}^T = 0$ to show that the equation from c) is equivalent to:

$$\dot{\widehat{\boldsymbol{\Omega}}}\mathbb{J} + \mathbb{J}\dot{\widehat{\boldsymbol{\Omega}}} + \widehat{\boldsymbol{\Omega}}^2\mathbb{J} - \mathbb{J}\widehat{\boldsymbol{\Omega}}^2 = \breve{\Xi}. \tag{7.12}$$

e) Use the identity $\mathbb{I} = \mathrm{tr}(\mathbb{J})\mathbf{I} - \mathbb{J}$ and the ***breve map*** given in Example 5.36 to show that the vector representation of the left-hand side of eqn (7.12) above is

$$2(\mathbb{I}\dot{\boldsymbol{\Omega}} - \mathbb{I}\boldsymbol{\Omega} \times \boldsymbol{\Omega}).$$

Conclude that the motion of a (non-free) rigid body in $\mathbb{R}^3$ is given by

$$\mathbb{I}\dot{\boldsymbol{\Omega}} = \mathbb{I}\boldsymbol{\Omega} \times \boldsymbol{\Omega} + \Xi. \tag{7.13}$$

## 7.2 Euler–Poincaré reduction: the rigid body

As we have seen, it is difficult to use the Euler–Lagrange equations directly to determine the motion of the rigid body. In this section, we follow an alternative path, taking advantage of the symmetry of the Lagrangian of the free rigid body. In this case, observe that the Lagrangian can be expressed in terms of the body angular velocity $\boldsymbol{\Omega}$ only, since

$$L(\mathrm{R}, \dot{\mathrm{R}}) = \frac{1}{2}\,\mathrm{tr}(\dot{\mathrm{R}}\,\mathbb{J}\dot{\mathrm{R}}^T) = \frac{1}{2}\,\mathrm{tr}(\widehat{\boldsymbol{\Omega}}\,\mathbb{J}\widehat{\boldsymbol{\Omega}}^T) = \frac{1}{2}\boldsymbol{\Omega}^T\mathbb{I}\boldsymbol{\Omega}.$$

Thus, we can introduce a ***reduced Lagrangian*** $l : \mathfrak{so}(3) \to \mathbb{R}$, defined by

$$l(\widehat{\mathbf{\Omega}}) = L(\mathbf{R}, \dot{\mathbf{R}}) = \frac{1}{2}\,\mathrm{tr}(\widehat{\mathbf{\Omega}}\,\mathbb{J}\widehat{\mathbf{\Omega}}^T) = \frac{1}{2}\mathbf{\Omega}^T\mathbb{I}\mathbf{\Omega}.$$

Note that

$$l(\widehat{\mathbf{\Omega}}) = L(\mathbf{I}, \widehat{\mathbf{\Omega}}),$$

so $l$ is actually the restriction of $L$ to $T_I SO(3) = \mathfrak{so}(3)$.

The equations of motion are provided by Hamilton's principle:

$$\delta \int_a^b L(\mathbf{R}, \dot{\mathbf{R}})\,\mathrm{d}t = 0, \tag{7.14}$$

where the variations $\delta\mathbf{R}$ are taken among paths $\mathbf{R}(t) \in SO(3)$, $t \in [a,b]$, with fixed endpoints, so that $\delta\mathbf{R}(a) = \delta\mathbf{R}(b) = 0$. By the definition of $l$, Hamilton's principle takes the equivalent form:

$$\delta \int_a^b l(\widehat{\mathbf{\Omega}})\,\mathrm{d}t = 0, \tag{7.15}$$

where the variations $\delta\widehat{\mathbf{\Omega}}$ are induced by the variations $\delta\mathbf{R}$. This is equivalent to:

$$\int_a^b \left\langle \frac{\delta l}{\delta\widehat{\mathbf{\Omega}}}, \delta\widehat{\mathbf{\Omega}} \right\rangle \mathrm{d}t = 0. \tag{7.16}$$

The pairing in the previous equation is the natural pairing of elements of $\mathfrak{so}(3)^*$ with elements of $\mathfrak{so}(3)$. When we represent $\delta l/\delta\widehat{\mathbf{\Omega}}$ as a matrix, its pairing with $\delta\widehat{\mathbf{\Omega}}$ will be the trace pairing, introduced in Example 3.74.

The variation in body angular velocity $\delta\widehat{\mathbf{\Omega}}$ is found by taking the variational derivative of its definition $\widehat{\mathbf{\Omega}} = \mathbf{R}^{-1}\dot{\mathbf{R}}$. This yields

$$\begin{aligned}\delta\widehat{\mathbf{\Omega}} &= -\mathbf{R}^{-1}\delta\mathbf{R}\mathbf{R}^{-1}\dot{\mathbf{R}} + \mathbf{R}^{-1}\delta\dot{\mathbf{R}} \\ &= -\left(\mathbf{R}^{-1}\delta\mathbf{R}\right)\left(\mathbf{R}^{-1}\dot{\mathbf{R}}\right) + \mathbf{R}^{-1}\delta\dot{\mathbf{R}},\end{aligned} \tag{7.17}$$

so that

$$\delta\widehat{\mathbf{\Omega}} = -\left(\mathbf{R}^{-1}\delta\mathbf{R}\right)\widehat{\mathbf{\Omega}} + \mathbf{R}^{-1}\delta\dot{\mathbf{R}}. \tag{7.18}$$

Define $\widehat{\mathbf{\Sigma}} \in \mathfrak{so}(3)$ by

$$\widehat{\mathbf{\Sigma}} = \mathbf{R}^{-1}\delta\mathbf{R}. \tag{7.19}$$

Note that $\widehat{\boldsymbol{\Sigma}}$ vanishes at the endpoints in time, since $\delta \mathrm{R}$ does. Differentiating the previous equation gives

$$\frac{\mathrm{d}\widehat{\boldsymbol{\Sigma}}}{\mathrm{d}t} = -\,\mathrm{R}^{-1}\dot{\mathrm{R}}\mathrm{R}^{-1}\delta\mathrm{R} + \mathrm{R}^{-1}\delta\dot{\mathrm{R}}$$

(using an identity in Exercise 3.12), so

$$\mathrm{R}^{-1}\delta\dot{\mathrm{R}} = \frac{\mathrm{d}\widehat{\boldsymbol{\Sigma}}}{\mathrm{d}t} + \mathrm{R}^{-1}\dot{\mathrm{R}}\widehat{\boldsymbol{\Sigma}}. \tag{7.20}$$

Substituting eqns (7.19) and (7.20) into eqn (7.18) gives

$$\delta\widehat{\boldsymbol{\Omega}} = -\,\widehat{\boldsymbol{\Sigma}}\widehat{\boldsymbol{\Omega}} + \frac{\mathrm{d}\widehat{\boldsymbol{\Sigma}}}{\mathrm{d}t} + \widehat{\boldsymbol{\Omega}}\widehat{\boldsymbol{\Sigma}}.$$

That is,

$$\delta\widehat{\boldsymbol{\Omega}} = \frac{\mathrm{d}\widehat{\boldsymbol{\Sigma}}}{\mathrm{d}t} + [\widehat{\boldsymbol{\Omega}}, \widehat{\boldsymbol{\Sigma}}], \tag{7.21}$$

where $[\,\cdot, \cdot\,]$ is the matrix commutator. Since $[\widehat{\boldsymbol{\Omega}}, \widehat{\boldsymbol{\Sigma}}] = \widehat{\boldsymbol{\Omega} \times \boldsymbol{\Sigma}}$, one finds the equivalent vector representation,

$$\delta\boldsymbol{\Omega} = \dot{\boldsymbol{\Sigma}} + \boldsymbol{\Omega} \times \boldsymbol{\Sigma}. \tag{7.22}$$

Substituting these manipulations into eqn (7.16) produces:

$$\begin{aligned}
0 &= \int_a^b \left\langle \frac{\delta l}{\delta\widehat{\boldsymbol{\Omega}}}, \delta\widehat{\boldsymbol{\Omega}} \right\rangle \mathrm{d}t = \int_a^b \left\langle \frac{\delta l}{\delta\boldsymbol{\Omega}}, \delta\boldsymbol{\Omega} \right\rangle \mathrm{d}t \\
&= \int_a^b \left\langle \frac{\delta l}{\delta\boldsymbol{\Omega}}, \dot{\boldsymbol{\Sigma}} + \boldsymbol{\Omega} \times \boldsymbol{\Sigma} \right\rangle \mathrm{d}t \\
&= \int_a^b \left\langle \frac{\delta l}{\delta\boldsymbol{\Omega}}, \frac{\mathrm{d}}{\mathrm{d}t}\boldsymbol{\Sigma} \right\rangle + \left\langle \frac{\delta l}{\delta\boldsymbol{\Omega}}, \boldsymbol{\Omega} \times \boldsymbol{\Sigma} \right\rangle \mathrm{d}t \\
&= \int_a^b \left\langle -\frac{\mathrm{d}}{\mathrm{d}t}\left(\frac{\delta l}{\delta\boldsymbol{\Omega}}\right), \boldsymbol{\Sigma} \right\rangle \mathrm{d}t + \int_a^b \left\langle -\boldsymbol{\Omega} \times \frac{\delta l}{\delta\boldsymbol{\Omega}}, \boldsymbol{\Sigma} \right\rangle \mathrm{d}t \\
&= \int_a^b \left\langle -\frac{\mathrm{d}}{\mathrm{d}t}\left(\frac{\delta l}{\delta\boldsymbol{\Omega}}\right) + \frac{\delta l}{\delta\boldsymbol{\Omega}} \times \boldsymbol{\Omega}, \boldsymbol{\Sigma} \right\rangle \mathrm{d}t
\end{aligned} \tag{7.23}$$

where, in integrating by parts, we used that $\boldsymbol{\Sigma}(t)$ vanishes at the endpoints, i.e. $\boldsymbol{\Sigma}(a) = \boldsymbol{\Sigma}(b) = 0$. Therefore, since

$$0 = \int_a^b \left\langle -\frac{\mathrm{d}}{\mathrm{d}t}\left(\frac{\delta l}{\delta \boldsymbol{\Omega}}\right) + \frac{\delta l}{\delta \boldsymbol{\Omega}} \times \boldsymbol{\Omega}, \boldsymbol{\Sigma} \right\rangle \mathrm{d}t$$

for *any* path $\boldsymbol{\Sigma}(t)$ in $\mathfrak{so}(3) \simeq \mathbb{R}^3$ that vanishes at the end points, the motion is given by the equation:

$$\frac{\mathrm{d}}{\mathrm{d}t}\left(\frac{\delta l}{\delta \boldsymbol{\Omega}}\right) = \frac{\delta l}{\delta \boldsymbol{\Omega}} \times \boldsymbol{\Omega}.$$

Since

$$\frac{\delta l}{\delta \boldsymbol{\Omega}} = \mathbb{I}\boldsymbol{\Omega},$$

we immediately retrieve Euler's equations:

$$\mathbb{I}\dot{\boldsymbol{\Omega}} = \mathbb{I}\boldsymbol{\Omega} \times \boldsymbol{\Omega}.$$

We have shown the following:

**Theorem 7.2 (Euler–Poincaré reduction for the free rigid body)** *For any curve* $\mathbf{R}(t)$ *in* $SO(3)$ *let*

$$\widehat{\boldsymbol{\Omega}}(t) := \mathbf{R}^{-1}(t)\dot{\mathbf{R}}(t).$$

*Consider the Lagrangian of the free rigid body* $L : TSO(3) \to \mathbb{R}$,

$$L(\mathbf{R}, \dot{\mathbf{R}}) = \frac{1}{2}\operatorname{tr}(\dot{\mathbf{R}}\,\mathbb{J}\dot{\mathbf{R}}^T) = \frac{1}{2}\operatorname{tr}(\widehat{\boldsymbol{\Omega}}\,\mathbb{J}\widehat{\boldsymbol{\Omega}}^T) = \frac{1}{2}\boldsymbol{\Omega}^T\mathbb{I}\boldsymbol{\Omega},$$

*and define the reduced Lagrangian* $l : \mathfrak{so}(3) \to \mathbb{R}$,

$$l(\widehat{\boldsymbol{\Omega}}) = L(\mathbf{R}, \dot{\mathbf{R}}) = \frac{1}{2}\operatorname{tr}(\widehat{\boldsymbol{\Omega}}\,\mathbb{J}\widehat{\boldsymbol{\Omega}}^T) = \frac{1}{2}\boldsymbol{\Omega}^T\mathbb{I}\boldsymbol{\Omega}.$$

*Then the following four statements are equivalent:*

(i) *The variational principle*

$$\delta \int_a^b L(\mathbf{R}(t), \dot{\mathbf{R}}(t))\mathrm{d}t = 0$$

*holds, for variations among paths with fixed endpoints.*

(ii) $\mathbf{R}(t)$ *satisfies the Euler-Lagrange equations for Lagrangian L defined on TSO(3).*

(iii) *The variational principle*

$$\delta \int_a^b l(\boldsymbol{\Omega}(t))\mathrm{d}t = 0 \tag{7.24}$$

*holds on* $\mathfrak{so}(3) \simeq \mathbb{R}^3$, *using variations of the form*

$$\delta\boldsymbol{\Omega} = \dot{\boldsymbol{\Sigma}} + \boldsymbol{\Omega} \times \boldsymbol{\Sigma}, \tag{7.25}$$

*where* $\boldsymbol{\Sigma}(t)$ *is an arbitrary path in* $\mathfrak{so}(3) \simeq \mathbb{R}^3$ *that vanishes at the endpoints, i.e.* $\boldsymbol{\Sigma}(a) = \boldsymbol{\Sigma}(b) = 0$.

(iv) *Euler's equation holds:*

$$\mathbb{I}\dot{\boldsymbol{\Omega}} = \mathbb{I}\boldsymbol{\Omega} \times \boldsymbol{\Omega}.$$

## Reconstruction of $\mathbf{R}(t) \in SO(3)$

Euler's equation determines the body angular velocity $\boldsymbol{\Omega}(t)$. The tangent vectors $\dot{\mathbf{R}}(t) \in T_{\mathbf{R}(t)}\mathrm{SO}(3)$ along the integral curve in the rotation group $\mathbf{R}(t) \in \mathrm{SO}(3)$ may be retrieved via the ***reconstruction (or attitude) formula***,

$$\dot{\mathbf{R}}(t) = \mathbf{R}(t)\widehat{\boldsymbol{\Omega}}(t).$$

This is a differential equation with time-dependent coefficients. Its solution yields the integral curve $\mathbf{R}(t) \in \mathrm{SO}(3)$ for the orientation of the rigid body.

## Non-free rigid body motion

For general Lagrangians on $SO(3)$ we can still deduce the equations of motion in coordinates $(\mathbf{R}, \widehat{\boldsymbol{\Omega}})$. Define

$$\tilde{L}(\mathbf{R}, \widehat{\boldsymbol{\Omega}}) := L(\mathbf{R}, \dot{\mathbf{R}}) = L(\mathbf{R}, \mathbf{R}\widehat{\boldsymbol{\Omega}}).$$

In $(\mathbf{R}, \widehat{\boldsymbol{\Omega}})$ coordinates Hamilton's principle takes the form:

$$\delta \int_a^b \tilde{L}(\mathbf{R}, \widehat{\boldsymbol{\Omega}})\, \mathrm{d}t = 0, \tag{7.26}$$

i.e.

$$\int_a^b \left\langle \frac{\delta \tilde{L}}{\delta \mathbf{R}}, \delta \mathbf{R} \right\rangle + \left\langle \frac{\delta \tilde{L}}{\delta \widehat{\boldsymbol{\Omega}}}, \delta \widehat{\boldsymbol{\Omega}} \right\rangle \mathrm{d}t = 0, \tag{7.27}$$

Substituting eqns (7.19) and (7.21) yields:

$$0 = \int_a^b \left\langle \frac{\delta \tilde{L}}{\delta \mathbf{R}}, \mathbf{R}\widehat{\boldsymbol{\Sigma}} \right\rangle + \left\langle \frac{\delta \tilde{L}}{\delta \widehat{\boldsymbol{\Omega}}}, \frac{d}{dt}\widehat{\boldsymbol{\Sigma}} + \left[\widehat{\boldsymbol{\Omega}}, \widehat{\boldsymbol{\Sigma}}\right] \right\rangle \, dt. \tag{7.28}$$

The derivative $\delta \tilde{L}/\delta \mathbf{R}$ can be represented as a matrix with $(i,j)$th entry $\delta \tilde{L}/\delta(\mathbf{R}_{ij})$. As explained in Example 3.74, this matrix is the gradient of $\tilde{L}$ with respect to the trace pairing in $\mathcal{M}(3,\mathbb{R})$, so the left-most pairing in the equation above must now be interpreted as the trace pairing. Recall that any matrix $\mathbf{P} \in \mathcal{M}(3,\mathbb{R})$ has symmetric and skew-symmetric parts given by:

$$\mathbf{P}_S := \frac{1}{2}(\mathbf{P} + \mathbf{P}^T), \qquad \mathbf{P}_A := \frac{1}{2}(\mathbf{P} - \mathbf{P}^T),$$

respectively. Since the trace pairing of any symmetric matrix with an antisymmetric matrix vanishes (see Exercise 3.26), we have

$$\left\langle \mathbf{P}, \widehat{\boldsymbol{\Sigma}} \right\rangle = \left\langle \mathbf{P}_A, \widehat{\boldsymbol{\Sigma}} \right\rangle = \left\langle \frac{1}{2}(\mathbf{P} - \mathbf{P}^T), \widehat{\boldsymbol{\Sigma}} \right\rangle \quad \text{for all } \widehat{\boldsymbol{\Sigma}} \in \mathfrak{so}(3).$$

Therefore,

$$\begin{aligned}
\left\langle \frac{\delta \tilde{L}}{\delta \mathbf{R}}, \mathbf{R}\widehat{\boldsymbol{\Sigma}} \right\rangle &= \operatorname{tr}\left( \frac{\delta \tilde{L}}{\delta \mathbf{R}} \left(\mathbf{R}\widehat{\boldsymbol{\Sigma}}\right)^T \right) = \operatorname{tr}\left( \mathbf{R}^T \frac{\delta \tilde{L}}{\delta \mathbf{R}} \widehat{\boldsymbol{\Sigma}}^T \right) \\
&= \left\langle \mathbf{R}^T \left( \frac{\delta \tilde{L}}{\delta \mathbf{R}} \right), \widehat{\boldsymbol{\Sigma}} \right\rangle \\
&= \left\langle \left( \mathbf{R}^T \left( \frac{\delta \tilde{L}}{\delta \mathbf{R}} \right) \right)_A, \widehat{\boldsymbol{\Sigma}} \right\rangle \\
&= \left\langle \frac{1}{2} \left[ \mathbf{R}^T \left( \frac{\delta \tilde{L}}{\delta \mathbf{R}} \right) - \left( \frac{\delta \tilde{L}}{\delta \mathbf{R}} \right)^T \mathbf{R} \right], \widehat{\boldsymbol{\Sigma}} \right\rangle \\
&=: \left\langle \breve{\boldsymbol{\Xi}}, \widehat{\boldsymbol{\Sigma}} \right\rangle,
\end{aligned}$$

in which the last equality defines $\breve{\boldsymbol{\Xi}} \in \mathfrak{so}^*(3)$, using the vector representation of $\mathfrak{so}^*(3)$ as introduced in Example 5.36. Recall that by definition,

$$\left\langle \breve{\boldsymbol{\Xi}}, \widehat{\boldsymbol{\Sigma}} \right\rangle = \boldsymbol{\Xi} \cdot \boldsymbol{\Sigma}.$$

Using the calculations above and the identity $[\widehat{\boldsymbol{\Omega}}, \widehat{\boldsymbol{\Sigma}}] = \widehat{\boldsymbol{\Omega} \times \boldsymbol{\Sigma}}$, relation (7.28) becomes:

$$
\begin{aligned}
0 &= \int_a^b \left\langle \frac{\delta \tilde{L}}{\delta \mathbf{R}}, \mathbf{R}\widehat{\boldsymbol{\Sigma}} \right\rangle + \left\langle \frac{\delta \tilde{L}}{\delta \widehat{\boldsymbol{\Omega}}}, \frac{\mathrm{d}}{\mathrm{d}t}\widehat{\boldsymbol{\Sigma}} + \left[\widehat{\boldsymbol{\Omega}}, \widehat{\boldsymbol{\Sigma}}\right] \right\rangle \mathrm{d}t \\
&= \int_a^b \left\langle \widetilde{\boldsymbol{\Xi}}, \widehat{\boldsymbol{\Sigma}} \right\rangle + \left\langle \frac{\delta \tilde{L}}{\delta \widehat{\boldsymbol{\Omega}}}, \frac{\mathrm{d}}{\mathrm{d}t}\widehat{\boldsymbol{\Sigma}} + \left[\widehat{\boldsymbol{\Omega}}, \widehat{\boldsymbol{\Sigma}}\right] \right\rangle \mathrm{d}t \\
&= \int_a^b \langle \boldsymbol{\Xi}, \boldsymbol{\Sigma} \rangle + \left\langle \frac{\delta \tilde{L}}{\delta \boldsymbol{\Omega}}, \dot{\boldsymbol{\Sigma}} + \boldsymbol{\Omega} \times \boldsymbol{\Sigma} \right\rangle \mathrm{d}t \\
&= \int_a^b \langle \boldsymbol{\Xi}, \boldsymbol{\Sigma} \rangle + \left\langle -\frac{\mathrm{d}}{\mathrm{d}t}\left(\frac{\delta \tilde{L}}{\delta \boldsymbol{\Omega}}\right) + \frac{\delta \tilde{L}}{\delta \boldsymbol{\Omega}} \times \boldsymbol{\Omega}, \boldsymbol{\Sigma} \right\rangle \mathrm{d}t,
\end{aligned}
$$

where, when integrating by parts, we used the condition that the variations vanish at the endpoints in time. Therefore, since

$$
0 = \int_a^b \left\langle \boldsymbol{\Xi} - \frac{\mathrm{d}}{\mathrm{d}t}\left(\frac{\delta \tilde{L}}{\delta \boldsymbol{\Omega}}\right) + \frac{\delta \tilde{L}}{\delta \boldsymbol{\Omega}} \times \boldsymbol{\Omega}, \boldsymbol{\Sigma} \right\rangle \mathrm{d}t
$$

for any curve $\boldsymbol{\Sigma}(t)$ in $\mathfrak{so}(3)$ with such that $\boldsymbol{\Sigma}(a) = \boldsymbol{\Sigma}(b) = 0$, we find that the non-free rigid body motion is governed by the equation:

$$
\frac{\mathrm{d}}{\mathrm{d}t}\left(\frac{\delta \tilde{L}}{\delta \boldsymbol{\Omega}}\right) = \frac{\delta \tilde{L}}{\delta \boldsymbol{\Omega}} \times \boldsymbol{\Omega} + \boldsymbol{\Xi}.
$$

The meaning of the last term of the right hand side is that of a torque exerted on the body as felt (or seen) on the body. As for the free rigid body, the equation above is accompanied by the ***reconstruction (or attitude) relation***:

$$
\dot{\mathbf{R}} = \mathbf{R}\widehat{\boldsymbol{\Omega}}.
$$

**Exercise 7.8**
State and prove the reduced variational principle for a ***right invariant*** 'rigid body' Lagrangian defined on $TSO(3)$.

**Exercise 7.9**
Consider the planar motion of a free flat rigid body. Find the Lagrangian of this system and show it is $SO(2)$ invariant. State and prove the reduced variational principle for an invariant Lagrangian defined on $TSO(2)$. Write

the equations of motion and solve them. Then write explicitly the reconstruction equation and deduce its general solution.

**Exercise 7.10**
Let $\boldsymbol{\Omega}(t) = \boldsymbol{\Omega}_0 = const.$ be an equilibrium solution of Euler's equation (7.1). Integrate the reconstruction equation. What about non-equilibrium solutions?

## 7.3 Euler–Poincaré reduction theorem

Theorem 7.2 generalizes naturally to any left (or right) invariant Lagrangian defined on the tangent bundle of a Lie group $G$.

Consider a Lie group $G$ together with the left multiplication action on itself,

$$G \times G \to G, \quad (g,h) \to L_g(h) := gh.$$

**Proposition 7.3 (Left trivialized motion on a Lie group)** *Let $G$ be a Lie group together with a Lagrangian $L : TG \to \mathbb{R}$. Then, in left trivialized coordinates $(g,\xi) := (g, g^{-1}\dot{g}) := (g, T_g L_{g^{-1}}\dot{g}) \in G \times \mathfrak{g}$ of $TG$, the equations of motion are given by*

$$\frac{\mathrm{d}}{\mathrm{d}t}\left(\frac{\delta L}{\delta \xi}\right) = \mathrm{ad}^*_\xi \frac{\delta L}{\delta \xi} + T_e^* L_g \left(\frac{\delta L}{\delta g}\right), \tag{7.29}$$

$$\dot{g} = g\xi. \tag{7.30}$$

*Proof* Consider the Lagrangian $L$ written in $(g,\xi)$ coordinates, that is, define $\tilde{L}(g,\xi) := L(g, g\xi)$, or equivalently, $L(g,\dot{g}) = \tilde{L}(g, g^{-1}\dot{g})$. Hamilton's variational principle becomes

$$0 = \delta \int_a^b L(g(t), \dot{g}(t))\mathrm{d}t = \delta \int_a^b \tilde{L}\left(g(t), \xi(t)\right) \mathrm{d}t$$

for variations $\delta g$ among paths $g(t) \in G$ with fixed endpoints. So we have:

$$0 = \delta \int_a^b \tilde{L}\left(g(t), \xi(t)\right) \mathrm{d}t = \int_a^b \left\langle \frac{\delta \tilde{L}}{\delta g}, \delta g \right\rangle + \left\langle \frac{\delta \tilde{L}}{\delta \xi}, \delta \xi \right\rangle \mathrm{d}t, \tag{7.31}$$

where the variations $\delta \xi$ are induced by the variations $\delta g$.

In the following computation we assume that $G$ is a matrix Lie group. For the general case, see Lemma 7.4. Define $g_\epsilon(t)$ to be a family of curves in $G$ such that $g_0(t) = g(t)$ and let

$$\delta g := \left.\frac{\mathrm{d}g_\epsilon(t)}{\mathrm{d}\epsilon}\right|_{\epsilon=0}.$$

For $\xi = g^{-1}\dot{g}$, the variation of $\xi$ is computed in terms of $\delta g$ as

$$\delta\xi = \left.\frac{\mathrm{d}}{\mathrm{d}\epsilon}\right|_{\epsilon=0}(g_\epsilon^{-1}\dot{g}_\epsilon) = -g^{-1}(\delta g)g^{-1}\dot{g} + g^{-1}\left.\frac{\mathrm{d}^2 g}{\mathrm{d}t\,\mathrm{d}\epsilon}\right|_{\epsilon=0}. \tag{7.32}$$

Set $\eta := g^{-1}\delta g$. That is, $\eta(t)$ is an arbitrary path in $\mathfrak{g}$ that vanishes at the endpoints. The time derivative of $\eta$ is computed as

$$\frac{\mathrm{d}\eta}{\mathrm{d}t} = \frac{\mathrm{d}}{\mathrm{d}t}\left(g^{-1}\left.\frac{\mathrm{d}}{\mathrm{d}\epsilon}\right|_{\epsilon=0} g_\epsilon\right) = -g^{-1}\dot{g}g^{-1}(\delta g) + g^{-1}\left.\frac{\mathrm{d}^2 g}{\mathrm{d}t\,\mathrm{d}\epsilon}\right|_{\epsilon=0}. \tag{7.33}$$

Taking the difference of eqn (7.32) and eqn (7.33) implies

$$\delta\xi - \frac{\mathrm{d}\eta}{\mathrm{d}t} = -g^{-1}(\delta g)g^{-1}\dot{g} + g^{-1}\dot{g}g^{-1}(\delta g) = \xi\eta - \eta\xi = [\xi, \eta].$$

That is:

$$\delta\xi = \dot{\eta} + [\xi, \eta] = \dot{\eta} + \mathrm{ad}_\xi \eta, \tag{7.34}$$

where $[\xi, \eta]$ is the matrix commutator and where we used eqn (6.5). Substituting the above into relation (7.31) and taking into account that

$$\delta g = g\eta,$$

we obtain:

$$\begin{aligned} 0 &= \int_a^b \left\langle \frac{\delta \tilde{L}}{\delta g}, \delta g \right\rangle + \left\langle \frac{\delta \tilde{L}}{\delta \xi}, \delta\xi \right\rangle \mathrm{d}t \\ &= \int_a^b \left\langle \frac{\delta \tilde{L}}{\delta g}, g\eta \right\rangle + \left\langle \frac{\delta \tilde{L}}{\delta \xi}, \frac{\mathrm{d}\eta}{\mathrm{d}t} + \mathrm{ad}_\xi \eta \right\rangle \mathrm{d}t \end{aligned}$$

$$= \int_a^b \left\langle \frac{\delta \tilde{L}}{\delta g}, T_e L_g(\eta) \right\rangle + \left\langle \frac{\delta \tilde{L}}{\delta \xi}, \frac{\mathrm{d}\eta}{\mathrm{d}t} \right\rangle + \left\langle \frac{\delta \tilde{L}}{\delta \xi}, \mathrm{ad}_\xi \eta \right\rangle \mathrm{d}t$$

$$= \int_a^b \left\langle T_e^* L_g \left( \frac{\delta \tilde{L}}{\delta g} \right), \eta \right\rangle + \left\langle -\frac{\mathrm{d}}{\mathrm{d}t} \left( \frac{\delta \tilde{L}}{\delta \xi} \right) + \mathrm{ad}_\xi^* \left( \frac{\delta \tilde{L}}{\delta \xi} \right), \eta \right\rangle \mathrm{d}t$$

$$= \int_a^b \left\langle -\frac{\mathrm{d}}{\mathrm{d}t} \left( \frac{\delta \tilde{L}}{\delta \xi} \right) + \mathrm{ad}_\xi^* \left( \frac{\delta \tilde{L}}{\delta \xi} \right) + T_e^* L_g \left( \frac{\delta \tilde{L}}{\delta g} \right), \eta \right\rangle \mathrm{d}t.$$

Since the equality above must be fulfilled for any curve $\eta(t) \in \mathfrak{g}$ with vanishing endpoints, the conclusion follows. ■

**Lemma 7.4** *Let $g : U \subset \mathbb{R}^2 \to G$ be a smooth map and denote its partial derivatives by*

$$\xi(t, \varepsilon) := T_{g(t,\varepsilon)} L_{g(t,\varepsilon)^{-1}} \frac{\partial g(t, \varepsilon)}{\partial t},$$
$$\eta(t, \varepsilon) := T_{g(t,\varepsilon)} L_{g(t,\varepsilon)^{-1}} \frac{\partial g(t, \varepsilon)}{\partial \varepsilon}. \tag{7.35}$$

*Then*

$$\frac{\partial \xi}{\partial \varepsilon} - \frac{\partial \eta}{\partial t} = [\xi, \eta], \tag{7.36}$$

*where $[\xi, \eta]$ is the Lie algebra bracket on $\mathfrak{g}$. Conversely, if $U \subset \mathbb{R}^2$ is simply connected and $\xi, \eta : U \to \mathfrak{g}$ are smooth functions satisfying (7.36), then there exists a smooth function $g : U \to G$ such that eqn (7.35) holds.*

*Proof* See [BKMR96]. ■

**Remark 7.5** *If we choose to use the* **right trivialization** *of $TG$, that is, identifying $TG$ to $G \times \mathfrak{g}$ via*

$$(g, \dot{g}) \to (g, \xi) = (g, \dot{g}g^{-1}) := (g, T_g R_{g^{-1}}(\dot{g})),$$

*the equations of motion are given by*

$$\frac{\mathrm{d}}{\mathrm{d}t} \left( \frac{\delta L}{\delta \xi} \right) = -\,\mathrm{ad}_\xi^* \frac{\delta L}{\delta \xi} + T_e^* R_g \left( \frac{\delta L}{\delta g} \right),$$
$$\dot{g} = \xi g.$$

Recall from Section 6.1 the definition of an invariant (or symmetric) function under a group action. For the reader's convenience, we repeat this definition in the present context:

**Definition 7.6** *Let $G$ act on $TG$ by left translation. A function $F : TG \to \mathbb{R}$ is called* ***left invariant*** *if and only if*

$$F(g(h,\dot{h})) = F(h,\dot{h}) \quad \textit{for all} \;\; (h,\dot{h}) \in TG,$$

*where*

$$g(h,\dot{h}) := (gh, T_e L_g(\dot{h})).$$

If the Lagrangian is left invariant, then:

$$L(g,\dot{g}) = L(g^{-1}g, g^{1}\dot{g}) = L(e, g^{-1}\dot{g}) = L(e,\xi) \quad \text{for all} \quad (g,\dot{g}) \in TG,$$

where $\xi := g^{-1}\dot{g}$. Note that in this case the left trivialized Lagrangian satisfies

$$\tilde{L}(g,\xi) = L(g,\dot{g}) = L(e,\xi),$$

so it is independent of $g$. Thus, in eqn (7.29) of Proposition 7.3 above, the last term vanishes. This equation can be re-expressed as

$$\frac{\mathrm{d}}{\mathrm{d}t}\left(\frac{\delta l}{\delta \xi}\right) = \mathrm{ad}^*_\xi \frac{\delta l}{\delta \xi},$$

where $l$ is defined to be the restriction of $L$ to $\mathfrak{g}$:

$$l : \mathfrak{g} \to \mathbb{R}, \quad l(\xi) := L(e,\xi) = \tilde{L}(g,\xi) \qquad \text{for all } g \in \xi.$$

The following theorem is now easily verified:

**Theorem 7.7 (Euler–Poincaré reduction)** *Let $G$ be a Lie group, $L : TG \to \mathbb{R}$ a* ***left-invariant*** *Lagrangian, and define the* ***reduced Lagrangian,***

$$l : \mathfrak{g} \to \mathbb{R}, \quad l(\xi) := L(e,\xi),$$

*as the restriction of $L$ to $\mathfrak{g}$. For a curve $g(t) \in G$, let*

$$\xi(t) = g(t)^{-1}\dot{g}(t) := T_{g(t)}L_{g(t)^{-1}}\dot{g}(t) \in \mathfrak{g}.$$

*Then, the following four statements are equivalent:*

(i) *The variational principle*

$$\delta \int_a^b L(g(t), \dot{g}(t))\mathrm{d}t = 0$$

*holds, for variations among paths with fixed endpoints.*

(ii) $g(t)$ *satisfies the Euler–Lagrange equations for Lagrangian* $L$ *defined on* $G$.

(iii) *The variational principle*

$$\delta \int_a^b l(\xi(t))\mathrm{d}t = 0$$

*holds on* $\mathfrak{g}$, *using variations of the form* $\delta\xi = \dot{\eta} + [\xi, \eta]$, *where* $\eta(t)$ *is an arbitrary path in* $\mathfrak{g}$ *that vanishes at the endpoints, i.e.* $\eta(a) = \eta(b) = 0$.

(iv) *The (left-invariant)* ***Euler–Poincaré equations*** *hold:*

$$\frac{\mathrm{d}}{\mathrm{d}t}\frac{\delta l}{\delta \xi} = \mathrm{ad}^*_\xi \frac{\delta l}{\delta \xi}.$$

**Remark 7.8** *A similar statement holds, with obvious changes for* ***right-invariant*** *Lagrangian systems on* $TG$. *In this case the Euler-Poincaré equations are given by:*

$$\frac{\mathrm{d}}{\mathrm{d}t}\frac{\delta l}{\delta \xi} = -\,\mathrm{ad}^*_\xi \frac{\delta l}{\delta \xi}. \tag{7.37}$$

## Reconstruction

The reconstruction of the solution $g(t)$ of the Euler–Lagrange equations, with initial conditions $g(0) = g_0$ and $\dot{g}(0) = v_0$, is as follows: first, solve the initial value problem for the left-invariant Euler–Poincaré equations:

$$\frac{\mathrm{d}}{\mathrm{d}t}\frac{\delta l}{\delta \xi} = \mathrm{ad}^*_\xi \frac{\delta l}{\delta \xi} \quad \text{with} \quad \xi(0) = \xi_0 := g_0^{-1} v_0.$$

Second, using the solution $\xi(t)$ of the above, find the curve $g(t) \in G$ by solving the ***reconstruction equation***

$$\dot{g}(t) = g(t)\xi(t) \quad \text{with} \quad g(0) = g_0,$$

which is a differential equation with time-dependent coefficients.

**Exercise 7.11**
Prove the Euler–Poincaré reduction Theorem 7.7.

**Exercise 7.12**
Write out the proof of the Euler–Poincaré reduction theorem for right-invariant Lagrangians.

**Exercise 7.13**
Describe the corresponding reconstruction procedure for right-invariant Lagrangians.

**Exercise 7.14 (Motion on $SE(3)$)**
Write Euler–Poincaré equations for motion on $SE(3)$.

**Exercise 7.15 (Motion on $SO(4)$)**
Write out the Euler–Poincaré equations in matrix form for a free rigid body fixed at its centre of mass in a 4-dimensional space. Then use the analogue of the 'hat' map for $\mathfrak{so}(4)$ (see Exercise 5.21) and write the $\mathbb{R}^6$ vector representation of the equations.

**Exercise 7.16**
Consider the following action of a Lie group $G$ on a product space $G \times Y$, where $Y$ is some manifold:

$$(g, (h, y)) \to (gh, y).$$

Let $L : T(G \times Y) \to \mathbb{R}$ be invariant with respect to this action. Define $l : \mathfrak{g} \times TY \to \mathbb{R}$ as the restriction of $L$, i.e.

$$l(\xi, y, \dot{y}) := L(e, \xi, y, \dot{y}).$$

Deduce the reduced Hamilton's principle for $l$ and show that the equations of motion are given by:

$$\frac{\mathrm{d}}{dt}\frac{\delta l}{\delta \xi} = \mathrm{ad}^*_\xi \frac{\delta l}{\delta \xi},$$
$$\frac{\mathrm{d}}{dt}\frac{\delta l}{\delta \dot{y}} = \frac{\delta l}{\delta y}.$$

## 7.4 Modelling heavy-top dynamics

The heavy top is a rigid body rotating with a fixed point of support (the 'pivot') in a constant gravitational field. Just as for the free rigid body in the absence of gravity, the kinetic energy of the heavy top is:

$$K := \frac{1}{2}\int_{\mathcal{B}} \rho(\mathbf{X})||\dot{\mathbf{R}}\mathbf{X}||^2 \mathrm{d}^3\mathbf{X} = \frac{1}{2}\mathrm{tr}(\dot{\mathbf{R}}\mathbb{J}\dot{\mathbf{R}}^T) = \frac{1}{2}\mathrm{tr}(\widehat{\boldsymbol{\Omega}}\mathbb{J}\widehat{\boldsymbol{\Omega}}^T) = \frac{1}{2}\boldsymbol{\Omega}^T\mathbb{I}\boldsymbol{\Omega},$$

where $\mathbb{J}$ is the coefficient of inertia matrix and $\mathbb{I}$ is the moment of inertia matrix. However, for the top, the matrices $\mathbb{J}$ and $\mathbb{I}$ must be calculated with respect to the pivot, which is not in general the centre of mass.

Let $m$ be the mass of the body, and let $\mathbf{k}$ be the vertical unit vector. Let $\chi$ be the vector from the point of support to the body's centre of mass, and note that $\chi$ is constant in body coordinates. With this notation, the potential is given by:

$$V_{\mathbf{k}}(\mathbf{R}) = mg\left\langle \mathbf{k}, \mathbf{R}\chi\right\rangle, \tag{7.38}$$

where $\langle\cdot,\cdot\chi\rangle$ is the standard dot product in $\mathbb{R}^3$. Thus, the dynamics is determined by the Lagrangian

$$L_{\mathbf{k}} : TSO(3) \to \mathbb{R},$$

given explicitly by

$$L_{\mathbf{k}}(\mathbf{R}, \dot{\mathbf{R}}) = K - V_{\mathbf{k}} = \frac{1}{2}\,\mathrm{tr}(\dot{\mathbf{R}}\mathbb{J}\dot{\mathbf{R}}^T) - mg\left\langle \mathbf{k}, \mathbf{R}\,\chi\right\rangle, \tag{7.39}$$

where $\mathbf{k}$ is a fixed parameter. We will identify the vector $\mathbf{k} \in \mathbb{R}^3$ with the map $\langle \mathbf{k}, \cdot\rangle$, which is a covector in $(\mathbb{R}^3)^*$.

*In the absence of gravity*, i.e. for $V = 0$, the Lagrangian models a free rigid body; in particular, it is $SO(3)$ left invariant. Gravity breaks this symmetry, leaving invariance only with respect the subgroup $SO(2)_{vert}$ of rotations around the vertical axis. While, in principle, one can apply reduction with respect to $SO(2)_{vert}$ only, we can take advantage of the full broken symmetry as follows.

First, we extend the configuration space $SO(3)$ to $SO(3) \times (\mathbb{R}^3)^*$ so that the direction of gravity $\mathbf{k}$ can be considered as a value of the new coordinate $\mathbf{v} \in (\mathbb{R}^3)^*$. The purpose of this is to obtain an extended Lagrangian that is left-invariant under an action of all of $SO(3)$, not just $SO(2)_{vert}$. Define

$$L_{ext} : TSO(3) \times T(\mathbb{R}^3)^* \to \mathbb{R},$$

as

$$L_{ext}(\mathbf{R}, \dot{\mathbf{R}}, \mathbf{v}, \dot{\mathbf{v}}) = \frac{1}{2} \operatorname{tr}(\dot{\mathbf{R}} \mathbb{J} \dot{\mathbf{R}}^T) - mg \langle \mathbf{v}, \mathbf{R} \boldsymbol{\chi} \rangle . \tag{7.40}$$

Note that

$$L_{ext}(\mathbf{R}, \dot{\mathbf{R}}, \mathbf{v}, \dot{\mathbf{v}}) \Big|_{\{\mathbf{v}=\mathbf{k}\}} = L_{\mathbf{k}}(\mathbf{R}, \dot{\mathbf{R}}). \tag{7.41}$$

The motion determined by $L_{\mathbf{k}}$ (via Hamilton's principle) corresponds to the motion determined by $L_{ext}$, *with the added constraint*

$$\mathbf{v}(t) = \mathbf{k} = const.$$

**Remark 7.9** *By construction, the extended Lagrangian $L_{ext}$ is a function of $(\mathbf{R}, \dot{\mathbf{R}}, \mathbf{v}) \in TSO(3) \times (\mathbb{R}^3)^*$ only. However, the* ***heavy-top variational principle*** *deduced in this section is the end result of the application of Hamilton's principle with constraints. This is a variational principle defined on a tangent bundle. Consequently, we choose to carry the domain of definition of $L_{ext}$ as $T\left(SO(3) \times (\mathbb{R}^3)^*\right)$ only to restrict it again later to $TSO(3) \times (\mathbb{R}^3)^*$.*

**Definition 7.10** *The (left) diagonal action of $SO(3)$ on $SO(3) \times (\mathbb{R}^3)^*$ is:*

$$(\mathbf{Q}, (\mathbf{R}, \mathbf{v})) \to (\mathbf{QR}, \mathbf{Qv}) , \quad \textit{for all } \mathbf{Q} \in SO(3). \tag{7.42}$$

*Its tangent lift is given by*

$$\left(\mathbf{Q}, \left(\mathbf{R}, \dot{\mathbf{R}}, \mathbf{v}, \dot{\mathbf{v}}\right)\right) \to \mathbf{Q}\left(\mathbf{R}, \dot{\mathbf{R}}, \mathbf{v}, \dot{\mathbf{v}}\right) := \left(\mathbf{QR}, \mathbf{Q}\dot{\mathbf{R}}, \mathbf{Qv}, \mathbf{Q}\dot{\mathbf{v}}\right) . \tag{7.43}$$

**Lemma 7.11** *Under the tangent-lifted diagonal action above, the Lagrangian $L_{ext}$ in eqn (7.40) is left invariant.*

*Proof*

$$
\begin{aligned}
L_{ext}\left(Q\left(\mathbf{R},\dot{\mathbf{R}},\mathbf{v},\dot{\mathbf{v}}\right)\right) &= \frac{1}{2}\mathrm{tr}\left((\mathbf{Q}\dot{\mathbf{R}})\mathbb{J}(\mathbf{Q}\dot{\mathbf{R}})^T\right) - mg\left\langle(\mathbf{Q}\mathbf{v}),(\mathbf{Q}\mathbf{R})\chi\right\rangle \\
&= \frac{1}{2}\mathrm{tr}\left(\dot{\mathbf{R}}\mathbb{J}\dot{\mathbf{R}}^T\right) - mg\left\langle(\mathbf{Q}\mathbf{R})^{-1}\mathbf{Q}\mathbf{v},\chi\right\rangle \\
&= \frac{1}{2}\mathrm{tr}\left(\dot{\mathbf{R}}\mathbb{J}\dot{\mathbf{R}}^T\right) - mg\left\langle\mathbf{R}^{-1}\mathbf{v},\chi\right\rangle \\
&= \frac{1}{2}\mathrm{tr}\left(\dot{\mathbf{R}}\mathbb{J}\dot{\mathbf{R}}^T\right) - mg\left\langle\mathbf{v},\mathbf{R}\chi\right\rangle \\
&= L_{ext}(\mathbf{R},\dot{\mathbf{R}},\mathbf{v},\dot{\mathbf{v}}).
\end{aligned}
\tag{7.44}
$$

■

## Body coordinates for the heavy top

As in Section 7.1, angular velocity in body coordinates is given by:

$$\widehat{\boldsymbol{\Omega}} := \mathbf{R}^{-1}\dot{\mathbf{R}}. \tag{7.45}$$

In addition, we transform $\mathbf{v}$ into body coordinates,

$$\boldsymbol{\Gamma} := \mathbf{R}^{-1}\mathbf{v}. \tag{7.46}$$

**Remark 7.12** *The gravity vector, which is the constant* $\mathbf{k}$ *in spatial coordinates, becomes time dependent when expressed in body coordinates. For a fixed path* $\mathbf{R}(t)$*, the gravity vector is represented in the body frame by the unit vector*

$$\boldsymbol{\Gamma}(t) = \mathbf{R}^{-1}(t)\mathbf{k}. \tag{7.47}$$

*(One may think of* $\boldsymbol{\Gamma}(t)$ *as* $\mathbf{k}$ *seen from the moving body.) Such 'carried' quantities are called* ***advected*** *and appear naturally in fluid modelling, as various characteristics (e.g. mass, heat) of the fluid carried along with the flow of each fluid element.*

From the definition of $\boldsymbol{\Gamma}$, it follows that:

$$\dot{\boldsymbol{\Gamma}} := \frac{\mathrm{d}\boldsymbol{\Gamma}}{\mathrm{d}t} = \frac{\mathrm{d}}{\mathrm{d}t}(\mathbf{R}^{-1}\mathbf{v}) = \dot{\mathbf{R}}^{-1}\mathbf{v} + \mathbf{R}^{-1}\dot{\mathbf{v}}$$
$$= -\mathbf{R}^{-1}\dot{\mathbf{R}}\mathbf{R}^{-1}\mathbf{v} + \mathbf{R}^{-1}\dot{\mathbf{v}}$$
$$= -\widehat{\boldsymbol{\Omega}}\boldsymbol{\Gamma} + \mathbf{R}^{-1}\dot{\mathbf{v}} = \boldsymbol{\Gamma} \times \boldsymbol{\Omega} + \mathbf{R}^{-1}\dot{\mathbf{v}},$$

and so

$$\dot{\boldsymbol{\Gamma}} = \boldsymbol{\Gamma} \times \boldsymbol{\Omega} + \mathbf{R}^{-1}\dot{\mathbf{v}}. \tag{7.48}$$

Taken together, these transformations give us a new set of coordinates:

$$(\mathbf{R}, \widehat{\boldsymbol{\Omega}}, \boldsymbol{\Gamma}, \dot{\boldsymbol{\Gamma}}) \in SO(3) \times \mathfrak{so}(3) \times T(\mathbb{R}^3)^*.$$

Note that the (left) action of $SO(3)$ defined above by eqn (7.43), in the new coordinates, is:

$$\left(\mathbf{Q}, \left(\mathbf{R}, \widehat{\boldsymbol{\Omega}}, \boldsymbol{\Gamma}, \dot{\boldsymbol{\Gamma}}\right)\right) \to \mathbf{Q}\left(\mathbf{R}, \widehat{\boldsymbol{\Omega}}, \boldsymbol{\Gamma}, \dot{\boldsymbol{\Gamma}}\right) = \left(\mathbf{QR}, \widehat{\boldsymbol{\Omega}}, \boldsymbol{\Gamma}, \dot{\boldsymbol{\Gamma}}\right). \tag{7.49}$$

We now define $\tilde{L}_{ext}$ to be $L_{ext}$ in body coordinates:

$$\tilde{L}_{ext}(\mathbf{R}, \widehat{\boldsymbol{\Omega}}, \boldsymbol{\Gamma}, \dot{\boldsymbol{\Gamma}}) := L_{ext}(\mathbf{R}, \mathbf{R}\widehat{\boldsymbol{\Omega}}, \mathbf{R}\boldsymbol{\Gamma}, \mathbf{R}\dot{\boldsymbol{\Gamma}}) = L_{ext}(\mathbf{R}, \dot{\mathbf{R}}, \mathbf{v}, \dot{\mathbf{v}}).$$

We noted earlier that the motion of the heavy top is determined by Hamilton's variational principle for $L_{ext}$ subject to the constraint $\mathbf{v} = \mathbf{k}$, that is (see Section 1.3),

$$\delta \int_a^b L_{ext}(\mathbf{R}, \dot{\mathbf{R}}, \mathbf{v}, \dot{\mathbf{v}})\mathrm{d}t = 0 \tag{7.50}$$

with respect to arbitrary variations $\delta\mathbf{R}$ that vanish at the end points *and* satisfy $\mathbf{v} = \mathbf{k}$. Note that the constraint allows only null variations $\delta\mathbf{v} = \mathbf{0}$ and $\delta\dot{\mathbf{v}} = \mathbf{0}$. The corresponding variational principle in the new coordinates is

$$\delta \int_a^b \tilde{L}_{ext}(\mathbf{R}, \widehat{\boldsymbol{\Omega}}, \boldsymbol{\Gamma}, \dot{\boldsymbol{\Gamma}})\mathrm{d}t = 0 \tag{7.51}$$

with respect to variations $\delta\mathbf{R}$, $\delta\boldsymbol{\Omega}$, $\delta\boldsymbol{\Gamma}$ and $\delta\dot{\boldsymbol{\Gamma}}$ *and* subject to $\mathbf{R}\boldsymbol{\Gamma} = \mathbf{k}$. The variations $\delta\mathbf{R}$ are still arbitrary except for vanishing at the endpoints. As shown in Section 7.2, the variations of $\mathbf{\Omega}$ are given by

$$\delta\mathbf{\Omega} = \dot{\mathbf{\Sigma}} + \mathbf{\Omega} \times \mathbf{\Sigma}, \tag{7.52}$$

where $\widehat{\mathbf{\Sigma}} := \mathbf{R}^{-1}\delta\mathbf{R}$ is an arbitrary path in $\mathfrak{so}(3)$ that also vanishes at the endpoints. The allowed variations of $\mathbf{\Gamma}$ follow from the calculation:

$$\begin{aligned} \delta\mathbf{\Gamma} = \delta\left(\mathbf{R}^{-1}\mathbf{k}\right) &= (\delta\mathbf{R}^{-1})\mathbf{k} \\ &= -(\mathbf{R}^{-1}\delta\mathbf{R}\mathbf{R}^{-1})\mathbf{k} \\ &= -(\mathbf{R}^{-1}\delta\mathbf{R})(\mathbf{R}^{-1}\mathbf{k}) \\ &= -\widehat{\mathbf{\Sigma}}\mathbf{\Gamma} = -\mathbf{\Sigma} \times \mathbf{\Gamma} = \mathbf{\Gamma} \times \mathbf{\Sigma}. \end{aligned} \tag{7.53}$$

The allowed variations of $\dot{\mathbf{\Gamma}}$ are,

$$\delta\dot{\mathbf{\Gamma}} = \dot{\mathbf{\Gamma}} \times \mathbf{\Sigma} + \mathbf{\Gamma} \times \dot{\mathbf{\Sigma}},$$

though this formula will not be needed for the heavy top, since $\tilde{L}_{ext}$ is independent of $\dot{\mathbf{\Gamma}}$.

Since $\tilde{L}_{ext}$ is independent of $\mathbf{R}$, the variations $\delta\mathbf{R}$ are relevant only through their relations to $\delta\mathbf{\Omega}, \delta\mathbf{\Gamma}$ and $\delta\dot{\mathbf{\Gamma}}$. Note that these variations only depend on $\delta\mathbf{R}$ via $\widehat{\mathbf{\Sigma}} := \mathbf{R}^{-1}\delta\mathbf{R}$. Hence, we can remove $\delta\mathbf{R}$ from consideration completely, keeping only $\widehat{\mathbf{\Sigma}}$. Since $\delta\mathbf{R}$ is arbitrary, $\widehat{\mathbf{\Sigma}}$ is an arbitrary path in $\mathfrak{so}(3)$ that vanishes at the endpoints.

The left invariance of $L_{ext}$ is inherited by $\tilde{L}_{ext}$ (see Exercise 7.17). In particular,

$$\tilde{L}_{ext}(\mathbf{R}, \widehat{\mathbf{\Omega}}, \mathbf{\Gamma}, \dot{\mathbf{\Gamma}}) = \tilde{L}_{ext}(\mathbf{R}^{-1}\mathbf{R}, \widehat{\mathbf{\Omega}}, \mathbf{\Gamma}, \dot{\mathbf{\Gamma}}) = \tilde{L}_{ext}(\mathbf{I}_3, \widehat{\mathbf{\Omega}}, \mathbf{\Gamma}, \dot{\mathbf{\Gamma}}).$$

**Definition 7.13** *The **reduced Lagrangian** is defined as*

$$l : \mathfrak{so}(3) \times (\mathbb{R}^3)^* \times (\mathbb{R}^3)^* \to \mathbb{R},$$

*by setting*

$$l(\widehat{\mathbf{\Omega}}, \mathbf{\Gamma}, \dot{\mathbf{\Gamma}}) := \tilde{L}_{ext}(\mathbf{I}, \widehat{\mathbf{\Omega}}, \mathbf{\Gamma}, \dot{\mathbf{\Gamma}}) = \tilde{L}_{ext}(\mathbf{R}, \widehat{\mathbf{\Omega}}, \mathbf{\Gamma}, \dot{\mathbf{\Gamma}}) \tag{7.54}$$

$$= \frac{1}{2}\,\mathrm{tr}(\widehat{\mathbf{\Omega}}\mathbb{J}\widehat{\mathbf{\Omega}}^T) - mg\,\langle \mathbf{\Gamma}, \boldsymbol{\chi} \rangle, \tag{7.55}$$

*for any* $\mathbf{R} \in SO(3)$. *In vector notation for* $\mathfrak{so}(3)$,

$$l(\mathbf{\Omega}, \mathbf{\Gamma}, \dot{\mathbf{\Gamma}}) = \frac{1}{2}\langle \mathbf{\Omega}, \mathbb{I}\mathbf{\Omega} \rangle - mg\,\langle \mathbf{\Gamma}, \boldsymbol{\chi} \rangle. \tag{7.56}$$

The variational principle for $\tilde{L}_{ext}$ in (7.51) is equivalent to:

$$\delta \int_a^b l(\mathbf{\Omega}, \mathbf{\Gamma}, \dot{\mathbf{\Gamma}})\mathrm{d}t = 0 \tag{7.57}$$

with respect to variations $\delta\mathbf{\Omega}$, $\delta\mathbf{\Gamma}$ and $\delta\dot{\mathbf{\Gamma}}$ deduced above. Since $l$ is independent of $\dot{\mathbf{\Gamma}}$, we can consider $l$ as a function of the variables $(\mathbf{\Omega}, \mathbf{\Gamma})$ only and state the variational principle as

$$\delta \int_a^b l(\mathbf{\Omega}, \mathbf{\Gamma})dt = 0, \tag{7.58}$$

with respect to variations

$$\delta\mathbf{\Omega} = \dot{\mathbf{\Sigma}} + \mathbf{\Omega} \times \mathbf{\Sigma} \quad \text{and} \quad \delta\mathbf{\Gamma} = \mathbf{\Gamma} \times \mathbf{\Sigma},$$

where $\mathbf{\Sigma}$ is an arbitrary path of displacements in $\mathbb{R}^3$ that vanish at the endpoints. We retain the constraint equation $\mathbf{R}\mathbf{\Gamma} = \mathbf{k}$, since it is not a consequence of the variational principle, even though it was used in the computation of the admissible variations. Note that this equation is not sufficient for reconstructing $\mathbf{R}$. For this, we must use the same reconstruction equation as in Section 7.2:

$$\dot{\mathbf{R}}(t) = \mathbf{R}(t)\widehat{\mathbf{\Omega}}(t),$$

which is simply a rearrangement of the definition of $\mathbf{\Omega}$.

## Heavy-top equations of motion

Now we derive the equations of motion for the heavy top. From the variational principle, we obtain:

$$\begin{aligned} 0 &= \delta \int_a^b l\left(\mathbf{\Omega}(t), \mathbf{\Gamma}(t)\right) \mathrm{d}t \\ &= \int_a^b \left\langle \frac{\delta l}{\delta \mathbf{\Omega}}, \delta\mathbf{\Omega} \right\rangle \mathrm{d}t + \int_a^b \left\langle \frac{\delta l}{\delta \mathbf{\Gamma}}, \delta\mathbf{\Gamma} \right\rangle \mathrm{d}t \end{aligned}$$

$$= \int_a^b \left\langle \frac{\delta l}{\delta \mathbf{\Omega}}, \dot{\mathbf{\Sigma}} + \mathbf{\Omega} \times \mathbf{\Sigma} \right\rangle \mathrm{d}t + \int_a^b \left\langle \frac{\delta l}{\delta \mathbf{\Gamma}}, \mathbf{\Gamma} \times \mathbf{\Sigma} \right\rangle \mathrm{d}t$$

$$= \int_a^b \left\langle -\frac{\mathrm{d}}{\mathrm{d}t}\left(\frac{\delta l}{\delta \mathbf{\Omega}}\right) + \frac{\delta l}{\delta \mathbf{\Omega}} \times \mathbf{\Omega} + \frac{\delta l}{\delta \mathbf{\Gamma}} \times \mathbf{\Gamma}, \mathbf{\Sigma} \right\rangle \mathrm{d}t.$$

Since $\mathbf{\Sigma}(t) \in \mathbb{R}^3$ is arbitrary except for vanishing at the endpoints, the equations of motion are:

$$\frac{\mathrm{d}}{\mathrm{d}t}\left(\frac{\delta l}{\delta \mathbf{\Omega}}\right) = \frac{\delta l}{\delta \mathbf{\Omega}} \times \mathbf{\Omega} + \frac{\delta l}{\delta \mathbf{\Gamma}} \times \mathbf{\Gamma}. \tag{7.59}$$

From the formula for $l$ in eqn (7.56),

$$\frac{\delta l}{\delta \mathbf{\Omega}} = \mathbb{I}\mathbf{\Omega} \quad \text{and} \quad \frac{\delta l}{\delta \mathbf{\Gamma}} = mg\boldsymbol{\chi},$$

so the equations of motion are:

$$\mathbb{I}\dot{\mathbf{\Omega}} = \mathbb{I}\mathbf{\Omega} \times \mathbf{\Omega} - mg\boldsymbol{\chi} \times \mathbf{\Gamma}.$$

Finally, we differentiate the constraint equation $\mathbf{R}(t)\mathbf{\Gamma}(t) = \mathbf{k}$ to give $\dot{\mathbf{R}}\mathbf{\Gamma} + \mathbf{R}\dot{\mathbf{\Gamma}} = 0$, which can be rearranged as

$$\dot{\mathbf{\Gamma}} = -\mathbf{R}^{-1}\dot{\mathbf{R}}\mathbf{\Gamma} = -\widehat{\mathbf{\Omega}}\mathbf{\Gamma} = -\mathbf{\Omega} \times \mathbf{\Gamma},$$

i.e.

$$\dot{\mathbf{\Gamma}} = \mathbf{\Gamma} \times \mathbf{\Omega}. \tag{7.60}$$

The evolutionary system is completed by the reconstruction equation,

$$\dot{\mathbf{R}} = \mathbf{R}\widehat{\mathbf{\Omega}}.$$

In conclusion, we have proven:

**Theorem 7.14 (Euler–Poincaré reduction for the heavy top)**
*For any curve* $\mathbf{R}(t) \in SO(3)$, *let*

$$\widehat{\mathbf{\Omega}}(t) := \mathbf{R}^{-1}(t)\dot{\mathbf{R}}(t)$$

*and*

$$\mathbf{\Gamma}(t) := \mathbf{R}^{-1}(t)\mathbf{k}.$$

*Then,* $\mathbf{\Gamma}(t)$ *is the unique solution of the non-autonomous initial value problem*

$$\dot{\mathbf{\Gamma}}(t) = \mathbf{\Gamma}(t) \times \mathbf{\Omega}(t), \quad \mathbf{\Gamma}(0) = \mathbf{R}^{-1}(0)\mathbf{k}.$$

*With* $L_{\mathbf{k}}$ *and* $l$ *defined in eqn (7.54), the following statements are equivalent:*

(i) *Hamilton's variational principle,*

$$\delta \int_a^b L_{\mathbf{k}}(\mathbf{R}(t), \dot{\mathbf{R}}(t))dt = 0,$$

*holds for variations* $\delta\mathbf{R}$ *of* $\mathbf{R}(t)$ *in* $SO(3)$ *vanishing at the end points.*

(ii) *The curve* $\mathbf{R}(t)$ *satisfies the Euler–Lagrange equations for* $L_{\mathbf{k}}$.

(iii) *The reduced Hamilton's variational principle,*

$$\delta \int_a^b l(\mathbf{\Omega}(t), \mathbf{\Gamma}(t))\mathrm{d}t = 0,$$

*holds for variations of* $\mathbf{\Omega}$ *and* $\mathbf{\Gamma}$ *of the form*

$$\delta\mathbf{\Omega} = \dot{\mathbf{\Sigma}} + \mathbf{\Omega} \times \mathbf{\Sigma}, \tag{7.61}$$

$$\delta\mathbf{\Gamma} = \mathbf{\Gamma} \times \mathbf{\Sigma}, \tag{7.62}$$

*where* $\mathbf{\Sigma}$ *is an arbitrary path of displacements in* $\mathbb{R}^3$ *vanishing at the end points.*

(iv) *The following extended Euler–Poincaré equations hold:*

$$\mathbb{I}\dot{\mathbf{\Omega}} = \mathbb{I}\mathbf{\Omega} \times \mathbf{\Omega} - mg\chi \times \mathbf{\Gamma}.$$

## Lagrangians on *SO*(3) with broken symmetry

Consider a family of Lagrangians $L_{\mathbf{k}} : TSO(3) \to \mathbb{R}$, parameterized by $\mathbf{k} \in (\mathbb{R}^3)^*$. Suppose there exists an extended Lagrangian

$$L_{ext} : TSO(3) \times T(\mathbb{R}^3)^* \to \mathbb{R},$$

such that

$$L_{ext}(\mathbf{R}, \dot{\mathbf{R}}, \mathbf{v}, \dot{\mathbf{v}})\Big|_{\{\mathbf{v}=\mathbf{k}\}} = L_{\mathbf{k}}(\mathbf{R}, \dot{\mathbf{R}}) \tag{7.63}$$

and $L_{ext}$ is invariant under the tangent-lifted diagonal left action of $SO(3)$, as in Definition 7.10. With $\mathbf{\Omega}$ and $\mathbf{\Gamma}$ as above, define the reduced Lagrangian as

$$\begin{aligned} l(\mathbf{\Omega}, \mathbf{\Gamma}, \dot{\mathbf{\Gamma}}) &:= L_{ext}(\mathbf{I}, \mathbf{\Omega}, \mathbf{\Gamma}, \dot{\mathbf{\Gamma}}) \\ &= L_{ext}(\mathbf{I}, \mathbf{R}^{-1}\dot{\mathbf{R}},\ \mathbf{R}^{-1}\mathbf{v}, \mathbf{R}^{-1}\dot{\mathbf{v}}) = L_{ext}(\mathbf{R}, \dot{\mathbf{R}}, \mathbf{v}, \dot{\mathbf{v}}). \end{aligned}$$

Then, by the same reasoning as for the heavy top, the variational principle for $L_{ext}$ subject to the constraint $\{\mathbf{v} = \mathbf{k}\}$ induces the reduced variational principle

$$\delta \int_a^b l(\mathbf{\Omega}, \mathbf{\Gamma}, \dot{\mathbf{\Gamma}})dt = 0, \tag{7.64}$$

with respect to variations

$$\delta\mathbf{\Omega} = \dot{\mathbf{\Sigma}} + \mathbf{\Omega} \times \mathbf{\Sigma}, \quad \delta\mathbf{\Gamma} = \mathbf{\Gamma} \times \mathbf{\Sigma} \quad \text{and} \quad \delta\dot{\mathbf{\Gamma}} = \dot{\mathbf{\Gamma}} \times \mathbf{\Sigma} + \mathbf{\Gamma} \times \dot{\mathbf{\Sigma}},$$

where $\mathbf{\Sigma}$ is an arbitrary path of displacements in $\mathbb{R}^3$ that vanishes at the end points in time. The dynamics is completed by

$$\dot{\mathbf{\Gamma}} = \mathbf{\Gamma} \times \mathbf{\Omega} \qquad \text{and} \qquad \dot{\mathbf{R}} = \mathbf{R}\widehat{\mathbf{\Omega}}.$$

An example is given in Exercise 7.18.

**Exercise 7.17**
Verify that $\tilde{L}_{ext}$ is left invariant, that is,

$$\tilde{L}_{ext}(\mathbf{R}, \widehat{\mathbf{\Omega}}, \mathbf{\Gamma}, \dot{\mathbf{\Gamma}}) = \tilde{L}_{ext}(\mathbf{QR}, \widehat{\mathbf{\Omega}}, \mathbf{\Gamma}, \dot{\mathbf{\Gamma}}),$$

for all $\mathbf{Q} \in SO(3)$.

**Exercise 7.18 (Charged heavy top in a magnetic field)**
Consider a heavy top with an electric charge at the centre of mass, in a magnetic field with vector potential $\mathbf{A}$ given by $A(\mathbf{q}) = \mathbf{k} \times \mathbf{q}$ for all

$\mathbf{q} \in \mathbb{R}^3$ (see Example 1.34). The Lagrangian is given by:

$$L_{\mathbf{k}}(\mathbf{R}, \dot{\mathbf{R}}) = \frac{1}{2}\mathrm{tr}(\dot{\mathbf{R}}\mathbb{J}\dot{\mathbf{R}}^T) - mg\left\langle \mathbf{k}, \mathbf{R}\,\chi \right\rangle + e\left\langle \mathbf{A}(\mathbf{R}\chi), \dot{\mathbf{R}}\chi \right\rangle. \tag{7.65}$$

Show that

$$l(\boldsymbol{\Omega}, \boldsymbol{\Gamma}, \dot{\boldsymbol{\Gamma}}) = \frac{1}{2}\left\langle \boldsymbol{\Omega}, \mathbb{I}\boldsymbol{\Omega} \right\rangle - mg\left\langle \boldsymbol{\Gamma}, \chi \right\rangle + \left\langle \boldsymbol{\Gamma} \times \chi, \boldsymbol{\Omega} \times \chi \right\rangle$$

and calculate the equations of motion.

**Exercise 7.19**
In place of $L_{ext}$ in eqn (7.50), consider a general Lagrangian $L : TSO(3) \times T(\mathbb{R}^3)^* \to \mathbb{R}$ that is invariant under the tangent-lifted diagonal left action of $SO(3)$. Suppose $(\mathbf{R}(t), \mathbf{v}(t))$ satisfies the variational principle

$$\delta \int_a^b L(\mathbf{R}, \dot{\mathbf{R}}, \mathbf{v}, \dot{\mathbf{v}})dt = 0 \tag{7.66}$$

with respect to arbitrary variations $(\delta\mathbf{R}, \delta\mathbf{v})$ with vanishing end points. Note that this is eqn (7.50), but for a general $L$ and *without any constraint*. Define $\widehat{\boldsymbol{\Omega}} := \mathbf{R}^{-1}\dot{\mathbf{R}}$, $\boldsymbol{\Gamma} := \mathbf{R}^{-1}\mathbf{v}$ and $\dot{\boldsymbol{\Gamma}} := \mathbf{R}^{-1}\dot{\mathbf{v}}$, and let $l : \mathfrak{so}(3) \times (\mathbb{R}^3)^* \times (\mathbb{R}^3)^* \to \mathbb{R}$ be given by

$$l(\widehat{\boldsymbol{\Omega}}, \boldsymbol{\Gamma}, \dot{\boldsymbol{\Gamma}}) := L(\mathbf{I}, \widehat{\boldsymbol{\Omega}}, \boldsymbol{\Gamma}, \dot{\boldsymbol{\Gamma}}) = L(\mathbf{R}, \dot{\mathbf{R}}, \mathbf{v}, \dot{\mathbf{v}}), \quad \text{for any } \mathbf{R}.$$

Deduce the corresponding variational principle for $l$. (HINT: let $\mathbf{W} = \mathbf{R}^{-1}\delta\mathbf{v}$, and express all variations in terms of $\boldsymbol{\Sigma}$ and $\mathbf{W}$. Note that, if the constraint $\mathbf{R}\boldsymbol{\Gamma} = \mathbf{k}$ is applied, $\mathbf{W} = 0$ always. )

## 7.5 Euler–Poincaré systems with advected parameters

Theorem 7.14 generalizes to the following setting. Consider a Lie group $G$ and a vector space $V$ on which $G$ acts on the left by linear maps, i.e. for all $g \in G$, the map $\Phi_g : V \to V$ is linear. This action induces a left action of

$G$ on $V^*$

$$G \times V^* \to V^* \quad (g, w) \to gw,$$

where

$$\langle gw, v \rangle := \left\langle w, g^{-1} v \right\rangle \qquad \text{for all} \qquad v \in V.$$

Note that if $G$ is a matrix Lie group, $gw = g^{-T} w := (g^T)^{-1} w$. For $\xi \in \mathfrak{g}$, let $\xi v$ be the infinitesimal generator corresponding to the $G$ action on $V$, that is

$$\xi v := \xi_V(v) = \left.\frac{\mathrm{d}}{\mathrm{d}t}\right|_{t=0} g(t) v,$$

where $g(t)$ is a curve in $G$ with $g(0) = g$ and $\dot{g} = \xi$. Further, for $\xi \in \mathfrak{g}$, let $\xi w$ be the infinitesimal generator for the $G$ action on $V^*$, that is

$$\xi w := \xi_{V^*}(w) = \left.\frac{\mathrm{d}}{\mathrm{d}t}\right|_{t=0} g(t) w.$$

Now, for every $v \in V$, consider the linear transformation

$$\rho_v : \mathfrak{g} \to V, \qquad \xi \to \rho_v(\xi) := \xi v.$$

Its dual, $\rho_v^* : V^* \to \mathfrak{g}^*$, defines the ***diamond operation***

$$V \times V^* \to \mathfrak{g}^*, \quad (v, w) \to v \diamond w := \rho_v^*(w).$$

Note that for any $\xi \in \mathfrak{g}$ and $(v, w) \in V \times V^*$ we have:

$$\langle v \diamond w, \xi \rangle = \left\langle \rho_v^*(w), \xi \right\rangle = \langle w, \rho_v(\xi) \rangle = \langle w, \xi v \rangle .$$

**Theorem 7.15 (Euler–Poincaré with advected parameters)** *Consider a Lie group $G$ and a left linear action of $G$ on a vector space $V$. For a given $a_0 \in V^*$, let $L_{a_0} : TG \to \mathbb{R}$ be a Lagrangian with parameter $a_0$. Suppose there exists a function $L : TG \times V^* \to \mathbb{R}$ such that*

$$L(g, \dot{g}, a_0) = L_{a_0}(g, \dot{g}), \qquad \textit{for all } (g, \dot{g}) \in TG,$$

*and $L$ is invariant under the diagonal left action of $G$ on $TG \times V^*$,*

$$G \times (TG \times V^*) \to (TG \times V^*), \quad (h, (g, \dot{g}, v)) \to (hg, h\dot{g}, hw).$$

*Define* $l : \mathfrak{g} \times V^* \to \mathbb{R}$ *by*

$$l(\xi, a) := L(e, \xi, a).$$

*For any given curve* $g(t) \in G$, *define*

$$\xi(t) := g(t)^{-1}\dot{g}(t)$$

*and*

$$a(t) = g(t)^{-1}a_0.$$

*Note that* $a(t)$ *is the unique solution of*

$$\dot{a}(t) = -\xi(t)a(t), \quad a(0) = a_0.$$

*Then the following are equivalent:*

(i) *Hamilton's variational principle,*

$$\delta \int_{t_1}^{t_2} L_{a_0}(g(t), \dot{g}(t))\mathrm{d}t = 0, \tag{7.67}$$

*holds for variations* $\delta g$ *of* $g(t)$ *in* $G$ *that vanish at the end points.*

(ii) *The curve* $g(t)$ *satisfies the Euler–Lagrange equations for* $L_{a_0}$.

(iii) *The constrained variational principle*

$$\delta \int_{t_1}^{t_2} l(\xi(t), a(t))\mathrm{d}t = 0 \tag{7.68}$$

*holds on* $\mathfrak{g} \times V$, *using variations of the form*

$$\delta\xi = \dot{\eta} + \mathrm{ad}_\xi \eta, \quad \delta a = -\eta a,$$

*where* $\eta(t)$ *is any curve in* $\mathfrak{g}$ *that vanishes at the endpoints.*

(iv) *The Euler–Poincaré equations,*

$$\frac{\mathrm{d}l}{\mathrm{d}t} = \mathrm{ad}^*_\xi \frac{\delta l}{\delta \xi} + \frac{\delta l}{\delta a} \diamond a, \tag{7.69}$$

*hold on* $\mathfrak{g} \times V^*$.

*Proof* The equivalence of (i) and (ii) holds for any configuration manifold, so, in particular, holds in this case.

To show the equivalence of (i) and (iii), first note that $G$-invariance of $L$ and the definition of $a(t)$ imply that the integrands (7.67) and (7.68) are equal. As we already have seen in Proposition 7.3, all variations $\delta g(t) \in TG$ of $g(t)$ that vanish at the endpoints induce and are induced by variations $\delta\xi(t) \in \mathfrak{g}$ of $\xi(t)$ of the form

$$\delta\xi = \dot{\eta} + [\xi, \eta]$$

with $\eta(t) \in \mathfrak{g}$ vanishing at endpoints. The variation of $a(t) = g(t)^{-1}a_0$ is given by:

$$\delta a(t) = \delta g(t)^{-1}a_0 = -g(t)^{-1}\delta g(t)g(t)^{-1}a_0 = -\eta(t)a(t).$$

Conversely, if the variation of $a(t)$ is defined by $\delta a(t) = -\eta(t)a(t)$, then the variation of $g(t)a(t) = a_0$ vanishes, which is consistent with the dependence of $L_{a_0}$ on $(g(t), \dot{g}(t))$.

We end the proof by showing the equivalence (iii) with (iv). Using the definitions and integrating by parts, the variation of the integral becomes:

$$\begin{aligned}
0 &= \delta \int_{t_1}^{t_2} l(\xi(t), a(t))\mathrm{d}t \\
&= \delta \int_{t_1}^{t_2} \left( \left\langle \frac{\delta l}{\delta \xi}, \delta\xi \right\rangle + \left\langle \frac{\delta l}{\delta a}, \delta a \right\rangle \right) \mathrm{d}t \\
&= \delta \int_{t_1}^{t_2} \left( \left\langle \frac{\delta l}{\delta \xi}, \dot{\eta} + \mathrm{ad}_\xi \eta \right\rangle + \left\langle \frac{\delta l}{\delta a}, (-\eta a) \right\rangle \right) \mathrm{d}t \\
&= \delta \int_{t_1}^{t_2} \left( \left\langle -\frac{\mathrm{d}}{\mathrm{d}t}\left(\frac{\delta l}{\delta \xi}\right) + \mathrm{ad}^*_\xi \left(\frac{\delta l}{\delta \xi}\right), \eta \right\rangle + \left\langle \frac{\delta l}{\delta a} \diamond a, \eta \right\rangle \right) \mathrm{d}t \\
&= \delta \int_{t_1}^{t_2} \left( \left\langle -\frac{\mathrm{d}}{\mathrm{d}t}\left(\frac{\delta l}{\delta \xi}\right) + \mathrm{ad}^*_\xi \left(\frac{\delta l}{\delta \xi}\right) + \frac{\delta l}{\delta a} \diamond a, \eta \right\rangle \right) \mathrm{d}t
\end{aligned}$$

and the result follows. ■

**Remark 7.16** *There are four versions of the preceding theorem, the given 'left-left' version (i.e. left action of G on itself together with a linear left* *action of G on V), a 'left-right', 'right-left' and a 'right-right' version. The most important ones are left-left and the right-right versions.*

**Remark 7.17** *In the preceding theorem, it suffices for L to be defined on $TG \times A$, where*

$$A := Orb(a_0) = \{g^{-1}a_0 : g \in G\}.$$

We conclude this chapter with a theoretical result that includes and generalizes a number of related conservation laws that will appear later in the book.

**Theorem 7.18 (Kelvin–Noether theorem)** *In the context of the previous theorem, let $\mathcal{C}$ be a smooth manifold on which G acts (on the left), fix $c_0 \in \mathcal{C}$, and let $c(t) = g(t)^{-1}c_0$. Suppose there exists a map $\mathcal{K} : \mathcal{C} \times V^* \to \mathfrak{g}^{**}$ that is equivariant with respect to the dual of the coadjoint action on $\mathfrak{g}^*$. Define the* ***Kelvin–Noether*** *quantity $I\colon \mathcal{C} \times \mathfrak{g} \times V^* \to \mathbb{R}$ by*

$$I(c,\xi,a) := \left\langle \mathcal{K}(c,a), \frac{\delta l}{\delta \xi}(\xi,a) \right\rangle$$

*and let $c(t) = g(t)^{-1}c_0$. Then*

$$\frac{\mathrm{d}I(t)}{\mathrm{d}t} = \left\langle \mathcal{K}(c(t),a(t)), \frac{\delta l}{\delta a}(\xi(t),a(t)) \diamond a(t) \right\rangle.$$

*Proof* Recall the identity (6.8):

$$\frac{\mathrm{d}}{\mathrm{d}t}(g(t)\cdot\mu(t)) = g(t)\cdot\left[\frac{\mathrm{d}\mu}{\mathrm{d}t} - \mathrm{ad}^*_{\xi(t)}\mu(t)\right].$$

Using the equivariance property of $\mathcal{K}$ :

$$\mathcal{K}(gc, ga) = g\mathcal{K}(c,a), \quad \text{for all} \quad g \in G, c \in \mathcal{C} \text{ and } a \in V^*,$$

and taking into account that $a(t) = g^{-1}(t)a_0$, we have:

$$\begin{aligned}
&\left\langle \mathcal{K}\left(c(t),a(t)\right), \left(\frac{\delta l}{\delta \xi}\left(\xi(t),a(t)\right)\right)\right\rangle \\
&\quad = \left\langle g(t)^{-1}\mathcal{K}(c_0,a_0), \left(\frac{\delta l}{\delta \xi}\left(\xi(t),a(t)\right)\right)\right\rangle \\
&\quad = \left\langle \mathcal{K}(c_0,a_0), g(t)\cdot\left(\frac{\delta l}{\delta \xi}\left(\xi(t),a(t)\right)\right)\right\rangle.
\end{aligned}$$

Differentiating the above and taking into account the identity (6.54) we have

$$\begin{aligned}
\frac{\mathrm{d}I(t)}{\mathrm{d}t} &= \frac{\mathrm{d}}{\mathrm{d}t}\left\langle \mathcal{K}(c_0,a_0), g(t)\cdot\left(\frac{\delta l}{\delta \xi}(\xi(t),a(t))\right)\right\rangle \\
&= \left\langle \mathcal{K}(c_0,a_0), \frac{\mathrm{d}}{\mathrm{d}t}\left\{g(t)\cdot\left(\frac{\delta l}{\delta \xi}(\xi(t),a(t))\right)\right\}\right\rangle \\
&= \left\langle \mathcal{K}(c_0,a_0), g(t)\cdot\left(\frac{\mathrm{d}}{\mathrm{d}t}\left(\frac{\delta l}{\delta \xi}(\xi(t),a(t))\right) - \mathrm{ad}^*_\xi\frac{\delta l}{\delta \xi}\right)\right\rangle \\
&= \left\langle \mathcal{K}(c_0,a_0), g(t)\cdot\left(\mathrm{ad}^*_\xi\frac{\delta l}{\delta \xi} + \frac{\delta l}{\delta a}\diamond a - \mathrm{ad}^*_\xi\frac{\delta l}{\delta \xi}\right)\right\rangle \\
&= \left\langle \mathcal{K}(c_0,a_0), g(t)\cdot\left(\frac{\delta l}{\delta a}\diamond a\right)\right\rangle \\
&= \left\langle g(t)^{-1}\mathcal{K}(c_0,a_0), \frac{\delta l}{\delta a}\diamond a\right\rangle = \left\langle \mathcal{K}\left(c(t),a(t)\right), \frac{\delta l}{\delta a}\diamond a\right\rangle .
\end{aligned}$$

■

In the absence of advected quantities, the right-hand side of the above formula for $\mathrm{d}I/\mathrm{d}t$ equals zero. This proves the following.

**Corollary 7.19** *In the context of the previous theorem, suppose there exists a map $\mathcal{K} : C \to \mathfrak{g}^{**}$ that is equivariant with respect to the dual of the coadjoint action on $\mathfrak{g}^*$. Define the* **Kelvin–Noether** *quantity $I\colon C\times \mathfrak{g} \to \mathbb{R}$ by*

$$I(c,\xi) := \left\langle \mathcal{K}(c), \frac{\delta l}{\delta \xi}(\xi)\right\rangle .$$

*Then,*

$$\frac{\mathrm{d}I(t)}{\mathrm{d}t} = 0.$$

**Remark 7.20** *Theorem 7.18 is associated with Kelvin and Noether because it arises from symmetry (Noether) and generalizes the Kelvin circulation theorem, as we will see in Chapters 10 and 17.*

**Exercise 7.20**
State and prove the right-right version of the Euler–Poincaré theorem for Lagrangians with advected parameters.

**Exercise 7.21**
For the Euler–Poincaré equations for motion on Lie groups, show that the Kelvin quantity $I(t)$, defined as above but with $I : \mathcal{C} \times \mathfrak{g} \to \mathbf{R}$, is conserved.

**Exercise 7.22 (Double heavy top)**
Consider two coupled rigid bodies in the gravitational field where the first body hangs from a fixed point and the bodies are coupled by an ideal ball and socket joint.

1) Deduce the Lagrangian of this system.
2) Deduce the Euler–Poincaré equations of motion.

## Solutions to selected exercises

**Solution to Exercise 7.3** Consider the expression

$$\langle \dot{\boldsymbol{R}}_1 \,,\, \mathbb{I}\dot{\boldsymbol{R}}_2 \rangle = \int_B (\dot{\boldsymbol{R}}_1 \boldsymbol{X}) \cdot (\dot{\boldsymbol{R}}_2 \boldsymbol{X}) \, \mathrm{d}^3 \boldsymbol{X} = \mathrm{tr}\left( \dot{\boldsymbol{R}}_1 \mathbb{I} \dot{\boldsymbol{R}}_2^T \right).$$

Under the right action, $\dot{\boldsymbol{R}} \to \dot{\boldsymbol{R}}\boldsymbol{Q}$ for some $\boldsymbol{Q} \in SO(3)$. Right invariance of the kinetic energy requires

$$\langle \dot{\boldsymbol{R}}_1 \boldsymbol{Q} \,,\, \mathbb{I}\dot{\boldsymbol{R}}_2 \boldsymbol{Q} \rangle = \mathrm{tr}\left( \dot{\boldsymbol{R}}_1 \boldsymbol{Q} \mathbb{I} \boldsymbol{Q}^T \dot{\boldsymbol{R}}_2^T \right) = \mathrm{tr}\left( \dot{\boldsymbol{R}}_1 \mathbb{I} \dot{\boldsymbol{R}}_2^T \right).$$

Thus, the condition for right invariance is $\boldsymbol{Q}\mathbb{I}\boldsymbol{Q}^T = \mathbb{I}$ for all $\boldsymbol{Q} \in SO(3)$. This is only true when $\mathbb{I} = \lambda \boldsymbol{I}$ for some $\lambda \in \mathbb{R}$. Therefore invariance of the kinetic energy under the right $SO(3)$ action is equivalent to the condition

$$\mathbb{I} = \lambda \boldsymbol{I}.$$

This condition is equivalent to saying our rigid body is a sphere.

**Solution to Exercise 7.9** The setup is formally the same as in 3 dimensions. The kinetic energy is

$$\begin{aligned}
K(\boldsymbol{R},\dot{\boldsymbol{R}}) &= \int_{\mathcal{B}} \frac{\|\dot{\boldsymbol{x}}\|^2}{2}\mathrm{d}V \\
&= \int_{\mathcal{B}} \frac{\|\dot{\boldsymbol{R}}\boldsymbol{X}\|^2}{2}\mathrm{d}^3\boldsymbol{X} \\
&= \frac{1}{2}\langle \dot{\boldsymbol{R}}, \mathbb{J}\dot{\boldsymbol{R}}\rangle.
\end{aligned}$$

Consequently, since $\|\boldsymbol{Q}\boldsymbol{x}\|^2 = \|\boldsymbol{x}\|^2$ for all $\boldsymbol{Q} \in SO(2)$ and $\boldsymbol{x} = (x, y) \in \mathbb{R}^2$, $K(\boldsymbol{R},\dot{\boldsymbol{R}})$ is left $SO(2)$-invariant. Thus, the reduced Lagrangian takes the form

$$l(\widehat{\boldsymbol{\Omega}}) = K(\boldsymbol{I},\widehat{\boldsymbol{\Omega}}) = \boldsymbol{\Omega}\cdot\mathbb{I}\boldsymbol{\Omega},$$

where, as usual, $\hat{\boldsymbol{\Omega}} = \big(\boldsymbol{R}^{-1}\dot{\boldsymbol{R}}\big)\widehat{\ }$. Since the symmetry group is $SO(2)$,

$$\widehat{\boldsymbol{\Omega}} = \begin{pmatrix} 0 & -\Omega \\ \Omega & 0 \end{pmatrix}.$$

The action of $\widehat{\boldsymbol{\Omega}}$ on $\mathbb{R}^2$ is

$$\widehat{\boldsymbol{\Omega}}\boldsymbol{x} = \Omega(-y, x),$$

which coincides with the definition of angular velocity in $\mathbb{R}^2$. Also,

$$\mathbb{I} = \|\boldsymbol{x}\|^2\boldsymbol{I} - \boldsymbol{x}\boldsymbol{x}^T,$$

just as in the 3-dimensional case only now $\boldsymbol{x} \in \mathbb{R}^2$. Consider variations of the form $\delta\widehat{\boldsymbol{\Omega}} = \dot{\widehat{\boldsymbol{\Sigma}}} + \mathrm{ad}_{\widehat{\boldsymbol{\Omega}}}\widehat{\boldsymbol{\Sigma}}$ where $\widehat{\boldsymbol{\Sigma}} = \big(\boldsymbol{R}^{-1}\delta\boldsymbol{R}\big)\widehat{\ }$. Now, in $SO(2)$, $\mathrm{ad}_{\widehat{\boldsymbol{\Omega}}} = 0$. Therefore, the variations are $\delta\widehat{\boldsymbol{\Omega}} = \dot{\widehat{\boldsymbol{\Sigma}}}$. Hamilton's Principle gives

$$\begin{aligned}
\delta S &= \delta \int_{t_0}^{t_1} l\,\mathrm{d}t \\
&= \int_{t_0}^{t_1} \left\langle \frac{\delta l}{\delta\widehat{\boldsymbol{\Omega}}}, \delta\widehat{\boldsymbol{\Omega}} \right\rangle \mathrm{d}t \\
&= -\int_{t_0}^{t_1} \left\langle \frac{\mathrm{d}}{\mathrm{d}t}\frac{\delta l}{\delta\widehat{\boldsymbol{\Omega}}}, \widehat{\boldsymbol{\Sigma}} \right\rangle \mathrm{d}t \\
&= 0.
\end{aligned}$$

Therefore the equations of motion are given by

$$\frac{\mathrm{d}}{\mathrm{d}t}\mathbb{I}\boldsymbol{\Omega} = 0.$$

This is solved by $\Omega(t) = \Omega(0)$ and the reconstruction equation is given by

$$\dot{\boldsymbol{R}} = \boldsymbol{R}\widehat{\boldsymbol{\Omega}}.$$

The solution to this equation is $\boldsymbol{R} = \exp(t\widehat{\boldsymbol{\Omega}})$ as is easily verified. Therefore the general solution for planar motion of a flat rigid body is given by

$$\boldsymbol{x}(t) = \exp(t\widehat{\boldsymbol{\Omega}})\boldsymbol{x}(0).$$

Explicitly, $\widehat{\boldsymbol{\Omega}}^2 = -\Omega^2 \boldsymbol{I}$ so that

$$\begin{aligned}
\exp(t\widehat{\boldsymbol{\Omega}}) &= \sum_{i=0}^{\infty} t^i \widehat{\boldsymbol{\Omega}}^i \\
&= \sum_{i=0}^{\infty}(-1)^i t^2 i \Omega^2 i \boldsymbol{I} + \sum_{i=0}^{\infty}(-1)^{i+1} t^{(2i+1)} \Omega^{(2i+1)} \begin{bmatrix} 0 & -1 \\ 1 & 0 \end{bmatrix} \\
&= \cos(t\Omega)\, \boldsymbol{I} + \sin(t\Omega) \begin{bmatrix} 0 & -1 \\ 1 & 0 \end{bmatrix} \\
&= \begin{bmatrix} \cos(t\Omega) & -\sin(t\Omega) \\ \sin(t\Omega) & \cos(t\Omega) \end{bmatrix}.
\end{aligned}$$

Therefore the solution to the rigid body in $\mathbb{R}^2$ is given by

$$\begin{bmatrix} x \\ y \end{bmatrix}(t) = \begin{bmatrix} x_0 \cos(\Omega t) - y_0 \sin(\Omega t) \\ x_0 \sin(\Omega t) + y_0 \cos(\Omega t) \end{bmatrix}.$$

**Solution to Exercise 7.10** The reconstruction equation is given by

$$\boldsymbol{R}^{-1}\dot{\boldsymbol{R}} = \boldsymbol{\Omega}, \qquad \dot{\boldsymbol{R}} = \boldsymbol{R}\boldsymbol{\Omega}.$$

Since $\boldsymbol{\Omega}(t) = \boldsymbol{\Omega}_0 = const.$ there is a simple solution given by $\boldsymbol{R} = \boldsymbol{R}_0 \exp(t\boldsymbol{\Omega})$, which follows as

$$\frac{\mathrm{d}}{\mathrm{d}t}\boldsymbol{R}_0 \exp(t\boldsymbol{\Omega}) = \boldsymbol{R}_0 \exp(t\boldsymbol{\Omega})\boldsymbol{\Omega} = \boldsymbol{R}\boldsymbol{\Omega}.$$

Note that away from the equilibrium solutions $\boldsymbol{\Omega}(t)$ is not constant so the reconstruction equation is not as simple as it is here. The problem is still integrable but the solution has to be expressed in terms of an elliptic integral or a series solution. This calculation is performed explicitly for $SO(3)$ in the last chapter of [MR03].

**Solution to Exercise 7.14** The Euler–Poincaré equations are

$$\frac{\mathrm{d}}{\mathrm{d}t}\frac{\delta l}{\delta \xi} = \mathrm{ad}^*_\xi \frac{\delta l}{\delta \xi}.$$

In $SE(3)$ the coadjoint operator is given by

$$\mathrm{ad}^*_{(\widehat{\boldsymbol{\Omega}},\boldsymbol{\Gamma})}(\widehat{\boldsymbol{\mu}}, \boldsymbol{m}) = -(\boldsymbol{\Omega}\times\boldsymbol{\mu} + \boldsymbol{\Gamma}\times \boldsymbol{m}, \boldsymbol{\Omega}\times\boldsymbol{m})\widehat{\ }.$$

Also, observe that

$$\frac{\delta l}{\delta\left(\widehat{\boldsymbol{\Omega}},\boldsymbol{\Gamma}\right)} = \left(\frac{\delta l}{\delta \widehat{\boldsymbol{\Omega}}}, \frac{\delta l}{\delta \boldsymbol{\Gamma}}\right).$$

Thus, the Euler–Poincaré equations in $SE(3)$ are given by

$$\begin{aligned}
\frac{\mathrm{d}}{\mathrm{d}t}\frac{\delta l}{\delta \boldsymbol{\Omega}} &= -\boldsymbol{\Omega}\times\frac{\delta l}{\delta \boldsymbol{\Omega}} - \boldsymbol{\Gamma}\times\frac{\delta l}{\delta \boldsymbol{\Gamma}},\\
\frac{\mathrm{d}}{\mathrm{d}t}\frac{\delta l}{\delta \boldsymbol{\Gamma}} &= -\boldsymbol{\Omega}\times\frac{\delta l}{\delta \boldsymbol{\Gamma}}.
\end{aligned}$$

**Solution to Exercise 7.15** Recall Exercise 5.21 where a basis, $\{\widehat{\boldsymbol{J}}, \widehat{\boldsymbol{K}}\}$, of $\mathfrak{so}(4)$ is derived. The hat map in this case can be regarded as mapping $\mathbb{R}^6$ onto $\mathfrak{so}(4)$ by $(\boldsymbol{a},\boldsymbol{b}) \to \widehat{(\boldsymbol{a},\boldsymbol{b})} := (\boldsymbol{a}\cdot\widehat{\boldsymbol{J}} + \boldsymbol{b}\cdot\widehat{\boldsymbol{K}})$. By a long calculation, the adjoint action can be shown to be

$$\mathrm{ad}_{\widehat{(\boldsymbol{a},\boldsymbol{b})}}\widehat{(\boldsymbol{c},\boldsymbol{d})} = (\boldsymbol{a}\times\boldsymbol{c} + \boldsymbol{b}\times\boldsymbol{d})\cdot\widehat{\boldsymbol{J}} + (\boldsymbol{a}\times\boldsymbol{d} - \boldsymbol{b}\times\boldsymbol{c})\cdot\widehat{\boldsymbol{K}}.$$

With this in mind we substitute $l = l(\Psi)$ with $\Psi = \boldsymbol{\Omega}\cdot\widehat{\boldsymbol{J}} + \boldsymbol{\Lambda}\cdot\widehat{\boldsymbol{K}}$ into the Euler-Poincaré equations for a left invariant Lagrangian and obtain

$$\begin{aligned}
\frac{\mathrm{d}}{\mathrm{d}t}\frac{\partial l}{\partial \boldsymbol{\Omega}} &= \frac{\partial l}{\partial \boldsymbol{\Omega}}\times\boldsymbol{\Omega} + \frac{\partial l}{\partial \boldsymbol{\Lambda}}\times\boldsymbol{\Lambda},\\
\frac{\mathrm{d}}{\mathrm{d}t}\frac{\partial l}{\partial \boldsymbol{\Lambda}} &= \frac{\partial l}{\partial \boldsymbol{\Lambda}}\times\boldsymbol{\Omega} - \frac{\partial l}{\partial \boldsymbol{\Omega}}\times\boldsymbol{\Lambda}.
\end{aligned}$$

The two equations correspond to the $\widehat{\boldsymbol{J}}$ and $\widehat{\boldsymbol{K}}$ components respectively. For more on this problem, see [Hol08].

**Solution to Exercise 7.18** The Lagrangian is,

$$L_{\mathrm{k}}(\boldsymbol{R},\dot{\boldsymbol{R}}) = \frac{1}{2}\langle \dot{\boldsymbol{R}}\,,\,\mathbb{I}\dot{\boldsymbol{R}}\rangle - mg\langle \mathrm{k}\,,\,\boldsymbol{R}\chi\rangle + e\langle \boldsymbol{A}(\boldsymbol{R}\chi)\,,\,\dot{\boldsymbol{R}}\chi\rangle.$$

This Lagrangian is left $SO(3)$ invariant. The reduced Lagrangian is calculated by

$$\begin{aligned}
L_{\mathrm{k}}(\boldsymbol{I},\boldsymbol{R}^{-1}\dot{\boldsymbol{R}}) &= \frac{1}{2}\Big\langle \boldsymbol{R}^{-1}\dot{\boldsymbol{R}}\,,\,\mathbb{I}\boldsymbol{R}^{-1}\dot{\boldsymbol{R}}\Big\rangle - mg\Big\langle \boldsymbol{R}^{-1}\mathrm{k}\,,\,\chi\Big\rangle \\
&\quad + e\Big\langle \boldsymbol{R}^{-1}\boldsymbol{A}(\boldsymbol{R}\chi)\,,\,\boldsymbol{R}^{-1}\dot{\boldsymbol{R}}\chi\Big\rangle \\
&= \frac{1}{2}\langle \widehat{\boldsymbol{\Omega}}\,,\,\mathbb{I}\widehat{\boldsymbol{\Omega}}\rangle - mg\langle \boldsymbol{\Gamma}\,,\,\chi\rangle + e\langle \boldsymbol{\Gamma}\times\chi\,,\,\widehat{\boldsymbol{\Omega}}\chi\rangle \\
&= \frac{1}{2}\langle \boldsymbol{\Omega}\,,\,\mathbb{I}\boldsymbol{\Omega}\rangle - mg\langle \boldsymbol{\Gamma}\,,\,\chi\rangle + e\langle \boldsymbol{\Gamma}\times\chi\,,\,\boldsymbol{\Omega}\times\chi\rangle \\
&= l(\boldsymbol{\Omega},\boldsymbol{\Gamma}).
\end{aligned}$$

Introducing the free variation $\widehat{\boldsymbol{\Sigma}} = \boldsymbol{R}^{-1}\delta\boldsymbol{R}$ enables the derivation of the extended Euler–Poincaré equations,

$$\begin{aligned}
\delta S &= \delta\int_{t_0}^{t_1} \frac{1}{2}\langle \boldsymbol{\Omega}\,,\,\mathbb{I}\boldsymbol{\Omega}\rangle - mg\langle \boldsymbol{\Gamma}\,,\,\chi\rangle + e\langle \boldsymbol{\Gamma}\times\chi\,,\,\boldsymbol{\Omega}\times\chi\rangle \mathrm{d}t \\
&= \int_{t_0}^{t_1} \langle \mathbb{I}\boldsymbol{\Omega} + e\chi\times(\boldsymbol{\Gamma}\times\chi)\,,\,\delta\boldsymbol{\Omega}\rangle + \langle e\chi\times(\boldsymbol{\Omega}\times\chi) - mg\chi\,,\,\delta\boldsymbol{\Gamma}\rangle \mathrm{d}t \\
&= \int_{t_0}^{t_1} \Big\langle \mathbb{I}\boldsymbol{\Omega} + e\chi\times(\boldsymbol{\Gamma}\times\chi)\,,\,\dot{\boldsymbol{\Sigma}} + \boldsymbol{\Omega}\times\boldsymbol{\Sigma}\Big\rangle \\
&\quad + \langle e\chi\times(\boldsymbol{\Omega}\times\chi) - mg\chi\,,\,\boldsymbol{\Gamma}\times\boldsymbol{\Sigma}\rangle \mathrm{d}t \\
&= \int_{t_0}^{t_1} \Big\langle -\left(\frac{\mathrm{d}}{\mathrm{d}t} + \boldsymbol{\Omega}\times\right)(\mathbb{I}\boldsymbol{\Omega} + e\chi\times(\boldsymbol{\Gamma}\times\chi)) \\
&\quad + (e\chi\times(\boldsymbol{\Omega}\times\chi) - mg\chi)\times\boldsymbol{\Gamma}, \boldsymbol{\Sigma}\Big\rangle \mathrm{d}t \\
&= 0.
\end{aligned}$$

Therefore, the Euler–Poincaré equation is given by

$$\left(\frac{\mathrm{d}}{\mathrm{d}t} + \boldsymbol{\Omega}\times\right)(\mathbb{I}\boldsymbol{\Omega} + e\chi\times(\boldsymbol{\Gamma}\times\chi)) = (e\chi\times(\boldsymbol{\Omega}\times\chi) - mg\chi)\times\boldsymbol{\Gamma},$$

which has to be augmented by the relation

$$\dot{\boldsymbol{\Gamma}} = \boldsymbol{\Gamma}\times\boldsymbol{\Omega}.$$

# 8 Momentum maps

Lie group symmetries of the Hamiltonian are associated with conserved quantities. For example, the flow of any SO(3)-invariant Hamiltonian vector field on $T^*\mathbb{R}^3$ conserves angular momentum, $\mathbf{q} \times \mathbf{p}$. More generally, given a Hamiltonian $H$ on a phase space $P$ that is invariant under a canonical action of a Lie group $G$, there often exists a ***momentum map*** $\mathbf{J} : P \to \mathfrak{g}^*$ that is conserved by the flow of the Hamiltonian vector field.

The definition of a momentum map is a geometric one, independent of the dynamics induced by a given Hamiltonian. The main idea is as follows. The group action generates a family of vector fields on the manifold $P$, each vector $\xi$ in the Lie algebra $\mathfrak{g}$ determining an infinitesimal generator vector field[1] $\xi_P$. If this vector field turns out to be Hamiltonian (as is often the case), then its Hamiltonian function is denoted $J_\xi$. These Hamiltonian functions are unique up to a choice of additive constants. By adjusting these constants if necessary, it is possible to 'bundle together' the family of Hamiltonians $\{J_\xi : \xi \in \mathfrak{g}\}$ (see Exercise 8.5) into a function $\mathbf{J} : P \to \mathfrak{g}^*$ that satisfies $\langle \mathbf{J}(z), \xi \rangle = J_\xi(z)$ for all $z \in P$ and all $\xi \in \mathfrak{g}$, where $<,>$ indicates the natural pairing of the Lie algebra with its dual.

## 8.1 Definition and examples

**Definition 8.1** *A* ***momentum map*** *for a canonical action of $G$ on a Poisson manifold $P$ is a map $\mathbf{J} : P \to \mathfrak{g}^*$ such that, for every $\xi \in \mathfrak{g}$, the Hamiltonian vector field of the map $J_\xi : P \to \mathbb{R}$ defined by*

$$J_\xi(z) = \langle \mathbf{J}(z), \xi \rangle$$

*satisfies*

$$X_{J_\xi} = \xi_P. \tag{8.1}$$

[1] Let $G$ act smoothly on $P$, and let $\xi \in \mathfrak{g}$. Recall from Definition 6.22 that the infinitesimal generator $\xi_P$ is the vector field on $P$ defined by

$$\xi_P(z) = \left.\frac{d}{dt}\right|_{t=0} (\exp(t\xi)z), \quad \text{for all } z \in P.$$

**Remark 8.2** *Not every canonical action on a Poisson manifold has a globally defined momentum map (see Exercise 8.3).*

**Example 8.3 (The momentum map for rotations on $\mathbb{R}^3$)** Consider the cotangent bundle of ordinary Euclidean space $\mathbb{R}^3$. This cotangent bundle has coordinates $(\mathbf{q}, \mathbf{p}) \in T^*\mathbb{R}^3 \simeq \mathbb{R}^6$, and is equipped with the canonical Poisson bracket. An element $\mathbf{R}$ of the rotation group $SO(3)$ acts on $T^*\mathbb{R}^3$ according to

$$\mathbf{R}(\mathbf{q}, \mathbf{p}) = (\mathbf{R}\mathbf{q}, \mathbf{R}\mathbf{p}).$$

Let $\widehat{\boldsymbol{\xi}} \in \mathfrak{so}(3)$ be an infinitesimal rotation. To calculate the infinitesimal generator $\widehat{\boldsymbol{\xi}}_{T^*\mathbb{R}^3}(\mathbf{q}, \mathbf{p})$, consider a path $\mathbf{R}(t) \in SO(3)$ such that $\mathbf{R}(0) = \mathbf{I}$ and $\mathbf{R}'(0) = \widehat{\boldsymbol{\xi}}$. Then

$$\begin{aligned}\widehat{\boldsymbol{\xi}}_{T^*\mathbb{R}^3}(\mathbf{q}, \mathbf{p}) &= \left.\frac{\mathrm{d}}{\mathrm{d}t}\right|_{t=0} \mathbf{R}(t)(\mathbf{q}, \mathbf{p}) = \left.\frac{\mathrm{d}}{\mathrm{d}t}\right|_{t=0} (\mathbf{R}(t)\mathbf{q}, \mathbf{R}(t)\mathbf{p}) \\ &= (\widehat{\boldsymbol{\xi}}\mathbf{q}, \widehat{\boldsymbol{\xi}}\mathbf{p}) = (\boldsymbol{\xi} \times \mathbf{q}, \boldsymbol{\xi} \times \mathbf{p}).\end{aligned}$$

Now we have to solve

$$\left(\frac{\partial J_{\widehat{\xi}}}{\partial \mathbf{p}}, -\frac{\partial J_{\widehat{\xi}}}{\partial \mathbf{q}}\right) = \widehat{\boldsymbol{\xi}}_{T^*\mathbb{R}^3}(\mathbf{q}, \mathbf{p}) = (\boldsymbol{\xi} \times \mathbf{q}, \boldsymbol{\xi} \times \mathbf{p})$$

for $J_{\widehat{\xi}}$. That is,

$$\frac{\partial J_{\widehat{\xi}}}{\partial \mathbf{p}} = \boldsymbol{\xi} \times \mathbf{q}, \qquad -\frac{\partial J_{\widehat{\xi}}}{\partial \mathbf{q}} = \boldsymbol{\xi} \times \mathbf{p}.$$

Hence,

$$J_{\widehat{\xi}}(\mathbf{q}, \mathbf{p}) = (\boldsymbol{\xi} \times \mathbf{q}) \cdot \mathbf{p} = (\mathbf{q} \times \mathbf{p}) \cdot \boldsymbol{\xi},$$

so

$$\left\langle \mathbf{J}(\mathbf{q}, \mathbf{p}), \widehat{\boldsymbol{\xi}} \right\rangle = J_{\widehat{\xi}}(\mathbf{q}, \mathbf{p}) = (\mathbf{q} \times \mathbf{p}) \cdot \boldsymbol{\xi}.$$

If $\mathbf{J}(\mathbf{q}, \mathbf{p}) \in \mathfrak{so}(3)^*$ is identified with a vector in $\mathbb{R}^3$ via the breve map (see Example 5.36), then the previous formula can be rewritten as

$$\left\langle \mathbf{J}(\mathbf{q}, \mathbf{p}), \boldsymbol{\xi} \right\rangle = (\mathbf{q} \times \mathbf{p}) \cdot \boldsymbol{\xi}.$$

Since this relation holds for all $\boldsymbol{\xi}$, it follows that the momentum map for the rotation group is the angular momentum

$$\mathbf{J} = \mathbf{q} \times \mathbf{p}.$$

**Example 8.4 (Momentum map for linear symplectic actions)** Let $(V, \omega)$ be a symplectic vector space, i.e. let $\omega$ be a constant symplectic bilinear form on $V$, represented as a matrix. Let $G$ be a Lie group acting linearly and symplectically on $V$. Without loss of generality, we think of $V$ as $\mathbb{R}^n$ and $G$ as a matrix Lie group, acting in the standard way (see Example 6.2 and also Remark 6.24). This action admits a momentum map $\mathbf{J} : V \to \mathfrak{g}^*$ given by

$$J_\xi(\mathbf{v}) := \langle \mathbf{J}(\mathbf{v}), \xi \rangle := \frac{1}{2}\omega(\xi_V(\mathbf{v}), \mathbf{v}) = \frac{1}{2}\left(\xi_V(\mathbf{v})\right)^T \omega \mathbf{v}, \quad \text{for all } \xi \in \mathfrak{g}, \mathbf{v} \in V,$$

as we will now show. We have to verify that

$$\mathrm{d}J_\xi(\mathbf{v})(\mathbf{w}) = \omega\left(\xi_V(\mathbf{v}), \mathbf{w}\right) \quad \text{for all } \mathbf{w} \in V.$$

We compute,

$$\mathrm{d}J_\xi(\mathbf{v})(\mathbf{w}) = \frac{1}{2}\left(\xi_V(\mathbf{v})\right)^T \omega \mathbf{w} + \frac{1}{2}\left(\xi_V(\mathbf{w})\right)^T \omega \mathbf{v}.$$

Since $G$ acts symplectically on $V$, we have

$$\omega(g\mathbf{w}, g\mathbf{v}) = \omega(\mathbf{w}, \mathbf{v}) = \mathbf{w}^T \omega \mathbf{v}, \quad \text{for all } g \in G,$$

and thus,

$$\left.\frac{\mathrm{d}}{\mathrm{d}t}\right|_{t=0} (g(t)\mathbf{w})^T \omega \; (g(t)\mathbf{v}) = 0,$$

for any path $g(t) \in G$. If $g(0) = I$ and $g'(0) = \xi$, then

$$\mathbf{w}^T \omega \left(\xi_V(\mathbf{v})\right) + \left(\xi_V(\mathbf{w})\right)^T \omega \mathbf{v} = 0.$$

Therefore,

$$\begin{aligned} \mathrm{d}J_\xi(\mathbf{v})(\mathbf{w}) &= \frac{1}{2}\left(\xi_V(\mathbf{v})\right)^T \omega \mathbf{w} - \frac{1}{2}\mathbf{w}^T \omega \left(\xi_V(\mathbf{v})\right) \\ &= \left(\xi_V(\mathbf{v})\right)^T \omega \mathbf{w} = \omega\left(\xi_V(\mathbf{v}), \mathbf{w}\right), \end{aligned}$$

where we have used the skew-symmetry of $\omega$.

**Remark 8.5 (The momentum map for actions on symplectic manifold)** *If $(P, \omega)$ is a symplectic manifold, recall that $P$ is also a Poisson manifold and that, for any $F \in \mathcal{F}(P)$, the Hamiltonian vector field $X_F$ satisfies (see eqn (4.19)):*

$$\omega_z \left( X_F(z), \cdot \right) = \mathrm{d}F(z).$$

*Therefore, the momentum map condition (8.1) becomes*

$$\omega_z \left( \xi_P(z), \cdot \right) = \mathrm{d}J_\xi(z).$$

*See Example eqn 8.4.*

**Remark 8.6 (The momentum map for actions on cotangent bundles)** *Consider $P = T^*Q$, for some manifold $Q$, with the canonical Poisson bracket. The Hamiltonian vector field $X_{J_\xi}$ has the following formula in any cotangent lifted coordinates:*

$$X_{J_\xi} = \left( \frac{\partial J_\xi}{\partial \mathbf{p}}, -\frac{\partial J_\xi}{\partial \mathbf{q}} \right).$$

*In this case, the momentum map condition (8.1) becomes*

$$\left( \frac{\partial J_\xi}{\partial \mathbf{p}}, -\frac{\partial J_\xi}{\partial \mathbf{q}} \right) = \xi_{T^*Q}\left( \mathbf{q}, \mathbf{p} \right). \tag{8.2}$$

**Theorem 8.7 (Noether's formula for cotangent bundles)** *Let $G$ act on $Q$, and by cotangent lifts on $T^*Q$. Then, the momentum map $\mathbf{J} : T^*Q \to \mathfrak{g}^*$ is obtained via the formula*

$$J_\xi(\mathbf{q}, \mathbf{p}) = \left\langle \mathbf{p}, \xi_Q\left( \mathbf{q} \right) \right\rangle, \tag{8.3}$$

*where, for every $\xi \in \mathfrak{g}$, the map $J_\xi : T^*Q \to \mathbb{R}$ satisfies $J_\xi\left( \mathbf{q}, \mathbf{p} \right) = \left\langle \mathbf{J}\left( \mathbf{q}, \mathbf{p} \right), \xi \right\rangle$.*

*Proof* In local coordinates, considering $\mathbf{p}$ to be a column vector, definition (8.3) reads $J_\xi(\mathbf{q}, \mathbf{p}) = \mathbf{p}^T \xi_Q(\mathbf{q})$, from which we calculate

$$\frac{\partial J_\xi}{\partial \mathbf{p}}(\mathbf{q}, \mathbf{p}) = \xi_Q(\mathbf{q}),$$

and

$$\frac{\partial J_\xi}{\partial \mathbf{q}}(\mathbf{q}, \mathbf{p}) \cdot \dot{\mathbf{q}} = \mathbf{p}^T \mathrm{D}\xi_Q(\mathbf{q})\, \dot{\mathbf{q}} = \left( \left( \mathrm{D}\xi_Q(\mathbf{q}) \right)^T \mathbf{p} \right)^T \dot{\mathbf{q}},$$

which implies

$$\frac{\partial J_\xi}{\partial \mathbf{q}}(\mathbf{q},\mathbf{p}) = \left(\mathrm{D}\xi_Q(\mathbf{q})\right)^T \mathbf{p}.$$

Thus, by Remark 8.6 above, we need to show that

$$\left(\xi_Q(\mathbf{q}), -\left(\mathrm{D}\xi_Q(\mathbf{q})\right)^T \mathbf{p}\right) = \xi_{T^*Q}(\mathbf{q},\mathbf{p}).$$

But this was proven in Proposition 6.31. ■

A coordinate-free proof is given in [MR02].

**Example 8.8 (Revisiting the momentum map for rotations on $\mathbb{R}^3$)**
In the notation of Example 8.3, we apply Noether's formula (8.3):

$$J_{\widehat{\xi}}(\mathbf{q},\mathbf{p}) = \left\langle \mathbf{p}, \widehat{\boldsymbol{\xi}}_{\mathbb{R}^3}(\mathbf{q})\right\rangle = \langle \mathbf{p}, \widehat{\boldsymbol{\xi}}\,\mathbf{q}\rangle = \left\langle \mathbf{p}, \boldsymbol{\xi}\times\mathbf{q}\right\rangle = \mathbf{p}\cdot(\boldsymbol{\xi}\times\mathbf{q}) = (\mathbf{q}\times\mathbf{p})\cdot\boldsymbol{\xi}.$$

With J identified with a vector via the tilde map, we get $\mathbf{J} = \mathbf{q}\times\mathbf{p}$.

**Example 8.9 (Left and right multiplicative actions of $GL(n)$ on itself)**
Consider the left, respectively right, action of $GL(n)$ on itself; that is, for any $g \in GL(n)$, we have:

$$L_g(Q) = gQ \quad \text{for all } Q \in G \quad \text{(left multiplicative action)},$$

$$R_g(Q) = Qg \quad \text{for all } Q \in G \quad \text{(right multiplicative action)}.$$

Consider also the corresponding cotangent lifted actions on $T^*GL(n)$. To compute the momentum maps, we use Noether's formula (8.3). Using the definition, the infinitesimal generators are calculated:

$$\xi^l_G(Q) = \frac{\mathrm{d}}{\mathrm{d}t} L_{(\exp t\xi)} Q\Big|_{t=0} = \frac{\mathrm{d}}{\mathrm{d}t}(\exp t\xi)\Big|_{t=0} = \xi Q \quad \text{(left generator)}$$

and

$$\xi^r_G(Q) = \frac{\mathrm{d}}{\mathrm{d}t} R_{(\exp t\xi)} Q\Big|_{t=0} = \frac{\mathrm{d}}{\mathrm{d}t} Q(\exp t\xi)\Big|_{t=0} = Q\xi \quad \text{(right generator)},$$

where $\xi \in \mathfrak{g}$. Further, for the left action,

$$\begin{aligned} J_\xi(Q,P) &= \left\langle P, \xi^l_G(Q)\right\rangle = \langle P, \xi\, Q\rangle = -\frac{1}{2}\mathrm{tr}\left(P^T(\xi\, Q)\right) \\ &= -\frac{1}{2}\mathrm{tr}\left(P(\xi\, Q)^T\right) = -\frac{1}{2}\mathrm{tr}\left((PQ^T)\,\xi^T\right) = \left\langle PQ^T, \xi\right\rangle. \end{aligned}$$

Thus $\mathbf{J}_L$ can be identified with $PQ^T$ via the trace pairing:

$$\mathbf{J}_L(Q, P) = P\,Q^T \quad \text{(left momentum map for matrix groups).} \tag{8.4}$$

Analogously, we have

$$\mathbf{J}_R(Q, P) = Q^T P \quad \text{(right momentum map for matrix groups).} \tag{8.5}$$

Using the formula for $\mathrm{Ad}^*_Q$ in Example 6.43, it follows that

$$\mathbf{J}_R(Q, P) = Q^T \mathbf{J}_L(Q, P) Q^{-T} = \mathrm{Ad}^*_Q \mathbf{J}_L(Q, P).$$

**Example 8.10 (Multiplicative actions of $SO(n)$ on a matrix group)**
Let $G$ be a matrix group containing $SO(n)$, and consider the left, respectively right multiplicative action of $SO(n)$ on $G$

$$SO(n) \times G \to G, \quad (g, Q) \to L_g(Q) = gQ \quad \text{(left multiplicative action),}$$

$$SO(n) \times G \to G, \quad (g, Q) \to R_g(Q) = Qg \quad \text{(right multiplicative action),}$$

The formulae for the infinitesimal generators are identical to those in the previous exercise

$$\xi^l_G(Q) = \frac{\mathrm{d}}{\mathrm{d}t} L_{(\exp t\xi)} Q\Big|_{t=0} = \frac{\mathrm{d}}{\mathrm{d}t} (\exp t\xi) Q\Big|_{t=0} = \xi Q \quad \text{(left generator)}$$

and

$$\xi^r_G(Q) = \frac{\mathrm{d}}{\mathrm{d}t} R_{(\exp t\xi)} Q\Big|_{t=0} = \frac{\mathrm{d}}{\mathrm{d}t} Q\,(\exp(-t\xi))\Big|_{t=0} = Q\xi \quad \text{(right generator),}$$

but now $\xi \in \mathfrak{so}(n)$. In particular, $\xi$ is a *skew-symmetric* $n \times n$ matrix. For a matrix $M$, denote by $M_S$ and $M_A$ its symmetric and skew-symmetric parts respectively. We have, for the left action,

$$\begin{aligned} J_\xi(Q, P) &= \left\langle P, \xi^l_G(Q) \right\rangle = \left\langle P^T, \xi Q \right\rangle = \mathrm{tr}\left(P^T \xi Q\right) = \mathrm{tr}\left(Q P^T \xi\right) \\ &= \mathrm{tr}\left((QP^T)_S\, \xi\right) + \mathrm{tr}\left((QP^T)_A\, \xi\right) \end{aligned}$$

$$
\begin{aligned}
&= \operatorname{tr}\left((QP^T)_A\,\xi\right) \\
&= \operatorname{tr}\left((PQ^T)_A^T\,\xi\right) \\
&= \left\langle (PQ^T)_A,\, \xi \right\rangle \\
&= \left\langle \frac{1}{2}\left(PQ^T - (PQ^T)^T\right), \xi \right\rangle = \left\langle \frac{1}{2}\left(PQ^T - QP^T\right), \xi \right\rangle .
\end{aligned}
$$

Thus, $\mathbf{J}_L$ can be identified with $\frac{1}{2}\left(PQ^T - QP^T\right)$ via the trace pairing:

$$
\mathbf{J}_L(Q,P) = \frac{1}{2}\left(P\,Q^T - Q\,P^T\right), \tag{8.6}
$$

for the left multiplicative action of $SO(n)$ on a matrix group. Analogously, the momentum map for the right action is:

$$
\mathbf{J}_R(Q,P) = \frac{1}{2}\left(Q^T P - P^T Q\right). \tag{8.7}
$$

This example applies to the left and right multiplicative actions of $SO(n)$ on $GL(n)$. It also applies to the left and right multiplicative actions of $SO(n)$ on itself. In the latter case, for any $(Q,P) \in T^*SO(n)$, we have $P\,Q^T = P\,Q^{-1} \in \mathfrak{so}(n)$ and, similarly, $Q^T P = Q^{-1}P \in \mathfrak{so}(n)$, so both $P\,Q^T$ and $Q^T P$ are skew-symmetric. Thus, the momentum maps for the actions of $SO(n)$ on itself are

$$
\begin{aligned}
\mathbf{J}_L(Q,P) &= P\,Q^T \qquad \text{(left multiplicative action)}, \\
\mathbf{J}_R(Q,P) &= Q^T P \qquad \text{(right multiplicative action)}.
\end{aligned} \tag{8.8}
$$

Note that these are the same formulae as for the multiplicative actions of $GL(n)$ on itself, given in the previous example. We also have:

$$
\mathbf{J}_R(Q,P) = Q^T \mathbf{J}_L(Q,P) Q = \mathrm{Ad}^*_Q \mathbf{J}_L(Q,P).
$$

**Remark 8.11 (Angular momentum of a rigid body)** *For a rigid body, the momentum $\mathbf{J}_L$, as computed in the previous example, can be interpreted as spatial angular momentum. Indeed, consider a rigid body with a fixed centre of mass, which has configuration space $SO(3)$ and a simple mechanical*

*Lagrangian of the form*

$$L(\mathbf{R}, \dot{\mathbf{R}}) = \frac{1}{2} \operatorname{tr}(\dot{\mathbf{R}} \mathbb{J} \dot{\mathbf{R}}^T) + V(\mathbf{R}),$$

*where* $\mathbb{J}$ *is the coefficient of the inertia matrix, introduced in Section 7.1. The Legendre transform for this Lagrangian is*

$$(\mathbf{R}, \mathbf{P}) = \left(\mathbf{R}, \frac{\partial L}{\partial \dot{\mathbf{R}}}\right) = \left(\mathbf{R}, \dot{\mathbf{R}} \mathbb{J}\right),$$

*where* $\frac{\partial L}{\partial \dot{\mathbf{R}}}$ *has been computed using the trace pairing as in Section 3.4. The left multiplicative action of* $SO(3)$ *on itself corresponds to spatial rotations, as noted earlier in Section 7.1. From Example 8.10, the momentum map of the cotangent lift of this action is* $\mathbf{J}_L(\mathbf{R}, \mathbf{P}) = \mathbf{P}\mathbf{R}^T$. *We now show that* $\mathbf{J}_L$ *corresponds to the spatial angular momentum vector* $\boldsymbol{\pi}$ *of the rigid body, which was introduced in Section 1.5, where it was shown that* $\boldsymbol{\pi} = \mathbf{R}\boldsymbol{\Pi} = \mathbf{R}\mathbb{I}\boldsymbol{\Omega}$. *Recall from Remark 7.1 that*

$$\operatorname{tr}\left(\widehat{\boldsymbol{\xi}}\, \mathbb{J}\, \widehat{\boldsymbol{\eta}}^T\right) = \boldsymbol{\xi}^T\, \mathbb{I}\, \boldsymbol{\eta},$$

*for all* $\widehat{\boldsymbol{\xi}}, \widehat{\boldsymbol{\eta}} \in \mathfrak{so}(3)$, *where* $\mathbb{I}$ *is the moment of inertia matrix. Thus, for all* $\widehat{\boldsymbol{\xi}} \in \mathfrak{so}(3)$,

$$\begin{aligned}
\left\langle \mathbf{J}_L(\mathbf{R}, \mathbf{P}), \widehat{\boldsymbol{\xi}} \right\rangle &= \operatorname{tr}\left(\dot{\mathbf{R}}\, \mathbb{J}\, \mathbf{R}^T \widehat{\boldsymbol{\xi}}^T\right) \\
&= \operatorname{tr}\left(\mathbf{R}^{-1} \dot{\mathbf{R}}\, \mathbb{J}\, \mathbf{R}^T \widehat{\boldsymbol{\xi}}^T \mathbf{R}\right) \\
&= \operatorname{tr}\left(\widehat{\boldsymbol{\Omega}}\, \mathbb{J}\, \left(\widehat{\mathbf{R}^T \boldsymbol{\xi}}\right)^T\right) \\
&= \boldsymbol{\Omega}^T \mathbb{I} \mathbf{R}^T \boldsymbol{\xi} \\
&= \boldsymbol{\pi}^T \boldsymbol{\xi}.
\end{aligned}$$

*By definition of the isomorphism* $\breve{\ } : \mathbb{R}^3 \to \mathfrak{so}(3)^*$ *defined in Chapter 5, we have*

$$\mathbf{J}_L = \breve{\boldsymbol{\pi}}.$$

*In this sense,* $\mathbf{J}_L$ *'is' spatial angular momentum for the rigid body. We know from Chapter 1 that spatial angular momentum is conserved, a fact that we deduced from Newton's laws (and the corresponding properties of solid bodies). For the rigid body, we now have another way of deducing this conservation law, namely Noether's theorem. Indeed, the Lagrangian* *above is clearly invariant under the tangent-lifted left multiplicative action, as was noted in Chapter 7, and from this it follows that the corresponding Hamiltonian is invariant under the the cotangent-lifted left multiplicative action of $SO(3)$ on $T^*SO(3)$ (see Chapter 4). Thus Noether's theorem (Theorem 8.14) implies that $\mathbf{J}_L$ is constant.*

*Recall also that the body angular momentum (i.e. angular momentum in body coordinates) is $\mathbf{\Pi} := \mathbf{R}^T\boldsymbol{\pi}$. From Example 6.44, it follows that*

$$\breve{\mathbf{\Pi}} = \left(\mathbf{R}^T\boldsymbol{\pi}\right)^{\breve{}} = \mathrm{Ad}^*_{\mathbf{R}}\mathbf{J}_L = \mathbf{R}^T\mathbf{J}_L(\mathbf{R},\mathbf{P})\mathbf{R} = \mathbf{J}_R(\mathbf{R},\mathbf{P}).$$

**Example 8.12 (Momentum maps for actions of a Lie group on itself)** Let $G$ be a Lie group and consider the left and right actions of $G$ on itself; that is, for any $h \in G$, we have:

$$\begin{aligned} G \times G \to G, &\quad (h,g) \to L_h(g) = hg \quad \text{(left multiplicative action)},\\ G \times G \to G, &\quad (h,g) \to R_h(g) = gh \quad \text{(right multiplicative action)}. \end{aligned}$$

It is useful to recall that the tangent lifts are:

$$\begin{aligned} T_gL_h(v) &= \frac{\mathrm{d}}{\mathrm{d}t}hg(t)\Big|_{t=0} = hv \quad \text{(left tangent action)},\\ T_gR_h(v) &= \frac{\mathrm{d}}{\mathrm{d}t}g(t)h\Big|_{t=0} = vh \quad \text{(right tangent action)}, \end{aligned}$$

where $g(t)$ is a path in $G$ with $g(0) = g$ and $g'(0) = v$. The infinitesimal actions on $G$ are

$$\begin{aligned} \xi^l_G(g) &= \frac{\mathrm{d}}{\mathrm{d}t}\exp(t\xi)\,g\Big|_{t=0} = \xi g = T_eR_g\xi \quad \text{(left infinitesimal action)},\\ \xi^r_G(g) &= \frac{\mathrm{d}}{\mathrm{d}t}g\exp(t\xi)\Big|_{t=0} = g\xi = T_eL_g\xi \quad \text{(right infinitesimal action)}. \end{aligned}$$

Let $\alpha \in T^*_gG$. For every $\xi \in \mathfrak{g}$, we have $\langle J_L(\alpha),\xi\rangle = \langle \alpha, \xi_G(g)\rangle = \langle \alpha, T_eR_g(\xi)\rangle = \langle T^*_eR_g(\alpha),\xi\rangle$ so

$$\mathbf{J}_L(\alpha) = T^*_eR_g(\alpha) \quad \text{(left momentum for a Lie group)}. \tag{8.9}$$

A similar calculation shows:

$$\mathbf{J}_R(\alpha) = T_e^* L_g(\alpha) \quad \text{(right momentum for a Lie group).} \tag{8.10}$$

Alternatively, writing $\alpha = T_g^* L_{g^{-1}}(\rho)$ for some $\rho \in \mathfrak{g}^*$ we have

$$\mathbf{J}_L\left(T_g^* L_{g^{-1}}(\rho)\right) = T_e^* R_g \circ T_g^* L_{g^{-1}}(\rho) = T_e^*(L_{g^{-1}} \circ R_g)(\rho) = \mathrm{Ad}^*_{g^{-1}}(\rho) \quad \text{and}$$
$$\mathbf{J}_R\left(T_g^* L_{g^{-1}}(\rho)\right) = T_e^* L_g \circ T_g^* L_{g^{-1}}(\rho) = T_e^*(L_{g^{-1}} \circ L_g)(\rho) = \rho.$$

It follows that $\mathbf{J}_R(\alpha) = \mathrm{Ad}^*_g \mathbf{J}_L(\alpha)$.

**Exercise 8.1 (Momentum map for the scaling group)**
Consider the action of the multiplicative group of real numbers on $\mathbb{R}^n$:

$$(\mathbb{R} \setminus \{0\}) \times \mathbb{R}^n \to \mathbb{R}^n \tag{8.11}$$
$$(\lambda, \mathbf{q}) \to \lambda \mathbf{q}. \tag{8.12}$$

Show that the momentum map $\mathbf{J} : T^*\mathbb{R}^n \to \mathbb{R}$ for the corresponding cotangent-lifted action is $\mathbf{J}(\mathbf{q}, \mathbf{p}) = \mathbf{p}^T \mathbf{q}$.

**Exercise 8.2**
Compute the momentum map for the cotangent lift of the standard action of $SE(3)$ on $\mathbb{R}^3$.

**Exercise 8.3**
Show that the $S^1$ action on the torus $\mathbb{T}^2$ given by $\theta \cdot (\alpha, \beta) = (\alpha + \theta, \beta)$ is canonical with respect to the area form $d\alpha \wedge d\beta$ but has no globally defined momentum map (see, for example, [Sin01]).

**Exercise 8.4**
Suppose that $G$ acts on $P_1$, with momentum map $J_1$. Let $P_2$ be another manifold, and define the $G$ action on $P_1 \times P_2$ by $g(p_1, p_2) = (gp_1, p_2)$. Define $J : P_1 \times P_2 \to \mathfrak{g}^*$ by $J(p_1, p_2) = J_1(p_1)$. Show that $J$ is a momentum map for the $G$ action on $P_1 \times P_2$.

**Exercise 8.5**
Let $G$ act on a connected symplectic manifold $P$ and suppose that, for every $\xi \in \mathfrak{g}$, the function $J_\xi$ is a Hamiltonian for the infinitesimal vector field $\xi_P$. Let $(\xi^1, \ldots, \xi^n)$ be a basis for $\mathfrak{g}$, and define $\mathbf{J} : P \to \mathfrak{g}^*$ by $\langle \mathbf{J}(p), a_i \xi^i \rangle = a_i J_{\xi^i}(p)$. Show that $\mathbf{J}$ is a momentum map for the group action, and that $\langle \mathbf{J}(p), \xi \rangle = J_\xi(p) + \text{constant}$, for every $\xi \in \mathfrak{g}$.

## 8.2 Properties of momentum maps

Momentum maps of symmetries are conserved quantities. This result, called Noether's theorem, is the most important property of momentum maps. To prove it, we will require the following,

**Proposition 8.13** *If $H : P \to \mathbb{R}$ is $G$-invariant, meaning that*

$$H(gx) = H(x) \qquad \textit{for all } g \in G$$

*and $x \in P$, then $\pounds_{\xi_P} H = 0$ for all $\xi \in \mathfrak{g}$. This property is called **infinitesimal invariance**.*

*Proof* Let $g(t)$ be some path in $G$ such that $g(0) = e$ and $g'(0) = \xi$. Then, for every $x \in P$,

$$\left(\pounds_{\xi_P} H\right)(x) = \mathrm{d}H(\xi_P(x)) = \frac{\mathrm{d}}{\mathrm{d}t} H\left(g(t)x\right)\bigg|_{t=0} = \frac{\mathrm{d}}{\mathrm{d}t} H\left(x\right)\bigg|_{t=0} = 0.$$

■

**Theorem 8.14 (Noether's theorem)** *Let $G$ act canonically on $(P, \{\cdot, \cdot\})$ with momentum map $\mathbf{J}$. If $H$ is $G$-invariant, then $\mathbf{J}$ is conserved by the flow of $X_H$.*

*Proof* The momentum map $\mathbf{J}$ is conserved along the flow of $X_H$ if, for every $\xi \in \mathfrak{g}$, the map $J_\xi$ is a constant of motion. That is, for every $\xi \in \mathfrak{g}$, the Lie derivative of $J_\xi$ along the vector field $X_H$ is zero. We have:

$$\pounds_{X_H} J_\xi = \left\{J_\xi, H\right\} = -\left\{H, J_\xi\right\} = -\pounds_{X_{J_\xi}} H = -\pounds_{\xi_P} H = 0,$$

where the last equality is given by the infinitesimal invariance of $H$. ■

**Definition 8.15** *Let $M$ and $N$ be manifolds together with a Lie group $G$ acting on $M$ and $N$. A map $f : M \to N$ is said to be* **equivariant** *with respect to these actions if, for all $g \in G$,*

$$f(gx) = g(f(x)).$$

If we denote the action of $G$ on $M$ and $N$ by $\Phi_g$ and $\Psi_g$, respectively, then the equivariance of $f$ may be stated as:

$$f \circ \Phi_g = \Psi_g \circ f \quad \text{for all } g \in G,$$

that is, the following diagram commutes:

$$\begin{array}{ccc} M & \xrightarrow{f} & N \\ \Phi_g \downarrow & & \downarrow \Psi_g \\ M & \xrightarrow{f} & N. \end{array}$$

Now specialize $f$ to be the momentum map $\mathbf{J} : P \to \mathfrak{g}^*$. Then the definition of equivariance reads $J(gz) = g(J(z))$. A momentum map is said to be ***equivariant*** when it is equivariant with respect to the given action on $P$ and the coadjoint action on $\mathfrak{g}^*$, i.e. for all $g \in G$,

$$\mathrm{J}(g\,z) = \mathrm{Ad}^*_{g^{-1}}\big(\mathrm{J}(z)\big) \tag{8.13}$$

for every $z \in P$, that is, the following diagram commutes:

$$\begin{array}{ccc} P & \xrightarrow{\mathrm{J}} & \mathfrak{g}^* \\ \Phi_g \downarrow & & \downarrow \mathrm{Ad}^*_{g^{-1}} \\ P & \xrightarrow{\mathrm{J}} & \mathfrak{g}^*. \end{array}$$

**Example 8.16 (Equivariance of angular momentum)** Recall from Example 8.8 that the momentum map for the standard action of the rotation group on $\mathbb{R}^3$ is $\mathbf{J} : SO(3) \to \mathfrak{so}(3)^* \simeq \mathbb{R}^3$, $\mathbf{J}(\mathbf{q}, \mathbf{p}) = \mathbf{q} \times \mathbf{p}$. The coadjoint action on $\mathfrak{so}(3)^*$, when expressed in vector notation, is $\boldsymbol{\pi} \mapsto \mathbf{R}\boldsymbol{\pi}$ (see Example 6.44). Thus the equivariance of the momentum map takes the form, $(\mathbf{Rq}) \times (\mathbf{Rp}) = \mathbf{R}(\mathbf{q} \times \mathbf{p})$, a well-known vector identity.

**Proposition 8.17** *The momentum map of a cotangent-lifted action is equivariant.*

*Proof* Exercise (Hint: use Noether's formula given in Theorem 8.7). ■

Two more important properties of momentum maps, namely that they are infinitesimally equivariant and Poisson, will be proven in Section 9.4.

**Exercise 8.6**
Show that the momentum map for linear symplectic actions in Example 8.4 is equivariant.

## Solutions to selected exercises

**Solution to Exercise 8.2** Consider the action of $SE(3)$ on $\mathbb{R}^3$ given by

$$(\boldsymbol{R}, \boldsymbol{r}) \cdot \boldsymbol{q} = \boldsymbol{R}\boldsymbol{q} + \boldsymbol{r}.$$

Then we can calculate a corresponding momentum map $\boldsymbol{J} : T^*\mathbb{R}^3 \to \mathfrak{se}^*(3)$. The infinitesimal action is given by

$$\left(\widehat{\boldsymbol{\xi}}, \boldsymbol{v}\right) \cdot \boldsymbol{q} = \boldsymbol{\xi} \times \boldsymbol{q} + \boldsymbol{v}.$$

Therefore, the cotangent-lifted momentum map is calculated as follows,

$$\begin{aligned}
\left\langle \boldsymbol{J}(\boldsymbol{q}, \boldsymbol{p}), \left(\widehat{\boldsymbol{\xi}}, \boldsymbol{v}\right)\right\rangle &= \boldsymbol{p} \cdot (\boldsymbol{\xi} \times \boldsymbol{q} + \boldsymbol{v}) \\
&= (\boldsymbol{q} \times \boldsymbol{p}) \cdot \boldsymbol{\xi} + \boldsymbol{p} \cdot \boldsymbol{v} \\
&= \left\langle \left((\boldsymbol{q} \times \boldsymbol{p})\widehat{\ }, \boldsymbol{p}\right), \left(\widehat{\boldsymbol{\xi}}, \boldsymbol{v}\right)\right\rangle.
\end{aligned}$$

Thus, the momentum map is given by,

$$\boldsymbol{J}(\boldsymbol{q}, \boldsymbol{p}) = \left((\boldsymbol{q} \times \boldsymbol{p})\widehat{\ }, \boldsymbol{p}\right).$$

The first component is the angular momentum and the second is linear momentum.

**Solution to Exercise 8.3** First, the pull-back of $\omega = \mathrm{d}\alpha \wedge \mathrm{d}\beta$ by $\Phi_\theta$ is computed to be

$$\Phi_\theta^* \omega = \mathrm{d}(\alpha + \theta) \wedge \mathrm{d}\beta = \mathrm{d}\alpha \wedge \mathrm{d}\beta.$$

Thus, the form $\omega$ is invariant under the action of $\theta$, so the action is symplectic. Now, to compute the infinitesimal generator, let $\theta(t)$ be a curve such that $\theta(0) = 0$ and $\dot{\theta}(0) = \xi$. Then, writing $\Phi_\theta(\alpha, \beta) = \theta \cdot (\alpha, \beta)$ for short,

$$\begin{aligned}\xi_{\mathbb{T}^2}(\alpha, \beta) &= \left.\frac{\mathrm{d}}{\mathrm{d}t}\right|_{t=0} (\theta(t) \cdot (\alpha, \beta)) \\ &= \left.\frac{\mathrm{d}}{\mathrm{d}t}\right|_{t=0} (\alpha + \theta(t), \beta) \\ &= (\xi, 0) \, .\end{aligned}$$

To find a momentum map associated with the symplectic form $d\alpha \wedge d\beta$ it is necessary to find a function $J^\xi : \mathbb{T}^2 \to \mathbb{R}$ such that

$$X_{J^\xi} = \xi_{\mathbb{T}^2},$$

where $X_{J^\xi}$ is the Hamiltonian vector field of $J^\xi$ with respect to $\omega$. Thus, such a $J^\xi$ must satisfy

$$\omega(\alpha, \beta)(\xi_{\mathbb{T}^2}, \cdot) = \xi \mathrm{d}\beta = \mathrm{d}J^\xi .$$

Note that $J^\xi$ is independent of $\alpha$, therefore $J^\xi : S^1 \to \mathbb{R}$. Suppose such a function does exist globally. $\xi \neq 0$ would give

$$\mathrm{d}\left(\frac{J^\xi}{\xi}\right) = \mathrm{d}\beta.$$

Recall from Exercise 3.35 that the form $\mathrm{d}\beta$ considered as a form on $S^1$ is not exact. Therefore, the function $J^\xi/\xi$ does not exist, so $J^\xi$ does not exist either, a contradiction. Thus, there is no globally defined momentum map.

# 9 Lie–Poisson reduction

In Chapter 7, the Euler equations of motion for the rigid body,

$$\mathbb{I}\dot{\mathbf{\Omega}} = \mathbb{I}\mathbf{\Omega} \times \mathbf{\Omega}, \tag{9.1}$$

were shown to be determined by a reduced variational principle applied to the reduced Lagrangian $l$: $\mathfrak{so}(3) \to \mathbb{R}$,

$$l(\mathbf{\Omega}) = \frac{1}{2} \langle \mathbf{\Omega}, \mathbb{I}\mathbf{\Omega} \rangle .$$

The Euler equations can be reformulated in the variable $\mathbf{\Pi} := \mathbb{I}\mathbf{\Omega}$ as

$$\dot{\mathbf{\Pi}} = \mathbf{\Pi} \times \mathbb{I}^{-1}\mathbf{\Pi}. \tag{9.2}$$

As shown in Example 4.24, these equations actually define a Hamiltonian vector field $X_h$ for the Hamiltonian

$$h(\mathbf{\Pi}) = \frac{1}{2}\left\langle \mathbf{\Pi}, \mathbb{I}^{-1}\mathbf{\Pi} \right\rangle$$

and the rigid body bracket

$$\{F, K\}(\mathbf{\Pi}) := -\langle \mathbf{\Pi}, \nabla F \times \nabla K \rangle. \tag{9.3}$$

The purpose of the present chapter is to generalize the rigid body example to symmetric mechanical systems on Lie groups, in two ways:

1. We introduce a *reduced Legendre transformation*, generalizing the change of coordinates $\mathbf{\Omega} \mapsto \mathbf{\Pi}$, which allows us to write equations on $\mathfrak{g}^*$, called Lie-Poisson equations, that correspond to the Euler–Poincaré equations on $\mathfrak{g}$.
2. We show that the Lie–Poisson equations have a Hamiltonian structure with respect to a suitable bracket.

## 9.1 The reduced Legendre transform

Consider a Lagrangian $L : TG \to \mathbb{R}$ that is invariant under the tangent-lifted left translation action of $G$ on itself. From Chapter 7 we know that the corresponding dynamics on $\mathfrak{g}$ are given by the Euler-Poincaré equations,

$$\frac{\mathrm{d}}{\mathrm{d}t}\frac{\delta l}{\delta \xi} = \mathrm{ad}^*_\xi \frac{\delta l}{\delta \xi}, \tag{9.4}$$

where $l : \mathfrak{g} \to \mathbb{R}$ is the reduced Lagrangian defined by $l(\xi) = L(e, \xi)$. Assuming $L$ is hyperregular, define the **reduced Legendre transform:**

$$fl : \mathfrak{g} \to \mathfrak{g}^*, \quad \langle fl(\xi), \eta\rangle := \frac{\mathrm{d}}{\mathrm{d}s}\Big|_{s=0} l(\xi + s\eta) = \left\langle \frac{\delta l}{\delta \xi}, \eta \right\rangle \tag{9.5}$$

for all $\xi, \eta \in \mathfrak{g}$. Note that, since

$$\mathbb{F}L(e, \xi) = (e, fl(\xi))\,, \tag{9.6}$$

the map $fl$ is the restriction of $\mathbb{F}L$ to $\mathfrak{g}$. Thus, if $L$ is hyperregular (that is, $\mathbb{F}L$ is a diffeomorphism), then $fl$ is a diffeomorphism.

**Lemma 9.1** *If $L : TG \to \mathbb{R}$ is $G$-invariant, then*

$$\frac{\delta L}{\delta \dot g}(g_0,\, g_0\,\xi_0) = g_0\left(e,\, \frac{\delta l}{\delta \xi}(\xi_0)\right),$$

*for every $g_0 \in G, \xi_0 \in \mathfrak{g}$, where the action of $G$ on the right hand side is the cotangent lift of the $G$ action on itself.*

*Proof*

$$\begin{aligned}\left\langle \frac{\delta L}{\delta \dot g}(g_0,\, g_0\,\xi_0), (g_0, g_0\eta)\right\rangle &= \frac{\mathrm{d}}{\mathrm{d}t}\Big|_{t=0} L(g_0, g_0\,\xi_0 + t\,g_0\,\eta) \\ &= \frac{\mathrm{d}}{\mathrm{d}t}\Big|_{t=0} L\left(g_0(e, \xi_0 + t\,\eta)\right)\end{aligned}$$

$$= \left.\frac{\mathrm{d}}{\mathrm{d}t}\right|_{t=0} L\,(e, \xi_0 + t\,\eta)$$

$$= \left.\frac{\mathrm{d}}{\mathrm{d}t}\right|_{t=0} l\,(\xi_0 + t\,\eta)$$

$$= \left\langle \frac{\delta l}{\delta \xi}(\xi_0), \eta \right\rangle$$

$$= \left\langle \left( e, \frac{\delta l}{\delta \xi}(\xi_0) \right), (e, \eta) \right\rangle$$

$$= \left\langle g_0 \left( e, \frac{\delta l}{\delta \xi}(\xi_0) \right), g_0(e, \eta) \right\rangle,$$

where in the second-to-last line we used the identification of $\mathfrak{g}^*$ with $\{e\} \times \mathfrak{g}^*$, and in the last line we used the definition of a cotangent lift (see eqn (6.2)). ■

A slight generalization of the previous lemma is the following, the proof of which is left as Exercise 9.5.

**Proposition 9.2** *Let $Q$ be a manifold. If $L : TQ \to \mathbb{R}$ is $G$-invariant, then $\mathbb{F}L$ is $G$-equivariant.*

We further define the **reduced energy function**

$$\tilde{e} : \mathfrak{g} \to \mathbb{R}, \quad \tilde{e}(\xi) := \langle fl(\xi), \xi \rangle - l(\xi),$$

and the **reduced Hamiltonian**

$$h : \mathfrak{g}^* \to \mathbb{R}, \quad h(\mu) := \tilde{e} \circ fl^{-1}. \tag{9.7}$$

Thus, if $\mu = fl(\xi)$, then

$$h(\mu) = \tilde{e} \circ fl^{-1}(\mu) = \langle \mu, \xi(\mu) \rangle - l(\xi(\mu)).$$

**Proposition 9.3** *If $E$ is the energy function corresponding to $L$ and if $\xi = g^{-1}\dot{g}$ then*

$$\tilde{e}(\xi) = E(e, \xi) = E(g, \dot{g}).$$

*Suppose $L$ is hyperregular and $H$ is the Hamiltonian corresponding to $L$, i.e. $H = E \circ (\mathbb{F}L)^{-1}$. If $\mu = g^{-1}\alpha$ then*

$$h(\mu) = (e, \xi) = H(g, \alpha).$$

*Proof* Exercise. ■

Note that:

$$\frac{\delta h}{\delta \mu} = \xi(\mu) + \left\langle \mu, \frac{\delta \xi}{\delta \mu} \right\rangle - \left\langle \frac{\delta l}{\delta \xi}, \frac{\delta \xi}{\delta \mu} \right\rangle = \xi(\mu) + \left\langle \mu, \frac{\delta \xi}{\delta \mu} \right\rangle - \left\langle \mu, \frac{\delta \xi}{\delta \mu} \right\rangle = \xi(\mu).$$

Thus, the Euler-Poincaré equations (9.4) can be written in the new variable $\mu$, leading to the following result:

**Proposition 9.4** *With $\mu$ and $h$ as defined above, the Euler–Poincaré equations (9.4) are equivalent to the* **Lie–Poisson** *equations:*

$$\dot{\mu} = \mathrm{ad}^*_{\delta h/\delta \mu} \mu.$$

**Remark 9.5** *Recall that for a right invariant Lagrangian, the Euler-Poincaré equations are*

$$\frac{\mathrm{d}}{\mathrm{d}t} \frac{\delta l}{\delta \xi} = -\mathrm{ad}^*_\xi \frac{\delta l}{\delta \xi}$$

*and thus the Lie-Poisson equations read:*

$$\dot{\mu} = -\mathrm{ad}^*_{\delta h/\delta \mu} \mu.$$

**Example 9.6** The reduced Legendre transform for the free rigid body is

$$fl : \mathfrak{so}(3) \to \mathfrak{so}^*(3), \qquad \mathbf{\Omega} \to \mathbf{\Pi} := fl(\mathbf{\Omega}) = \frac{\partial l}{\partial \mathbf{\Omega}} = \mathbb{I}\mathbf{\Omega},$$

with inverse $(fl)^{-1}(\mathbf{\Pi}) = \mathbb{I}^{-1}\mathbf{\Pi}$. The reduced energy function is

$$\tilde{e} : \mathfrak{so}(3) \to \mathbb{R}, \quad \tilde{e}(\mathbf{\Omega}) := \langle \mathbb{I}\mathbf{\Omega}, \mathbf{\Omega} \rangle - \frac{1}{2} \langle \mathbf{\Omega}, \mathbb{I}\mathbf{\Omega} \rangle = \frac{1}{2} \langle \mathbf{\Omega}, \mathbb{I}\mathbf{\Omega} \rangle,$$

so the reduced Hamiltonian is

$$h : \mathfrak{g}^* \to \mathbb{R}, \quad h(\mathbf{\Pi}) := \tilde{e} \circ fl^{-1}(\mathbf{\Pi}) = \frac{1}{2} \left\langle \mathbb{I}^{-1}\mathbf{\Pi}, \mathbf{\Pi} \right\rangle.$$

Using the formula for ad* in eqn (6.53), the Lie–Poisson equations are

$$\dot{\mathbf{\Pi}} = \mathrm{ad}^*_{\delta h/\delta \mathbf{\Pi}} \mathbf{\Pi} = \mathrm{ad}^*_{\mathbb{I}^{-1}\mathbf{\Pi}} \mathbf{\Pi} = \mathbf{\Pi} \times \mathbb{I}^{-1}\mathbf{\Pi}.$$

Having obtained the Lie–Poisson equations, the next question is: are these equations Hamiltonian? We will see in the following sections that they *are* Hamiltonian, with respect to a bracket on $\mathfrak{g}^*$ called the *Lie–Poisson bracket*. The theory relating canonical dynamics on $T^*G$ with reduced Hamiltonian dynamics on $\mathfrak{g}^*$ is called ***Lie–Poisson reduction***. This is a Hamiltonian counterpart to Euler–Poincaré reduction, in which reduced Lagrangian dynamics on $\mathfrak{g}$ are shown to be determined by a reduced variational principle.

Lie–Poisson reduction is a special case of the theory of Poisson reduction, presented in Section 6.4. Suppose $L$ is hyperregular, and let $H : T^*G \to \mathbb{R}$ be the corresponding Hamiltonian, which is left $G$-invariant, by Proposition 6.35. Poisson reduction theory shows that the canonical Hamiltonian flow on $T^*G$ corresponds to a Hamiltonian flow on $(T^*G)/G$. By Proposition 6.61, the following map is a diffeomorphism,

$$\begin{array}{ccccc} (T^*G)/G & \simeq & (G \times \mathfrak{g}^*)/G & \simeq & \mathfrak{g}^*, \\ [g, \alpha] & \longmapsto & [g, g^{-1}\alpha] & \longmapsto & \mu := g^{-1}\alpha. \end{array}$$

This map can be used to push-forward the reduced Poisson bracket and reduced Hamiltonian on $(T^*G)/G$ to a bracket and Hamiltonian on $\mathfrak{g}^*$. The push-forward of the bracket will be done in the following section. The push-forward of the Hamiltonian is straightforward. Indeed, the reduced Hamiltonian $h^{red}$ on $(T^*G)/G$ is defined by $h^{red}([g, \alpha]) = H(g, \alpha)$. Let $h^-$ be the push-forward of $h^{red}$ to $\mathfrak{g}^*$. Then, by definition,

$$h^-(\mu) = h^{red}([e, \mu]) = H(e, \mu).$$

**Proposition 9.7 (Consistency with the reduced Legendre transform)** *In the above context, the Hamiltonian $h^-$ coincides with the reduced Hamiltonian $h$ defined in eqn (9.7).*

*Proof* Since $FL(e, \xi) = (e, fl(\xi))$ for every $\xi \in \mathfrak{g}$,

$$\begin{aligned} h^-(fl(\xi)) &= H(e, fl(\xi)) \\ &= E \circ (\mathbb{F}L)^{-1}(e, fl(\xi)) = E(e, \xi) = \tilde{e}(\xi) = h(fl(\xi)). \end{aligned}$$

■

**Remark 9.8** *Lie–Poisson reduction is a general theory that applies to any symmetric Hamiltonian $H$ on $T^*G$. In particular, $H$ need not to correspond to a hyperregular Lagrangian.*

**Exercise 9.1**
Consider the left trivializations of $TG$ and $T^*G$ given by the maps

$$\lambda : TG \to G \times \mathfrak{g}, \quad (h,\dot{h}) \to \lambda(h,\dot{h}) := (h, h^{-1}\dot{h}) = (h, T_h L_{h^{-1}}\dot{h}),$$
$$\lambda^* : T^*G \to G \times \mathfrak{g}^*, \quad (h,\alpha) \to (h, h^{-1}\alpha) := (h, T_e^* L_h \alpha),$$

(first defined in eqn (6.11) and eqn (6.12), respectively).

Let $\lambda_2$ and $\lambda_2^*$ be the second components of these maps. Show that the following diagram commutes,

$$\begin{array}{ccc} TG \simeq G \times \mathfrak{g} & \stackrel{\mathbb{F}L}{\longrightarrow} & G \times \mathfrak{g}^* \simeq T^*G \\ \lambda_2 \downarrow & & \downarrow \lambda_2^* \\ \mathfrak{g} & \stackrel{fl}{\longrightarrow} & \mathfrak{g}^* \end{array}$$

**Exercise 9.2**
Show that if $(g(t), \dot{g}(t))$ is an integral curve for a left invariant Lagrangian $L$, then $E\,(g(t), \dot{g}(t)) = \tilde{e}\,(\xi(t)) = const.$, where $\xi = g^{-1}\dot{g}$.

**Exercise 9.3**
Consider a product Lie group $G \times K$ acting on itself by

$$\Big((\tilde{g}, \tilde{k}), (g, k)\Big) \to (\tilde{g}g, k\tilde{k}).$$

Use Proposition 9.4 and Remark 9.5 to deduce Lie–Poisson equations for a reduced Hamiltonian $h : \mathfrak{g}^* \times \mathfrak{k}^* \to \mathbb{R}$.

**Exercise 9.4**
Consider the following action of a Lie group $G$ on a product space $G \times Y$, where $Y$ is some manifold:

$$(g, (h, y)) \to (gh, y).$$

Let $L : T(G \times Y) \to \mathbb{R}$ be invariant with respect to this action. Define $l : \mathfrak{g} \times TY \to \mathbb{R}$ to be the restriction of $L$, i.e.

$$l(\xi, y, \dot{y}) := L(e, \xi, y, \dot{y}).$$

The equations of motion for this $l$ appear in Exercise 7.16. Define a reduced Legendre transformation and reduced Hamiltonian, by analogy with eqns (9.5) and (9.7), and show that the equations of motion on $\mathfrak{g}^* \times T^*Y$ are:

$$\dot{\mu} = \mathrm{ad}^*_{\delta h/\delta \mu}\, \mu,$$
$$\dot{y} = \frac{\partial h}{\partial p},$$
$$\dot{p} = -\frac{\partial h}{\partial y},$$

where $p$ is the conjugate momentum corresponding to $y$.

Exercise 9.5
Prove Proposition 9.2.

## 9.2 Lie–Poisson reduction: geometry

Consider the cotangent bundle $T^*G$ of a Lie group $G$ under the cotangent-lifted left multiplication action of $G$. We know from Section 6.3 that the quotient space $P/G = T^*G/G$ is diffeomorphic to $\mathfrak{g}^*$, the dual of the Lie algebra of $G$.

Recall from Section 6.3 the left trivialization map for $T^*G$ :

$$\lambda : T^*G \to G \times \mathfrak{g}^*, \quad \lambda(g, \alpha) = (\lambda_1(g, \alpha), \lambda_2(g, \alpha)) = (g, T_e^* L_g \alpha), \tag{9.8}$$

and the diffeomorphism

$$\psi : (T^*G)/G \longmapsto \mathfrak{g}^*, \qquad [g, \alpha] \longmapsto g^{-1}\alpha := T_e^* L_g(\alpha).$$

Note that $\psi \circ \pi = \lambda_2$, where $\pi : T^*G \to (T^*G)/G$ is projection and $\lambda_2$ is the second component of $\lambda$, defined above. The map $\psi$, together with the reduced Poisson bracket on $(T^*G)/G$ from Theorem 6.63, induces a

Poisson bracket on $\mathfrak{g}^*$ that makes $\psi$ a Poisson map. The induced Poisson bracket on $\mathfrak{g}^*$ will be denoted $\{\cdot,\cdot\}_{\mathfrak{g}^*}^{left}$ to indicate that it is associated with the left translation.

By definition of this bracket, $\psi$ is a Poisson map. By definition of the reduced Poisson bracket on $(T^*G)/G$, the projection $\pi$ is a Poisson map. Since $\lambda_2 = \psi \circ \pi$, it follows that $\lambda_2$ is a Poisson map, i.e.

$$\{F,K\}_{\mathfrak{g}^*}^{left} \circ \lambda_2 := \{F \circ \lambda_2, K \circ \lambda_2\}_{T^*G}, \tag{9.9}$$

where the bracket on the right is the canonical Poisson bracket.

For any $F \in \mathcal{F}(\mathfrak{g}^*)$, the corresponding function $F \circ \lambda_2 \in \mathcal{F}(T^*G)$ is left $G$–invariant and it is called the ***left extension*** of $F$. We denote by $\mathcal{F}_L(T^*G)$ the space of smooth left invariant functions on $T^*G$. The map

$$\mathcal{F}(\mathfrak{g}^*) \to \mathcal{F}_L(T^*G), \qquad F \to F \circ \lambda_2,$$

is in fact a bijection, with the inverse given by restriction to $\mathfrak{g}^*$ (this is Exercise 9.6).

**Theorem 9.9** *Let $F_L$ and $K_L$ be $G$-invariant smooth functions on $T^*G$, and let $F$ and $K$ be their restrictions (respectively) to $\mathfrak{g}^*$. For every $(g_0,\alpha_0) \in T^*G$, let $\mu_0 = T_e^* L_g(\alpha_0)$. Then*

$$\{F_L, K_L\}(g_0,\alpha_0) = -\left\langle \mu_0, \left[\frac{\delta F}{\delta \mu}, \frac{\delta K}{\delta \mu}\right]\Big|_{\mu_0}\right\rangle, \tag{9.10}$$

*where the bracket on the left is the canonical Poisson bracket.*

*Proof* Recall from Section 4.3 that, given a Poisson structure on a manifold P, the Poisson bracket $\{F_1,F_2\}$ *at a point* $p \in P$ of two functions $F_1, F_2 \in \mathcal{F}(P)$ depends *only* on the first derivatives of $F_1$ and $F_2$ at $p$. Thus the value of $\{F_L,K_L\}(g,\alpha)$ depends only on the first derivatives of $F_L$ and $K_L$ at $(g,\alpha)$, and so we can assume, without loss of generality, that $F_L$ and $K_L$ are linear in the cotangent fibres, i.e., that they depend linearly on $\alpha$. Thus, in particular, $\frac{\delta F}{\delta \mu}$ and $\frac{\delta K}{\delta \mu}$ are constant elements of $\mathfrak{g}$. Let $X = \left(\frac{\delta F}{\delta \mu}\right)_L$, i.e. the left-invariant vector field on $G$ corresponding to $\frac{\delta F}{\delta \mu}$, defined by $X(h) = T_e L_h\left(\frac{\delta F}{\delta \mu}\right)$, for all $h \in G$. By left-invariance,

$$\frac{\delta F_L}{\delta \alpha}(g,\alpha) = TL_g\left(\frac{\delta F}{\delta \mu}\right) = X(g), \tag{9.11}$$

for all $(g,\alpha) \in T^*G$ (see Exercise 9.8). Similarly, let $Y = \left(\frac{\delta K}{\delta\mu}\right)_L$, so $Y(h) = T_eL_h\left(\frac{\delta K}{\delta\mu}\right)$, and note that $\frac{\delta K_L}{\delta\alpha}(g,\alpha) = Y(g)$. Using the assumption that $F_L$ and $K_L$ are linear in $\alpha$, we deduce that

$$F_L(g,\alpha) = \langle \alpha, X(g)\rangle\,, \quad \text{and} \quad K_L(g,\alpha) = \langle \alpha, Y(g)\rangle\,. \tag{9.12}$$

By definition of the Lie bracket on $\mathfrak{g}^*$, $\left[\frac{\delta F}{\delta\mu}, \frac{\delta K}{\delta\mu}\right] = [X,Y](e)$, where the bracket on the right is the Jacobi-Lie bracket. By the left invariance of $X$ and $Y$, and the invariance of the Jacobi–Lie bracket under pull-backs,

$$TL_{g_0{}^{-1}}\left([X,Y](g_0)\right) = [X,Y](e) = \left[\frac{\delta F}{\delta\mu}, \frac{\delta K}{\delta\mu}\right] \tag{9.13}$$

(see Exercise 9.9).

Let $(q,p)$ be a cotangent-lifted local coordinate system for $T^*G$, so in these coordinates, $g_0$ is a particular value of $q$ and $\alpha_0$ is a particular value of $p$. In these coordinates, eqn (9.12) becomes

$$F_L(q,p) = \sum_j p_j X^j, \qquad \text{and} \qquad K_L(q,p) = \sum_j p_j Y^j,$$

and so

$$\frac{\delta F_L}{\partial q^i} = \sum_j p_j \frac{\partial X^j}{\partial q^i}, \qquad \frac{\delta F_L}{\partial p_i} = X^i,$$

and similarly for $K_L$. Therefore,

$$\begin{aligned}
\{F_L, K_L\}(g_0,\alpha_0) &= \sum_i \left(\frac{\partial F_L}{\partial q^i}\frac{\partial K_L}{\partial p_i} - \frac{\partial K_L}{\partial q^i}\frac{\partial F_L}{\partial p_i}\right)\Bigg|_{(q,p)=(g_0,\alpha_0)} \\
&= \sum_i\sum_j \left(p_j\left(\frac{\partial X^j}{\partial q^i}Y^i - \frac{\partial Y^j}{\partial q^i}X^i\right)\right)\Bigg|_{(q,p)=(g_0,\alpha_0)} \\
&= -\langle (g_0,\alpha_0), [X,Y](g_0)\rangle \\
&= -\langle T^*L_{g_0}(g_0,\alpha_0), TL_{g_0{}^{-1}}\left([X,Y](g_0)\right)\rangle \\
&= -\left\langle \mu_0, \left[\frac{\delta F}{\delta\mu}, \frac{\delta K}{\delta\mu}\right]\right\rangle,
\end{aligned}$$

where in the last line we have used eqn (9.13). ■

Combining eqns (9.9) and (9.10), and taking into account that the restriction of $F \circ \lambda_2$ and $G \circ \lambda_2$ to $\mathfrak{g}^*$ are just $F$ and $G$, respectively, we have that

$$\{F, K\}_{\mathfrak{g}^*}^{left} \circ \lambda_2(g_0, \alpha_0) = \{F \circ \lambda_2, K \circ \lambda_2\}_{T^*G}(g_0, \alpha_0)$$
$$= -\left\langle \mu_0, \left[\frac{\delta F}{\delta \mu}, \frac{\delta K}{\delta \mu}\right]\bigg|_{\mu=\mu_0} \right\rangle.$$

Since $\mu_0 = T_e^* L_{g_0}(\alpha_0) = \lambda_2(g_0, \alpha_0)$, we can rewrite this as

$$\{F, K\}_{\mathfrak{g}^*}^{left}(\mu_0) = -\left\langle \mu_0, \left[\frac{\delta F}{\delta \mu}, \frac{\delta K}{\delta \mu}\right]\bigg|_{\mu=\mu_0} \right\rangle, \tag{9.14}$$

or, more concisely,

$$\{F, K\}_{\mathfrak{g}^*}^{left} = -\left\langle \mu, \left[\frac{\delta F}{\delta \mu}, \frac{\delta K}{\delta \mu}\right] \right\rangle. \tag{9.15}$$

This formula defines a bracket operation on $\mathfrak{g}^*$ called the ***left*** (or ***minus***) ***Lie–Poisson bracket***. This bracket satisfies the axioms of a Poisson bracket, as can be checked either directly or by using its relation to the canonical Poisson bracket, stated in the previous theorem; this is left as Exercise 9.10. We have shown:

**Theorem 9.10 (Lie-Poisson reduction theorem, geometry part)**
*Consider a Lie group $G$, and let $\mathfrak{g}^*$ be the dual of its Lie algebra. If $\mathcal{F}(\mathfrak{g}^*)$ is identified with $\mathcal{F}_L(T^*G)$, as above, then the canonical Poisson bracket on $T^*G$ induces a Poisson bracket on $\mathfrak{g}^*$, called the 'left' (or the 'minus')* ***Lie–Poisson bracket****, given by*

$$\{F, K\}_{\mathfrak{g}^*}^{left} = -\left\langle \mu, \left[\frac{\partial F}{\partial \mu}, \frac{\partial K}{\partial \mu}\right] \right\rangle. \tag{9.16}$$

*Further, $\{\cdot,\cdot\}_{\mathfrak{g}^*}^{left}$ corresponds to the reduced Poisson bracket $\{\cdot,\cdot\}^{red}$ on $(T^*G)/G$, via the isomorphism*

$$\psi : (T^*G)/G \longmapsto \mathfrak{g}^*,$$
$$[g, \alpha] \longmapsto g^{-1}\alpha := T_e^* L_g(\alpha).$$

**Example 9.11** The left bracket on $\mathfrak{so}^*(3)$ is given by

$$\{F, K\}^{left} = -\left\langle \mathbf{\Pi}, \left[\frac{\partial F}{\partial \mathbf{\Pi}}, \frac{\partial K}{\partial \mathbf{\Pi}}\right]\right\rangle = -\mathbf{\Pi} \cdot (\nabla F \times \nabla K).$$

**Remark 9.12** *If instead we use the* ***right*** *action and thus identify $\mathcal{F}(\mathfrak{g}^*)$ with $\mathcal{F}_R(T^*G)$, the space of right invariant smooth functions on $T^*G$, we obtain the same formula but with opposite sign, that is*

$$\{F, K\}^{right}_{\mathfrak{g}^*} = +\left\langle \mu, \left[\frac{\partial F}{\partial \mu}, \frac{\partial K}{\partial \mu}\right]\right\rangle.$$

*This bracket is called the 'right' (or the 'plus') Lie–Poisson bracket on $\mathfrak{g}^*$.*

**Remark 9.13** *Recall from the previous chapter that the momentum map corresponding to the cotangent-lifted right translation is $\mathbf{J_R} : T^*G \to \mathfrak{g}^*$, $\mathbf{J_R}(g, \alpha) = T_e^* L_g(\alpha)$ and thus it coincides with $\lambda_2$. Therefore, an alternative way to write eqn (9.9), which we emphasize describes the left Poisson bracket, is:*

$$\{F, K\}^{left}_{\mathfrak{g}^*} \circ \mathbf{J_R} = \{F \circ \mathbf{J_R}, K \circ \mathbf{J_R}\}_{T^*G}. \tag{9.17}$$

*In particular, we have that $\mathbf{J_R}$ is a Poisson map with respect to the left Poisson bracket on $\mathfrak{g}^*$.*

The following proposition is useful in applications where the configuration space is a Lie group product.

**Proposition 9.14** *Consider the product group $G \times H$, where $G$ and $H$ are given Lie groups. Identify $\mathcal{F}(\mathfrak{g}^*)$ with $\mathcal{F}_L(T^*G)$, (respectively, $\mathcal{F}_R(T^*G)$) and $\mathcal{F}(\mathfrak{h}^*)$ with $\mathcal{F}_L(T^*H)$, (respectively $\mathcal{F}_R(T^*H)$). Then, the dual of its Lie algebra $\mathfrak{g}^* \times \mathfrak{h}^*$ has a Poisson bracket given by*

$$\begin{aligned}\{F, K\}_{\mathfrak{g}^* \times \mathfrak{h}^*} &= \{F, K\}^{left(right)}_{\mathfrak{g}^*} + \{F, K\}^{left(right)}_{\mathfrak{h}^*} \\ &= \underset{(+)}{-} \left\langle \mu, \left[\frac{\partial F}{\partial \mu}, \frac{\partial K}{\partial \mu}\right]\right\rangle \underset{(+)}{-} \left\langle \nu, \left[\frac{\partial F}{\partial \nu}, \frac{\partial K}{\partial \nu}\right]\right\rangle\end{aligned} \tag{9.18}$$

*which corresponds to the reduced Poisson bracket on $T^*(G \times H)/(G \times H)$.*

*Proof* Exercise. ■

**Corollary 9.15** *If $P = G \times H \times Q$ where $Q$ is a smooth manifold, then $\mathfrak{g}^* \times \mathfrak{h}^* \times T^*Q$ is a Poisson manifold endowed with the bracket*

$$\{F, K\}_{\mathfrak{g}^* \times \mathfrak{h}^* \times T^*Q} = \{F, K\}_{\mathfrak{g}^*}^{left(right)} + \{F, K\}_{\mathfrak{h}^*}^{left(right)} + \{F, K\}_{T^*Q} \quad (9.19)$$

*which corresponds to the reduced Poisson bracket on $T^*(G \times H \times Q)/(G \times H)$.*

**Exercise 9.6**
Show that the following is a bijection:

$$\mathcal{F}(\mathfrak{g}^*) \to \mathcal{F}_L(T^*G), \qquad F \to F \circ \lambda_2,$$

where the inverse is given by restriction to $\mathfrak{g}^*$.

**Exercise 9.7**
Prove Proposition 9.14.

**Exercise 9.8**
Verify eqn (9.11). Note that $\frac{\delta F}{\delta \alpha}$ is the *fibre derivative* of $F$, introduced earlier in Section 4.2 in the context of the Legendre transformation.

**Exercise 9.9**
Verify eqn (9.13). (HINT: use the left invariance of $X$ and $Y$, and Theorem 3.25.)

**Exercise 9.10**
Check that the Lie Poisson brackets (both left and right) are indeed Poisson brackets on $\mathfrak{g}^*$.

**Exercise 9.11**
Write the Lie–Poisson bracket on $\mathfrak{su}(2)^*$ in terms of the tilde map defined in Exercise 5.20.

## 9.3 Lie–Poisson reduction: dynamics

**Proposition 9.16 (Lie-Poisson reduction, dynamics part)** *Let $H : T^*G \to \mathbb{R}$ be a $G$ left invariant Hamiltonian and let $h^- := H\big|_{\mathfrak{g}^*}$ be its restriction to $\mathfrak{g}^*$. Then,*

$$H = h^- \circ \lambda_2$$

*where $\lambda_2 : T^*G \to \mathfrak{g}^*$ is the second component of the left trivialization map (9.8), and the flow $\phi_t$ of the Hamiltonian vector field $X_H$ and the Lie–Poisson flow $\phi_t^-$ of $X_{h^-}$ are related by*

$$\lambda_2 \circ \phi_t = \phi_t^- \circ \lambda_2.$$

*Proof* For any $(g, \alpha) \in TG$ we have

$$\begin{aligned} H(g,\alpha) &= H(e, T_e^* L_g(g,\alpha)) \\ &= h^-(T_e^* L_g(g,\alpha)) = h^-(\lambda_2(g,\alpha)) = (h^- \circ \lambda_2)(g,\alpha), \end{aligned}$$

and thus the first statement follows.

The second statement follows from general Poisson reduction, which is Theorem 6.63, and the geometric part of Lie–Poisson reduction, which is stated in Theorem 9.10. Indeed, from general Poisson reduction, we know that

$$\pi \circ \phi_t = \phi_t^{red} \circ \pi, \tag{9.20}$$

for all $t$, where $\pi : P \to P/G$ is the standard projection and $\phi^{red}$ is the Hamiltonian flow on $P/G$ corresponding to the reduced Hamiltonian $h^{red}$ and the reduced Poisson bracket $\{\cdot,\cdot\}_{red}$. From Theorem 9.10, we know that there is a diffeomorphism $\psi : (T^*G)/G \to \mathfrak{g}^*$, satisfying $\lambda_2 = \psi \circ \pi$, that is a Poisson map with respect to $\{\cdot,\cdot\}_{\mathfrak{g}^*}^{left}$ on $\mathfrak{g}^*$ and $\{\cdot,\cdot\}_{red}$ on $(T^*G)/G$. Since $\psi$ is a Poisson map,

$$\psi \circ \phi_t^{red} = \phi_t^- \circ \psi, \tag{9.21}$$

for all $t$. Applying first eqn (9.20) and then eqn (9.21), we have

$$\lambda_2 \circ \phi_t = \psi \circ \pi \circ \phi_t = \psi \circ \phi_t^{red} \circ \pi = \phi_t^- \circ \psi \circ \pi = \phi_t^- \circ \lambda_2.$$

■

**Proposition 9.17 (The Lie–Poisson equations)** *The equations of motion for a Hamiltonian $h : \mathfrak{g}^* \to \mathbb{R}$ with respect to the left Lie–Poisson bracket on $\mathfrak{g}^*$ are*

$$\frac{\mathrm{d}\mu}{\mathrm{d}t} = \mathrm{ad}^*_{\delta h/\delta\mu}\,\mu. \tag{9.22}$$

*That is, the Hamiltonian vector field for h is given by*

$$X_h(\mu) = \mathrm{ad}^*_{\delta h/\delta\mu}\,\mu.$$

*Proof* Let $F \in \mathcal{F}(\mathfrak{g}^*)$. Then

$$\frac{\mathrm{d}F}{\mathrm{d}t} = \mathrm{D}F(\mu)\cdot\dot{\mu} = \left\langle \dot{\mu}, \frac{\delta F}{\delta\mu} \right\rangle.$$

Further,

$$\begin{aligned}
\{F,h\}^{left}_{\mathfrak{g}^*}(\mu) &= -\left\langle \mu, \left[\frac{\delta F}{\delta\mu}, \frac{\delta h}{\delta\mu}\right]\right\rangle \\
&= -\left\langle \mu, \mathrm{ad}_{\frac{\delta F}{\delta\mu}} \frac{\delta h}{\delta\mu}\right\rangle \\
&= -\left\langle \mu, -\mathrm{ad}_{\frac{\delta h}{\delta\mu}} \frac{\delta F}{\delta\mu}\right\rangle = \left\langle \mathrm{ad}^*_{\frac{\delta h}{\delta\mu}}\mu, \frac{\delta F}{\delta\mu}\right\rangle.
\end{aligned}$$

Since $F$ is arbitrary, the conclusion follows. ■

**Proposition 9.18 (The reconstruction equation)** *Let $H : T^*G \to \mathbb{R}$ be a left invariant Hamiltonian, $h^- := H\big|_{\mathfrak{g}^*}$, and $\mu(t)$ the integral curve of the Lie-Poisson equations*

$$\frac{\mathrm{d}\mu}{\mathrm{d}t} = \mathrm{ad}^*_{\delta h^-/\delta\mu}\,\mu \tag{9.23}$$

*with initial condition $\mu(0) = \mu_0$. Then the integral curve $(g(t), \alpha(t)) \in T^*G$ of $X_H$ with initial condition $(g(0), \alpha(0)) = (g_0, \alpha_0)$, where $\mu_0 = T_e^* L_{g_0}(g_0, \alpha_0)$, is determined by the non-autonomous ODE*

$$\dot{g}(t) = g(t)\frac{\delta h^-}{\delta\mu}, \qquad g(0) = g_0 \tag{9.24}$$

*and the relation*

$$\alpha(t) = T^*_{g(t)} L_{g^{-1}(t)} \mu(t).$$

*Proof* Recall from the previous chapter, Example 8.12, that the left and right momentum maps are related by

$$\mathbf{J}_L = \mathrm{Ad}^*_{g^{-1}} \mathbf{J}_R. \tag{9.25}$$

Since $H$ is left invariant, Noether's theorem guarantees that $\mathbf{J}_L\,(g(t), \alpha(t)) = const.$ along any $(g(t), \alpha(t))$ integral curve of $X_H$. Differentiating eqn (9.25) along an integral curve and using Proposition 6.54 we obtain:

$$0 = \mathrm{Ad}^*_{g^{-1}(t)} \left[ \frac{\mathrm{d}\mu}{\mathrm{d}t} - \mathrm{ad}^*_{\xi(t)} \mu(t) \right]$$

where $\xi(t) = g^{-1}(t)\dot{g}(t)$. Since $\mu(t)$ satisfies eqn (9.23), we must have

$$0 = \left[ \mathrm{ad}^*_{\frac{\delta h^-}{\delta \mu}} \mu(t) - \mathrm{ad}^*_{\xi(t)} \mu(t) \right]$$

or

$$0 = \mathrm{ad}^*_{\left( \frac{\delta h^-}{\delta \mu} - \xi(t) \right)} \mu(t)$$

from where it is sufficient that

$$\frac{\delta h^-}{\delta \mu} = \xi(t).$$

Since $g(0) = g_0$, it follows that $g(t)$ verifies eqn (9.24). The last relation is just the inverse of $\mu(t) = T_e^* L_{g(t)}(g(t), \alpha(t))$. ■

**Example 9.19 (Hamiltonian form of the rigid body)** The reduced Hamiltonian for the rigid body is given by

$$h(\mathbf{\Pi}) = \frac{1}{2} < \mathbf{\Pi}, \mathbb{I}^{-1} \mathbf{\Pi} >$$

and the Lie–Poisson equations of motion for the rigid body are

$$\dot{\mathbf{\Pi}} = \mathbf{\Pi} \times \nabla h = \mathbf{\Pi} \times \mathbb{I}^{-1} \mathbf{\Pi}.$$

If $\mathbf{\Pi}(t)$ solves the above, than the non-autonomous ODE

$$\dot{\mathbf{R}} = \mathbf{R}(t) \mathbb{I}^{-1} \mathbf{\Pi}(t), \quad \mathbf{R}(0) = \mathbf{R}_0$$

describes the orientation in time of the body.

## 9.4 Momentum maps revisited

In this section we present further computations and properties of momentum maps that rely on the Lie–Poisson bracket.

**Example 9.20 (The momentum map for the coadjoint action)**
Consider the coadjoint action of a Lie group $G$ on its Lie algebra dual $\mathfrak{g}^*$, that is,

$$\mathrm{CoAd}_g(\mu) = \mathrm{Ad}^*_{g^{-1}}(\mu) \quad \text{for all } g \in G, \mu \in \mathfrak{g}^*.$$

This action is canonical with respect to the plus (or minus) Lie–Poisson bracket – see Exercise 9.12. Recall from Remark 6.51 that the infinitesimal generator for the coadjoint action is given by

$$\xi_{\mathfrak{g}^*}(\mu) = -\mathrm{ad}^*_\xi(\mu).$$

By the definition (8.1) of the momentum map, we are looking for a Hamiltonian function $J_\xi$ such that $X_{J_\xi}(\mu) = \xi_{\mathfrak{g}^*}(\mu)$ for all $\mu \in \mathfrak{g}^*$. Since

$$X_F(\mu) = -\mathrm{ad}^*_{\frac{\delta F}{\delta \mu}}(\mu) \quad \text{for all } F \in \mathcal{F}(\mathfrak{g}^*)$$

we have

$$-\mathrm{ad}^*_{\frac{\delta J_\xi}{\delta \mu}}(\mu) = -\mathrm{ad}^*_\xi(\mu) \quad \text{for all } \mu \in \mathfrak{g}^*,$$

i.e. $\dfrac{\delta J_\xi}{\delta \mu} = \xi$. Thus we can take $J_\xi(\mu) = \langle \mu, \xi \rangle$ and, since $J_\xi(\mu) = \langle \mathbf{J}(\mu), \xi \rangle$, we conclude that $\mathbf{J}$ is the identity map.

Consider the left action of a group $G$ on a Poisson manifold $P$, and suppose it has a momentum map $\mathbf{J}$. Recall from Section 8.2 that $\mathbf{J}$ is equivariant if and only if for all $g \in G$, $p \in P$,

$$\mathbf{J}(g p) = \mathrm{Ad}^*_{g^{-1}}\big(\mathbf{J}(p)\big),$$

that is,

$$\langle \mathbf{J}(g p), \xi \rangle = \left\langle \mathrm{Ad}^*_{g^{-1}}\big(\mathbf{J}(p)\big), \xi \right\rangle \quad \text{for all } \xi \in \mathfrak{g}.$$

Moving the "$\mathrm{Ad}^*_{g^{-1}}$" to the right hand side of the pairing, the above takes the form

$$J_\xi(g\,p) = J_{\mathrm{Ad}_{g^{-1}}\xi}(p) \quad \text{for all } \xi \in \mathfrak{g}. \tag{9.26}$$

For any $\eta \in \mathfrak{g}$, by taking the derivative along a curve $g(t) \in G$ with $g(0) = e$ and $g'(0) = \eta$, we obtain the following property, called ***infinitesimal equivariance of the momentum map***:

$$J_{[\xi,\eta]} = \{J_\xi, J_\eta\}. \tag{9.27}$$

Indeed, the derivative of the left-hand side of eqn (9.26) is

$$\begin{aligned}
&\frac{\mathrm{d}}{\mathrm{d}t}\bigg|_{t=0} J_\xi\,(g(t)\,p)\\
&\quad= \frac{\mathrm{d}}{\mathrm{d}t}\bigg|_{t=0} \langle \mathbf{J}\,(g(t)\,p)\,,\xi\rangle\\
&\quad= \left\langle T_p\mathbf{J}\circ \eta_P(p),\xi\right\rangle\\
&\quad= \left\langle T_p\mathbf{J}\circ X_{J_\eta}(p),\xi\right\rangle \quad \text{(by Definition 8.1 of momentum maps)}\\
&\quad= \mathrm{d}J_\xi(p)X_{J_\eta}(p)\\
&\quad= \mathcal{L}_{X_{J_\eta}(p)}J_\xi(p) \quad \text{(by Definition 3.15 of the Lie derivative)}\\
&\quad= \{J_\xi, J_\eta\}(p) \quad \text{(by Definition 4.22 of the Poisson bracket).}
\end{aligned}$$

The derivative of the right-hand side of eqn (9.26) is

$$\begin{aligned}
\frac{\mathrm{d}}{\mathrm{d}t}\bigg|_{t=0} J_{\left(\mathrm{Ad}_{g(t)^{-1}}\xi\right)}(p) &= \frac{\mathrm{d}}{\mathrm{d}t}\bigg|_{t=0} \left\langle \mathrm{Ad}^*_{g(t)^{-1}}(\mathbf{J}(p)),\xi\right\rangle\\
&= \left\langle -\mathrm{ad}^*_\eta(\mathbf{J}(p)),\xi\right\rangle \quad \text{(using Remark 6.51)}\\
&= \left\langle \mathbf{J}(p), -\mathrm{ad}_\eta\xi\right\rangle\\
&= \left\langle \mathbf{J}(p), \mathrm{ad}_\xi\eta\right\rangle\\
&= \langle \mathbf{J}(p), [\xi,\eta]\rangle = J_{[\xi,\eta]}(p).
\end{aligned}$$

**Remark 9.21** *If one considers instead a right action on $G$, the arguments above must be modified with the corresponding signs, but the outcome identity is the same as eqn (9.27).*

Thus we have proven:

**Proposition 9.22** *Equivariant momentum maps are infinitesimally equivariant.*

**Proposition 9.23** *Consider the left action of a Lie group $G$ on a Poisson manifold. If the associated momentum map $\mathbf{J}$ is infinitesimally equivariant, then $\mathbf{J}$ is a Poisson map with respect to the right Lie-Poisson bracket, that is:*

$$\{F, K\}_{\mathfrak{g}^*}^{right} \circ \mathbf{J} = \{F \circ \mathbf{J}, K \circ \mathbf{J}\}, \tag{9.28}$$

*for all $F, K \in \mathcal{F}(\mathfrak{g}^*)$.*

**Remark 9.24** *If instead one considers a right action, we have*

$$\{F, K\}_{\mathfrak{g}^*}^{left} \circ \mathbf{J} = \{F \circ \mathbf{J}, K \circ \mathbf{J}\},$$

*for all $F, K \in \mathcal{F}(\mathfrak{g}^*)$. This identity appeared in the context of actions on cotangent bundles of Lie groups $T^*G$, in Section 9.2, relation (9.17).*

*Proof* Recall that

$$\{F, K\}_{\mathfrak{g}^*}^{right} \circ \mathbf{J}(p) = \left\langle \mathbf{J}(p), \left[\frac{\delta F}{\delta \mu}, \frac{\delta K}{\delta \mu}\right]\bigg|_{\mu=\mathbf{J}(p)} \right\rangle.$$

Denoting

$$\xi := \frac{\delta F}{\delta \mu}\bigg|_{\mu=\mathbf{J}(p)} \qquad \eta := \frac{\delta K}{\delta \mu}\bigg|_{\mu=\mathbf{J}(p)},$$

we calculate:

$$\{F, K\}_{\mathfrak{g}^*}^{right} \circ \mathbf{J}(p) = \langle \mathbf{J}(p), [\xi, \eta]\rangle = J_{[\xi,\eta]}(p) = \{J_\xi, J_\eta\}(p),$$

where in the last equality we used the infinitesimal equivariance property (9.27) of the momentum map. It remains to show that

$$\{J_\xi, J_\eta\}(p) = \{F \circ \mathbf{J}, K \circ \mathbf{J}\}(p) \quad \text{for all } p \in P. \tag{9.29}$$

For this, recall from Section 4.3 that, given a Poisson structure on a manifold, the Poisson bracket $\{F_1, F_2\}$ *at a point* $p \in P$ of two functions depends only on the first derivatives of $F_1$ and $F_2$ at $p$. Thus, to show eqn (9.29) above, it suffices to verify that the first derivatives of $J_\xi$ and $J_\eta$ at $p \in P$ coincide with the first derivatives of $F \circ \mathbf{J}$ and $K \circ \mathbf{J}$ at $p \in P$, respectively.

Now, for any $p \in P$ and $v_p \in T_pP$, we have:

$$\begin{aligned} \mathrm{d}J_\xi(p)(v_p) &= \left\langle T_p\mathbf{J}(v_p), \xi \right\rangle \\ &= \left\langle T_p\mathbf{J}(v_p), \left.\frac{\delta F}{\delta \mu}\right|_{\mu=\mathbf{J}(p)} \right\rangle \\ &= \mathrm{d}F\left(\mathbf{J}(p)\right) \circ T_p\mathbf{J}(v_p) = \mathrm{d}(F \circ \mathbf{J})(p)(v_p). \end{aligned}$$

So, $J_\xi$ and $F \circ \mathbf{J}$ and, analogously, $J_\eta$ and $K \circ \mathbf{J}$, have identical derivatives at $p \in P$. ■

**Remark 9.25** *The set of* ***collective Hamiltonians*** *is defined as*

$$\{F \circ \mathbf{J} : P \to \mathbb{R} \mid F \in \mathcal{F}(\mathfrak{g}^*)\}.$$

*The equality of the first derivatives of* $J_{(\delta F/\delta\mu)|_{\mu=\mathbf{J}(p)}}$ *and* $F \circ \mathbf{J}$ *at* $p \in P$ *allows one to prove that*

$$X_{F\circ\mathbf{J}}(p) = X_{J_{(\delta F/\delta\mu)|_{\mu=\mathbf{J}(p)}}}(p) = \left( \left.\frac{\delta F}{\delta \mu}\right|_{\mu=\mathbf{J}(p)} \right)_P (p) \quad \textit{for all } p \in P.$$

*Note that if* $F$ *is a linear function of* $\xi \in \mathfrak{g}$, *then* $(\delta F/\delta\mu)|_{\mu=\mathbf{J}(p)} = \xi$ *and the last equality above becomes*

$$X_{F\circ\mathbf{J}}(p) = X_{J_\xi}(p) = \xi_P(p).$$

*Observe that the last equality is the definition of the momentum map. For more on collective Hamiltonians, see [MR02].*

**Remark 9.26** *By Exercise 8.6 the momentum map for linear symplectic actions is equivariant and therefore, by the previous proposition, is a Poisson map.*

**Example 9.27 (Cayley–Klein parameters and the Hopf fibration)** Consider the symplectic form on $\mathbb{C}^2$ given by minus the imaginary part of the standard Hermitian inner product. (Recall that the standard Hermitian inner product $\mathbb{C}^n$ is given by $\mathbf{z}\cdot\mathbf{w} := \sum_{j=1}^n z_j\overline{w}_j$, where $\mathbf{z} = (z_1, \ldots, z_n)$, $\mathbf{w} = (w_1, \ldots, w_n) \in \mathbb{C}^n$.) The symplectic form is given by

$$\Omega(\mathbf{z}, \mathbf{w}) := -\operatorname{Im}(\mathbf{z} \cdot \mathbf{w})$$

and it is identical to the one given before on $\mathbb{R}^{2n}$ by identifying $\mathbf{z} = \mathbf{u} + i\mathbf{v} \in \mathbb{C}^n$ with $(\mathbf{u}, \mathbf{v}) \in \mathbb{R}^{2n}$ and $\mathbf{w} = \mathbf{u}' + i\mathbf{v}' \in \mathbb{C}^n$ with $(\mathbf{u}', \mathbf{v}') \in \mathbb{R}^{2n}$.

Consider the natural action of $SU(2)$ on $\mathbb{C}^2$. It is easily checked that $SU(2)$ acts by isometries of the Hermitian metric, and hence that this action is symplectic (Exercise). Thus, Example 8.4 applies here, showing that there is a momentum map $\mathbf{J} : \mathbb{C}^2 \to \mathfrak{su}(2)^*$ given by

$$\langle \mathbf{J}(z, w), \xi \rangle = \frac{1}{2}\Omega(\xi \cdot (z, w), (z, w)),$$

where $z, w \in \mathbb{C}$ and $\xi \in \mathfrak{su}(2)$.

The Lie algebra $\mathfrak{su}(2)$ of $SU(2)$ consists of $2 \times 2$ skew Hermitian matrices of trace zero. This Lie algebra is isomorphic to $\mathfrak{so}(3)$ and therefore to $(\mathbb{R}^3, \times)$ by the isomorphism given by the tilde map,

$$\mathbf{x} = (x^1, x^2, x^3) \in \mathbb{R}^3 \mapsto \widetilde{\mathbf{x}} := \frac{1}{2}\begin{bmatrix} -ix^3 & -ix^1 - x^2 \\ -ix^1 + x^2 & ix^3 \end{bmatrix} \in \mathfrak{su}(2).$$

Thus we have $[\widetilde{\mathbf{x}}, \widetilde{\mathbf{y}}] = (\mathbf{x} \times \mathbf{y})^{\sim}$ for any $\mathbf{x}, \mathbf{y} \in \mathbb{R}^3$. Other useful relations are $\det(2\widetilde{\mathbf{x}}) = \|\mathbf{x}\|^2$ and $\operatorname{trace}(\widetilde{\mathbf{x}}\widetilde{\mathbf{y}}) = -\frac{1}{2}\mathbf{x} \cdot \mathbf{y}$. Identify $\mathfrak{su}(2)^*$ with $\mathbb{R}^3$ by the map $\mu \in \mathfrak{su}(2)^* \mapsto \check{\mu} \in \mathbb{R}^3$ defined by

$$\check{\mu} \cdot \mathbf{x} := -2\langle \mu, \widetilde{\mathbf{x}} \rangle$$

for any $\mathbf{x} \in \mathbb{R}^3$. With these notations, the momentum map $\check{\mathbf{J}} : \mathbb{C}^2 \to \mathbb{R}^3$ can be explicitly computed in coordinates: for any $\mathbf{x} \in \mathbb{R}^3$ we have

$$\begin{aligned}
\check{\mathbf{J}}(z, w) \cdot \mathbf{x} &= -2\langle \mathbf{J}(z, w), \widetilde{\mathbf{x}} \rangle = (-2)\frac{1}{2}\Omega\big(\widetilde{\mathbf{x}} \cdot (z, w), (z, w)\big) \\
&= \frac{1}{2}\operatorname{Im}\left(\begin{bmatrix} -ix^3 & -ix^1 - x^2 \\ -ix^1 + x^2 & ix^3 \end{bmatrix}\begin{bmatrix} z \\ w \end{bmatrix} \cdot \begin{bmatrix} z \\ w \end{bmatrix}\right) \\
&= -\frac{1}{2}(2\operatorname{Re}(w\overline{z}), 2\operatorname{Im}(w\overline{z}), |z|^2 - |w|^2) \cdot \mathbf{x}.
\end{aligned}$$

Therefore,

$$\check{\mathbf{J}}(z, w) = -\frac{1}{2}(\operatorname{Re}(w\overline{z}), 2\operatorname{Im}(w\overline{z}), |z|^2 - |w|^2) \in \mathbb{R}^3.$$

By Remark 9.26, $\check{\mathbf{J}}$ is a Poisson map from $\mathbb{C}^2$, endowed with the canonical symplectic structure, to $\mathfrak{su}(2)^*$, endowed with the plus Lie–Poisson bracket. Therefore, $-\check{\mathbf{J}} : \mathbb{C}^2 \to \mathfrak{su}(2)^*$ is a canonical map with respect to the minus Lie–Poisson bracket. Identifying $\mathfrak{su}(2)^*$ with $\mathbb{R}^3$, the map $-\check{\mathbf{J}} : \mathbb{C}^2 \to \mathbb{R}^3$ is canonical with respect to the rigid body bracket, relative to which the free rigid body equations are Hamiltonian. Pulling back the free rigid body Hamiltonian $H(\mathbf{\Pi}) = \mathbf{\Pi} \cdot \mathbb{I}^{-1}\mathbf{\Pi}/2$ to $\mathbb{C}^2$ gives a Hamiltonian function (called collective) $H \circ (-\check{\mathbf{J}})$ on $\mathbb{C}^2$. The classical Hamilton equations for this function are therefore projected by $-\check{\mathbf{J}}$ to the rigid body equations $\dot{\boldsymbol{\Pi}} = \boldsymbol{\Pi} \times \mathbb{I}^{-1}\boldsymbol{\Pi}$. In this context, the variables $(z, w)$ are called the ***Cayley–Klein parameters***.

Now notice that if $(z, w) \in S^3 := \{(z, w) \in \mathbb{C}^2 \mid |z|^2 + |w|^2 = 1\}$, then $\| -\check{\mathbf{J}}(z, w)\| = 1/2$, so that $-\check{\mathbf{J}}|_{S^3} : S^3 \to S^2_{1/2}$, where $S^2_{1/2}$ is the sphere in $\mathbb{R}^3$ of radius 1/2. For any $(x^1, x^2, x^3) = (x^1 + ix^2, x^3) = (re^{i\psi}, x^3) \in S^2_{1/2}$, the inverse image of this point is

$$-\check{\mathbf{J}}^{-1}(re^{i\psi}, x^3) = \left\{ \left( e^{i\theta}\sqrt{\frac{1}{2} + x^3}, e^{i\varphi}\sqrt{\frac{1}{2} - x^3} \right) \in S^3 \mid e^{i(\theta - \varphi + \psi)} = 1 \right\}.$$

Thus, $-\check{\mathbf{J}}|_{S^3}$ is surjective and its fibers are circles. One recognizes now that $-\check{\mathbf{J}}|_{S^3} : S^3 \to S^2$ is the ***Hopf fibration***. (For more on the Hopf fibration see, for instance, [CB97].) In other words, *the momentum map of the $SU(2)$-action on $\mathbb{C}^2$, the Cayley–Klein parameters and the family of Hopf fibrations on concentric three-spheres in $\mathbb{C}^2$ are all the same map.*

**Exercise 9.12**
Show that the coadjoint action of a Lie group $G$ on its Lie algebra dual $\mathfrak{g}^*$ is canonical (see Definition 6.19) with respect to the plus (or minus) Lie–Poisson bracket. HINT: Let $\nu = \mathrm{Ad}^*_{g^{-1}}\mu$, and show that

$$\frac{\delta F}{\delta \nu} = \mathrm{Ad}_g \frac{\delta \left( F \circ \mathrm{Ad}^*_{g^{-1}} \right)}{\delta \mu}.$$

## 9.5 Co-Adjoint orbits

Given $\mu \in \mathfrak{g}$, recall from Chapter 6 the definition of the Co-Adjoint orbit:

$$\mathcal{O}_\mu := \{g \cdot \mu \mid g \in G\} = \{\mathrm{Ad}^*_{g^{-1}}\mu \mid g \in G\}.$$

In this section we present some properties of Lie–Poisson systems related to CoAdjoint orbits.

**Proposition 9.28** *Let $h \in \mathcal{F}(\mathfrak{g}^*)$ and let $\mu(t)$ be the solution of the initial value problem*

$$\dot{\mu} = \mathrm{ad}^*_{\delta h/\delta\mu}\,\mu,$$
$$\mu(0) = \mu_0.$$

*Then, $\mu(t) \in \mathcal{O}_{\mu_0}$.*

*Proof* Given $\mu(t)$, the solution of the above initial-value problem, we have to show that for every $t$ there is a $g(t) \in G$ such that $\mu(t) \in \mathcal{O}_{\mu_0} = \{\mathrm{Ad}^*_{g^{-1}}\mu_0 \mid g \in G\} = \{\mathrm{Ad}^*_g\,\mu_0 \mid g \in G\}$. We will show that $\mu(t) = \mathrm{Ad}^*_{g(t)}\mu_0$, where $g(t)$ is the solution of the initial value problem:

$$\dot{g} = g\,\frac{\delta h}{\delta\mu},$$
$$g(0) = e.$$

Since $g^{-1}(t)\dot{g}(t) = (\delta h/\delta\mu)(t)$ for all $t$, applying Proposition 6.54, we obtain:

$$\begin{aligned}\frac{\mathrm{d}}{\mathrm{d}t}\left(\mathrm{Ad}^*_{g(t)^{-1}}\mu(t)\right) &= \mathrm{Ad}^*_{g(t)^{-1}}\mu(t)\left[\frac{\mathrm{d}\mu(t)}{\mathrm{d}t} - \mathrm{ad}^*_{g^{-1}(t)\dot{g}(t)}\mu(t)\right]\\ &= \mathrm{Ad}^*_{g(t)^{-1}}\mu(t)\left[\frac{\mathrm{d}\mu(t)}{\mathrm{d}t} - \mathrm{ad}^*_{\frac{\delta h}{\delta\mu}(t)}\mu(t)\right] = 0.\end{aligned}$$

Thus, $\mathrm{Ad}^*_{g(t)^{-1}}\mu(t) = const. = \mathrm{Ad}^*_{g(0)}\mu(0) = \mathrm{Ad}^*_e\mu_0 = \mu_0$. ■

**Definition 9.29** *The Co-Adjoint isotropy group of $\mu \in \mathfrak{g}^*$ is*

$$G_\mu := \{g \in G \mid \mathrm{Ad}^*_{g^{-1}}\mu = \mu\}.$$

**Definition 9.30** *The coadjoint isotropy algebra of $\mu \in \mathfrak{g}^*$ is*

$$\mathfrak{g}_\mu := \{\xi \in \mathfrak{g} \mid \mathrm{ad}^*_\xi\mu = 0\}.$$

Recall from Chapter 4 that a function $C \in \mathcal{F}(P)$ is a Casimir of

$$\{C, F\} = 0 \quad \text{for all } F \in \mathcal{F}(P).$$

We have the following

**Proposition 9.31** *A function $C \in \mathcal{F}(\mathfrak{g}^*)$ is a Casimir function if and only if $\delta C/\delta\mu \in \mathfrak{g}_\mu$ for all $\mu \in \mathfrak{g}^*$.*

*Proof* Since $C \in \mathcal{F}(\mathfrak{g}^*)$ is a Casimir, we have $\{C, F\} = 0$ for all $F \in \mathcal{F}(\mathfrak{g}^*)$. We have

$$0 = \{C, F\}(\mu) = -\left\langle \mu, \left[\frac{\delta C}{\delta \mu}, \frac{\delta F}{\delta \mu}\right]\right\rangle = -\left\langle \mu, \mathrm{ad}_{\frac{\delta C}{\delta \mu}} \frac{\delta F}{\delta \mu}\right\rangle = -\left\langle \mathrm{ad}^*_{\frac{\delta C}{\delta \mu}} \mu, \frac{\delta F}{\delta \mu}\right\rangle$$

for $F$ arbitrary. Thus,

$$\mathrm{ad}^*_{\frac{\delta C}{\delta \mu}} \mu = 0,$$

and the conclusion follows. ∎

**Proposition 9.32** *If $C \in \mathcal{F}(\mathfrak{g}^*)$ is* Ad$^*$ *invariant, then $C$ is a Casimir function.*

**Remark 9.33** *The converse of this statement can be found in [MR02].*

*Proof* Since $C$ is Ad$^*$ invariant, we have $C(\mathrm{Ad}^*_{g^{-1}}\mu) = C(\mu)$ for all $\mu \in \mathfrak{g}^*$. Taking the derivative with respect to $g$ at $g = e$ we have

$$\mathrm{d}C(\mu)\left(-\mathrm{ad}^*_\xi \mu\right) = 0$$

where $\xi = \dot{g}(0)$. Thus

$$0 = \left\langle -\mathrm{ad}^*_\xi \mu, \frac{\delta C}{\delta \mu}\right\rangle = \left\langle \mu, -\mathrm{ad}_\xi \frac{\delta C}{\delta \mu}\right\rangle = \left\langle \mu, \mathrm{ad}_{\frac{\delta C}{\delta \mu}} \xi\right\rangle = \left\langle \mathrm{ad}^*_{\frac{\delta C}{\delta \mu}} \mu, \xi\right\rangle.$$

Since $\xi$ is arbitrary, we have

$$\mathrm{ad}^*_{\frac{\delta C}{\delta \mu}} \mu = 0,$$

and thus $\delta C/\delta \mu \in \mathfrak{g}_\mu$. The conclusion follows by applying the previous proposition. ∎

**Example 9.34 (Co-Adjoint orbits on $SO(3)$ and rigid body Casimirs)** For $\mathfrak{so}(3)^*$ the Co-Adjoint orbits are the sets (see Example 6.44)

$$\mathcal{O}_{\mathbf{\Pi}} = \{\mathbf{R}\mathbf{\Pi} \mid \mathbf{R} \in SO(3)\}.$$

Given $\mathbf{\Pi} \in \mathfrak{so}(3)^*$, the coadjoint isotropy group is given by

$$(\mathfrak{so}(3))_{\mathbf{\Pi}} = \{\mathbf{\Omega} \in \mathfrak{so}(3) \mid \mathbf{\Pi} \times \mathbf{\Omega} = 0\}$$

i.e. it contains all angular velocities parallel to $\mathbf{\Pi}$. The rigid body Casimirs are functions of the form

$$C(\mathbf{\Pi}) = \Phi\left(\frac{1}{2}||\mathbf{\Pi}||^2\right).$$

Indeed, since

$$\frac{\delta C}{\delta \mathbf{\Pi}} = \left(\Phi'\left(\frac{1}{2}||\mathbf{\Pi}||^2\right)\right)\mathbf{\Pi}$$

belongs to the coadjoint isotropy group $(\mathfrak{so}(3))_{\mathbf{\Pi}}$, by Proposition 9.31, $\Phi$ is a Casimir.

**Exercise 9.13**
Recall the left and right momentum maps from Example 8.12. Show that $J_L(J_R^{-1}(\mu)) = \mathcal{O}_\mu$.

## 9.6 Lie–Poisson brackets on semidirect products

In Chapter 5, Example 5.14 we introduced the special Euclidian group $SE(3)$, a Lie group that describes rigid motions and coordinate transformations of the three-dimensional space. Recall that $SE(3) \simeq SO(3) \times \mathbb{R}^3$ and that it may be described as the set of $(4 \times 4)$-dimensional matrices of the form

$$\left\{ \begin{bmatrix} \mathrm{R} & \mathrm{v} \\ 0 & 1 \end{bmatrix} \;\middle|\; \mathrm{R} \in SO(3), \quad \mathrm{v} \in \mathbb{R}^3 \right\}$$

The group operation is the usual multiplication of matrices. Equivalently, we can write the multiplication as the operation $\star : \left(SO(3) \times \mathbb{R}^3\right) \times \left(SO(3) \times \mathbb{R}^3\right) \to SO(3) \times \mathbb{R}^3$ :

$$(\mathrm{R}_1, \mathrm{v}_1) \star (\mathrm{R}_2, \mathrm{v}_2) = (\mathrm{R}_1\mathrm{R}_2, \mathrm{v}_1 + \mathrm{R}_1\mathrm{v}_2).$$

This kind of product structure generalizes naturally as follows:

**Definition 9.35 (Semidirect products)** *Consider a Lie group $G$ and vector space $V$ on which $G$ acts on the left by linear maps. That is, for $g \in G$,*

$\Phi_g : V \to V$ *is a linear map. The semidirect product* $S := G\circledS V$ *is a Lie group structure on the set* $G \times V$ *where the multiplication is given by*

$$(g_1, v_1) \star (g_2, v_2) := (g_1 g_2, v_1 + g_1 v_2).$$

Note that the identity element of $G\circledS V$ is $(e, 0)$, where $e$ is the identity of $G$ and $0$ is the zero vector on $V$.

The Lie algebra of $G\circledS V$ is sometimes denoted $\mathfrak{g}\circledS V$ or just $\mathfrak{s}$. It has a bracket given by

$$[(\xi_1, v_1), (\xi_2, v_2)] := ([\xi_1, \xi_2], \xi_1 v_2 - \xi_2 v_1), \tag{9.30}$$

where $\xi v$ denotes the induced infinitesimal action of $\mathfrak{g}$ on $V$:

$$\mathfrak{g} \times V \to V, \quad (\xi, v) \to \xi v := \xi_V(v) = \left.\frac{\mathrm{d}}{\mathrm{d}t}\right|_{t=0} g(t)v,$$

where $g(t)$ is a curve in $G$ with $g(0) = e$ and $\dot{g}(0) = \xi$.

For the reader's convenience, we repeat certain definition and notations from Section 7.5 related to the $G$ action on $V$ and $V^*$. For every $v \in V$ define the map

$$\rho_v : \mathfrak{g} \to V, \quad \rho_v(\xi) := \xi v. \tag{9.31}$$

It is immediate that $\rho_v$ is linear. The dual map $\rho_v^* : V^* \to \mathfrak{g}^*$, defines the diamond operator $v \diamond a := \rho_v^*(a)$. Further, for any $\xi \in \mathfrak{g}$ and $(v, a) \in V \times V^*$ we have:

$$\langle v \diamond a, \xi \rangle = \left\langle \rho_v^*(a), \xi \right\rangle = \langle a, \rho_v(\xi) \rangle = \langle a, \xi v \rangle. \tag{9.32}$$

The left action of $G$ on $V$ induces a left action of $G$ on $V^*$ by

$$G \times V^* \to V^* \quad (g, a) = ga \quad \text{where } \langle ga, u \rangle = \left\langle a, g^{-1}u \right\rangle \quad \text{for all } u \in V. \tag{9.33}$$

The infinitesimal generator for this action is given by:

$$\xi a := \xi_{V^*}(a) := \frac{\mathrm{d}}{\mathrm{d}t} g(t)a,$$

where $g(t)$ is a curve in $G$ with $g(0) = e$ and $\dot{g}(0) = \xi$.

The dual of the Lie algebra $\mathfrak{s} = \mathfrak{g}\circledS V$ is denoted $\mathfrak{s}^*$. Direct calculations show that

$$\mathrm{Ad}_{(g,v)}(\xi, u) = \left(\mathrm{Ad}_g \xi, gu - (\mathrm{Ad}_g \xi)\, v\right) \quad (\textit{adjoint action}), \tag{9.34}$$

and

$$\mathrm{Ad}^*_{(g,v)^{-1}}(\mu,a) = \left(\mathrm{Ad}^*_{g^{-1}}\,\mu + v \diamond (ga), (ga)\right) \quad (\textit{coadjoint action}). \quad (9.35)$$

We now apply the general Lie–Poisson formula (9.16) in the semidirect products setting. Consider a *left* $G$ action on $V$ and a $\mp$ Lie–Poisson bracket on $\mathfrak{g}^*$. the Lie–Poisson bracket on $\mathfrak{s}^*$ is given by

$$\{F,K\}_{\mathfrak{s}^*} = \mp\left\langle(\mu,a), \left[\frac{\delta F}{\delta(\mu,a)}, \frac{\delta K}{\delta(\mu,a)}\right]\right\rangle \quad (9.36)$$

$$= \mp\left\langle(\mu,a), \left[\left(\frac{\delta F}{\delta\mu}, \frac{\delta F}{\delta a}\right), \left(\frac{\delta K}{\delta\mu}, \frac{\delta K}{\delta a}\right)\right]\right\rangle, \quad (9.37)$$

where as usual, $F,K \in C(\mathfrak{s}^*$. In the above formula, we identify $\mathfrak{g}^{**} \simeq \mathfrak{g}$ and $V^{**} \simeq V$ such that for any function $F : \mathfrak{s}^* \to \mathbb{R}$, we have $(\delta F/\delta\mu) \in \mathfrak{g}$ and $(\delta F/\delta a) \in V$. Taking into account the Lie bracket formula (9.30) we calculate:

$$\{F,K\}_{\mathfrak{s}^*} = \mp\left\langle\mu, \left[\frac{\delta F}{\delta\mu}, \frac{\delta K}{\delta\mu}\right]\right\rangle \mp \left\langle a, \frac{\delta F}{\delta\mu}\frac{\delta K}{\delta a} - \frac{\delta K}{\delta\mu}\frac{\delta F}{\delta a}\right\rangle. \quad (9.38)$$

Further, by applying the formulae (9.22), given a Hamiltonian $h : \mathfrak{s}^* \to \mathbb{R}$, the Lie–Poisson equations of motion are

$$\dot{\mu} = \pm\mathrm{ad}^*_{\frac{\delta h}{\delta\mu}}\,\mu \mp \frac{\delta h}{\delta a} \diamond a, \quad (9.39)$$

$$\dot{a} = \mp\frac{\delta h}{\delta\mu}a. \quad (9.40)$$

**Remark 9.36** *The construction above applies to right actions of $G$ on the vector space $V$, as well. In this case the group multiplication is given by*

$$(g_1, v_1) \star (g_2, v_2) := (g_1 g_2, v_2 + v_1 g_2),$$

*and the Lie bracket on $\mathfrak{g}$Ⓢ$V$ is*

$$[(\xi_1, v_1), (\xi_2, v_2)] := ([\xi_1, \xi_2], v_2\xi_1 - v_1\xi_2)\,. \quad (9.41)$$

*Taking the $\mp$ Lie–Poisson bracket on $\mathfrak{g}^*$, the Lie–Poisson bracket on $\mathfrak{s}^*$ reads:*

$$\{F,K\}_{\mathfrak{s}^*} = \mp\left\langle\mu, \left[\frac{\delta F}{\delta\mu}, \frac{\delta K}{\delta\mu}\right]\right\rangle \pm \left\langle a, \frac{\delta F}{\delta\mu}\frac{\delta K}{\delta a} - \frac{\delta K}{\delta\mu}\frac{\delta F}{\delta a}\right\rangle \quad (9.42)$$

*and, given a Hamiltonian $h : \mathfrak{s}^* \to \mathbb{R}$, the Lie–Poisson equations of motion are*

$$\dot{\mu} = \pm \mathrm{ad}^*_{\frac{\delta h}{\delta \mu}} \mu \pm \frac{\delta h}{\delta a} \diamond a, \tag{9.43}$$

$$\dot{a} = \pm \frac{\delta h}{\delta \mu} a. \tag{9.44}$$

**Example 9.37 (Motion on $\mathfrak{se}(3)^*$)** Consider the the minus Lie–Poisson bracket on $\mathfrak{so}(3)^*$. Then, on $\mathfrak{se}(3)^*$, the Lie–Poisson bracket is given by

$$\begin{aligned} &\{F, K\}_{\mathfrak{se}(3)^*}(\boldsymbol{\Pi}, \boldsymbol{\Gamma}) \\ &\quad := -\langle \boldsymbol{\Pi}, \nabla_{\boldsymbol{\Pi}} F \times \nabla_{\boldsymbol{\Pi}} K \rangle - \langle \boldsymbol{\Gamma}, \nabla_{\boldsymbol{\Pi}} F \times \nabla_{\boldsymbol{\Gamma}} K - \nabla_{\boldsymbol{\Pi}} K \times \nabla_{\boldsymbol{\Gamma}} F \rangle . \end{aligned}$$

Given a Hamiltonian $h : \mathfrak{se}(3)^* \to \mathbb{R}$, $h = h(\boldsymbol{\Pi}, \boldsymbol{\Gamma})$ with $\boldsymbol{\Pi} \in \mathfrak{so}(3)$ and $\boldsymbol{\Gamma} \in \mathbb{R}^3$, the equations of motion are:

$$\dot{\boldsymbol{\Pi}} = \boldsymbol{\Pi} \times (\nabla_{\boldsymbol{\Pi}} h) + (\nabla_{\boldsymbol{\Gamma}} h) \diamond \boldsymbol{\Gamma}, \tag{9.45}$$

$$\dot{\boldsymbol{\Gamma}} = -(\nabla_{\boldsymbol{\Gamma}} h) \times \boldsymbol{\Gamma}, \tag{9.46}$$

where we used the usual identification

$$\hat{\boldsymbol{\Omega}} \mathbf{v} = \boldsymbol{\Omega} \times \mathbf{v} \quad \text{for all } \hat{\boldsymbol{\Omega}} \in \mathfrak{so}(3) \text{ and } \mathbf{v} \in \mathbb{R}^3.$$

For any $\mathbf{v} \in \mathbb{R}^3$ and $\mathbf{a} \in (\mathbb{R}^3)^* \simeq \mathbb{R}^3$, we compute $\mathbf{v} \diamond \mathbf{a}$ using eqn (9.32):

$$\left\langle \mathbf{v} \diamond \mathbf{a}, \hat{\boldsymbol{\Omega}} \right\rangle = \left\langle \mathbf{a}, \hat{\boldsymbol{\Omega}}\, \mathbf{v} \right\rangle = \langle \mathbf{a}, \boldsymbol{\Omega} \times \mathbf{v} \rangle = \langle \mathbf{v} \times \mathbf{a}, \boldsymbol{\Omega} \rangle \quad \text{for all } \hat{\boldsymbol{\Omega}} \in \mathfrak{so}(3).$$

So,

$$\mathbf{v} \diamond \mathbf{a} = \mathbf{v} \times \mathbf{a}$$

and the equations of motions on $\mathfrak{se}(3)^*$ read:

$$\dot{\boldsymbol{\Pi}} = \boldsymbol{\Pi} \times (\nabla_{\boldsymbol{\Pi}} h) + (\nabla_{\boldsymbol{\Gamma}} h) \times \boldsymbol{\Gamma} \tag{9.47}$$

$$\dot{\boldsymbol{\Gamma}} = -(\nabla_{\boldsymbol{\Gamma}} h) \times \boldsymbol{\Gamma}. \tag{9.48}$$

**Lie–Poisson formulation of the heavy top** Recall from Chapter 7 the heavy top reduced Lagrangian

$$l : \mathfrak{so}(3) \times \mathbb{R}^3 \to \mathbb{R}$$

$$l(\boldsymbol{\Omega}, \boldsymbol{\Gamma}) = \frac{1}{2} \langle \boldsymbol{\Omega}, \mathbb{I} \boldsymbol{\Omega} \rangle - \langle mg\boldsymbol{\Gamma}, \boldsymbol{\chi} \rangle ,$$

together with the equations of motion:

$$\frac{\mathrm{d}}{\mathrm{d}t}(\mathbb{I}\boldsymbol{\Omega}) = \mathbb{I}\boldsymbol{\Omega} \times \boldsymbol{\Omega} + mg\boldsymbol{\Gamma} \times \boldsymbol{\chi} \tag{9.49}$$

$$\frac{\mathrm{d}\boldsymbol{\Gamma}}{\mathrm{d}t} = \boldsymbol{\Gamma} \times \boldsymbol{\Omega}. \tag{9.50}$$

Applying the reduced Legendre transformation in variable $\boldsymbol{\Omega}$ only, we have

$$\boldsymbol{\Pi} := \frac{\delta l}{\delta \boldsymbol{\Omega}} = \mathbb{I}\boldsymbol{\Omega}.$$

In the variables $(\boldsymbol{\Pi}, \boldsymbol{\Gamma}) \in \mathfrak{so}(3)^* \times \mathbb{R}^3$ the equations above become

$$\frac{\mathrm{d}\boldsymbol{\Pi}}{\mathrm{d}t} = \boldsymbol{\Pi} \times \mathbb{I}^{-1}\boldsymbol{\Omega} + mg\boldsymbol{\Gamma} \times \boldsymbol{\chi} \tag{9.51}$$

$$\frac{\mathrm{d}\boldsymbol{\Gamma}}{\mathrm{d}t} = \boldsymbol{\Gamma} \times \boldsymbol{\Omega}. \tag{9.52}$$

Defining the Hamiltonian by

$$h : \mathfrak{so}(3)^* \times \mathbb{R}^3 \to \mathbb{R}$$

$$h(\boldsymbol{\Pi}, \boldsymbol{\Gamma}) = \langle \boldsymbol{\Pi}, \boldsymbol{\Gamma}(\boldsymbol{\Pi}) \rangle - l\,(\boldsymbol{\Omega}(\boldsymbol{\Pi}), \boldsymbol{\Gamma}) = \frac{1}{2}\langle \boldsymbol{\Pi}, \mathbb{I}\boldsymbol{\Pi} \rangle + \langle mg\boldsymbol{\Gamma}, \boldsymbol{\chi} \rangle,$$

the system of eqns (9.51)+(9.52) above is Hamiltonian with respect to the Lie–Poisson bracket on $\mathfrak{se}(3)^*$.

**Exercise 9.14 (Semi-direct products facts).**

a) Show that the inverse of $(g, v) \in G \circledS V$ is given by $(g^{-1}, -g^{-1}v)$.
b) Show that eqn (9.30) is indeed a Lie algebra bracket.
c) Verify that for any $v \in V$, the map $\rho_v$ is linear.
d) Calculate and retrieve formulae (9.34) and (9.35).

**Exercise 9.15 (Semidirect bracket)**
For right $G$ actions on $V$ and a $\mp$ Lie–Poisson bracket on $\mathfrak{g}^*$, deduce the Lie–Poisson bracket (9.42) and the equation of motion (9.43).

**Exercise 9.16 (Semidirect bracket)**
Verify that the system of eqns (9.51)+(9.52) is Hamiltonian with respect to the Lie–Poisson structure on $\mathfrak{se}(3)^*$.

**Exercise 9.17 (Semidirect bracket)**
Verify that functions of the form $C(\mathbf{\Pi}, \mathbf{\Gamma}) = \Phi(\mathbf{\Pi}\cdot\mathbf{\Gamma}, ||\mathbf{\Pi}||^2)$ are Casimirs for the heavy top.

## Solutions to selected exercises

**Solution to Exercise 9.3** Suppose a Hamiltonian $H : T^*G \times T^*K \to \mathbb{R}$ is invariant under the action $(\tilde{g}, \tilde{k}) \to (\tilde{g}g, k\tilde{k})$. Then the Lagrangian is both right $G$-invariant and left $K$-invariant. Therefore, the right Lie–Poisson equations for variations of $G$ and the left equations for variations in $K$ hold on the respective factors of $T^*G \times T^*K$. These considerations yield the equations

$$\frac{\mathrm{d}}{\mathrm{d}t}(\mu, \eta) = (\dot{\mu}, \dot{\eta}) = \left(-\operatorname{ad}^*_{\frac{\delta h}{\delta \mu}} \mu, \operatorname{ad}^*_{\frac{\delta h}{\delta \eta}} \eta\right),$$

for $\mu \in \mathfrak{g}^*$ and $\eta \in \mathfrak{k}^*$.

**Solution to Exercise 9.8** Let $F_L : T^*G \to \mathbb{R}$ be a $G$-invariant function. Then $F_L \circ T^*L_{g^{-1}} = F_L$. Taking the fiber derivative of this relationship at $(g, \alpha)$ while supposing $(g, \alpha) = T_e^*L_{g^{-1}}(e, \mu)$ for some $\mu \in \mathfrak{g}^*$ gives

$$\begin{aligned}
\left\langle\left(g, \frac{\delta F_L}{\delta \alpha}\right), (g, \delta\alpha)\right\rangle &= \left\langle\left(g, \frac{\delta F_L}{\delta \alpha}\right), T_e^*L_{g^{-1}}(e, \delta\mu)\right\rangle \\
&= \left\langle T_g L_{g^{-1}}\left(g, \frac{\delta F_L}{\delta \alpha}\right), (e, \delta\mu)\right\rangle \\
&= \left\langle\left(e, \frac{\delta F_L}{\delta \mu}\right), (e, \delta\mu)\right\rangle.
\end{aligned}$$

Here, the first step follows since $T^*L_{g^{-1}}$ is linear, so

$$\delta\left(T_e^*L_{g^{-1}}(e, \mu)\right) = T_e^*L_{g^{-1}}\delta(e, \mu).$$

Consequently,

$$\left(g, \frac{\delta F_L}{\delta \alpha}\right) = T_e L_g \left(e, \frac{\delta F_L}{\delta \mu}\right).$$

**Solution to Exercise 9.9** Consider the left invariant vector fields

$$X(g) = T_e L_g \frac{\delta F}{\delta \mu}, \qquad Y(g) = T_e L_g \frac{\delta K}{\delta \mu}.$$

Theorem 3.25 (regarding push-forwards and Lie brackets) yields

$$\begin{aligned} T_g L_{g^{-1}} [X, Y](g) &= \left(\left(L_{g^{-1}}\right)_* [X, Y]\right)(e) \\ &= \left[\left(L_{g^{-1}}\right)_* X, \left(L_{g^{-1}}\right)_* Y\right](e) \\ &= [X, Y](e) \\ &= \left[\frac{\delta F}{\delta \mu}, \frac{\delta K}{\delta \mu}\right]. \end{aligned}$$

**Solution to Exercise 9.11** The Poisson bracket for $\mathfrak{su}(2)^*$ is given by the Lie–Poisson bracket.

$$\{F, K\}(\mu) = \left\langle \mu, \left[\frac{\delta F}{\delta \mu}, \frac{\delta K}{\delta \mu}\right]\right\rangle.$$

The tilde map on $\mathfrak{su}(2)^*$ that we studied in Exercise 5.20 allows an alternative expression,

$$\{F, K\}(\tilde{\boldsymbol{\mu}}) = -\frac{1}{2}\boldsymbol{\mu} \cdot \left(\frac{\partial F}{\partial \boldsymbol{\mu}} \times \frac{\partial K}{\partial \boldsymbol{\mu}}\right).$$

# 10 Pseudo-rigid bodies

Consider a body that can stretch and shear as well as rotate, so that it can undergo arbitrary orientation-preserving linear transformations around some fixed point. The configuration of the object at a certain time is determined by an element of $GL^+(3)$ that maps a given reference configuration to the current one. This system is called the ***pseudo-rigid*** body, or ***affine rigid body***. It is very similar to the rigid body, modelled in Sections 1.5 and 7.1, except that the the configuration space is larger. Pseudo-rigid bodies are used to model solid objects undergoing small elastic deformations, and also some fluid motions, for example rotating gas planets and liquid drops [Cha87].

We will assume in this chapter that, in the reference configuration, the body is spherically symmetric, meaning that the moment of inertia tensor is rotationally invariant. A sufficient condition for this symmetry is that the body's density function is spherically symmetric, which implies that the body is spherical in the reference configuration and ellipsoidal at all times. For most of the chapter, we will study *free ellipsoidal motion*, meaning that the pseudo-rigid body evolves according to a Lagrangian that equals kinetic energy only, with no potential energy.

## 10.1 Modelling

We fix a reference configuration of the body. We use a fixed inertial *spatial coordinate system*, and a moving *body coordinate system*, both with origin at the fixed point of the body. When the body is in its reference position, the two systems coincide. The body coordinate system moves with the body, which means that the position of any given particle, in the body coordinate system, remains fixed. The position in body coordinates is called the particle's *label*. The spatial position of a particle at a given time is also called its *Eulerian* position, while its label is called its *Lagrangian* position. We denote by $GL^+(3)$ the set of all real-valued matrices with positive determinant, i.e. matrices corresponding to orientation-preserving transformations. We assume that the configuration of the body at time $t$ is determined by a matrix $\mathbf{Q}(t) \in GL^+(3)$ that takes the label $\mathbf{X}$ of any particle in the body to its spatial position $\mathbf{x}(t)$. Thus the configuration space of the pseudo-rigid body is $GL^{+}(3)$. The position and velocity at time $t$ of the particle with label $\mathbf{X}$ are

$$\mathbf{x}(t,\mathbf{X}) = \mathbf{Q}(t)\mathbf{X}, \qquad \text{and } \dot{\mathbf{x}}(t,\mathbf{X}) = \dot{\mathbf{Q}}(t)\mathbf{X} = \dot{\mathbf{Q}}(t)\mathbf{Q}^{-1}(t)\mathbf{x}(t,\mathbf{X}). \quad (10.1)$$

Let $\rho(\mathbf{X})$ be the density of the body at position $\mathbf{X}$ in body coordinates, which we assume to be constant. Let $\mathcal{B}$ be the region occupied by the body in its reference configuration. The ***coefficient of inertia matrix***, with respect to the origin, is

$$\mathbb{J} := \int_{\mathcal{B}} \rho(\mathbf{X})\mathbf{X}\mathbf{X}^{T}\, \mathrm{d}^{3}\mathbf{X}.$$

Clearly, $\mathbb{J}$ is symmetric and therefore can be diagonalized by a change of basis.

The ***kinetic energy*** of the pseudo-rigid body is:

$$\begin{aligned} K &= \frac{1}{2}\int_{\mathcal{B}} \rho(\mathbf{X})\,\|\dot{\mathbf{x}}\|^{2}\, \mathrm{d}^{3}\mathbf{X} \\ &= \frac{1}{2}\int_{\mathcal{B}} \rho(\mathbf{X})\,\|\dot{\mathbf{Q}}\mathbf{X}\|^{2}\, \mathrm{d}^{3}\mathbf{X} \\ &= \frac{1}{2}\int_{\mathcal{B}} \rho(\mathbf{X})\,\mathrm{tr}\left((\dot{\mathbf{Q}}\mathbf{X})(\dot{\mathbf{Q}}\mathbf{X})^{T}\right) \mathrm{d}^{3}\mathbf{X} \\ &= \frac{1}{2}\,\mathrm{tr}\left(\dot{\mathbf{Q}}\left(\int_{\mathcal{B}} \rho(\mathbf{X})\mathbf{X}\mathbf{X}^{T}\, d^{3}\mathbf{X}\right)\dot{\mathbf{Q}}^{T}\right) \\ &= \frac{1}{2}\,\mathrm{tr}\left(\dot{\mathbf{Q}}\mathbb{J}\dot{\mathbf{Q}}^{T}\right). \end{aligned}$$

Note that this expression for kinetic energy is left-invariant under $\mathbf{Q} \to \mathbf{R}\mathbf{Q}$ for any choice of rotation $\mathbf{R} \in SO(3)$ and any distribution of mass in the body represented by $\mathbb{J}$.

We will assume from now on that $\mathbb{J}$ is ***spherically symmetric***, meaning that it is invariant under any change of basis by a rotation. This implies that $\mathbb{J}$ is a multiple of the identity matrix,

$$\mathbb{J} = k\,\mathbf{I},$$

for some constant $k \in \mathbb{R}$. Without loss of generality, we will assume that $k = 1$. It is convenient to assume that the region $\mathcal{B}$ is spherical, with the density distributed in a spherically-symmetric manner, which implies that $\mathbb{J}$ is spherically symmetric, but this assumption is not strictly necessary. If $\mathcal{B}$ is spherical, then the shape of the body at time $t$, which is $\mathbf{Q}(t)(\mathcal{B})$, is always ellipsoidal. The motion of the body is called ***ellipsoidal motion*** whenever $\mathbb{J}$

is spherically symmetric (even if $\mathcal{B}$ is not spherical). Using the assumption of spherical symmetry, the kinetic energy of ellipsoidal motion is:

$$K = \frac{1}{2}\operatorname{tr}\left(\dot{\mathbf{Q}}\dot{\mathbf{Q}}^T\right). \tag{10.2}$$

We consider ***free ellipsoidal motion***, which is the motion determined by the Lagrangian function $L : T\,\mathrm{GL}^+(3) \to \mathbb{R}$ that equals the kinetic energy:

$$L(\mathbf{Q}, \dot{\mathbf{Q}}) = K = \frac{1}{2}\operatorname{tr}\left(\dot{\mathbf{Q}}\dot{\mathbf{Q}}^T\right). \tag{10.3}$$

Note that $L$ is invariant under the tangent lifts of both the left and right translation actions of $SO(3)$, since

$$L(\mathbf{gQh}, \mathbf{g}\dot{\mathbf{Q}}\mathbf{h}) = \frac{1}{2}\operatorname{tr}\left(\mathbf{g}\dot{\mathbf{Q}}\mathbf{h}\left(\mathbf{g}\dot{\mathbf{Q}}\mathbf{h}\right)^T\right) = \frac{1}{2}\operatorname{tr}\left(\mathbf{g}\dot{\mathbf{Q}}\mathbf{h}\mathbf{h}^T\dot{\mathbf{Q}}^T\mathbf{h}^T\right) = \frac{1}{2}\operatorname{tr}\left(\dot{\mathbf{Q}}\dot{\mathbf{Q}}^T\right), \tag{10.4}$$

for all $\mathbf{g}, \mathbf{h} \in SO(3)$. The Lagrangian is also invariant under the tangent lift of the discrete symmetry $\mathbf{Q} \mapsto \mathbf{Q}^T$, which is called the ***Dedekind duality*** principle.

Every $\mathbf{Q} \in GL^+(3)$ can be decomposed as

$$\mathbf{Q} = \mathbf{RAS},$$

where **R** and **S** are rotation matrices and **A** is diagonal with positive entries.[1] This is a minor variation on the ***singular value decomposition***, also called the ***bipolar decomposition***. In fact, the singular value decomposition gives $\mathbf{R}, \mathbf{S} \in O(3)$, but these matrices can be easily modified to be in $SO(3)$ (see Exercise 10.1). This decomposition $\mathbf{Q} = \mathbf{RAS}$ is not unique. However it is known (see [Kat76]) that if $\mathbf{Q}(t)$ is an analytic path in $GL(3)$, then there exist analytic paths $\mathbf{R}(t), \mathbf{S}(t) \in O(3)$ and $\mathbf{A}(t) \in Diag^+(3)$ such that $\mathbf{Q}(t) = \mathbf{R}(t)\mathbf{A}(t)\mathbf{S}(t)$ for all $t$. We will assume that this is the case.

If $\mathbf{Q}(t)$ is a path in $GL^+(3)$ then $\mathbf{R}(0)$ and $\mathbf{S}(0)$ may be assumed to be in $SO(3)$, as noted above. By continuity, this implies that $\mathbf{R}(t), \mathbf{S}(t) \in SO(3)$ for all $t$. The various components of the motion can be interpreted as follows:

- $\mathbf{S} \in SO(3)$ rotates the **X**-coordinates in the reference configuration
- **A** stretches the body along the instantaneous principal axes of $\mathbf{S}(\mathcal{B})$.
- $\mathbf{R} \in SO(3)$ rotates the **x**-coordinates.

[1] Some authors prefer to write the decomposition as $\mathbf{Q} = \mathbf{R}^T\mathbf{AS}$.

Thus $\mathbf{S}$ is a rotation in body coordinates, while $\mathbf{R}$ is a rotation in spatial coordinates. Note that $\mathbf{A}(t)$ completely determines the shape of the body at time $t$. If the reference configuration of the body is spherically-symmetric, then the transformation $\mathbf{X} \mapsto \mathbf{SX}$ changes neither the region occupied by the body nor the mass distribution of the body. Physically, it is only directly visible if the particles have some observable quality such as colour. This kind of motion is called ***internal circulation*** of material.

**Example 10.1 (Jacobi and Dedekind ellipsoids)** Consider a body whose reference configuration is the solid unit sphere. Let $\mathbf{A}(t) = \text{diag}(a_1, a_2, a_3)$ for all $t$. For any motion $\mathbf{Q}(t) = \mathbf{R}(t)\mathbf{A}(t)\mathbf{S}(t)$ as above, the shape of the body at time $t$ is an ellipsoid with principal axes of lengths $2a_1, 2a_2, 2a_3$. If $\mathbf{S}(t) = \mathbf{I}$, for all $t$, then the resulting motion is a rotating ellipsoid with no internal circulation, called a ***Jacobi ellipsoid***. On the other hand, if $\mathbf{R}(t) = \mathbf{I}$, for all $t$, then the resulting motion is a body whose orientation is fixed, but that is internally circulating. Such a motion is called a ***Dedekind ellipsoid***. Note that the Dedekind duality $\mathbf{Q} \mapsto \mathbf{Q}^T$ interchanges Jacobi and Dedekind ellipsoids. (See Chandrasekhar [Cha87] for further discussion of Dedekind duality in the context of self-gravitating figures of equilibrium.)

We define an extended configuration space $Q_{ext} := SO(3) \times Diag(3)^+ \times SO(3)$, and a non-injective map $\Phi : Q \to GL^+(3)$ by

$$\Phi(\mathbf{R}, \mathbf{A}, \mathbf{S}) = \mathbf{RAS}.$$

As we noted above, every smooth path $\mathbf{Q}(t)$ in $GL^+(3)$ can be expressed as

$$\mathbf{Q}(t) = \mathbf{R}(t)\mathbf{A}(t)\mathbf{S}(t) = \Phi(\mathbf{R}(t), \mathbf{A}(t), \mathbf{S}(t)),$$

for some smooth path $(\mathbf{R}(t), \mathbf{A}(t), \mathbf{S}(t))$ in $Q_{ext}$. In particular, $\Phi$ is surjective, and one can also check that it is a submersion. We define the extended Lagrangian $L_{ext} : TQ_{ext} \to \mathbb{R}$ by $L_{ext} = L \circ T\Phi$. From the invariance property of $L$ in eqn (10.4), it follows that

$$L_{ext}(\mathbf{gR}, \mathbf{A}, \mathbf{Sh}, \mathbf{g}\dot{\mathbf{R}}, \dot{\mathbf{A}}, \dot{\mathbf{S}}\mathbf{h}) = L_{ext}(\mathbf{R}, \mathbf{A}, \mathbf{S}, \dot{\mathbf{R}}, \dot{\mathbf{A}}, \dot{\mathbf{S}}). \tag{10.5}$$

**Proposition 10.2** *Let* $(\mathbf{R}(t), \mathbf{A}(t), \mathbf{S}(t))$ *be a path in* $Q_{ext}$, *and let* $\mathbf{Q}(t) := \mathbf{R}(t)\mathbf{A}(t)\mathbf{S}(t)$ *be the corresponding path in* $GL^+(3)$. *The path* $(\mathbf{R}(t), \mathbf{A}(t), \mathbf{S}(t))$ *satisfies Hamilton's principle for* $L_{ext}$ *if and only if* $\mathbf{Q}(t)$ *satisfies Hamilton's principle for* $L$.

*Proof* All deformations $\mathbf{Q}(t,s)$ are of the form $\Phi(\mathbf{R}(t,s), \mathbf{A}(t,s), \mathbf{S}(t,s))$ for some deformation $(\mathbf{R}(t,s), \mathbf{A}(t,s), \mathbf{S}(t,s)$ in $Q_{ext}$. By the chain rule, this implies that

$$\left(\mathbf{Q}(t,s),\dot{\mathbf{Q}}(t,s)\right) = T\Phi\left(\mathbf{R}(t,s),\mathbf{A}(t,s),\mathbf{S}(t,s),\dot{\mathbf{R}}(t,s),\dot{\mathbf{A}}(t,s),\dot{\mathbf{S}}(t,s)\right), \tag{10.6}$$

and hence that

$$L\left(\mathbf{Q}(t,s),\dot{\mathbf{Q}}(t,s)\right) = L_{ext}\left(\mathbf{R}(t,s),\mathbf{A}(t,s),\mathbf{S}(t,s),\dot{\mathbf{R}}(t,s),\dot{\mathbf{A}}(t,s),\dot{\mathbf{S}}(t,s)\right),$$

for all $(t,s)$ in the domain of the deformation. Hamilton's principle for $L_{ext}$ is

$$\delta\int_a^b L_{ext}\left(\mathbf{R}(t),\mathbf{A}(t),\mathbf{S}(t),\dot{\mathbf{R}}(t),\dot{\mathbf{A}}(t),\dot{\mathbf{S}}(t)\right)\,dt = 0$$

for all variations that vanish at the endpoints. This is equivalent to

$$\left.\frac{d}{ds}\right|_{s=0}\int_a^b L_{ext}\left(\mathbf{R}(t,s),\mathbf{A}(t,s),\mathbf{S}(t,s),\dot{\mathbf{R}}(t,s),\dot{\mathbf{A}}(t,s),\dot{\mathbf{S}}(t,s)\right)\,\mathrm{d}t = 0$$

for all deformations leaving the endpoints fixed. Taking into account eqn (10.6), this is equivalent to

$$\left.\frac{\mathrm{d}}{\mathrm{d}s}\right|_{s=0}\int_a^b L\left(\mathbf{Q}(t,s),\dot{\mathbf{Q}}(t,s)\right)\,\mathrm{d}t = 0,$$

for all deformations leaving the endpoints fixed, i.e.

$$\delta\int_a^b L\left(\mathbf{Q}(t),\dot{\mathbf{Q}}(t)\right)\,\mathrm{d}t = 0,$$

for all variations that vanish at the endpoints, which is Hamilton's principle for $L$. ■

**Remark 10.3** *Given an initial condition* $(\mathbf{Q}_0,\dot{\mathbf{Q}}_0)$, *suppose we choose*

$$\left(\mathbf{R}_0,\mathbf{A}_0,\mathbf{S}_0,\dot{\mathbf{R}}_0,\dot{\mathbf{A}}_0,\dot{\mathbf{S}}_0\right),\in T\Phi^{-1}(\mathbf{Q}_0,\dot{\mathbf{Q}}_0).$$

*Let* $\left(\mathbf{R}(t),\mathbf{A}(t),\mathbf{S}(t)\right)$ *be a solution to Hamilton's principle for* $L_{ext}$ *with initial conditions*

$$\left(\mathbf{R}(0),\mathbf{A}(0),\mathbf{S}(0),\dot{\mathbf{R}}(0),\dot{\mathbf{A}}(0),\dot{\mathbf{S}}(0)\right) = \left(\mathbf{R}_0,\mathbf{A}_0,\mathbf{S}_0,\dot{\mathbf{R}}_0,\dot{\mathbf{A}}_0,\dot{\mathbf{S}}_0\right).$$

*Then by the previous proposition,* $\Phi\left(\mathbf{R}(t),\mathbf{A}(t),\mathbf{S}(t)\right)$ *satisfies Hamilton's principle for* $L$ *with initial conditions* $(\mathbf{Q}_0,\dot{\mathbf{Q}}_0)$. *Assuming that* $L$ *is regular, there is a unique solution* $\mathbf{Q}(t)$ *with this initial condition. Thus, regardless of the choice of initial condition* $\left(\mathbf{R}_0, \mathbf{A}_0, \mathbf{S}_0, \dot{\mathbf{R}}_0, \dot{\mathbf{A}}_0, \dot{\mathbf{S}}_0\right) \in T\Phi^{-1}(\mathbf{Q}_0, \dot{\mathbf{Q}}_0)$*, the path* $\mathbf{Q}(t) := \Phi\left(\mathbf{R}(t), \mathbf{A}(t), \mathbf{S}(t)\right)$ *will be the same.*

The previous proposition and remark show that, to study the motion of the pseudo-rigid body, it suffices to study the extended Lagrangian $L_{ext} : TQ_{ext} \to \mathbb{R}$, given by $L_{ext} = L \circ \Phi$. Since

$$\dot{\mathbf{Q}} = \frac{\mathrm{d}}{\mathrm{d}t}(\mathbf{RAS}) = \dot{\mathbf{R}}\mathbf{AS} + \mathbf{R}\dot{\mathbf{A}}\mathbf{S} + \mathbf{RA}\dot{\mathbf{S}}, \tag{10.7}$$

which we denote $(\mathbf{RAS})^{\cdot}$, and similarly $\dot{\mathbf{Q}}^T = (\mathbf{S}^T\mathbf{AR}^T)^{\cdot}$, the definition of $L$ in eqn (10.3) implies that

$$L_{ext}(\mathbf{R}, \mathbf{A}, \mathbf{S}, \dot{\mathbf{R}}, \dot{\mathbf{A}}, \dot{\mathbf{S}}) = \frac{1}{2}\operatorname{tr}\left((\mathbf{RAS})^{\cdot}\left(\mathbf{S}^{\mathrm{T}}\mathbf{AR}^{\mathrm{T}}\right)^{\cdot}\right). \tag{10.8}$$

In the remainder of this chapter, we drop the subscript "ext". No confusion will arise if we use the same letter '$L$' to denote both the original Lagrangian $L$ on $TGL^+(3)$ and the extended Lagrangian $L_{ext}$ on $TSO(3) \times TDiag^+(3) \times TSO(3)$. Which of these is meant should be clear from the context.

**Exercise 10.1**
Let $\mathbf{Q} = \mathbf{RAS}$ with $\mathbf{R}, \mathbf{S} \in O(3)$ and $\mathbf{A} \in Diag^+(3)$, which is a singular value decomposition of a non-singular matrix $\mathbf{Q}$ with real entries. Show that there exists matrices $\mathbf{R}', \mathbf{S}' \in SO(3)$ and $\mathbf{A}' \in Diag^+(3)$ such that $\mathbf{Q} = \mathbf{R}'\mathbf{A}'\mathbf{S}'$.

## 10.2 Euler–Poincaré reduction

By the invariance property (10.5) of the Lagrangian on $TSO(3) \times TDiag^+(3) \times TSO(3)$ we can write:

$$\begin{aligned} L\left(\mathbf{R}, \dot{\mathbf{R}}, \mathbf{A}, \dot{\mathbf{A}}, \mathbf{S}, \dot{\mathbf{S}}\right) &= L(\mathbf{R}^{-1}\mathbf{R}, \mathbf{R}^{-1}\dot{\mathbf{R}}, \mathbf{A}, \dot{\mathbf{A}}, \mathbf{S}\mathbf{S}^{-1}, \dot{\mathbf{S}}\mathbf{S}^{-1}) \\ &= L(\mathbf{I}, \widehat{\boldsymbol{\Omega}}, \mathbf{A}, \dot{\mathbf{A}}, \mathbf{I}, \widehat{\boldsymbol{\Lambda}}) \end{aligned}$$

where

$$\widehat{\boldsymbol{\Omega}} := \mathbf{R}^{-1}\dot{\mathbf{R}} \qquad \text{and} \qquad \widehat{\boldsymbol{\Lambda}} := \dot{\mathbf{S}}\mathbf{S}^{-1}.$$

Thus, we can define the *reduced* Lagrangian

$$l : \mathfrak{so}(3) \times T\mathbb{R}^3 \times \mathfrak{so}(3) \to \mathbb{R},$$

$$l(\widehat{\mathbf{\Omega}}, \mathbf{A}, \dot{\mathbf{A}}, \mathbf{S}, \mathbf{\Lambda}) := L(\mathbf{I}, \widehat{\mathbf{\Omega}}, \mathbf{A}, \dot{\mathbf{A}}, \mathbf{I}, \widehat{\mathbf{\Lambda}}) = L\left(\mathbf{R}, \dot{\mathbf{R}}, \mathbf{A}, \dot{\mathbf{A}}, \mathbf{S}, \dot{\mathbf{S}}\right),$$

for all $\mathbf{R}, \dot{\mathbf{R}}, \mathbf{S}, \dot{\mathbf{S}}$ such that $\widehat{\mathbf{\Omega}} := \mathbf{R}^{-1}\dot{\mathbf{R}}$ and $\widehat{\mathbf{\Lambda}} := \dot{\mathbf{S}}\mathbf{S}^{-1}$. Hamilton's principle for $L$,

$$0 = \delta \int_a^b L\left(\mathbf{R}(t), \dot{\mathbf{R}}(t), \mathbf{A}(t), \dot{\mathbf{A}}(t), \mathbf{S}(t), \dot{\mathbf{S}}(t)\right) \mathrm{d}t,$$

becomes

$$0 = \delta \int_a^b l\left(\widehat{\mathbf{\Omega}}(t), \mathbf{A}(t), \dot{\mathbf{A}}(t), \mathbf{S}(t), \widehat{\mathbf{\Lambda}}(t)\right) \mathrm{d}t,$$

where the variations $\delta\widehat{\mathbf{\Omega}}$ and $\delta\widehat{\mathbf{\Lambda}}$ are induced by the variations $\delta\mathbf{R}, \delta\mathbf{S}$. By the calculations (7.17) – (7.21) in Section 7.2 we have:

$$\delta\widehat{\mathbf{\Omega}} = \dot{\widehat{\mathbf{\Sigma}}} + [\widehat{\mathbf{\Omega}}, \widehat{\mathbf{\Sigma}}],$$

where $\widehat{\mathbf{\Sigma}} := \mathbf{R}^{-1}\delta\mathbf{R}$ is an arbitrary path in $\mathfrak{so}(3)$ that vanishes at the endpoints, i.e., $\widehat{\mathbf{\Sigma}}(a) = \widehat{\mathbf{\Sigma}}(b) = 0$. Analogously,

$$\delta\widehat{\mathbf{\Lambda}} = \dot{\widehat{\mathbf{\Xi}}} - [\widehat{\mathbf{\Lambda}}, \widehat{\mathbf{\Xi}}]$$

where $\widehat{\mathbf{\Xi}} := \delta\mathbf{S}\,\mathbf{S}^{-1}$ is an arbitrary path in $\mathfrak{so}(3)$ that vanishes at the endpoints, i.e., $\widehat{\mathbf{\Xi}}(a) = \widehat{\mathbf{\Xi}}(b) = 0$. (The minus sign in the relation above is due to the fact that the variations are calculated for a right action.) The equations of motion follow from a direct calculation (see Exercise 10.2). Thus, we have:

**Theorem 10.4 (Euler–Poincaré reduction for the pseudo-rigid body)** *Consider a Lagrangian*

$$L\colon TSO(3) \times TDiag^{+}(3) \times TSO(3) \to \mathbb{R}, \quad L = L(\mathbf{R}, \dot{\mathbf{R}}, \mathbf{A}, \dot{\mathbf{A}}, \mathbf{S}, \dot{\mathbf{S}}),$$

*that is invariant under the tangent lift of the action of $SO(3) \times SO(3)$ given by*

$$(\mathbf{g}, \mathbf{h})(\mathbf{R}, \mathbf{A}, \mathbf{S}) \to (\mathbf{g}\mathbf{R}, \mathbf{A}, \mathbf{S}\mathbf{h}).$$

*Let*

$$l(\widehat{\boldsymbol{\Omega}}, \mathbf{A}, \dot{\mathbf{A}}, \widehat{\boldsymbol{\Lambda}}) := L\left(\mathbf{I}, \widehat{\boldsymbol{\Omega}}, \mathbf{A}, \dot{\mathbf{A}}, \mathbf{I}, \widehat{\boldsymbol{\Lambda}}\right)$$

*be the restriction of* $L$ *to* $\mathfrak{so}(3) \times T\,Diag^{+}(3) \times \mathfrak{so}(3)$, *and for any curves* $\mathbf{R}(t)$ *and* $\mathbf{S}(t)$ *let*

$$\widehat{\boldsymbol{\Omega}}(t) := \mathbf{R}^{-1}(t)\dot{\mathbf{R}}(t),$$

*and*

$$\widehat{\boldsymbol{\Lambda}}(t) := \dot{\mathbf{S}}(t)\mathbf{S}^{-1}(t).$$

*Then the following four statements are equivalent:*

(i) *The variational principle*

$$\delta \int_a^b L\left(\mathbf{R}(t), \dot{\mathbf{R}}(t), \mathbf{A}(t), \dot{\mathbf{A}}(t), \mathbf{S}(t), \delta\dot{\mathbf{S}}(t)\right) \mathrm{d}t = 0$$

*holds, for arbitrary variations* $\delta\mathbf{R}, \delta\mathbf{A}$, *and* $\delta\mathbf{S}$ *that vanish at the endpoints.*

(ii) $\mathbf{R}(t), \mathbf{A}(t)$ *and* $\mathbf{S}(t)$ *satisfy the Euler-Lagrange equations for Lagrangian* $L$.

(iii) *The variational principle*

$$\delta \int_a^b l\left(\widehat{\boldsymbol{\Omega}}(t), \mathbf{A}(t), \dot{\mathbf{A}}(t), \widehat{\boldsymbol{\Lambda}}(t)\right) dt = 0$$

*holds on* $\mathfrak{so}(3) \times T\,Diag^{+}(3) \times \mathfrak{so}(3)$, *for arbitrary variations* $\delta\mathbf{A}$ *vanishing at the endpoints, and variations* $\delta\widehat{\boldsymbol{\Omega}}$, $\delta\widehat{\boldsymbol{\Lambda}}$ *of the form*

$$\delta\widehat{\boldsymbol{\Omega}} = \dot{\widehat{\boldsymbol{\Sigma}}} + [\widehat{\boldsymbol{\Omega}}\,\widehat{\boldsymbol{\Sigma}}],$$

$$\delta\widehat{\boldsymbol{\Lambda}} = \dot{\widehat{\boldsymbol{\Xi}}} - [\widehat{\boldsymbol{\Lambda}}, \widehat{\boldsymbol{\Xi}}],$$

*with* $\widehat{\boldsymbol{\Sigma}}$ *and* $\widehat{\boldsymbol{\Xi}}$ *arbitrary paths in* $\mathfrak{so}(3)$ *vanishing at the endpoints.*

(iv) *The **Euler–Poincaré–Lagrange equations** hold:*

$$\frac{\mathrm{d}}{\mathrm{d}t}\left(\frac{\delta l}{\delta \widehat{\boldsymbol{\Omega}}}\right) = \left[\frac{\delta l}{\delta \widehat{\boldsymbol{\Omega}}}, \widehat{\boldsymbol{\Omega}}\right], \tag{10.9}$$

$$\frac{\mathrm{d}}{\mathrm{d}t}\left(\frac{\delta l}{\delta \dot{\mathbf{A}}}\right) = \frac{\delta l}{\delta \mathbf{A}}, \tag{10.10}$$

$$\frac{\mathrm{d}}{\mathrm{d}t}\left(\frac{\delta l}{\delta \widehat{\boldsymbol{\Lambda}}}\right) = -\left[\frac{\delta l}{\delta \widehat{\boldsymbol{\Lambda}}}, \widehat{\boldsymbol{\Lambda}}\right]. \tag{10.11}$$

If $\mathfrak{so}(3)$ is identified with $\mathbb{R}^3$ via the 'hat' map, then the Euler–Poincaré equations take the form:

$$\frac{\mathrm{d}}{\mathrm{d}t}\frac{\delta l}{\delta \boldsymbol{\Omega}} = \frac{\delta l}{\delta \boldsymbol{\Omega}} \times \boldsymbol{\Omega}, \tag{10.12}$$

$$\frac{\mathrm{d}}{\mathrm{d}t}\frac{\delta l}{\delta \dot{\mathbf{A}}} = \frac{\delta l}{\delta \mathbf{A}}, \tag{10.13}$$

$$\frac{\mathrm{d}}{\mathrm{d}t}\frac{\delta l}{\delta \boldsymbol{\Lambda}} = -\frac{\delta l}{\delta \boldsymbol{\Lambda}} \times \boldsymbol{\Lambda}. \tag{10.14}$$

For free ellipsoidal motion expressed using the decomposition $\mathbf{Q} = \mathbf{RAS}$, one can check that

$$\dot{\mathrm{Q}} = \mathrm{R}(\boldsymbol{\Omega}\mathrm{A} + \dot{\mathrm{A}} + \mathrm{A}\boldsymbol{\Lambda})\mathrm{S},$$

(see Exercise 10.3). Thus the Lagrangian can be expressed as:

$$\begin{aligned}
L(\mathrm{Q}, \dot{\mathrm{Q}}) &= \frac{1}{2}\mathrm{tr}\left(\dot{\mathrm{Q}}\dot{\mathrm{Q}}^T\right) \\
&= \frac{1}{2}\mathrm{tr}\left(\left(\mathrm{R}(\widehat{\boldsymbol{\Omega}}\mathrm{A} + \dot{\mathrm{A}} + \mathrm{A}\widehat{\boldsymbol{\Lambda}})\mathrm{S}\right)\left(\mathrm{R}(\widehat{\boldsymbol{\Omega}}\mathrm{A} + \dot{\mathrm{A}} + \mathrm{A}\widehat{\boldsymbol{\Lambda}})\mathrm{S}\right)^T\right) \\
&= \frac{1}{2}\mathrm{tr}\left((\widehat{\boldsymbol{\Omega}}\mathrm{A} + \dot{\mathrm{A}} + \mathrm{A}\widehat{\boldsymbol{\Lambda}})(\widehat{\boldsymbol{\Omega}}\mathrm{A} + \dot{\mathrm{A}} + \mathrm{A}\widehat{\boldsymbol{\Lambda}})^T\right) \\
&= \frac{1}{2}\mathrm{tr}\Big((\widehat{\boldsymbol{\Omega}}\mathrm{A})(\widehat{\boldsymbol{\Omega}}\mathrm{A})^T + (\widehat{\boldsymbol{\Omega}}\mathrm{A})\dot{\mathrm{A}}^T + (\widehat{\boldsymbol{\Omega}}\mathrm{A})(\mathrm{A}\widehat{\boldsymbol{\Lambda}})^T + \dot{\mathrm{A}}(\widehat{\boldsymbol{\Omega}}\mathrm{A})^T + \dot{\mathrm{A}}\dot{\mathrm{A}}^T \\
&\quad + \dot{\mathrm{A}}(\mathrm{A}\widehat{\boldsymbol{\Lambda}})^T + (\mathrm{A}\widehat{\boldsymbol{\Lambda}})(\widehat{\boldsymbol{\Omega}}\mathrm{A})^T + (\mathrm{A}\widehat{\boldsymbol{\Lambda}})\dot{\mathrm{A}}^T + (\mathrm{A}\widehat{\boldsymbol{\Lambda}})(\mathrm{A}\widehat{\boldsymbol{\Lambda}})^T\Big) \\
&= \frac{1}{2}\mathrm{tr}\left[-\widehat{\boldsymbol{\Omega}}^2\mathrm{A}^2 - \widehat{\boldsymbol{\Lambda}}^2\mathrm{A}^2 - \underbrace{2\widehat{\boldsymbol{\Omega}}\mathrm{A}\widehat{\boldsymbol{\Lambda}}\mathrm{A}}_{\text{Coriolis coupling}} + \dot{\mathrm{A}}^2\right],
\end{aligned}$$

and so the reduced Lagrangian is

$$l\left(\widehat{\boldsymbol{\Omega}}, \mathbf{A}, \dot{\mathbf{A}}, \widehat{\boldsymbol{\Lambda}}\right) = \frac{1}{2}\mathrm{tr}\left[-\widehat{\boldsymbol{\Omega}}^2\mathbf{A}^2 - \widehat{\boldsymbol{\Lambda}}^2\mathbf{A}^2 - 2\widehat{\boldsymbol{\Omega}}\mathbf{A}\widehat{\boldsymbol{\Lambda}}\mathbf{A} + \dot{\mathbf{A}}^2\right].$$

A direct calculation shows:

$$\frac{\delta l}{\delta\widehat{\boldsymbol{\Omega}}} = \frac{1}{2}\left(\widehat{\boldsymbol{\Omega}}\mathbf{A}^2 + \mathbf{A}^2\widehat{\boldsymbol{\Omega}}\right) + \mathbf{A}\widehat{\boldsymbol{\Lambda}}\mathbf{A}, \tag{10.15}$$

$$\frac{\delta l}{\delta\mathbf{A}} = \left(\widehat{\boldsymbol{\Omega}}\mathbf{A}\widehat{\boldsymbol{\Lambda}} + \widehat{\boldsymbol{\Lambda}}\mathbf{A}\widehat{\boldsymbol{\Omega}}\right), \tag{10.16}$$

$$\frac{\delta l}{\delta\dot{\mathbf{A}}} = \dot{\mathbf{A}}, \tag{10.17}$$

$$\frac{\delta l}{\delta\widehat{\boldsymbol{\Lambda}}} = \frac{1}{2}\left(\widehat{\boldsymbol{\Lambda}}\mathbf{A}^2 + \mathbf{A}^2\widehat{\boldsymbol{\Lambda}}\right) + \mathbf{A}\widehat{\boldsymbol{\Omega}}\mathbf{A}, \tag{10.18}$$

which can be substituted in eqns (10.9)–(10.11) to give the equations of motion in matrix form.

Alternatively, the equations of motion can be expressed in terms of the components of $\widehat{\boldsymbol{\Omega}}, \widehat{\boldsymbol{\Lambda}}$ and $\mathbf{A}$. Let us denote the (diagonal) entries of $\mathbf{A}$ and $\dot{\mathbf{A}}$ by $(d_1, d_2, d_3)$ and $(\dot{d}_1, \dot{d}_2, \dot{d}_3)$, respectively, and let $\Omega_1, \Omega_2, \Omega_3$ and $\Lambda_1, \Lambda_2, \Lambda_3$ be as in the definition of the hat map. The Lagrangian $l$ becomes:

$$\begin{aligned} l = \frac{1}{2}\Big[&(\Omega_3 d_2 + d_1\Lambda_3)^2 + (\Omega_2 d_3 + d_1\Lambda_2)^2 + (\Omega_3 d_1 + d_2\Lambda_3)^2 \\ &+ (\Omega_1 d_3 + d_2\Lambda_1)^2 + (\Omega_2 d_1 + d_3\Lambda_2)^2 + (\Omega_1 d_2 + d_3\Lambda_1)^2 \\ &+ (\dot{d}_1^2 + \dot{d}_2^2 + \dot{d}_3^2)\Big]. \end{aligned}$$

Further, we have:

$$\frac{\partial l}{\partial\Omega_k} = (d_i^2 + d_j^2)\Omega_k + 2d_i d_j\Lambda_k, \tag{10.19}$$

$$\frac{\partial l}{\partial\dot{d}_k} = \dot{d}_k \tag{10.20}$$

$$\frac{\partial l}{\partial d_k} = d_k(\Omega_i^2 + \Omega_j^2 + \Lambda_i^2 + \Lambda_j^2) + 2d_i(\Omega_j\Lambda_j) + 2d_j(\Omega_i\Lambda_i) \tag{10.21}$$

$$\frac{\partial l}{\partial\Lambda_k} = (d_i^2 + d_j^2)\Lambda_k + 2d_i d_j\Omega_k, \tag{10.22}$$

with $(i, j, k)$ cyclic permutations of $(1, 2, 3)$. The relations above can be substituted in equations (10.12) – (10.14) to obtain the equations of motion in vector representation.

**Exercise 10.2**
Deduce Euler–Poincaré equations (10.9)–(10.11).

**Exercise 10.3**
Show that if $Q = RAS$, then

$$\dot{Q} = R(\widehat{\Omega}A + \dot{A} + A\widehat{\Lambda})S$$

where $\widehat{\Omega} := R^{-1}\dot{R}$ and $\widehat{\Lambda} := \dot{S}S^{-1}$.

## 10.3 Lie–Poisson reduction

In this section we deduce Lie–Poisson equations of motion for the free pseudo-rigid body with symmetric reference configuration by applying results from Chapter 9. Section 9.1 is especially relevant, in particular Exercises 9.3 and 9.4, which concern the definition of a reduced Hamiltonian and reduced Legendre transformation for product spaces.

Consider the dual space $\mathfrak{so}^*(3) \times T^*Diag^+(3) \times \mathfrak{so}^*(3)$ of the reduced space $\mathfrak{so}(3) \times TDiag^+(3) \times \mathfrak{so}(3)$, and let $\mathbf{M} \in \mathfrak{so}^*(3)$, $\mathbf{B} \in T^*_{\mathbf{A}}Diag^+(3)$ and $\mathbf{N} \in \mathfrak{so}^*(3)$ be the conjugate momenta corresponding to the coordinates $\widehat{\Omega}$, $\mathbf{A}$ and $\widehat{\Lambda}$, respectively. They are obtained by applying the reduced Legendre transformation:

$$M := \frac{\delta l}{\delta \widehat{\Omega}} = \frac{1}{2}\left(\widehat{\Omega}A^2 + A^2\widehat{\Omega}\right) + A\widehat{\Lambda}A, \tag{10.23}$$

$$B := \frac{\delta l}{\delta \dot{A}} = \dot{A}, \tag{10.24}$$

$$N := \frac{\delta l}{\delta \widehat{\Lambda}} = \frac{1}{2}\left(\widehat{\Lambda}A^2 + A^2\widehat{\Lambda}\right) + A\widehat{\Omega}A, \tag{10.25}$$

where we used relations (10.15), (10.17), and (10.18). Inverting the above we obtain

$$\Omega_k = \frac{M_k}{(d_i - d_j)^2} - \frac{2d_i d_j}{(d_i - d_j)^2(d_i + d_j)^2}(M_k + N_k),$$

$$B_k = \dot{a}_k,$$

$$\Lambda_k = \frac{N_k}{(d_i - d_j)^2} - \frac{2d_i d_j}{(d_i - d_j)^2 (d_i + d_j)^2}(M_k + N_k),$$

with $(i, j, k)$ cyclic permutations of $(1, 2, 3)$. With

$$\widehat{\mathbf{\Omega}} = \widehat{\mathbf{\Omega}}(\mathbf{M}, \mathbf{A}, \mathbf{N}),$$
$$\dot{\mathbf{A}} = \mathbf{B},$$
$$\widehat{\mathbf{\Lambda}} = \widehat{\mathbf{\Lambda}}(\mathbf{M}, \mathbf{A}, \mathbf{N}),$$

determined above, the reduced Hamiltonian is:

$$h : \mathfrak{so}^*(3) \times T^*Diag^+(3) \times \mathfrak{so}^*(3) \to \mathbb{R}^3,$$

$$h(\mathbf{M}, \mathbf{A}, \mathbf{B}, \mathbf{N}) = \left\langle \mathbf{M}, \widehat{\mathbf{\Omega}} \right\rangle + \left\langle \mathbf{B}, \dot{\mathbf{A}} \right\rangle + \left\langle \mathbf{N}, \widehat{\mathbf{\Lambda}} \right\rangle - l\left(\widehat{\mathbf{\Omega}}, \mathbf{A}, \dot{\mathbf{A}}, \widehat{\mathbf{\Lambda}}\right). \qquad (10.26)$$

The equations of motion corresponding to eqns (10.9) – (10.11) are

$$\dot{\mathbf{M}} = [\mathbf{M}, \widehat{\mathbf{\Omega}}], \qquad (10.27)$$

$$\dot{\mathbf{A}} = \mathbf{B}, \ \ \dot{\mathbf{B}} = \frac{\delta h}{\delta \mathbf{A}}, \qquad (10.28)$$

$$\dot{\mathbf{N}} = -[\mathbf{N}, \widehat{\mathbf{\Lambda}}], \qquad (10.29)$$

where we used that for matrix dual algebras $\mathrm{ad}^*{}_{\mathbf{U}}\mathbf{V} = [\mathbf{V}, \mathbf{U}]$ (see formula (6.6)).

By Lie–Poisson reduction (see Theorems 9.10 and 9.16, Proposition 6.66 and Corollary 9.15), the reduced Poisson bracket on $\mathfrak{so}^*(3) \times T^*Diag^+(3) \times \mathfrak{so}^*(3)$ is:

$$\{F, K\} = \{F, K\}^{left}_{so^*(3)} + \{F, K\}_{T^*\mathbf{R}^3} + \{F, K\}^{right}_{so^*(3)} \qquad (10.30)$$

$$= -\left\langle \mathbf{M}, \left[\frac{\delta F}{\delta \mathbf{M}}, \frac{\delta K}{\delta \mathbf{M}}\right]\right\rangle + \left(\frac{\delta F}{\delta \mathbf{A}}\frac{\delta K}{\delta \mathbf{B}} - \frac{\delta F}{\delta \mathbf{B}}\frac{\delta F}{\delta \mathbf{A}}\right) + \left\langle \mathbf{N}, \left[\frac{\delta F}{\delta \mathbf{N}}, \frac{\delta K}{\delta \mathbf{N}}\right]\right\rangle, \qquad (10.31)$$

and the equations of motion (10.27) – (10.29) are the Hamiltonian equations corresponding to the reduced Hamiltonian $h$ and the the reduced Poisson bracket.

**Exercise 10.4**
Write an essay containing all theoretical details skipped in the presentation. Explain how the story changes when the coefficient of inertia matrix $\mathbb{J}$ is diagonal, but is not a multiple of the identity.

**Exercise 10.5**
Compute the reduced Hamiltonian (10.26) explicitly and verify that

$$\frac{\delta h}{\delta \mathbf{M}} = \widehat{\boldsymbol{\Omega}}(\mathbf{M}, \mathbf{A}, \mathbf{N}), \qquad \text{and} \qquad \frac{\delta h}{\delta \mathbf{N}} = \widehat{\boldsymbol{\Lambda}}(\mathbf{M}, \mathbf{A}, \mathbf{N}).$$

(A computer algebra system is helpful here!) This confirms that the equations of motion (10.27) – (10.29) are the Hamiltonian equations corresponding to the reduced Hamiltonian $h$ and the the reduced Poisson bracket in eqn (10.30).

## 10.4 Momentum maps: angular momentum and circulation

The Lagrangian for free ellipsoidal motion on $GL^+(3)$, given in eqn (10.3), is

$$L = \frac{1}{2}\,\mathrm{tr}(\dot{\mathbf{Q}}\dot{\mathbf{Q}}^T).$$

The Legendre transform is $(\mathbf{Q}, \dot{\mathbf{Q}}) \mapsto (\mathbf{Q}, \mathbf{P})$, where

$$\mathbf{P} = \frac{\delta L}{\delta \dot{\mathbf{Q}}} = \dot{\mathbf{Q}}.$$

The corresponding Hamiltonian is $H(\mathbf{Q}, \mathbf{P}) = \frac{1}{2}\,\mathrm{tr}(\mathbf{P}\dot{\mathbf{P}}^T)$. Since $L$ is invariant under the tangent lifts of the left and right translation actions of $SO(3)$ on $GL^+(3)$, as shown in eqn (10.4), it follows that $H$ is invariant under the cotangent lifts of these actions, which are given by

$$\begin{aligned} &\text{(left translation)} && \mathbf{g}(\mathbf{Q}, \mathbf{P}) = (\mathbf{gQ}, \mathbf{gP}), \\ &\text{(right translation)} && \mathbf{h}(\mathbf{Q}, \mathbf{P}) = (\mathbf{Qh}, \mathbf{Ph}). \end{aligned}$$

As shown in Example 8.10, the momentum maps for these actions are

$$\text{(left translation)} \qquad \mathbf{J}_L(\mathbf{Q},\mathbf{P}) = \frac{1}{2}\left(\mathbf{P}\mathbf{Q}^T - \mathbf{Q}\mathbf{P}^T\right),$$

$$\text{(right translation)} \qquad \mathbf{J}_R(\mathbf{Q},\mathbf{P}) = \frac{1}{2}\left(\mathbf{Q}^T\mathbf{P} - \mathbf{P}^T\mathbf{Q}\right). \tag{10.32}$$

(There is an implicit identification of $\mathbf{J}_L$ and $\mathbf{J}_R$, which are elements of $\mathfrak{so}(3)^*$, with elements of $\mathfrak{so}(3)$, via the trace pairing.) By substituting the formulae for **M** and **N** in eqn (10.23) into eqn (10.32), and using Exercise 10.3, we find:

$$\begin{aligned}
\mathbf{J}_L &= \frac{1}{2}\left(\mathbf{P}\mathbf{Q}^T - \mathbf{Q}\mathbf{P}^T\right) = \frac{1}{2}\left(\dot{\mathbf{Q}}\mathbf{Q}^T - \mathbf{Q}\dot{\mathbf{Q}}^T\right) \\
&= \frac{1}{2}\left(\mathbf{R}(\widehat{\boldsymbol{\Omega}}\mathbf{A} + \dot{\mathbf{A}} + \mathbf{A}\widehat{\boldsymbol{\Lambda}})\mathbf{A}\mathbf{R}^T - \mathbf{R}\mathbf{A}(-\mathbf{A}\widehat{\boldsymbol{\Omega}} + \dot{\mathbf{A}} - \widehat{\boldsymbol{\Lambda}}\mathbf{A})\mathbf{R}^T\right) \\
&= \frac{1}{2}\mathbf{R}\left(\widehat{\boldsymbol{\Omega}}\mathbf{A}^2 + \mathbf{A}^2\widehat{\boldsymbol{\Omega}} + 2\mathbf{A}\widehat{\boldsymbol{\Lambda}}\mathbf{A}\right)\mathbf{R}^T = \mathbf{R}\mathbf{M}\mathbf{R}^T.
\end{aligned}$$

Similarly, one can check that

$$\mathbf{J}_R = \mathbf{S}^T\mathbf{N}\mathbf{S}. \tag{10.33}$$

An alternative method to obtain these expressions for $\mathbf{J}_L$ and $\mathbf{J}_R$ in terms of $\mathbf{R}, \mathbf{M}, \mathbf{S}, \mathbf{N}$, is to begin with the expression for $L$ in terms of the decomposition $\mathbf{Q} = \mathbf{R}\mathbf{A}\mathbf{S}$. The Legendre transform is

$$(\mathbf{R}, \mathbf{A}, \mathbf{S}, \dot{\mathbf{R}}, \dot{\mathbf{A}}, \dot{\mathbf{S}}) \to (\mathbf{R}, \mathbf{A}, \mathbf{S}, \mathbf{P}_\mathbf{R}, \mathbf{P}_\mathbf{A}, \mathbf{P}_\mathbf{S})$$

where, using Remark 9.1,

$$\begin{aligned}
\mathbf{P}_\mathbf{R} &= \frac{\delta L}{\delta \dot{\mathbf{R}}} = \mathbf{R}\frac{\delta l}{\delta \widehat{\boldsymbol{\Omega}}} = \mathbf{R}\mathbf{M}, \\
\mathbf{P}_\mathbf{A} &= \frac{\delta L}{\delta \dot{\mathbf{A}}} = \mathbf{R}\frac{\delta l}{\delta \dot{\mathbf{A}}}, \\
\mathbf{P}_\mathbf{S} &= \frac{\delta L}{\delta \dot{\mathbf{S}}} = \frac{\delta l}{\delta \widehat{\boldsymbol{\Lambda}}}\mathbf{S} = \mathbf{N}\mathbf{S}.
\end{aligned}$$

The corresponding Hamiltonian is invariant under the following cotangent-lifted actions

$$\text{(left translation)} \qquad \mathbf{g}(\mathbf{R}, \mathbf{A}, \mathbf{S}, \mathbf{P}_\mathbf{R}, \mathbf{P}_\mathbf{A}, \mathbf{P}_\mathbf{S}) = (\mathbf{g}\mathbf{R}, \mathbf{A}, \mathbf{S}, \mathbf{g}\mathbf{P}_\mathbf{R}, \mathbf{P}_\mathbf{A}, \mathbf{P}_\mathbf{S}),$$

$$\text{(right translation)} \qquad \mathbf{h}(\mathbf{R}, \mathbf{A}, \mathbf{S}, \mathbf{P}_\mathbf{R}, \mathbf{P}_\mathbf{A}, \mathbf{P}_\mathbf{S}) = (\mathbf{R}, \mathbf{A}, \mathbf{S}\mathbf{h}, \mathbf{P}_\mathbf{R}, \mathbf{P}_\mathbf{A}, \mathbf{P}_\mathbf{S}\mathbf{h}).$$

As shown in Example 8.10, the momentum maps for these actions are

$$\text{(left translation)} \qquad \mathbf{J}_L = \mathbf{P}_\mathbf{R}\mathbf{R}^T = \mathbf{R}\mathbf{M}\mathbf{R}^T,$$

$$\text{(right translation)} \qquad \mathbf{J}_R = \mathbf{S}^T\mathbf{P}_\mathbf{S} = \mathbf{S}^T\mathbf{N}\mathbf{S}.$$

**Proposition 10.5** *The momentum maps* $\mathbf{J}_L$ *and* $\mathbf{J}_R$ *are conserved quantities.*

*Proof* This is immediate because the Hamiltonian $H(\mathbf{Q},\mathbf{P}) = \frac{1}{2}\operatorname{tr}(\mathbf{P}\dot{\mathbf{P}}^T)$ is both left- and right-invariant under the cotangent lift of $SO(3)$ action. ■

**Remark 10.6** *The discrete symmetry* $\mathbf{Q} \mapsto \mathbf{Q}^T$ *(Dedekind duality) reverses the roles and senses of Eulerian and Lagrangian rotations, and interchanges the angular momentum and the circulation,* $\mathbf{J}_L \leftrightarrow \mathbf{J}_R$.

## Spatial angular momentum for the pseudo-rigid body

We saw in Remark 8.11 that, for the $SO(3)$ action on the rigid body, $\mathbf{J}_L$ corresponds to the spatial angular momentum vector $\boldsymbol{\pi}$, via the "breve" map (defined in Example 5.36). The same arguments apply in the case of the pseudo-rigid body. An alternative argument begins with original definition of angular momentum in eqn (1.41),

$$\boldsymbol{\pi} := \int_{\mathcal{B}} \rho(\mathbf{X})\, \mathbf{x} \times \dot{\mathbf{x}}\, \mathrm{d}^3\mathbf{X},$$

and proceeds by direct computation. Recall that $\breve{\boldsymbol{\pi}}$ can be identified with $\frac{1}{2}\hat{\boldsymbol{\pi}}$ via the trace pairing. The following identity can be verified directly,

$$\widehat{\mathbf{x} \times \dot{\mathbf{x}}} = \dot{\mathbf{x}}\mathbf{x}^T - \mathbf{x}\dot{\mathbf{x}}^T.$$

Thus,

$$\begin{aligned}
\breve{\boldsymbol{\pi}} \leftrightarrow \frac{1}{2}\hat{\boldsymbol{\pi}} &= \frac{1}{2}\int_{\mathcal{B}(t)} \rho(\mathbf{x})\widehat{\mathbf{x} \times \dot{\mathbf{x}}}\, \mathrm{d}^3x \\
&= \frac{1}{2}\int_{\mathcal{B}(t)} \rho(\mathbf{x})\Big(\dot{\mathbf{x}}\mathbf{x}^T - \mathbf{x}\dot{\mathbf{x}}^T\Big)\, \mathrm{d}^3x \\
&= \frac{1}{2}\int_{\mathcal{B}} \rho(\mathbf{X})\Big((\dot{\mathbf{Q}}\mathbf{X})(\mathbf{Q}\mathbf{X})^T - (\mathbf{Q}\mathbf{X})(\dot{\mathbf{Q}}\mathbf{X})^T\Big)\, \mathrm{d}^3X \\
&= \frac{1}{2}\Big((\mathbb{J}\dot{\mathbf{Q}})\mathbf{Q}^T - \mathbf{Q}(\mathbb{J}\dot{\mathbf{Q}})^T\Big) \\
&= \frac{1}{2}\Big(\mathbf{P}\mathbf{Q}^T - \mathbf{Q}\mathbf{P}^T\Big) \leftrightarrow \mathbf{J}_L,
\end{aligned}$$

where "$\leftrightarrow$" indicates correspondence under the trace pairing. Thus $\breve{\pi} = \mathbf{J}_L$, and conservation of $\mathbf{J}_L$ is just conservation of spatial angular momentum in yet another guise.

**Remark 10.7** *In the absence of external torques, left-invariance implies that conservation of spatial angular momentum holds for any initial mass distribution, not just for the spherically symmetric reference configuration.*

## Circulation for pseudo-rigid motion

The other momentum map, $\mathbf{J}_R$ for right translations in eqn (10.32), is related to *circulation*, which is defined for pseudo-rigid motion as follows. Consider a smooth closed loop $\gamma$ in the reference coordinates of the pseudo-rigid body. Parametrize the loop $\gamma$ by $s \in \mathbb{R}$ and at time $t$ define $c(t) = \mathbf{Q}(t) \circ \gamma$, so that the curve in space $c(t)$ represents the evolution under pseudo-rigid motion of the reference loop $\gamma$. In this notation, the circulation is defined, as follows.

**Definition 10.8** *The* ***circulation*** *around the loop c at time t is the integral:*

$$I_{c(t)} := \oint_{c(t)} \dot{\mathbf{x}}(t, \mathbf{x}) \cdot \mathrm{d}\mathbf{x} := \oint_{\gamma} \dot{\mathbf{x}}(t, s) \cdot \frac{\mathrm{d}\mathbf{x}}{\mathrm{d}s} \, \mathrm{d}s. \tag{10.34}$$

**Remark 10.9** *The circulation (10.34) around a path summarizes the degree to which the velocity $\dot{\mathbf{x}}(t)$ at time t projects along the direction of* $\mathrm{d}\mathbf{x}/\mathrm{d}s$ *tangent to the loop $\gamma$; see Figure 10.1. If the velocity field is the gradient of a smooth function with respect to the spatial coordinate, then the circulation around any closed loop vanishes identically.*

The smooth closed loop $\gamma$ in body coordinates determines a family of smooth closed loops $c(t) := \mathbf{Q}(t) \circ \gamma$ parameterized by time $t$, which may be regarded as the path of the loop $\gamma$ carried along in the space of smooth closed loops by the pseudo-rigid flow. Since $\dot{\mathbf{x}}(t, X) = \dot{\mathbf{Q}}(t)\mathbf{X}$, the circulation

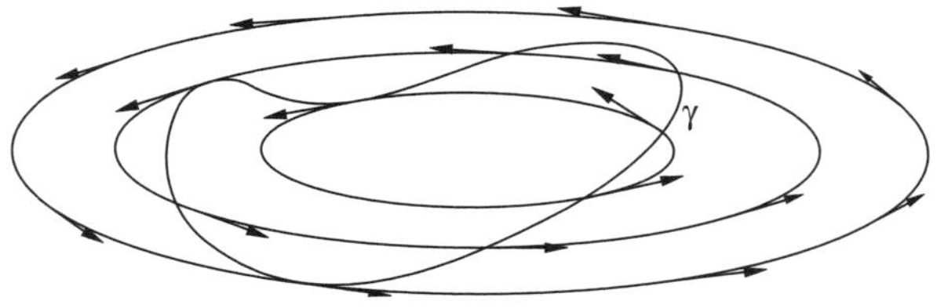


**Figure 10.1** The circulation of a flow around a path.

around such a loop $c(t)$ is defined as

$$I_{c(t)} = \oint_c \dot{\mathbf{x}} \cdot d\mathbf{x} = \oint_\gamma \left(\dot{\mathrm{Q}}\mathrm{X}\right) \cdot (\mathrm{Q}\, d\mathrm{X}) = \oint_\gamma \left(\mathrm{Q}^T\dot{\mathrm{Q}}\mathrm{X}\right) \cdot d\mathrm{X}. \qquad (10.35)$$

Following [RdSD99], we transform this integral by using an elementary identity, whose proof is left to Exercise 10.6.

**Lemma 10.10** *If* $\mathrm{L}^A = \frac{1}{2}(\mathrm{L} - \mathrm{L}^T)$, *then* $\oint_\gamma \mathrm{L}\mathrm{X} \cdot d\mathrm{X} = \oint_\gamma \mathrm{L}^A\mathrm{X} \cdot d\mathrm{X}$.

It follows that the circulation in eqn (10.35) around the loop $c(t)$ equals

$$I_{c(t)} = \frac{1}{2}\oint_\gamma \left(\left(\mathrm{Q}^T\dot{\mathrm{Q}} - \mathrm{Q}^T\dot{\mathrm{Q}}\right)\mathrm{X}\right) \cdot d\mathrm{X}.$$

The skew-symmetric $3 \times 3$ matrix in this circulation integral,

$$\widehat{\mathrm{K}} := \frac{1}{2}\left(\mathrm{Q}^T\dot{\mathrm{Q}} - \dot{\mathrm{Q}}^T\mathrm{Q}\right), \qquad (10.36)$$

defines, via the hat map, a vector $\mathbf{K}$ called the ***Kelvin vector.***

According to Lemma 10.10, the circulation around the loop $c$ may be expressed in factored form as

$$I_c = \oint_\gamma \left(\widehat{\mathrm{K}}\mathrm{X}\right) \cdot d\mathrm{X} = \oint_\gamma (\mathrm{K} \times \mathrm{X}) \cdot d\mathrm{X} = \mathrm{K} \cdot \left(\oint_\gamma \mathrm{X} \times d\mathrm{X}\right). \qquad (10.37)$$

The factor in parentheses in the last line depends only on the reference curve $\gamma$ and not on the pseudo-rigid motion. Thus, by eqn (10.37), the evolution of the circulation $I_c(t)$ is determined by the evolution of the Kelvin vector $\mathrm{K}$ defined by $\widehat{\mathrm{K}}$ in (10.36). It so happens that, for a pseudo-rigid body with a spherically symmetric reference configuration, $\widehat{\mathrm{K}}$ equals the momentum map $\mathbf{J}_R$ (more precisely, they correspond via the trace pairing), and $\mathbf{J}_R$ is a conserved quantity; see Proposition 10.5. Consequently, $\mathrm{K}$ is constant and therefore $I_c$ is also constant. This calculation proves the following.

**Proposition 10.11 (Kelvin circulation theorem: pseudo-rigid bodies)** *Pseudo-rigid motion starting initially from spherical symmetry conserves the circulation (10.34).*

**Remark 10.12** *The Kelvin circulation theorem can be seen as a special case of the Kelvin–Noether theorem, in particular the simplified version in Corollary 7.19, in which there are no advected quantities. Indeed, take* $G = SO(3)$, *acting on the right on* $GL(3)^+$. *Fix a closed loop* $\gamma$ *and let* $C$ *be* *the space of all closed loops of the form* $c = \mathbf{S} \circ \gamma$ *for some* $\mathbf{S} \in SO(3)$, *with left* $SO(3)$ *action given by* $\mathbf{S}c = \mathbf{S} \circ c$. *Define* $\mathcal{K} : \mathcal{C} \to \mathfrak{so}(3)^{**}$ *by setting*

$$\langle \mathcal{K}(\gamma), \mathbf{N} \rangle = \oint_\gamma (\mathbf{N}\mathbf{X}) \cdot \mathrm{d}\mathbf{X}$$

*and requiring that* $\mathcal{K}$ *be* $SO(3)$*-equivariant:*

$$\begin{aligned}\langle \mathcal{K}(\mathbf{S} \circ \gamma), \mathbf{N} \rangle &= \left\langle \mathbf{S}\, \mathcal{K}(\gamma)\, \mathbf{S}^T, \mathbf{N} \right\rangle \\ &= \left\langle \mathcal{K}(\gamma), \mathbf{S}^T \mathbf{N} \mathbf{S} \right\rangle \\ &= \oint_\gamma \left( \mathbf{S}^T \mathbf{N} \mathbf{S} \mathbf{X} \right) \cdot \mathrm{d}\mathbf{X}.\end{aligned}$$

*The associated* ***Kelvin-Noether*** *quantity* $I : \mathcal{C} \times \mathfrak{g} \to \mathbb{R}$ *is given by*

$$\begin{aligned}I(c, \widehat{\boldsymbol{\Lambda}}) = I(\mathbf{S} \circ \gamma, \widehat{\boldsymbol{\Lambda}})) &= \left\langle \mathcal{K}(\mathbf{S} \circ \gamma), \frac{\delta l}{\delta \widehat{\boldsymbol{\Lambda}}}(\widehat{\boldsymbol{\Lambda}}) \right\rangle \\ &= \oint_\gamma \left( \mathbf{S}^T \frac{\delta l}{\delta \widehat{\boldsymbol{\Lambda}}}(\widehat{\boldsymbol{\Lambda}}) \mathbf{S}\, \mathbf{X} \right) \cdot \mathrm{d}\mathbf{X} \\ &= \oint_\gamma (\mathbf{J}_R \mathbf{X}) \cdot \mathrm{d}\mathbf{X},\end{aligned}$$

*using the formula for* $\mathbf{J}_R$ *in eqn (10.33). Corollary 7.19 implies that* $I$ *is a conserved quantity. As noted earlier,* $\mathbf{J}_R = \widehat{\mathbf{K}}$ *when the reference configuration is spherically symmetric, so comparison with eqn (10.37) shows that* $I_c$ *is a conserved quantity.*

**Remark 10.13**

- *Geometrically, the quantity in parentheses in the factored form of the circulation* $I_c$ *in eqn (10.37) is given by*

$$\oint_\gamma \mathbf{X} \times \mathrm{d}\mathbf{X} = 2 \iint_{S(\gamma)} \mathrm{d}\mathbf{S}^\gamma, \tag{10.38}$$

  *where* $S(\gamma)$ *is any surface in* ***Stokes' theorem*** *whose boundary is the curve* $\gamma = \partial S(\gamma)$ *in the fluid reference coordinates and*

$$\mathrm{d}\mathbf{S}_i^\gamma = \frac{1}{2} \epsilon_{ijk} \mathrm{d}X^j \wedge \mathrm{d}X^k$$

*is its surface element. In components,* $\mathrm{d}S_1^\gamma = \mathrm{d}X^2 \wedge \mathrm{d}X^3$ *holds for cyclic permutations of the integers* $\{1, 2, 3\}$. *Thus, the circulation* $I_c$ *around the loop* $c(t) = \mathbf{Q}(t) \circ \gamma$ *is equal to twice the projection of the Kelvin vector* $\mathbf{K}$ *onto the 'net normal vector' of a surface bounded by the circulation loop* $\gamma$ *in the reference configuration. For example, for a circular loop, this surface may be taken as the disk bounded by the loop.*

- *The Stokes theorem allows the circulation* $I_c$ *defined in eqn (10.34) to be rewritten as*

$$I_{c(t)} = \iint_{S(c)} \operatorname{curl} \dot{\mathbf{x}} \cdot \mathrm{d}\mathbf{S}^c. \tag{10.39}$$

*Substituting* $\dot{\mathbf{x}} = \dot{\mathbf{Q}}\mathbf{Q}^{-1}\mathbf{x}$ *into (10.39) and transforming to body coordinates recovers the previous expressions in eqns (10.37 and (10.38) as*

$$I_{c(t)} = \left(\mathbf{Q}^T\dot{\mathbf{Q}} - \dot{\mathbf{Q}}^T\mathbf{Q}\right) \iint_{S(\gamma)} \mathrm{d}\mathbf{S}^\gamma = 2\mathbf{K} \cdot \iint_{S(\gamma)} d\mathbf{S}^\gamma .$$

- *Kelvin's circulation theorem for Euler fluid motion will be discussed in Part II. Specifically, Chapter 17 of Part II will use Theorem 7.18 (Kelvin–Noether theorem) to show that the circulation integral* $I_c$ *is conserved by all solutions of Euler's equations for any initial loop* $\gamma$. *Thus, conservation of circulation* $I_c$ *is a general fact, by no means limited only to pseudo-rigid motion for ellipsoidal bodies that are initially spherically symmetric.*

**Exercise 10.6**
Prove Lemma 10.10.

**Exercise 10.7 (Elliptical motions with potential energy on $GL(2)$)**
Compute the Euler–Poincaré–Lagrange equations, for elliptical motion in the plane, with

$$L = T(\boldsymbol{\Omega}, \boldsymbol{\Lambda}, \boldsymbol{D}, \dot{\boldsymbol{D}}) - V(\boldsymbol{D})$$

for some potential $V(\boldsymbol{D})$. Show that, if

$$V(\boldsymbol{D}) = V\left(\operatorname{tr}\boldsymbol{D}^2, \det(\boldsymbol{D})\right),$$

the equations become homogeneous in $r^2(t)$, and can be reduced to separated Newtonian forms,

$$\frac{d^2 r^2}{dt^2} = -\frac{dV(r)}{dr^2}, \qquad \frac{d^2\alpha}{dt^2} = -\frac{dW(\alpha)}{d\alpha},$$

for $r^2 = x^2 + y^2$ and $\alpha = \tan^{-1}(y/x)$ with $x(t)$ and $y(t)$ in two dimensions.

**Exercise 10.8 (Ellipsoidal motions with potential energy on $GL(3)$)**
Choose the Lagrangian in 3D,

$$L = \frac{1}{2}\mathrm{tr}\Big(\dot{Q}^T\dot{Q}\Big) - V\Big(\mathrm{tr}\,(Q^T\,Q), \det(Q)\Big),$$

where $Q(t) \in GL(3)$ is a $3 \times 3$ matrix function of time and the potential energy $V$ is an arbitrary function of $\mathrm{tr}\,(Q^T\,Q)$ and $\det(Q)$.

1. Legendre transform this Lagrangian. That is, find the momenta $P_{ij}$ canonically conjugate to $Q_{ij}$, construct the Hamiltonian $H(Q, P)$ and write Hamilton's canonical equations of motion for this problem.
2. Show that the Hamiltonian is invariant under $Q \to O\,Q$ where $O \in SO(3)$. Construct the cotangent lift and the momentum map of this action.
3. Construct another distinct action of $SO(3)$ on this system that also leaves its Hamiltonian $H(Q, P)$ invariant. Calculate its momentum map. Do the two momentum maps Poisson commute? Why?

**Exercise 10.9 ($GL(n,\mathbb{R})$-invariant motions)**
Begin with the Lagrangian

$$L = \frac{1}{2}\,\mathrm{tr}\Big(\dot{S}S^{-1}\dot{S}S^{-1}\Big) + \frac{1}{2}\dot{\mathrm{Q}}^T S^{-1}\dot{\mathrm{Q}},$$

where $S$ is an $n \times n$ symmetric matrix and $\mathbf{q} \in \mathbb{R}^n$ is an $n$ component column vector.

1. Legendre transform to construct the corresponding Hamiltonian and canonical equations.

2. Show that the system is invariant under the group action

$$\mathbf{q} \to A\mathbf{q} \quad \text{and} \quad S \to ASA^T$$

for any constant invertible $n \times n$ matrix, $A$.
3. Compute the infinitesimal generator for this group action and construct its corresponding momentum map. Is this momentum map equivariant?
4. Verify directly that this momentum map is a conserved $n \times n$ matrix quantity by using the equations of motion.
5. Is this system completely integrable for any value of $n > 2$?

## Solutions to selected exercises

**Solution to Exercise 10.1** Consider the decomposition $\boldsymbol{Q} = \boldsymbol{RAS} \in GL^+(n, \mathbb{R})$ with $\boldsymbol{R}, \boldsymbol{S} \in O(n)$ and $\boldsymbol{A} \in Diag^+(3)$. Since $\det \boldsymbol{Q} > 0$ and $\det \boldsymbol{A} > 0$, either $\det \boldsymbol{R} = \det \boldsymbol{S} = 1$ or $\det \boldsymbol{R} = \det \boldsymbol{S} = -1$. The first case requires $\boldsymbol{R}, \boldsymbol{S} \in SO(3)$ as was to be shown. For the second case, let

$$\boldsymbol{J} = \begin{bmatrix} -1 & 0 & 0 \\ 0 & 1 & 0 \\ 0 & 0 & 1 \end{bmatrix}.$$

Then $\boldsymbol{Q} = \boldsymbol{R}'(\boldsymbol{JAJ})\boldsymbol{S}'$ where $\boldsymbol{R}' = \boldsymbol{RJ} \in SO(3)$ and $\boldsymbol{S}' = \boldsymbol{SJ} \in SO(3)$. Also $\boldsymbol{JAJ} \in Diag^+(3)$. Therefore, the decomposition transforms to

$$\boldsymbol{Q} = \boldsymbol{R}'\boldsymbol{A}'\boldsymbol{S}'.$$

such that $\boldsymbol{R}', \boldsymbol{S}' \in SO(3)$ and $\boldsymbol{A}' \in Diag^+(3)$.

**Solution to Exercise 10.2** The variations are given by $\delta\widehat{\boldsymbol{\Omega}} = \dot{\widehat{\boldsymbol{\Sigma}}} + [\widehat{\boldsymbol{\Omega}}, \widehat{\boldsymbol{\Sigma}}]$, $\delta\widehat{\Lambda} = \dot{\widehat{\Xi}} - [\widehat{\boldsymbol{\Lambda}}, \widehat{\Xi}]$ and $\delta \boldsymbol{A}$. Therefore the terms proportional to $\widehat{\boldsymbol{\Sigma}}$ in the variational principle are

$$\left\langle \frac{\delta l}{\delta \widehat{\boldsymbol{\Omega}}}, \dot{\widehat{\boldsymbol{\Sigma}}} + [\widehat{\boldsymbol{\Omega}}, \widehat{\boldsymbol{\Sigma}}] \right\rangle = \left\langle -\frac{\mathrm{d}}{\mathrm{d}t}\frac{\delta l}{\delta \widehat{\boldsymbol{\Omega}}} + \left[\frac{\delta l}{\delta \widehat{\boldsymbol{\Omega}}}, \widehat{\boldsymbol{\Omega}}\right], \widehat{\boldsymbol{\Sigma}} \right\rangle.$$

Meanwhile, the terms proportional to $\delta \boldsymbol{A}$ are

$$\left\langle \frac{\delta l}{\delta \dot{\boldsymbol{A}}}, \delta \dot{\boldsymbol{A}} \right\rangle + \left\langle \frac{\delta l}{\delta \boldsymbol{A}}, \delta \boldsymbol{A} \right\rangle = \left\langle -\frac{\mathrm{d}}{\mathrm{d}t} \frac{\delta l}{\delta \dot{\boldsymbol{A}}}, \delta \boldsymbol{A} \right\rangle.$$

Finally, the terms proportional to $\widehat{\boldsymbol{\Xi}}$ are

$$\left\langle \frac{\delta l}{\delta \widehat{\boldsymbol{\Lambda}}}, \dot{\widehat{\boldsymbol{\Xi}}} - [\widehat{\boldsymbol{\Lambda}}, \widehat{\boldsymbol{\Xi}}] \right\rangle = \left\langle -\frac{\mathrm{d}}{\mathrm{d}t} \frac{\delta l}{\delta \widehat{\boldsymbol{\Lambda}}} - \left[ \frac{\delta l}{\delta \widehat{\boldsymbol{\Lambda}}}, \widehat{\boldsymbol{\Lambda}} \right], \widehat{\boldsymbol{\Xi}} \right\rangle.$$

Combining these terms yields the following set of equations.

$$\frac{\mathrm{d}}{\mathrm{d}t} \frac{\delta l}{\delta \widehat{\boldsymbol{\Omega}}} = \left[ \frac{\delta l}{\delta \widehat{\boldsymbol{\Omega}}}, \widehat{\boldsymbol{\Omega}} \right],$$

$$\frac{\mathrm{d}}{\mathrm{d}t} \frac{\delta l}{\delta \dot{\boldsymbol{A}}} = \frac{\delta l}{\delta \boldsymbol{A}},$$

$$\frac{\mathrm{d}}{\mathrm{d}t} \frac{\delta l}{\delta \widehat{\boldsymbol{\Lambda}}} = -\left[ \frac{\delta l}{\delta \widehat{\boldsymbol{\Lambda}}}, \widehat{\boldsymbol{\Lambda}} \right].$$

**Solution to Exercise 10.6** Consider a symmetric matrix

$$\boldsymbol{S} = \begin{bmatrix} a & q & r \\ q & b & s \\ r & s & c \end{bmatrix}.$$

In order to evaluate the integral

$$\oint_{\gamma} \boldsymbol{S X} \cdot \mathrm{d}\boldsymbol{X},$$

note that

$$\begin{aligned} \boldsymbol{S X} \cdot d\boldsymbol{X} &= \begin{bmatrix} ax + qy + rz \\ qx + by + sz \\ rx + sy + cz \end{bmatrix} \cdot \begin{bmatrix} dx \\ dy \\ dz \end{bmatrix}, \\ &= (ax + qy + rz)\, dx + (qx + by + sz)\, dy + (rx + sy + cz)\, dz, \\ &= d\left( \frac{a}{2}x^2 + \frac{b}{2}y^2 + \frac{c}{2}z^2 \right) + d\,(qxy + rzx + syz)\,. \end{aligned}$$

Therefore,

$$\oint_{\gamma} \boldsymbol{SX} \cdot \mathrm{d}\boldsymbol{X} = \left[\frac{a}{2}x^2 + \frac{b}{2}y^2 + \frac{c}{2}z^2 + qxy + rzx + syz\right]_{\gamma},$$
$$= 0.$$

**Solution to Exercise 10.8** The Lagrangian is

$$L = \frac{1}{2}\mathrm{tr}\left(\dot{\boldsymbol{Q}}^T \dot{\boldsymbol{Q}}\right) - V\left(\mathrm{tr}\left(\boldsymbol{Q}^T \boldsymbol{Q}\right), \det(\boldsymbol{Q})\right).$$

1. To perform the Legendre transform, take the fiber derivative,

$$\boldsymbol{P} = \frac{\partial L}{\partial \dot{\boldsymbol{Q}}},$$
$$= \dot{\boldsymbol{Q}}.$$

   Therefore the corresponding Hamiltonian is

$$H(\boldsymbol{Q}, \boldsymbol{P}) = \frac{1}{2}\mathrm{tr}\left(\boldsymbol{P}^T \boldsymbol{P}\right) + V\left(\mathrm{tr}\left(\boldsymbol{Q}^T \boldsymbol{Q}\right), \det(\boldsymbol{Q})\right).$$

   Hamilton's equations are given by

$$\dot{\boldsymbol{P}} = -\frac{\partial H}{\partial \boldsymbol{Q}} = 2\frac{\partial V}{\partial \mathrm{tr}\left(\boldsymbol{Q}^T \boldsymbol{Q}\right)}\boldsymbol{Q} + \frac{\partial V}{\partial \det(\boldsymbol{Q})}\frac{\mathrm{d}\det(\boldsymbol{Q})}{\mathrm{d}\boldsymbol{Q}},$$

   and,

$$\dot{\boldsymbol{Q}} = \frac{\partial H}{\partial \boldsymbol{P}},$$
$$= \boldsymbol{P}.$$

   Note that the derivative of det was calculated in Example 5.10.

2. Observe that $(\boldsymbol{O}\boldsymbol{Q})^T \boldsymbol{O}\boldsymbol{Q} = \boldsymbol{Q}^T \boldsymbol{O}^T \boldsymbol{O}\boldsymbol{Q} = \boldsymbol{Q}^T \boldsymbol{Q}$. Also,

$$\det(\boldsymbol{O}\boldsymbol{Q}) = \det \boldsymbol{O} \det \boldsymbol{Q} = \det \boldsymbol{Q}.$$

   These two calculations show that $L$ is left invariant under $SO(3)$. The tangent lift of the action is given by $\dot{\boldsymbol{Q}} \to \boldsymbol{O}\dot{\boldsymbol{Q}}$, therefore the cotangent lift of the action is calculated as

$$\begin{aligned}\left\langle \boldsymbol{O}\cdot(\boldsymbol{Q},\boldsymbol{P})\,,\,(\boldsymbol{O}\,\boldsymbol{Q},\dot{\boldsymbol{Q}})\right\rangle &= \left\langle(\boldsymbol{Q},\boldsymbol{P})\,,\,\boldsymbol{O}^T\cdot(\boldsymbol{O}\,\boldsymbol{Q},\dot{\boldsymbol{Q}})\right\rangle,\\ &= \left\langle \boldsymbol{P}\,,\,\boldsymbol{O}^T\dot{\boldsymbol{Q}}\right\rangle,\\ &= \left\langle \boldsymbol{O}\boldsymbol{P}\,,\,\dot{\boldsymbol{Q}}\right\rangle.\end{aligned}$$

Now the cotangent lift of the infinitesimal action yields the momentum map. Let $\boldsymbol{\xi}\in\mathfrak{so}(3)$ and consider

$$\begin{aligned}\langle \boldsymbol{P}\,,\,\xi\,\boldsymbol{Q}\rangle &= \mathrm{tr}\left(\boldsymbol{P}^T\xi\,\boldsymbol{Q}\right),\\ &= \mathrm{tr}\left(\boldsymbol{Q}\boldsymbol{P}^T\xi\right),\\ &= \mathrm{tr}\left(\left(\boldsymbol{P}\,\boldsymbol{Q}^T\right)^T\xi\right).\end{aligned}$$

Since $\boldsymbol{\xi}$ is skew-symmetric, the trace of $\boldsymbol{\xi}$ with a symmetric matrix is zero, so

$$\begin{aligned}\mathrm{tr}\left(\left(\boldsymbol{P}\,\boldsymbol{Q}^T\right)^T\xi\right) &= \mathrm{tr}\left(\frac{1}{2}\left(\boldsymbol{P}\,\boldsymbol{Q}^T+\boldsymbol{Q}\boldsymbol{P}^T\right)^T\xi\right)\\ &\quad+\mathrm{tr}\left(\frac{1}{2}\left(\boldsymbol{P}\,\boldsymbol{Q}^T-\boldsymbol{Q}\boldsymbol{P}^T\right)^T\xi\right)\\ &= \mathrm{tr}\left(\frac{1}{2}\left(\boldsymbol{P}\,\boldsymbol{Q}^T-\boldsymbol{Q}\boldsymbol{P}^T\right)^T\xi\right)\\ &= \left\langle\frac{1}{2}\left(\boldsymbol{P}\,\boldsymbol{Q}^T-\boldsymbol{Q}\boldsymbol{P}^T\right),\,\xi\right\rangle\\ &= \langle \boldsymbol{J}_L(\boldsymbol{Q},\boldsymbol{P})\,,\,\xi\rangle.\end{aligned}$$

Thus, the momentum map for the left action of $SO(3)$ on $GL(3,\mathbb{R})$ is given by

$$\boldsymbol{J}_L(\boldsymbol{Q},\boldsymbol{P}) = \frac{1}{2}\left(\boldsymbol{P}\,\boldsymbol{Q}^T-\boldsymbol{Q}\boldsymbol{P}^T\right).$$

3. Observe that $H$ is invariant under the right action by $SO(3)$ given by $(\boldsymbol{Q},\boldsymbol{P}) \to (\boldsymbol{QO},\boldsymbol{PO})$. This is because

$$
\begin{aligned}
\operatorname{tr}\left((\boldsymbol{QO})^T\,\boldsymbol{QO}\right) &= \operatorname{tr}\left(\boldsymbol{O}^T\,\boldsymbol{Q}^T\,\boldsymbol{QO}\right),\\
&= \operatorname{tr}\left(\boldsymbol{O}\boldsymbol{O}^T\,\boldsymbol{Q}^T\,\boldsymbol{Q}\right),\\
&= \operatorname{tr}\left(\boldsymbol{Q}^T\,\boldsymbol{Q}\right).
\end{aligned}
$$

Also, $\det(\boldsymbol{QO}) = \det\boldsymbol{Q}\det\boldsymbol{O} = \det\boldsymbol{Q}$ just as before. The right action has infinitesimal action given by $\xi_{GL(3,\mathbb{R})}(\boldsymbol{Q}) = \boldsymbol{Q}\xi$. Therefore the momentum map can be computed as

$$
\begin{aligned}
\langle \boldsymbol{P}\,,\,\boldsymbol{Q}\xi\rangle &= \operatorname{tr}\left(\boldsymbol{P}^T\,\boldsymbol{Q}\xi\right),\\
&= \operatorname{tr}\left(\left(\boldsymbol{Q}^T\,\boldsymbol{P}\right)^T\,\xi\right),\\
&= \operatorname{tr}\left(\frac{1}{2}\left(\boldsymbol{Q}^T\,\boldsymbol{P} - \boldsymbol{P}^T\,\boldsymbol{Q}\right)^T\,\xi\right),\\
&= \left\langle\frac{1}{2}\left(\boldsymbol{Q}^T\,\boldsymbol{P} - \boldsymbol{P}^T\,\boldsymbol{Q}\right),\,\xi\right\rangle,\\
&= \langle \boldsymbol{J}_R(\boldsymbol{Q},\boldsymbol{P})\,,\,\xi\rangle.
\end{aligned}
$$

Thus the right action momentum map is given by

$$
\boldsymbol{J}_R(\boldsymbol{Q},\boldsymbol{P}) = \frac{1}{2}\left(\boldsymbol{Q}^T\,\boldsymbol{P} - \boldsymbol{P}^T\,\boldsymbol{Q}\right).
$$

The Poisson bracket of the two momentum maps is given by,

$$
\begin{aligned}
\{\boldsymbol{J}_L,\boldsymbol{J}_R\} &= \frac{1}{4}\left\{\left(\boldsymbol{P}\,\boldsymbol{Q}^T - \boldsymbol{Q}\boldsymbol{P}^T\right),\left(\boldsymbol{Q}^T\,\boldsymbol{P} - \boldsymbol{P}^T\,\boldsymbol{Q}\right)\right\},\\
&= \frac{1}{4}\left(\left(\boldsymbol{P} - \boldsymbol{P}^T\right)\left(\boldsymbol{Q}^T - \boldsymbol{Q}\right) - \left(\boldsymbol{Q}^T - \boldsymbol{Q}\right)\left(\boldsymbol{P} - \boldsymbol{P}^T\right)\right),\\
&= \frac{1}{4}\left(\left(\boldsymbol{P}\,\boldsymbol{Q}^T - \boldsymbol{Q}\boldsymbol{P}^T\right) - \left(\boldsymbol{Q}^T\,\boldsymbol{P} - \boldsymbol{P}^T\,\boldsymbol{Q}\right)\right.\\
&\qquad \left. + \left(\boldsymbol{Q}\boldsymbol{P} - \boldsymbol{P}^T\,\boldsymbol{Q}^T\right) - \left(\boldsymbol{P}\,\boldsymbol{Q} - \boldsymbol{Q}^T\,\boldsymbol{P}^T\right)\right),\\
&= \frac{1}{2}\left(\boldsymbol{J}_L - \boldsymbol{J}_R\right) + \frac{1}{4}\left(\boldsymbol{Q}\boldsymbol{P} - \boldsymbol{P}^T\,\boldsymbol{Q}^T\right) - \frac{1}{4}\left(\boldsymbol{P}\,\boldsymbol{Q} - \boldsymbol{Q}^T\,\boldsymbol{P}^T\right).
\end{aligned}
$$

Therefore the momentum maps do not Poisson commute. $\boldsymbol{J}_L$ and $\boldsymbol{J}_R$ are not in involution. Jacobi's identity shows that $\{\boldsymbol{J}_L, \boldsymbol{J}_R\}$ is also a conserved quantity. Although the difference of the momentum maps occurs, there are also two other terms, indicating that the left and right $SO(3)$ symmetry has not captured all of the symmetry of the Hamiltonian, i.e. there are more conserved quantities than those given to us by $SO(3)$ invariance.

# Part II

I shall speak of things ...so singular in their oddity as in some manner to instruct, or at least entertain, without wearying.

– LORENZO DA PONTE (1749–1838)
*Memoirs of Lorenzo Da Ponte* (2000). Translated by E. Abbott; edited and annotated by A. Livingston; introduction by C. Rosen. The New York Review of Books, New York.

# 11 EPDiff

## 11.1 Brief history of geometric ideal continuum motion

Arnold [Arn66] showed that the Euler equations for incompressible motion of an ideal fluid may be cast into Lagrangian and Hamiltonian forms that are similar to those for the rigid body, but are suitably generalized to allow for continuum motions, instead of only rigid rotations. Rigid rotations preserve the distances between any pair of particles. However, in continuum motions, no metric relation is imposed on the distances between the particles. In his generalization of dynamics from rigid-body rotations to continuum motions, Arnold applied the Lagrangian and Hamiltonian theories of geometric mechanics presented in Part I to rederive the Euler fluid equations. In particular, Arnold began with a configuration space $Q$ and formed a Lagrangian $L$ on the velocity space $TQ$, then passed to a Hamiltonian description on the momentum phase space $T^*Q$. In this approach, the methods of geometric mechanics systematically generated the Euler–Poincaré variational principle for the Euler fluid equations, as well as their associated Lie–Poisson Hamiltonian structure, Noether theorem, momentum maps, etc. For incompressible motion of an ideal fluid, the configuration space $Q = G$ is the group $G = \mathrm{Diff}_{Vol}(\mathcal{D})$ of volume preserving ***smooth diffeomorphisms*** (smooth invertible maps with smooth inverses) of the region $\mathcal{D}$ occupied by the fluid. The tangent vectors in $TG$ for the maps in $G = \mathrm{Diff}_{Vol}(\mathcal{D})$ represent the space of fluid velocities, which must satisfy appropriate physical conditions at the boundary of the region $\mathcal{D}$. Group multiplication in $G = \mathrm{Diff}_{Vol}(\mathcal{D})$ is composition of the smooth invertible volume-preserving maps.

This chapter begins by explaining how the Euler equations of ideal incompressible fluid motion may be recognized as Euler–Poincaré equations $\mathrm{EPDiff}_{Vol}$ defined on the dual of the tangent space at the identity $T_eG = T_e\,\mathrm{Diff}_{Vol}(\mathcal{D})$ of the ***right invariant vector fields*** over the domain $\mathcal{D}$. Then it sets about developing these ideas to apply to EPDiff, the Euler–Poincaré equations on the full diffeomorphism group.

Arnold's geometric approach for incompressible ideal fluid motion has been extended and applied to many other cases of ***ideal continuum motion***. See, e.g., [HMRW85, HMR98a] for references, discussion and progress in these more general applications. This wide range of applications of the geometric mechanics approach in fluids, plasmas and other continua has emerged for two reasons. First, all models of ideal continuum motion admit equivalent descriptions in either the ***material (or Lagrangian) picture*** or the ***spatial (or Eulerian) picture***. Second, the potential energy of a continuum flow depends upon material properties that are carried along (or *advected*) by the flow. These material properties evolve by the action of the flow as if each fluid parcel being carried along by the flow were a closed thermodynamic system. However, these advected material properties need not be passive. They may also influence the flow, because fluid dynamics allows an exchange between potential and kinetic energy.

In Hamilton's principle for the Eulerian description of ideal continuum dynamics, the kinetic energy may be expressed in terms of the ***right invariant spatial fluid velocity*** defined on smooth vector fields $T_eG = \mathfrak{X}(\mathcal{D})$. When *only* kinetic energy is present, the reduction of Hamilton's principle in the Lagrangian picture for fluids by the symmetries of its Eulerian picture produces the ***Euler–Poincaré (EP) theorem***. When the kinetic energy Lagrangian is right invariant under the diffeomorphisms, the dynamical equation resulting from the symmetry reduction of Hamilton's principle is called ***EPDiff***.

**Definition 11.1 (EPDiff)**
*EPDiff is the family of equations governing geodesic motion on the full diffeomorphism group with respect to whatever right invariant metric is chosen on the tangent space of the diffeomorphisms.*

**Remark 11.2** *Arnold's result [Ar1966] that the Euler fluid equations describe geodesic motion on the volume-preserving diffeomorphisms may then be understood by interpreting these equations as $EPDiff_{Vol}$ with respect to the right invariant $L^2$ metric of the Eulerian fluid velocity supplied by the fluid's kinetic energy.*

If potential energy due to advected quantities is present in an ideal continuum flow, then reduction by right invariance of Hamilton's principle under the diffeomorphisms produces the ***EP theorem with advected quantities*** discussed later in Theorem 17.8 in Chapter 17. The latter theorem encompasses most ideal continuum theories. This theorem was first stated and proved in [HMR98a], to which one may refer for additional details, as well as for abstract definitions and proofs. We shall also follow [HMR98a] in stating the conventions and terminology for the standard quantities in continuum mechanics treated in the present chapter.

## 11.2 Geometric setting of ideal continuum motion

**Definition 11.3** *Points in a domain $\mathcal{D}$ represent the positions of material particles of the system in its* ***reference configuration****. These points are denoted by $X \in \mathbb{R}^n$ and called the* ***particle labels****.*

- *A* ***configuration****, which we typically denote by $g$, is an element of* Diff($\mathcal{D}$), *the space of diffeomorphisms from $\mathcal{D}$ to itself.*
- *A* ***fluid motion****, denoted as $g_t$ or alternatively as $g(t)$, is a time-dependent curve in* Diff($\mathcal{D}$)*, providing an evolutionary sequence of diffeomorphism from the reference configuration to the current configuration in $\mathcal{D}$.*

The configuration space Diff($\mathcal{D}$) is a group, with the group operation being composition and the group identity being the identity map. This group acts on $\mathcal{D}$ in the obvious way: $g \cdot X := g(X)$, where we are using the 'dot' notation for the group action.

**Definition 11.4** *During a motion $g_t$ or $g(t)$, the particle labelled by $X$ describes a path in $\mathcal{D}$ along a locus of points*

$$x(X,t) := g_t(X) = g(t) \cdot X, \tag{11.1}$$

*which are called the* ***Eulerian*** *or* ***spatial points*** *of the path. This locus of points in $\mathbb{R}^n$ is also called the* ***Lagrangian****, or* ***material, trajectory****, because a Lagrangian fluid parcel follows this path in space.*

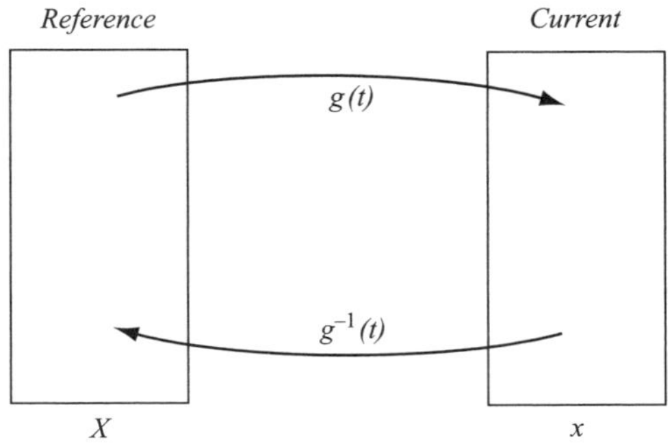


**Figure 11.1** The map from Lagrange reference coordinates $X$ in the fluid to the current Eulerian spatial position $x$ is performed by the time-dependent diffeomorphism $g(t)$, so that $x(t,X) = g(t) \cdot X$.

**Definition 11.5** *The* ***Lagrangian,*** *or* ***material, velocity*** *$U$ of the system along the motion $g_t$ or $g(t)$ is defined by taking the time derivative of the Lagrangian trajectory (11.1) keeping the particle labels $X$ fixed:*

$$U(X,t) := \frac{\partial}{\partial t} g_t \cdot X = \frac{\partial}{\partial t} x(X,t). \tag{11.2}$$

*Thus $U(X,t)$ is the velocity of the particle with label $X$ at time $t$.*

*The* ***Eulerian,*** *or* ***spatial, velocity*** *$u$ of the system is velocity expressed as a function of spatial position and time, meaning that if $x = x(X,t) = g_t(X)$ then*

$$u(x,t) := U(X,t) = U(g_t^{-1}(x),t). \tag{11.3}$$

*Thus, $u(x,t)$ is the velocity at time $t$ of the particle currently in position $x$.*

The Eulerian velocity $u$ can also be regarded as a time-dependent vector field $u_t \in \mathfrak{X}(\mathcal{D})$, where $u_t(x) := u(x,t)$. Similarly, we write $U_t(X) := U(X,t)$, though this is not really a vector field since $U_t(X)$ is a vector based at $x = x(X,t)$ rather than $X$ (cf. Definition 3.1). It follows from eqn (11.3) that

$$U_t = u_t \circ g_t. \tag{11.4}$$

In this sense, the Lagrangian velocity field at a particular time is a *right translation* of the Eulerian velocity field. This observation leads to consideration of the Lie-group structure of $\mathrm{Diff}(\mathcal{D})$.

The configuration space $\mathrm{Diff}(\mathcal{D})$ is an infinite-dimensional Lie group, but the precise meaning of this (there is more than one interpretation) is beyond the scope of this book.[1] Various subsets of $\mathrm{Diff}(\mathcal{D})$ may be chosen as configuration spaces, for example the space $\mathrm{Diff}_{vol}(\mathcal{D})$ of volume-preserving diffeomorphisms discussed in the previous section. As mentioned earlier, if $\mathcal{D}$ has a boundary, then the diffeomorphisms are required to respect the boundary conditions appropriate to the problem at hand. Thus, significant functional-analytic issues remain to be addressed before one could confidently regard $\mathrm{Diff}(\mathcal{D})$ as an infinite-dimensional Lie group. For a sense of the level of difficulty of some of these issues, see [EM70]. Acknowledging these important functional-analytic aspects, but not addressing them here, we will trust the approach of earlier successful endeavours in this field, by relying on a formal analogy to hold with the finite-dimensional theory from Part I, then checking the results of this assumption.

[1] If $\mathcal{D}$ is compact and without boundary, and if we specify that $\mathrm{Diff}(\mathcal{D})$ contains only $C^\infty$ diffeomorphisms, then $\mathrm{Diff}(\mathcal{D})$ is a *Fréchet manifold*, and in fact a Fréchet Lie group, since the group operations of composition and inversion are $C^\infty$ [Lee03].

**Definition 11.6** *Given a path $g(t)$ in $\mathrm{Diff}(\mathcal{D})$, the corresponding Lagrangian velocity fields $U_t$ are also denoted $\dot{g}(t)$ or $\frac{\partial}{\partial t}g(t)$. We use the 'dot' notation,*

$$\dot{g}(t)\cdot X := \dot{g}(t)(X) = U_t(X) = \frac{\partial}{\partial t}\, g_t \cdot X.$$

*For a given $t$, the velocity field $\dot{g}(t)$ is called a* ***tangent vector*** *to $\mathrm{Diff}(\mathcal{D})$ at $g(t)$. The* ***tangent space*** *of $\mathrm{Diff}(\mathcal{D})$ at $g$, denoted $T_g\,\mathrm{Diff}(\mathcal{D})$, is the set of all tangent vectors to $\mathrm{Diff}(\mathcal{D})$ at $g$, i.e. all possible Lagrangian velocity fields $U_t$ (for a fixed $t$) such that $U_t(X)\in T_{g(X)}\mathcal{D}$ for all $X$. The union of all of these tangent spaces is the* ***tangent bundle*** *$T\,\mathrm{Diff}(\mathcal{D})$.*

Note that $\dot{g}(t)\cdot X \in T_{g(t)\cdot X}\mathcal{D}$, so $\dot{g}(t)$ is not, in general, a vector field. However, if $g(t_0) = e$, then $\dot{g}(t_0)$ *is* a vector field. In fact, any smooth vector field on $\mathcal{D}$ can be expressed as $\dot{g}(0)$ for $g(t)$ equal to the flow of the vector field, so

$$T_e\,\mathrm{Diff}(\mathcal{D}) = \mathfrak{X}(\mathcal{D}),$$

where $\mathfrak{X}(\mathcal{D})$ is the set of smooth vector fields on $\mathcal{D}$. In general, we have

$$U_t = u_t \circ g_t,$$

and $u_t$ (for a fixed $t$) is a vector field, so general tangent vectors are right translations of vector fields, and

$$\begin{aligned} T_g\,\mathrm{Diff}(\mathcal{D}) &= \{u\circ g : u \in \mathfrak{X}(\mathcal{D})\} \\ &= \{\text{smooth } U : \mathcal{D}\to T\mathcal{D} \mid U(X)\in T_{g(X)}\mathcal{D} \text{ for all } X\}. \end{aligned}$$

Tangent lifts are defined as in Chapter 2. In particular,

**Remark 11.7 (Tangent lift of right translation)**
*Let $\varphi\in\mathrm{Diff}(\mathcal{D})$ and let $R_\varphi$ be the right translation map $g\mapsto g\circ\varphi$. The tangent lift of $R_\varphi$ is the map $TR_\varphi : T\,\mathrm{Diff}(\mathcal{D}) \to T\,\mathrm{Diff}(\mathcal{D})$ defined as follows. Let $U = \dot{g}(t_0)$. Then*

$$TR_\varphi(U) = TR_\varphi\left(\left.\frac{\mathrm{d}}{\mathrm{d}t}\right|_{t_0} g_t\right) := \left.\frac{\mathrm{d}}{\mathrm{d}t}\right|_{t_0}(g_t\circ\varphi) = U\circ\varphi,$$

*since for all $X\in\mathcal{D}$,*

$$\left.\frac{\mathrm{d}}{\mathrm{d}t}\right|_{t_0}(g_t\circ\varphi)(X) = \left.\frac{\mathrm{d}}{\mathrm{d}t}\right|_{t_0}(g_t\circ\varphi(X)) = \left(\left.\frac{\mathrm{d}}{\mathrm{d}t}\right|_{t_0} g_t\right)\cdot\varphi(X)) = U\circ\varphi(X).$$

*We use the notation* $U\varphi = TR_\varphi(U)$. *With this notation, the* ***Eulerian velocity*** *corresponding to a flow* $g(t)$ *is*

$$u_t = \dot{g}(t)g^{-1}(t).$$

The ***Lie algebra*** of Diff($\mathcal{D}$) is $\mathfrak{X}(\mathcal{D})$ with the ***Lie bracket*** defined by

$$[u, v]_L := [X_u^L, X_v^L](e),$$

for all $u, v \in \mathfrak{X}(\mathcal{D})$ (see Definition 5.27). This definition is hard to work with directly in the present context, since the left extension $X_v^L$ is a vector field on $T\,\mathrm{Diff}(\mathcal{D})$, and not on $\mathcal{D}$, so instead we use that fact that $[u, v]_L = \mathrm{ad}_u\, v$ (see Section 6.2). Let $\Phi_u(t)$ and $\Phi_v(t)$ be the flows of vector fields $u$ and $v$, respectively. The adjoint action of Diff($\mathcal{D}$) on $\mathfrak{X}(\mathcal{D})$ is

$$\mathrm{Ad}_g v = \frac{\mathrm{d}}{\mathrm{d}t}\bigg|_{t=0} g \circ \Phi_v(t) \circ g^{-1} = TL_g \circ v \circ g^{-1} = g_* v,$$

the push-forward of $v$ by $g$. It follows that the adjoint action of $\mathfrak{X}(\mathcal{D})$ on itself is

$$\mathrm{ad}_u v = \frac{\mathrm{d}}{\mathrm{d}t}\bigg|_{t=0} (\Phi_u(t))_* v = -\frac{\mathrm{d}}{\mathrm{d}t}\bigg|_{t=0} (\Phi_u(t))^* v = -\,\mathcal{L}_u\, v = -[u, v], \tag{11.5}$$

where the bracket on the right is the standard ***Jacobi–Lie bracket*** of the vector fields (see Section 3.1 and in particular Exercise 3.11). In components (summing on repeated indices),

$$-(\mathrm{ad}_u v)^i = [u, v]^i = u^j \frac{\partial v^i}{\partial x^j} - v^j \frac{\partial u^i}{\partial x^j},$$

$$\text{or} \quad -\mathrm{ad}_{\mathbf{u}}\mathbf{v} = [\mathbf{u}, \mathbf{v}] = \mathbf{u} \cdot \nabla \mathbf{v} - \mathbf{v} \cdot \nabla \mathbf{u}. \tag{11.6}$$

Thus, the Lie bracket on $\mathfrak{X}(\mathcal{D})$, considered as the Lie algebra of Diff($\mathcal{D}$), is minus the standard Jacobi-Lie bracket.

**Remark 11.8 (A matter of signs)**
*While the Lie bracket on* $T_eG$, *for any Lie group* $G$, *is conventionally defined using left extensions, it is also possible to define a bracket by right extensions, with the resulting bracket being the same except for a sign change (see Exercise 5.12). From the above calculation, it would seem that the definition by right extension is more natural in the present context, especially given the central role that right invariance plays in continuum motion. However, we will keep to the 'left extension' convention, and hence we're stuck with the minus sign.*

## 11.3 Euler–Poincaré reduction for continua

As discussed in Chapter 7, Euler–Poincaré reduction starts with a $G$-invariant Lagrangian

$$L : TG \to \mathbb{R}$$

defined on the tangent bundle of a Lie group $G$.

**Definition 11.9** *A Lagrangian $L : TG \to \mathbb{R}$ is said to be right $G$-invariant if $L(TR_h(v)) = L(v)$, for all $v \in T_gG$ and for all $g, h \in G$. In shorter notation,* ***right invariance*** *of the Lagrangian may be written as*

$$L(g(t)h, \dot{g}(t)h) = L(g(t), \dot{g}(t)),$$

*for all $h \in G$.*

For a $G$-invariant Lagrangian defined on $TG$, reduction by symmetry takes Hamilton's principle from $TG$ to $TG/G \simeq \mathfrak{g}$. Stationarity of the symmetry-reduced Hamilton's principle yields the Euler–Poincaré equations on $\mathfrak{g}^*$ discussed in Chapter 7. As we shall discuss later, the corresponding reduced Legendre transformation yields the now-standard ***Lie–Poisson bracket*** for the Hamiltonian formulation of these equations.

**Theorem 11.10 (Euler–Poincaré reduction)**
*Let $G$ be a Lie group and $L : TG \to \mathbb{R}$ be a* ***right invariant Lagrangian.*** *Let $\ell := L|_{\mathfrak{g}} : \mathfrak{g} \to \mathbb{R}$ be its restriction to $\mathfrak{g}$. For a curve $g(t) \in G$, let*

$$u(t) = \dot{g}(t) \cdot g(t)^{-1} := T_{g(t)}R_{g(t)^{-1}}\dot{g}(t) \in \mathfrak{g}.$$

*Then the following four statements are equivalent:*

(i) *$g(t)$ satisfies the* ***Euler–Lagrange equations*** *for Lagrangian $L$ defined on $G$.*

(ii) *The variational principle*

$$\delta \int_a^b L(g(t), \dot{g}(t))\mathrm{d}t = 0, \qquad (11.7)$$

*holds, for variations with fixed endpoints.*

(iii) *The (right invariant)* ***Euler–Poincaré equations*** *hold:*

$$\frac{\mathrm{d}}{\mathrm{d}t}\frac{\delta \ell}{\delta u} = -\,\mathrm{ad}^*_u \frac{\delta \ell}{\delta u}. \tag{11.8}$$

(iv) *The variational principle*

$$\delta \int_a^b \ell(u(t))\,\mathrm{d}t = 0, \tag{11.9}$$

*holds on* $\mathfrak{g}$*, using variations of the form*

$$\delta u = \dot{v} + [u, v], \tag{11.10}$$

*where* $u(t)$ *is an arbitrary path in* $\mathfrak{g}$ *that vanishes at the endpoints, i.e.* $u(a) = u(b) = 0$.

We identify the Lie group $G$ with the smooth invertible maps with smooth inverses; that is, we identify $G$ with Diff($\mathcal{D}$) the group of diffeomorphisms acting on the domain $\mathcal{D}$. The corresponding Lie algebra will be the algebra of smooth vector fields $\mathfrak{X}(\mathcal{D})$ endowed with the ad-operation given by (minus) the Jacobi–Lie bracket. We will forego any analytical technicalities that may arise in making this identification. The interested reader may consult Ebin and Marsden [EM70] for an approach to the analytical issues that arise in the volume-preserving case. The corresponding issues for the full diffeomorphism group remain an active field of current research.

## 11.4 EPDiff: Euler–Poincaré equation on the diffeomorphisms

### 11.4.1 The $n$-dimensional EPDiff equation

Eulerian geodesic motion of a fluid in $n$ dimensions is generated as an EP equation via Hamilton's principle, when the Lagrangian is given by the kinetic energy. The kinetic energy defines a norm $\|\mathbf{u}\|^2$ for the Eulerian fluid velocity, represented by the contravariant vector function $\mathbf{u}(\mathbf{x}, t) : \mathbb{R}^n \times \mathbb{R} \to \mathbb{R}^n$. The choice of the kinetic energy as a positive functional of fluid velocity $\mathbf{u}$ is a modelling step that depends upon the physics of the problem being studied. We shall choose the kinetic-energy Lagrangian,

$$\ell = L_{\mathfrak{g}} = \frac{1}{2}\|u\|^2_{Q_{op}} = \frac{1}{2}\int \mathbf{u}\cdot\mathbf{m}\,\mathrm{d}V \quad \text{with} \quad \mathbf{m} := Q_{op}\mathbf{u}. \tag{11.11}$$

This Lagrangian may also be expressed as the $L^2$ pairing,

$$\ell = \frac{1}{2}\big\langle u,\, m\big\rangle = \frac{1}{2}\int \mathbf{u}\cdot Q_{op}\mathbf{u}\,\mathrm{d}V, \tag{11.12}$$

where, in a coordinate basis, the components of the vector field $u$ and the 1-form density $m$ are defined by

$$u = u^j\frac{\partial}{\partial x^j} = \mathbf{u}\cdot\nabla \quad \text{and} \quad m = m_i\mathrm{d}x^i \otimes \mathrm{d}V = \mathbf{m}\cdot\mathrm{d}\mathbf{x}\otimes\mathrm{d}V.$$

We use the same font for a quantity and its dual. In particular, italic font denotes vector field $u$ and 1-form density $m$, and bold denotes vector $\mathbf{u}$ and covector $\mathbf{m}$. In eqns (11.11) and (11.12), the positive-definite, symmetric operator $Q_{op}$ defines the norm $\|\mathbf{u}\|$, for appropriate (homogeneous, say, or periodic) boundary conditions. Conversely, the spatial velocity vector $\mathbf{u}$ is obtained by convolution of the momentum covector $\mathbf{m}$ with the ***Green's function*** for the operator $Q_{op}$. This Green's function $G$ is defined by the vector equation

$$Q_{op}G = \delta(\mathbf{x}),$$

in which $\delta(\mathbf{x})$ is the Dirac measure and $G$ satisfies appropriate boundary conditions. Consequently,

$$\mathbf{u}(\mathbf{x}) = (G * \mathbf{m})(\mathbf{x}) = \int G(\mathbf{x},\mathbf{x}')\mathbf{m}(\mathbf{x}')\,\mathrm{d}\mathbf{x}'. \tag{11.13}$$

For more discussion of Green's functions for linear differential operators, see [Tay96].

**Remark 11.11** *An analogy exists between the kinetic energy in eqn (11.11) based on the norm $\|u\|_{Q_{op}}$ and the kinetic energy for the rigid body. In this analogy, the spatial velocity vector field $u$ corresponds to body angular velocity, the operator $Q_{op}$ to moment of inertia, and $G$ to its inverse.*

**Remark 11.12** *As defined earlier, the* ***EPDiff equation*** *is the Euler–Poincaré equation (11.8) for the Eulerian geodesic motion of a fluid with* *respect to norm* $\|\mathbf{u}\|$. *Its explicit form is given in the notation of Hamilton's principle by*

$$\frac{\mathrm{d}}{\mathrm{d}t}\frac{\delta \ell}{\delta u} + \mathrm{ad}^*_u \frac{\delta \ell}{\delta u} = 0, \quad \textit{in which} \quad \ell[u] = \frac{1}{2}\|\mathbf{u}\|^2. \tag{11.14}$$

**Definition 11.13** *The variational derivative of* $\ell$ *is defined by using the* $L^2$ ***pairing*** *between vector fields and 1-form densities as*

$$\delta \ell[u] = \left\langle \frac{\delta \ell}{\delta u}, \delta u \right\rangle = \int \frac{\delta \ell}{\delta \mathbf{u}} \cdot \delta \mathbf{u}\, \mathrm{d}V. \tag{11.15}$$

Consequently, the ***variational derivative*** with respect to the vector field $u$ is the ***one-form density*** of momentum given as in eqn (11.11),

$$\frac{\delta \ell}{\delta u} = \frac{\delta \ell}{\delta \mathbf{u}} \cdot d\mathbf{x} \otimes \mathrm{d}V = m, \tag{11.16}$$

which has vector components given by

$$\frac{\delta \ell}{\delta \mathbf{u}} = Q_{op}\mathbf{u} = \mathbf{m}. \tag{11.17}$$

In addition, $\mathrm{ad}^*$ is the dual of the vector-field ad-operation (minus the vector-field commutator) with respect to the $L^2$ pairing,

$$\langle \mathrm{ad}^*_u m, v\rangle = \langle m, \mathrm{ad}_u v\rangle, \tag{11.18}$$

where $u$ and $v$ are vector fields. The notation $\mathrm{ad}_u\, v$ from eqn (11.5) denotes the adjoint action of the ***right Lie algebra*** of $\mathrm{Diff}(\mathcal{D})$ on itself. The pairing in eqn (11.18) is the $L^2$ pairing. Hence, upon integration by parts, one finds

$$\begin{aligned}
\langle \mathrm{ad}^*_u m, v\rangle &= \langle m, \mathrm{ad}_u v\rangle \\
&= -\int m_i \left( u^j \frac{\partial v^i}{\partial x^j} - v^j \frac{\partial u^i}{\partial x^j}\right) \mathrm{d}V \\
&= \int \left( \frac{\partial}{\partial x^j}(m_i u^j) + m_j \frac{\partial u^j}{\partial x^j}\right) v^i\, \mathrm{d}V,
\end{aligned}$$

for homogeneous boundary conditions. In a coordinate basis, the preceding formula for $\mathrm{ad}^*_u m$ has the ***coordinate expression*** in $\mathbb{R}^n$,

$$\left(\mathrm{ad}^*_u m\right)_i \mathrm{d}x^i \otimes \mathrm{d}V = \left( \frac{\partial}{\partial x^j}(m_i u^j) + m_j \frac{\partial u^j}{\partial x^i}\right) \mathrm{d}x^i \otimes \mathrm{d}V. \tag{11.19}$$

In this notation, the abstract EPDiff equation (11.14) may be written explicitly in Euclidean coordinates as a partial differential equation for a covector function $\mathbf{m}(\mathbf{x}, t) : R^n \times R^1 \to R^n$. Namely, the ***EPDiff equation*** is given explicitly in Euclidean coordinates as

$$\frac{\partial}{\partial t}\mathbf{m} + \underbrace{\mathbf{u}\cdot\nabla\mathbf{m}}_{\text{Convection}} + \underbrace{(\nabla\mathbf{u})^T\cdot\mathbf{m}}_{\text{Stretching}} + \underbrace{\mathbf{m}(\operatorname{div}\mathbf{u})}_{\text{Expansion}} = 0. \tag{11.20}$$

Here, one denotes $(\nabla\mathbf{u})^T \cdot \mathbf{m} = \sum_j m_j \nabla u^j$. To explain the terms in underbraces, we rewrite EPDiff as preservation of the one-form density of momentum along the characteristic curves of the velocity. In vector coordinates, this is

$$\frac{\mathrm{d}}{\mathrm{d}t}\Big(\mathbf{m}\cdot \mathrm{d}\mathbf{x}\otimes \mathrm{d}V\Big) = 0 \quad \text{along} \quad \frac{\mathrm{d}\mathbf{x}}{\mathrm{d}t} = \mathbf{u} = G * \mathbf{m}. \tag{11.21}$$

This form of the EPDiff equation also emphasizes its non-locality, since the velocity is obtained from the momentum density by convolution against the Green's function $G$ of the operator $Q_{op}$, as in eqn (11.13). One may check that the ***characteristic form*** of EPDiff in eqn (11.21) recovers its Eulerian form by computing directly the result that

$$\begin{aligned}
&\frac{\mathrm{d}}{\mathrm{d}t}\Big(\mathbf{m}\cdot \mathrm{d}\mathbf{x}\otimes \mathrm{d}V\Big)\\
&\quad = \frac{\mathrm{d}\mathbf{m}}{\mathrm{d}t}\cdot \mathrm{d}\mathbf{x}\otimes \mathrm{d}V + \mathbf{m}\cdot \mathrm{d}\frac{\mathrm{d}\mathbf{x}}{\mathrm{d}t}\otimes \mathrm{d}V + \mathbf{m}\cdot \mathrm{d}\mathbf{x}\otimes\Big(\frac{\mathrm{d}}{\mathrm{d}t}\mathrm{d}V\Big)\\
&\quad = \Big(\frac{\partial}{\partial t}\mathbf{m} + \mathbf{u}\cdot\nabla\mathbf{m} + \nabla\mathbf{u}^T\cdot\mathbf{m} + \mathbf{m}(\operatorname{div}\mathbf{u})\Big)\cdot \mathrm{d}\mathbf{x}\otimes \mathrm{d}V = 0,
\end{aligned} \tag{11.22}$$

along

$$\frac{d\mathbf{x}}{\mathrm{d}t} = \mathbf{u} = G * \mathbf{m}.$$

This calculation explains the terms convection, stretching and expansion in the under-braces in eqn (11.20).

**Remark 11.14** *In 2D and 3D, the EPDiff equation (11.20) may also be written equivalently in terms of the operators div, grad and curl as,*

$$\frac{\partial}{\partial t}\mathbf{m} - \mathbf{u}\times \operatorname{curl}\mathbf{m} + \nabla(\mathbf{u}\cdot\mathbf{m}) + \mathbf{m}(\operatorname{div}\mathbf{u}) = 0. \tag{11.23}$$

*Thus, for example, the numerical solution of EPDiff would require an algorithm that has the capability to deal with the distinctions and relationships among the operators div, grad and curl.*

### 11.4.2 Variational derivation of EPDiff

The EPDiff equation (11.20) may be derived by following the proof of the EP reduction theorem leading to the Euler–Poincaré equations for right-invariance in the form of eqn (11.14). Following this calculation in Chapter 7 for the present right invariant case in the continuum notation yields

$$\begin{aligned}\delta \int_a^b l(u)\mathrm{d}t &= \int_a^b \left\langle \frac{\delta l}{\delta u}, \delta u \right\rangle \mathrm{d}t = \int_a^b \left\langle \frac{\delta l}{\delta u}, \frac{\mathrm{d}v}{\mathrm{d}t} - \mathrm{ad}_u v \right\rangle \mathrm{d}t \\ &= \int_a^b \left\langle \frac{\delta l}{\delta u}, \frac{\mathrm{d}v}{\mathrm{d}t} \right\rangle \mathrm{d}t - \int_a^b \left\langle \frac{\delta l}{\delta u}, \mathrm{ad}_u v \right\rangle \mathrm{d}t \\ &= -\int_a^b \left\langle \frac{\mathrm{d}}{\mathrm{d}t}\frac{\delta l}{\delta u} + \mathrm{ad}^*_u \frac{\delta l}{\delta u}, v \right\rangle \mathrm{d}t,\end{aligned}$$

where, as in (11.10), we have set

$$\delta u = \frac{\mathrm{d}v}{\mathrm{d}t} - \mathrm{ad}_u v, \tag{11.24}$$

for the variation of the right invariant vector field $u$ and $\langle \cdot, \cdot \rangle$ is the pairing between elements of the Lie algebra and its dual In our case, this is the $L^2$ pairing between vector fields and 1-form densities in eqn (11.15), written in components as

$$\left\langle \frac{\delta l}{\delta u}, \delta u \right\rangle = \int \frac{\delta l}{\delta u^i} \delta u^i \, \mathrm{d}V.$$

This $L^2$ pairing yields the component form of the EPDiff equation explicitly, as

$$\begin{aligned}\int_a^b \left\langle \frac{\delta l}{\delta u}, \delta u \right\rangle \mathrm{d}t &= \int_a^b \mathrm{d}t \int \frac{\delta l}{\delta u^i} \left( \frac{\partial v^i}{\partial t} + u^j \frac{\partial v^i}{\partial x^j} - v^j \frac{\partial u^i}{\partial x^j} \right) \mathrm{d}V \\ &= -\int_a^b \mathrm{d}t \int \left\{ \frac{\partial}{\partial t}\frac{\delta l}{\delta u^i} + \frac{\partial}{\partial x^j}\left( \frac{\delta l}{\delta u^i} u^j \right) + \frac{\delta l}{\delta u^j}\frac{\partial u^j}{\partial x^i} \right\} v^i \, \mathrm{d}V \\ &\quad + \int_a^b \mathrm{d}t \int \left\{ \frac{\partial}{\partial t}\left( \frac{\delta l}{\delta u^i} v^i \right) + \frac{\partial}{\partial x^j}\left( \frac{\delta l}{\delta u^i} v^i u^j \right) \right\} \mathrm{d}V.\end{aligned} \tag{11.25}$$

Invoking $v^i = 0$ at the endpoints in time and taking the fluid velocity vector $\mathbf{u}$ to be tangent to the (fixed) boundary in space, then substituting the definition $m = \delta l/\delta u$ recovers the coordinate forms in Euclidean components for the coadjoint action of vector fields in eqn (11.19) and the EPDiff equation itself in eqn (11.20). When $\ell[u] = \frac{1}{2}\|u\|^2$, EPDiff describes geodesic motion on the diffeomorphisms with respect to the norm $\|u\|$.

### 11.4.3 Noether's theorem for EPDiff

Noether's theorem associates conservation laws to continuous symmetries of a Lagrangian. See, e.g., [Olv00] for a clear discussion of the classical theory. Momentum and energy conservation for the EPDiff equation in eqn (11.20) readily emerge from Noether's theorem, since the Lagrangian in eqn (11.11) admits space and time translations. That is, the action for EPDiff,

$$S = \int \ell[\mathbf{u}]\mathrm{d}t = \int \frac{1}{2}\|\mathbf{u}\|^2\mathrm{d}t,$$

is invariant under the following transformations,

$$x^j \to x'^j = x^j + c^j \quad \text{and} \quad t \to t' = t + \tau, \tag{11.26}$$

for constants $\tau$ and $c^j$, with $j = 1, 2, 3$. Noether's theorem then implies conservation of corresponding momentum components $m_j$, with $j = 1, 2, 3$, and energy $E$ of the expected forms,

$$m_j = \frac{\delta \ell}{\delta u^j} \quad \text{and} \quad E = \frac{\delta \ell}{\delta u^j}u^j - \ell[\mathbf{u}], \tag{11.27}$$

which may be readily verified.

**Exercise 11.1**
Show that the EPDiff equation (11.14) may be written as

$$\Big(\frac{\partial}{\partial t} + \mathcal{L}_{\mathbf{u}}\Big)\Big(\mathbf{m}\cdot \mathrm{d}\mathbf{x}\otimes \mathrm{d}V\Big) = 0, \tag{11.28}$$

where $\mathcal{L}_{\mathbf{u}}$ is the Lie derivative with respect to the vector field with components $\mathbf{u} = G * \mathbf{m}$. How does the Lie-derivative form of EPDiff in eqn (11.28) differ from its characteristic form (11.21)? HINT: compare the coordinate expression obtained from the dynamical definition of the Lie derivative with the corresponding expression obtained from its definition via Cartan's formula.

**Exercise 11.2**
Show that EPDiff in 1D may be written as

$$m_t + um_x + 2mu_x = 0. \tag{11.29}$$

How does the factor of 2 arise in this equation? HINT: Take a look at eqn (11.20).

**Exercise 11.3**
Write the EPDiff equation in coordinate form (11.20) for (a) the $L^2$ norm and (b) the $H^1$ norm ($L^2$ norm of the gradient) of the spatial fluid velocity.

**Exercise 11.4**
Verify that the EPDiff equation (11.20) conserves the spatially integrated momentum and energy in eqn (11.27). HINT: for momentum conservation look at eqn (11.25) when $v^j = c^j$ for spatial translations.

# 12 EPDiff solution behaviour

This chapter discusses the coherent particle-like properties of the unidirectional singular solutions of the EPDiff equation (11.29). These singular solutions emerge from any smooth spatially confined initial velocity profile $u(x,0)$. After emerging, they dominate the evolution in interacting fully nonlinearly by exchanging momentum in elastic collisions. The mechanism for their emergence is proven to be pulse steepening due to nonlinearity. Several examples of the dynamics among singular solutions are given.

## 12.1 Introduction

Consider the following particular case of the EPDiff equation (11.29) in one spatial dimension,

$$m_t + um_x + 2mu_x = 0 \quad \text{with} \quad m = (1 - \alpha^2 \partial_x^2)u, \qquad (12.1)$$

in which the fluid velocity $u$ is a function of position $x$ on the real line and time $t$. This equation governs geodesic motion on the smooth invertible maps (diffeomorphisms) of the real line with respect to the metric associated with the $H^1$ Sobolev norm of the fluid velocity given by

$$\|u\|_{H^1}^2 = \int (u^2 + \alpha^2 u_x^2)\, \mathrm{d}x. \qquad (12.2)$$

The ***peakon*** is the solitary travelling wave solution for the EPDIff equation (12.1),

$$u(x,t) = c\, \mathrm{e}^{-|x-ct|/\alpha}. \qquad (12.3)$$

The peakon travelling wave moves at a speed equal to its maximum height, at which it has a sharp peak (jump in derivative). The spatial velocity profile $\mathrm{e}^{-|x|/\alpha}$ is the ***Green's function*** for the Helmholtz operator $(1 - \alpha^2 \partial_x^2)$

on the real line with vanishing boundary conditions at spatial infinity. In particular, it satisfies

$$(1 - \alpha^2 \partial_x^2) e^{-|x-ct|/\alpha} = 2\alpha \delta(x - ct). \tag{12.4}$$

A novel feature of the EPDiff equation (12.1) is that it admits solutions representing a ***wave train of peakons***

$$u(x,t) = \sum_{a=1}^{N} p_a(t) e^{-|x-q_a(t)|/\alpha}. \tag{12.5}$$

By eqn (12.4), this corresponds to a sum over delta functions representing the ***singular solution*** in momentum,

$$m(x,t) = 2\alpha \sum_{a=1}^{N} p_a(t)\, \delta(x - q_a(t)), \tag{12.6}$$

in which the ***delta function*** $\delta(x - q)$ is defined by

$$f(q) = \int f(x) \delta(x - q)\, \mathrm{d}x, \tag{12.7}$$

for an arbitrary smooth function $f$. Such a sum is an *exact solution* of the EPDiff equation (12.1) provided the time-dependent parameters $\{p_a\}$ and $\{q_a\}$, $a = 1, \ldots, N$, satisfy certain canonical Hamiltonian equations that will be discussed later.

**Remark 12.1** *The peakon-train solutions of EPDiff are an* ***emergent phenomenon****. A wave train of peakons emerges in solving the initial-value problem for the EPDiff equation (12.1) for essentially any spatially confined initial condition. An example of the emergence of a wave train of peakons from a Gaussian initial condition is shown in Figure 12.1.*

### 12.1.1 Steepening lemma: the peakon-formation mechanism

We may understand the mechanism for the emergent formation of the peakons seen in Figure 12.1, by showing that initial conditions exist for which the solution of the EPDiff equation (12.24) can develop a vertical slope in its velocity $u(t,x)$, in finite time. The mechanism turns out to be associated with ***inflection points of negative slope***, such as occur on the leading edge of a rightward-propagating velocity profile. In particular,

**Lemma 12.2 (Steepening lemma [CH93])**
*Suppose the initial profile of velocity $u(0,x)$ has an inflection point at $x = \overline{x}$*

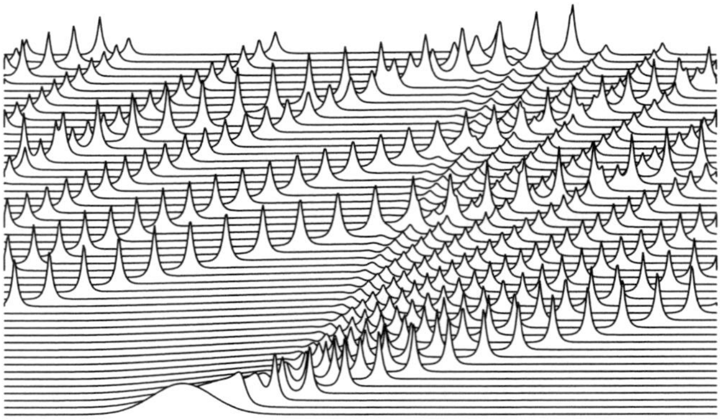

**Figure 12.1** Under the evolution of the EPDiff equation (12.1), an ordered ***wave train of peakons*** emerges from a smooth localized initial condition (a Gaussian). The spatial profiles at successive times are offset in the vertical to show the evolution. The peakon wave train eventually wraps around the periodic domain, thereby allowing the leading peakons to overtake the slower peakons from behind in collisions that conserve momentum and preserve the peakon shape but cause phase shifts in the positions of the peaks, as discussed in [CH93].

*to the right of its maximum, and otherwise it decays to zero in each direction sufficiently rapidly for the $H^1$ Sobolev norm of the fluid velocity in eqn (12.2) to be finite. Then, the negative slope at the inflection point will become vertical in finite time.*

*Proof* Consider the evolution of the slope at the inflection point. Define $s = u_x(\overline{x}(t), t)$. Then the EPDiff equation (12.1), rewritten as,

$$(1 - \alpha^2 \partial^2)(u_t + uu_x) = -\,\partial\Big(u^2 + \frac{\alpha^2}{2}u_x^2\Big), \tag{12.8}$$

yields an equation for the evolution of $s$. Namely, using $u_{xx}(\overline{x}(t), t) = 0$ leads to

$$\frac{ds}{dt} = -\frac{1}{2}s^2 + \frac{1}{2}\int_{-\infty}^{\infty} \operatorname{sgn}(\overline{x} - y)e^{-|\overline{x}-y|}\partial_y\left(u^2 + \frac{1}{2}u_y^2\right)dy. \tag{12.9}$$

Integrating by parts and using the inequality $A^2 + B^2 \geq 2AB$, for any two real numbers $A$ and $B$, leads to

$$\begin{aligned}\frac{ds}{dt} &= -\frac{1}{2}s^2 - \frac{1}{2}\int_{-\infty}^{\infty} e^{-|\overline{x}-y|}\left(u^2 + \frac{1}{2}u_y^2\right)dy + u^2(\overline{x}(t), t)\\ &\leq -\frac{1}{2}s^2 + 2u^2(\overline{x}(t), t).\end{aligned} \tag{12.10}$$

Then, provided $u^2(\overline{x}(t), t)$ remains finite, say less than a number $M/4$, we have

$$\frac{\mathrm{d}s}{\mathrm{d}t} = -\frac{1}{2}s^2 + \frac{M}{2}, \tag{12.11}$$

which implies, for negative slope initially $s \le -\sqrt{M}$, that

$$s \le \sqrt{M} \coth\left(\sigma + \frac{t}{2}\sqrt{M}\right), \tag{12.12}$$

where $\sigma$ is a negative constant that determines the initial slope, also negative. Hence, at time $t = -2\sigma/\sqrt{M}$ the slope becomes negative and vertical. The assumption that $M$ in eqn (12.11) exists is verified in general by a Sobolev inequality. In fact, $M = 8H_1$, since

$$\max_{x\in\mathbb{R}} u^2(x,t) \le \int_{-\infty}^{\infty} \left(u^2 + u_x^2\right) \mathrm{d}x = 2H_1 = const. \tag{12.13}$$

■

**Remark 12.3** *Suppose the initial condition is anti-symmetric, so the inflection point at $u = 0$ is fixed and $\mathrm{d}\overline{x}/\mathrm{d}t = 0$, due to the symmetry $(u,x) \to (-u,-x)$ admitted by eqn (13.1). In this case, $M = 0$ and no matter how small $|s(0)|$ (with $s(0) < 0$) verticality $s \to -\infty$ develops at $\overline{x}$ in finite time.*

*The steepening lemma indicates that travelling wave solutions of the EPDiff equation (12.1) cannot have the sech$^2$ shape that appears for KdV solitons, since inflection points with sufficiently negative slope can lead to unsteady changes in the shape of the profile if inflection points are present. In fact, numerical simulations show that the presence of an inflection point of negative slope in any confined initial velocity distribution triggers the steepening lemma as the* ***mechanism*** *for the formation of the peakons. Namely. the initial (positive) velocity profile "leans" to the right and steepens, then produces a peakon that is taller than the initial profile, so it propagates away to the right. This process leaves a profile behind with an inflection point of negative slope; so it repeats, thereby producing a wave train of peakons with the tallest and fastest ones moving rightward in order of height. Remarkably, this recurrent process produces only peakons.*

The EPDiff equation (12.1) arises from a shallow water wave equation in the limit of zero linear dispersion in one dimension. As we shall see, the peakon solutions (12.6) for EPDiff generalize to higher dimensions and other kinetic energy norms.

**Exercise 12.1**
Verify that the EPDIff equation (12.1) preserves the $H^1$ norm (12.2).

**Exercise 12.2**
Verify that the peakon formula (12.3) provides the solitary travelling wave solution for the EPDIff equation (12.1).

**Exercise 12.3**
Verify formula (12.4) for the Green's function. Why is this formula useful in representing the travelling-wave solution of the EPDIff equation (12.1)?

## 12.2 Shallow-water background for peakons

The EPDiff equation (12.1) whose solutions admit peakon wave trains (12.5) may be derived by taking the zero-dispersion limit of another equation obtained from Euler's fluid equations by using asymptotic expansions for shallow water waves [CH93]. Euler's equations for irrotational incompressible ideal fluid motion under gravity with a free surface have an asymptotic expansion for shallow water waves that involves two small parameters, $\epsilon$ and $\delta^2$, with ordering $\epsilon \geq \delta^2$. These small parameters are $\epsilon = a/h_0$ (the ratio of wave amplitude to mean depth) and $\delta^2 = (h_0/l_x)^2$ (the squared ratio of mean depth to horizontal length, or wavelength).

In one spatial dimension, EPDiff is the zero-dispersion limit of the Camassa–Holm (CH) equation for shallow water waves, which is the $b = 2$ case of the following ***b-equation***, that results from the asymptotic expansion for shallow water waves,

$$m_t + c_0 u_x + u m_x + b\, m u_x - \gamma u_{xxx} = 0. \qquad (12.14)$$

Here, $m = u - \alpha^2 u_{xx}$ is the momentum variable, and the constants $\alpha^2$ and $\gamma/c_0$ are squares of length scales. At *linear* order in the asymptotic expansion for shallow water waves in terms of the small parameters $\epsilon$ and $\delta^2$, one finds $\alpha^2 \to 0$, so that $m \to u$ in (12.14). In this case, the famous ***Korteweg–de Vries*** (KdV) soliton equation is recovered for $b = 2$,

$$u_t + 3uu_x = -c_0 u_x + \gamma u_{xxx}. \qquad (12.15)$$

Any value of the parameter *except* $b = -1$ may be achieved in eqn (12.14) by an appropriate near-identity (normal form) transformation of the solution [DGH04]. The value $b = -1$ is disallowed in (12.14) because it cancels the leading-order nonlinearity and, thus, breaks the asymptotic ordering.

Because of the relation $m = u - \alpha^2 u_{xx}$, the b-equation (12.14) is ***non-local.*** In other words, it is an integral-partial differential equation. In fact, after writing eqn (12.14) equivalently as,

$$(1 - \alpha^2 \partial_x^2)(u_t + uu_x) = -\,\partial_x \left( \frac{b}{2} u^2 + \frac{3-b}{2} \alpha^2 u_x^2 \right) - c_0 u_x + \gamma u_{xxx}. \tag{12.16}$$

The b-equation may be expressed in ***hydrodynamic form*** as

$$u_t + uu_x = -\,p_x, \tag{12.17}$$

with a 'pressure' $p$ given by

$$p = G * \left( \frac{b}{2} u^2 + \frac{3-b}{2} \alpha^2 u_x^2 + c_0 u - \gamma u_{xx} \right), \tag{12.18}$$

in which the convolution kernel is the Green's function $G(x, y) = (2\alpha)^{-1} e^{-|x-y|/\alpha}$ for the Helmholtz operator $(1 - \alpha^2 \partial_x^2)$.

One sees the interplay between local and non-local linear dispersion in the b-equation by linearizing eqn (12.16) around $u = 0$ to find its phase-velocity relation,

$$\frac{\omega}{k} = \frac{c_0 + \gamma k^2}{1 + \alpha^2 k^2}, \tag{12.19}$$

obtained for waves with frequency $\omega$ and wave number $k$. For $\gamma / c_0 > 0$, short waves and long waves travel in the same direction. Long waves travel faster than short ones (as required in shallow water) provided $\gamma / c_0 < \alpha^2$. Then, the phase velocity lies in the interval $\omega / k \in (\gamma / \alpha^2, c_0]$. The parameters $c_0$ and $\gamma$ represent linear wave dispersion, which modifies and may eventually balance the tendency for nonlinear waves to steepen and break. The parameter $\alpha$, which introduces non-locality, also allows a balance leading to a stable wave shape, even in the absence of $c_0$ and $\gamma$.

The nonlinear effects of the parameter $b$ on the solutions of eqn (12.14) were investigated in Holm and Staley [HS03], where $b$ was treated as a bifurcation parameter. In the limiting case when the linear dispersion coefficients are absent, peakon solutions of eqn (12.14) are allowed theoretically for any value of $b$. However, they were found numerically to be stable only for $b > 1$. These coherent solutions are allowed, because the two nonlinear terms in eqn (12.14) may balance each other, even in the *absence* of linear dispersion. However, the instability of the peakons found numerically for $b < 1$ indicates that the relative strengths of the two nonlinearities will determine whether this balance can be maintained.

**Proposition 12.4** *A solution $u$ of the b-equation (12.14) with $c_0 = 0$ and $\gamma = 0$ vanishing at spatial infinity blows up in $H^1$ if and only if its first-order derivative blows up, that is, if wave breaking occurs.*

*Proof* This result is implied by Exercise 12.6. ■

**Lemma 12.5 (Steepening lemma for the b-equation with $b > 1$)**
*Suppose the initial profile of velocity $u(0,x)$ has an inflection point at $x = \overline{x}$ to the right of its maximum, and otherwise it decays to zero in each direction. Assume that the velocity at the inflection point remains finite. Then, the negative slope at the inflection point will become vertical in finite time, provided $b > 1$.*

*Proof* Consider the evolution of the slope at the inflection point $x = \overline{x}(t)$. Define $s = u_x(\overline{x}(t), t)$. Then, the b-equation (12.14) with $c_0 = 0$ and $\gamma = 0$ may be rewritten in hydrodynamic form as, cf. eqn (12.17),

$$u_t + uu_x = -\,\partial_x G * \left(\frac{b}{2}u^2 + \frac{3-b}{2}\,\alpha^2 u_x^2\right). \tag{12.20}$$

The spatial derivative of this yields an equation for the evolution of $s$. Namely, using $u_{xx}(\overline{x}(t), t) = 0$ leads to

$$\begin{aligned}
\frac{ds}{dt} + s^2 &= -\,\partial_x^2 (G * p) \quad \text{with} \quad p := \left(\frac{b}{2}u^2(\overline{x}(t), t) + \frac{3-b}{2}\,\alpha^2 s^2\right) \\
&= \frac{1}{\alpha^2}(1 - \alpha^2\partial_x^2) G * p - \frac{1}{\alpha^2} G * p \\
&= \frac{1}{\alpha^2} p - \frac{1}{\alpha^2} G * p.
\end{aligned} \tag{12.21}$$

This calculation implies

$$\begin{aligned}
\frac{ds}{dt} &= \frac{1-b}{2}s^2 - \frac{1}{2\alpha}\int_{-\infty}^{\infty} e^{-|\overline{x}-y|/\alpha}\left(\frac{b}{2}u^2 + \frac{3-b}{2}\alpha^2 u_y^2\right) dy \\
&\quad + \frac{b}{2\alpha^2}u^2(\overline{x}(t), t) \\
&\le \frac{1-b}{2}s^2 + \frac{b}{2\alpha^2}u^2(\overline{x}(t), t),
\end{aligned} \tag{12.22}$$

where we have dropped the negative middle term in the last step. Then, provided $u^2(\overline{x}(t),t)$ remains finite, say less than a number $M$, we have

$$\frac{\mathrm{d}s}{\mathrm{d}t} \le \frac{1-b}{2}s^2 + \frac{bM}{2\alpha^2}, \tag{12.23}$$

which implies, for negative slope initially and $b > 1$, that the slope remains negative and becomes vertical in finite time. This proves the steepening lemma for the b-equation and identifies $b = 1$ as a special value. ■

**Remark 12.6** *One might wonder whether the dispersionless CH equation is the only shallow water b-equation that both possesses peakon solutions and is completely integrable as a Hamiltonian system. Mikhailov and Novikov [MN02] showed that among the b-equations only the cases $b = 2$ and $b = 3$ are completely integrable as Hamiltonian systems. The case $b = 3$ is the Degasperis–Processi equation, whose peakon solutions are studied in [DHH03].*

**Remark 12.7** *Hereafter, we specialize the b-equation (12.14) to the case $b = 2$. If, in addition, $c_0 = 0$ and $\gamma = 0$, then the b-equation specializes to EPDiff.*

### 12.2.1 Hamiltonian dynamics of EPDiff peakons

Upon substituting the peakon solution expressions (12.5) for velocity $u$ and eqn (12.6) for momentum $m$ into the EPDiff equation,

$$m_t + um_x + 2mu_x = 0, \quad \text{with} \quad m = u - \alpha^2 u_{xx}, \tag{12.24}$$

one finds ***Hamilton's canonical equations*** for the dynamics of the discrete set of peakon parameters $p_a(t)$ and $q_a(t)$. Namely,

$$\dot{q}_a(t) = \frac{\partial H_N}{\partial p_a} \quad \text{and} \quad \dot{p}_a(t) = -\frac{\partial H_N}{\partial q_a}, \tag{12.25}$$

for $a = 1, 2, \ldots, N$, with Hamiltonian given by [CH93],

$$H_N = \tfrac{1}{2} \sum_{a,b=1}^{N} p_a\, p_b\, \mathrm{e}^{-|q_a - q_b|/\alpha}. \tag{12.26}$$

The first canonical equation in eqn (12.25) implies that the peaks at the positions $x = q^a(t)$ in the peakon-train solution (12.5) move with the flow of the fluid velocity $u$ at those positions, since $u(q^a(t),t) = \dot{q}^a(t)$. This means the positions $q^a(t)$ are ***Lagrangian coordinates*** frozen into the flow of EPDiff. Thus, the singular momentum solution ansatz (12.6) is the map from Lagrangian coordinates to Eulerian coordinates (that is, the ***Lagrange-to-Euler map***) for the momentum.

**Remark 12.8** *The peakon wave train (12.6) forms a finite-dimensional invariant manifold of solutions of the EPDiff equation. On this invariant manifold of solutions for the partial differential equation (12.24), the dynamics turns out to be canonically Hamiltonian as in eqn (12.25). Chapter 14 will explain that the canonical Hamiltonian structure of the peakon solutions arises because the solution ansatz (12.6) for momentum m is a momentum map.*

### 12.2.2 Pulsons: Singular solutions of EPDiff for other Green's functions

The Hamiltonian $H_N$ in eqn (12.26) depends on $G$, the Green's function for the relation $u = G * m$ between velocity $u$ and momentum $m$. For the Helmholtz operator on the real line this Green's function is given by eqn (12.4) as $G(x) = \mathrm{e}^{-|x|/\alpha}/2\alpha$. However, the singular momentum solution ansatz (12.6) is *independent* of this Green's function. Thus, we may conclude the following [FH01].

**Proposition 12.9** *The singular momentum solution ansatz*

$$m(x,t) = \sum_{a=1}^{N} p_a(t)\,\delta(x - q_a(t)), \tag{12.27}$$

*for EPDiff,*

$$m_t + um_x + 2mu_x = 0, \quad \textit{with} \quad u = G * m, \tag{12.28}$$

*provides a **finite-dimensional invariant manifold** of solutions governed by canonical Hamiltonian dynamics, for any choice of the Green's function G relating velocity u and momentum m.*

*Proof* The singular momentum solution ansatz (12.27) is *independent* of the Green's function $G$. ■

**Remark 12.10** *The pulson singular solutions (12.27) of the EPDiff equation (12.28) form a finite-dimensional invariant symplectic manifold, on which the EPDiff solution dynamics is governed by a canonical Hamiltonian system for the conjugate pairs of variables* $(q_a, p_a)$ *with*

$a = 1, 2, \ldots, N$. *Perhaps surprisingly, these singular solutions will turn out to emerge from any smooth confined initial distribution of momentum.*

The fluid velocity solutions corresponding to the singular momentum ansatz (12.27) for eqn (12.28) are the ***pulsons***. A pulson wave train is defined by the sum over $N$ velocity profiles determined by the Green's function $G$, as

$$u(x,t) = \sum_{a=1}^{N} p_a(t) G\big(x, q_a(t)\big). \tag{12.29}$$

A solitary travelling wave solution for the pulson is given by

$$u(x,t) = cG(x, ct) = cG(x - ct) \quad \text{with} \quad G(0) = 1, \tag{12.30}$$

where one finds $G(x, ct) = G(x - ct)$, provided the Green's function $G$ is translation-invariant.

For EPDiff (12.28) with any choice of the Green's function $G$, the singular momentum solution ansatz (12.27) results in a finite-dimensional invariant manifold of exact solutions. The $2N$ parameters $p_a(t)$ and $q_a(t)$ in these pulson-train solutions of EPDiff satisfy Hamilton's canonical equations

$$\frac{\mathrm{d}q_a}{\mathrm{d}t} = \frac{\partial H_N}{\partial p_a} \quad \text{and} \quad \frac{\mathrm{d}p_a}{\mathrm{d}t} = -\frac{\partial H_N}{\partial q_a}, \tag{12.31}$$

with $N$-particle Hamiltonian,

$$H_N = \frac{1}{2} \sum_{a,b=1}^{N} p_a \, p_b \, G(q_a, q_b). \tag{12.32}$$

The canonical equations for the parameters in the pulson train define an invariant manifold of singular momentum solutions and provide a phase-space description of geodesic motion with respect to the cometric (inverse metric) given by the Green's function $G$. Mathematical analysis and numerical results for the dynamics of these pulson solutions are given in [FH01] whose results show how the results of collisions of pulsons (12.29) depend upon the *shape* of their travelling wave profile. The effects of the travelling-wave pulse shape

$$u(x - ct) = cG(x - ct)$$

on the multipulson collision dynamics are reflected in the Hamiltonian (12.32) that governs this dynamics. For example, see Figure 12.2, in which the pulsons are *triangular*.

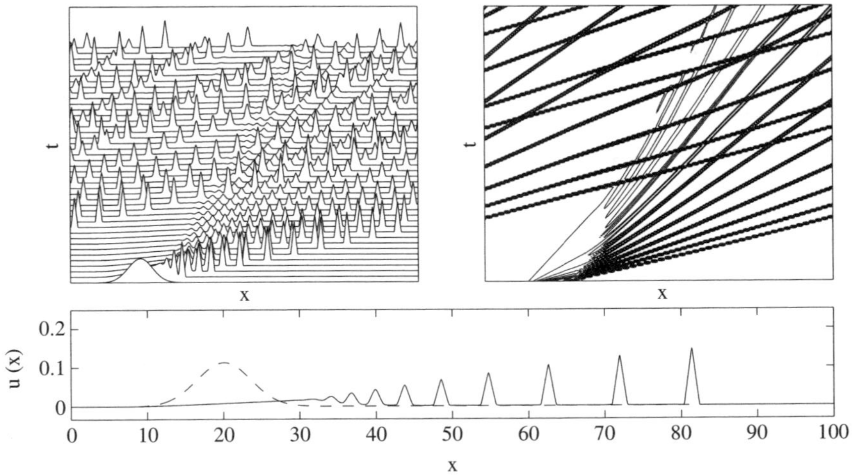


**Figure 12.2** When the Green's function $G$ has a triangular profile, a train of triangular pulsons emerges from a Gaussian initial velocity distribution as it evolves under the EPDiff equation (12.1). The upper panels show the collisions that occur as the faster triangular pulsons overtake the slower ones as they cross and re-cross the periodic domain. The upper left panel shows the progress of the pulsons by by showing offsets of the velocity profile at equal time intervals. The upper right panel shows the pulson paths obtained by plotting their elevation topography.

**Exercise 12.4**
Verify the hydrodynamic form of the b-equation in eqn (12.16).

**Exercise 12.5**
Verify that the b-equation (12.14) with $c_0 = 0$ and $\gamma = 0$ admits peakon-train solutions of the form (12.5) for any value of $b$.

**Exercise 12.6**
Verify that the b-equation (12.14) with $c_0 = 0$ and $\gamma = 0$ satisfies

$$\frac{d}{dt}\|u\|^2_{H^1} = (b-2)\int u_x^3\, dx,$$

for any value of $b$ and for solutions that vanish sufficiently rapidly at spatial infinity that no endpoint contributions arise upon integration by parts.

**Exercise 12.7**
Prove a steepening lemma for the b-equation (12.14) with $c_0 = 0$ and $\gamma = 0$ that avoids the assumption that $u^2(\overline{x}(t), t)$ remains finite. That is, establish a necessary and sufficient condition depending only on the initial data for blow-up to occur in finite time. How does this condition depend on the value of $b$? Does this steepening lemma hold for every value of $b > 1$?

**Exercise 12.8**
Are the equations of peakon dynamics for the b-equation (12.14) with $c_0 = 0$ and $\gamma = 0$ canonically Hamiltonian for every value of $b$? HINT: try $b = 3$.

## 12.3 Peakons and pulsons

### 12.3.1 Pulson–Pulson interactions

The solution of EPDiff in 1D

$$\partial_t m + u m_x + 2u_x m = 0, \tag{12.33}$$

with $u = G * m$ for the momentum $m = Q_{op} u$ is given for the interaction of only two pulsons by the sum of delta functions in eqn (12.27) with $N = 2$,

$$m(x,t) = \sum_{i=1}^{2} p_i(t)\, \delta(x - q_i(t)). \tag{12.34}$$

The parameters satisfy the finite dimensional geodesic canonical Hamiltonian equations (12.25), in which the Hamiltonian for $N = 2$ is given by

$$H_{N=2}(q_1, q_2, p_1, p_2) = \frac{1}{2}(p_1^2 + p_2^2) + p_1 p_2\, G(q_1 - q_2). \tag{12.35}$$

#### Conservation laws and reduction to quadrature

Provided the Green's function $G$ is symmetric under spatial reflections, $G(-x) = G(x)$, the two-pulson Hamiltonian system conserves the total momentum

$$P = p_1 + p_2. \tag{12.36}$$

Conservation of $P$ ensures integrability, by Liouville's theorem, and reduces the 2-pulson system to quadratures. To see this, we introduce sum and difference variables as

$$P = p_1 + p_2, \quad Q = q_1 + q_2, \quad p = p_1 - p_2, \quad q = q_1 - q_2. \tag{12.37}$$

In these variables, the Hamiltonian (12.35) becomes

$$H(q, p, P) = \frac{1}{4}(P^2 - p^2)\big(1 - G(q)\big). \tag{12.38}$$

Likewise, the 2-pulson equations of motion transform to sum and difference variables as

$$\frac{\mathrm{d}P}{\mathrm{d}t} = -2\frac{\partial H}{\partial Q} = 0, \frac{\mathrm{d}Q}{\mathrm{d}t} = 2\frac{\partial H}{\partial P} = P(1 + G(q)),$$
$$\frac{\mathrm{d}p}{\mathrm{d}t} = -2\frac{\partial H}{\partial q} = \frac{1}{2}(p^2 - P^2)G'(q), \frac{\mathrm{d}q}{\mathrm{d}t} = 2\frac{\partial H}{\partial p} = -\,p(1 - G(q)).$$

Eliminating $p^2$ between the formula for $H$ and the equation of motion for $q$ yields

$$\begin{aligned}\left(\frac{\mathrm{d}q}{\mathrm{d}t}\right)^2 &= P^2\big(1 - G(q)\big)^2 - 4H\big(1 - G(q)\big)\\ &=: Z(G(q); P, H) \geq 0,\end{aligned} \tag{12.39}$$

which rearranges into the following quadrature,

$$\mathrm{d}t = \frac{\mathrm{d}G(q)}{G'(q)\sqrt{Z(G(q); P, H)}}. \tag{12.40}$$

For the peakon case, we have $G(q) = \mathrm{e}^q$ so that $G'(q) = G(q)$ and the quadrature (12.40) simplifies to an elementary integral. Having obtained $q(t)$ from the quadrature, the momentum difference $p(t)$ is found from eqn (12.38) via the algebraic expression

$$p^2 = P^2 - \frac{4H}{1 - G(q)}, \tag{12.41}$$

in terms of $q$ and the constants of motion $P$ and $H$. Finally, the sum $Q(t)$ is found by a further quadrature.

Upon writing the quantities $H$ and $P$ as

$$H = c_1 c_2, \quad P = c_1 + c_2, \quad \frac{1}{2}c_1^2 + \frac{1}{2}c_2^2 = \frac{1}{2}P^2 - H, \tag{12.42}$$

in terms of the asymptotic speeds of the pulsons, $c_1$ and $c_2$, we find the relative momentum relation,

$$p^2 = (c_1 + c_2)^2 - \frac{4c_1c_2}{1 - G(q)}. \tag{12.43}$$

This equation has several implications for the qualitative properties of the 2-pulson collisions.

**Definition 12.11** *Overtaking, or rear-end, pulson collisions satisfy* $c_1c_2 > 0$, *while head-on pulson collisions satisfy* $c_1c_2 < 0$.

The pulson order $q_1 < q_2$ is preserved in an overtaking, or rear-end, collision. This follows, as

**Proposition 12.12 (Preservation of pulson order)** *For overtaking, or rear-end, collisions, the 2-pulson dynamics preserves the sign condition*

$$q = q_1 - q_2 < 0.$$

*Proof* Suppose the peaks were to overlap in an overtaking collision with $c_1c_2 > 0$, thereby producing $q = 0$ during a collision. The condition $G(0) = 1$ implies the second term in eqn (12.43) would diverge if this overlap were to occur. However, such a divergence would contradict $p^2 \geq 0$. ■

Consequently, seen as a collision between two 'particles' with initial speeds $c_1$ and $c_2$ that are initially well separated, the separation $q(t)$ reaches a non-zero distance of closest approach $q_{min}$ in an overtaking, or rear-end, collision that may be expressed in terms of the pulse shape, as follows.

**Corollary 12.13 (Minimum separation distance)**
*The minimum separation distance reachable in two-pulson collisions with* $c_1c_2 > 0$ *is given by,*

$$1 - G(q_{min}) = \frac{4c_1c_2}{(c_1 + c_2)^2}. \tag{12.44}$$

*Proof* Set $p^2 = 0$ in eqn (12.43). ■

**Proposition 12.14 (Head-on collisions admit $q \to 0$)**
*The 2-pulson dynamics allows the overlap $q \to 0$ in head-on collisions.*

*Proof* Because $p^2 \geq 0$, the overlap $q \to 0$ implying $g \to 1$ is only possible in eqn (12.43) for $c_1 c_2 < 0$. That is, for the head-on collisions. ∎

**Remark 12.15 (Divergence of head-on momentum)**
*Equation (12.43) implies that $p^2 \to \infty$ diverges when $q \to 0$ in head-on collisions. As we shall discuss, this signals the development of a vertical slope in the velocity profile of the solution at the moment of collision.*

### 12.3.2 Pulson–anti-pulson interactions

#### Head-on pulson–anti-pulson collision

In a ***completely anti-symmetric*** head-on collision of a pulson and anti-pulson, one has $p_1 = -p_2 = p/2$ and $q_1 = -q_2 = q/2$ (so that $P = 0$ and $Q = 0$). In this case, the quadrature formula (12.40) reduces to

$$\pm(t - t_0) = \frac{1}{\sqrt{-4H}} \int_{q(t_0)}^{q(t)} \frac{dq\,'}{\left(1 - G(q\,')\right)^{1/2}}, \tag{12.45}$$

and the second constant of motion in eqn (12.38) satisfies

$$-4H = p^2\left(1 - G(q)\right) \geq 0. \tag{12.46}$$

After the collision, the pulson and anti-pulson separate and travel apart in opposite directions; so that asymptotically in time $g(q) \to 0$, $p \to 2c$, and $H \to -c^2$, where $c$ (or $-c$) is the asymptotic speed (and amplitude) of the pulson (or anti-pulson). Setting $H = -c^2$ in eqn (12.46) gives a relation for the pulson–anti-pulson $(p, q)$ phase trajectories for any kernel,

$$p = \pm \frac{2c}{\left(1 - G(q)\right)^{1/2}}. \tag{12.47}$$

Notice that $p$ diverges (and switches branches of the square root) when $q \to 0^+$, because $G(0) = 1$. The convention of switching branches of the square root allows one to keep $q > 0$ throughout, so the particles retain their order. That is, the particles 'bounce' elastically at the moment when $q \to 0^+$ in the perfectly anti-symmetric head-on collision.

**Remark 12.16 (Preservation of particle identity in collisions)** *The relative separation distance $q(t)$ in pulson–anti-pulson collisions is determined*

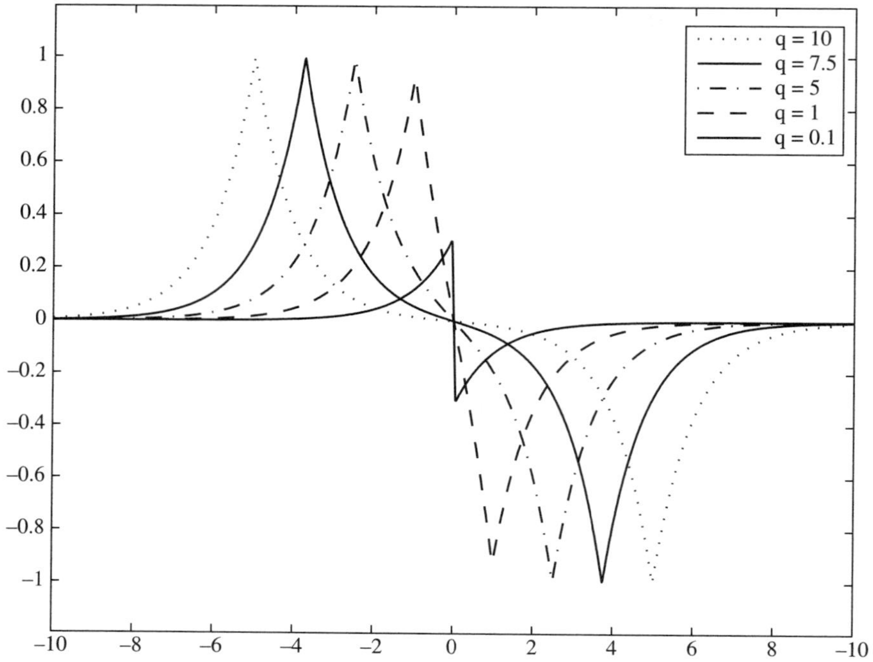


**Figure 12.3** This is the velocity profile (12.48) for the peakon-antipeakon head-on collision as a function of separation between the peaks [FH01].

*by following a phase point along a level surface of the Hamiltonian H in the phase space with coordinates $(q, p)$. Because H is quadratic, the relative momentum p has two branches on such a level surface, as indicated by the $\pm$ sign in eqn (12.47). At the pulson–anti-pulson collision point, both $q \to 0^+$ and either $1/p \to 0^+$ or $p \to 0^+$, so following a phase point through a collision requires that one must choose a convention for which branch of the level surface is taken after the collision. Taking the convention that p changes sign (corresponding to a* ***bounce****), but q does not change sign (so the* ***particles keep their identity****) is convenient, because it allows the phase points to be followed more easily through multiple collisions. This choice is also consistent with the pulson–pulson and anti-pulson–anti-pulson collisions. In these other* ***rear-end collisions****, as implied by eqn (12.43), the separation distance always remains positive and again the particles retain their identity.*

## Theorem 12.17 (Pulson–anti-pulson exact solution)

*The exact analytical solution for the pulson–anti-pulson collision for any symmetric G may be written as a function of position x and the separation*

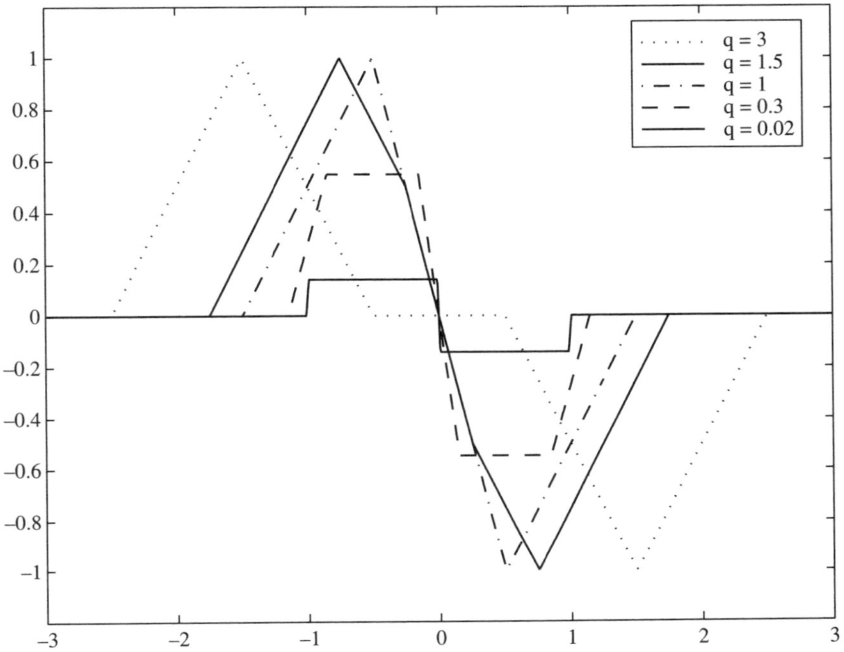


**Figure 12.4** Velocity profile (12.48) for the head-on collision of the triangular peakon–antipeakon pair as a function of separation between the peaks [FH01].

*between the pulses q for any pulse shape or kernel G(x) as*

$$u(x,q) = \frac{c}{\big(1 - G(q)\big)^{1/2}}\Big[G(x+q/2) - G(x-q/2)\Big], \tag{12.48}$$

*where c is the pulson speed at sufficiently large separation and the dynamics of the separation q(t) is given by the quadrature (12.45) with* $\sqrt{-4H} = 2c$.

*Proof* The solution for the velocity $u(x,t)$ in the head-on pulson–anti-pulson collision may be expressed in this notation as

$$u(x,t) = \frac{p}{2}G(x+q/2) - \frac{p}{2}G(x-q/2). \tag{12.49}$$

In using eqn (12.47) to eliminate $p$ this solution becomes eqn (12.48). ■

**Exercise 12.9**
According to eqn (12.45), how much time is required for the head-on pulson–anti-pulson collision, when $G(q) = e^{-q^2/2}$ is a Gaussian?

**Exercise 12.10**
For the case that $G(x) = e^{-|x|}$, which is Green's function for the Helmholtz operator in 1D with $\alpha = 1$, show that solution (12.49) for the peakon–anti-peakon collision yields

$$q = -\log \operatorname{sech}^2(ct), \quad p = \frac{\pm 2c}{\tanh(ct)}, \tag{12.50}$$

so the peakon–anti-peakon collision occurs at time $t = 0$ and eqn (12.49) results in

$$\begin{aligned} m(x,t) &= u - \alpha^2 u_{xx} \\ &= \frac{2c}{\tanh(ct)} \left[ \delta\left(x - \frac{1}{2}q(t)\right) - \delta\left(x + \frac{1}{2}q(t)\right) \right]. \end{aligned} \tag{12.51}$$

Discuss the behaviour of this solution. What happens to the slope and amplitude of the peakon velocity just at the moment of impact?

# 13 Integrability of EPDiff in 1D

In the previous chapter, we discussed the CH equation for unidirectional shallow-water waves derived in [CH93], as a special case of the $b$-equation (12.14) with $b = 2$,

$$m_t + um_x + 2mu_x = -c_0 u_x + \gamma u_{xxx}, \qquad m = u - \alpha^2 u_{xx}. \tag{13.1}$$

This partial differential equation (PDE) describes shallow-water dynamics at quadratic order in the asymptotic expansion for unidirectional shallow-water waves on a free surface under gravity. The previous chapter discussed its elastic particle-collision solution properties in the dispersionless case for which the linear terms on the right side of eqn (13.1) are absent. These elastic-collision solution properties hold for any Green's function $G(x)$ in the convolution relation $u = G * m$ between velocity $u$ and momentum $m$. For the CH equation $G(x) = \mathrm{e}^{-|x|/\alpha}$ is the Green's function for the 1D Helmholtz operator on the real line with homogeneous boundary conditions.

This chapter explains the noncanonical Hamiltonian properties of the CH equation (13.1) in one spatial dimension. In fact, the CH equation has two compatible Hamiltonian structures, so it is ***bi-Hamiltonian***. In this situation, Magri's lemmas for bi-Hamiltonian PDE in 1D imply systematically that CH arises as a different compatibility condition for an ***isospectral eigenvalue problem*** and a linear evolution equation for the corresponding eigenfunctions in the case when $G(x) = \mathrm{e}^{-|x|/\alpha}$. The properties of being bi-Hamiltonian and possessing an associated isospectral problem are ingredients for proving the one-dimensional CH equation (13.1) is ***completely integrable*** as a Hamiltonian system and is solvable by the ***inverse scattering transform (IST) method***.

## 13.1 The CH equation is bi-Hamiltonian

The CH equation is ***bi-Hamiltonian***. This means that eqn (13.1) may be written in two compatible Hamiltonian forms, namely as

$$m_t = -B_2 \frac{\delta H_1}{\delta m} = -B_1 \frac{\delta H_2}{\delta m}, \tag{13.2}$$

where $B_1$ and $B_2$ are Poisson operators. For the CH equation, the pairs of Hamiltonians and Poisson operators are given by

$$\begin{aligned}
H_1 &= \frac{1}{2}\int (u^2 + \alpha^2 u_x^2)\, \mathrm{d}x, \\
B_2 &= \partial_x m + m\partial_x + c_0\partial_x + \gamma\, \partial_x^3, \qquad (13.3)\\
H_2 &= \frac{1}{2}\int u^3 + \alpha^2 u\, u_x^2 + c_0 u^2 - \gamma\, u_x^2\, \mathrm{d}x, \\
B_1 &= \partial_x - \alpha^2 \partial_x^3. \qquad (13.4)
\end{aligned}$$

These bi-Hamiltonian forms restrict properly to those for KdV when $\alpha^2 \to 0$, and to those for EPDiff when $c_0, \gamma \to 0$. Compatibility of $B_1$ and $B_2$ is assured, because $(\partial_x m + m\partial_x)$, $\partial_x$ and $\partial_x^3$ are all mutually compatible Hamiltonian operators. That is, any linear combination of these operators defines a Poisson bracket,

$$\{f, h\}(m) = -\int \frac{\delta f}{\delta m}(c_1 B_1 + c_2 B_2)\frac{\delta h}{\delta m}\, \mathrm{d}x, \tag{13.5}$$

as a bilinear skew-symmetric operation that satisfies the Jacobi identity. (In general, the sum of the Poisson brackets would fail to satisfy the Jacobi identity.) Moreover, no further deformations of these Hamiltonian operators involving higher-order partial derivatives would be compatible with $B_2$, as shown in [Olv00]. This fact was already known in the literature for KdV, see [Fuc96].

### 13.1.1 Magri's lemmas

The property of ***compatibility*** of the two Hamiltonian operators for a bi-Hamiltonian equation enables the construction under certain conditions of an infinite hierarchy of Poisson-commuting Hamiltonians. The property of compatibility was used by Magri [Mag78] in proving the following important pair of lemmas (see also [Olv00] for a clear discussion of Magri's lemmas):

**Lemma 13.1 (Magri 1978)** *If $B_1$ and $B_2$ are compatible Hamiltonian operators, with $B_1$ non-degenerate, and if*

$$B_2 \frac{\delta H_1}{\delta m} = B_1 \frac{\delta H_2}{\delta m} \quad \textit{and} \quad B_2 \frac{\delta H_2}{\delta m} = B_1 \mathcal{K}, \tag{13.6}$$

*for Hamiltonians $H_1$, $H_2$, and some function $\mathcal{K}$, then there exists a third Hamiltonian functional $H$ such that $\mathcal{K} = \delta H / \delta m$.*

To prove the existence of an infinite hierarchy of Hamiltonians, $H_n$, $n = 1, 2, \ldots$, related to the two compatible Hamiltonian operators $B_1$, $B_2$, we need to check that the following two conditions hold:

(i) There exists an infinite sequence of functions $\mathcal{K}_1, \mathcal{K}_2, \ldots$ satisfying

$$B_2 \mathcal{K}_n = B_1 \mathcal{K}_{n+1}\,; \tag{13.7}$$

(ii) There exist two functionals $H_1$ and $H_2$ such that

$$\mathcal{K}_1 = \frac{\delta H_1}{\delta m}, \qquad \mathcal{K}_2 = \frac{\delta H_2}{\delta m}. \tag{13.8}$$

It then follows from Lemma 13.1 that there exist functionals $H_n$ such that

$$\mathcal{K}_n = \frac{\delta H_n}{\delta m}, \quad \text{for all} \quad n \geq 1. \tag{13.9}$$

**Lemma 13.2 (Magri 1978)** *Let $\{\cdot, \cdot\}_1$ and $\{\cdot, \cdot\}_2$ denote the Poisson brackets defined, respectively, by $B_1$ and $B_2$, which are assumed to be compatible Hamiltonian operators. Let $H_1, H_2, \ldots$ be an infinite sequence of Hamiltonian functionals constructed from eqns (13.7) and (13.9). Then, these Hamiltonian functionals mutually commute under both Poisson brackets:*

$$\{H_m, H_n\}_1 = \{H_m, H_n\}_2 = 0, \quad \textit{for all} \quad m, n \geq 1. \tag{13.10}$$

**Definition 13.3** *A set of Hamiltonians that Poisson-commute among themselves is said to be* ***in involution****.*

**Remark 13.4** *The condition for a canonical Hamiltonian system with N degrees of freedom to be* ***completely integrable*** *is that it possess N constants* *of motion in involution. The bi-Hamiltonian property is important because it produces the corresponding condition for an infinite-dimensional system. The infinite-dimensional case introduces additional questions, such as the completeness of the infinite set of independent constants of motion in involution. However, such questions are beyond our present scope.*

### 13.1.2 Applying Magri's lemmas

The bi-Hamiltonian property of eqn (13.1) allows one to construct an infinite number of Poisson-commuting conservation laws for it by applying Magri's lemmas. According to [Mag78], these conservation laws may be constructed for non-degenerate $B_1$ by defining the transpose operator $R^T = B_1^{-1}B_2$ that leads from the variational derivative of one conservation law to the next,

$$\frac{\delta H_n}{\delta m} = R^T \frac{\delta H_{n-1}}{\delta m}, \quad n = -1, 0, 1, 2, \ldots. \tag{13.11}$$

The operator $R^T = B_1^{-1}B_2$ recursively takes the variational derivative of $H_{-1}$ to that of $H_0$, to that of $H_1$, then to that of $H_2$, etc. The next steps are not so easy for the integrable CH hierarchy, because each application of the recursion operator introduces an additional convolution integral into the sequence. Correspondingly, the ***recursion operator*** $R = B_2B_1^{-1}$ leads to a hierarchy of commuting flows, defined by $\mathcal{K}_{n+1} = R\mathcal{K}_n$, for $n = 0, 1, 2, \ldots$,

$$\begin{aligned} m_t^{(n+1)} &= \mathcal{K}_{n+1}[m] = -B_1 \frac{\delta H_n}{\delta m} \\ &= -\, B_2 \frac{\delta H_{n-1}}{\delta m} = B_2 B_1^{-1} \mathcal{K}_n[m]. \end{aligned} \tag{13.12}$$

The first three flows in the 'positive hierarchy' when $c_0, \gamma \to 0$ are

$$m_t^{(1)} = 0, \quad m_t^{(2)} = -\, m_x, \quad m_t^{(3)} = -\,(m\partial + \partial m)u, \tag{13.13}$$

the third being EPDiff. The next flow is too complicated to be usefully written here. However, by Magri's construction, all of these flows commute with the other flows in the hierarchy, so they each conserve $H_n$ for $n = 0, 1, 2, \ldots$.

The recursion operator can also be continued for negative values of $n$. The conservation laws generated in this way do not introduce convolutions, but care must be taken to ensure the conserved densities are integrable. All the Hamiltonian densities in the negative hierarchy are expressible in terms of $m$ only and do not involve $u$. Thus, for instance, the second Hamiltonian in the negative hierarchy of EPDiff is given by

$$m_t = B_1 \frac{\delta H_{-1}}{\delta m} = B_2 \frac{\delta H_{-2}}{\delta m}, \tag{13.14}$$

which gives

$$H_{-2} = \frac{1}{2} \int_{-\infty}^{\infty} \left[ \frac{\alpha^2}{4} \frac{m_x^2}{m^{5/2}} - \frac{2}{\sqrt{m}} \right]. \tag{13.15}$$

The flow defined by eqn (13.14) is

$$m_t = -(\partial - \alpha^2 \partial^3) \left( \frac{1}{2\sqrt{m}} \right). \tag{13.16}$$

For $m = u - \alpha^2 u_{xx}$, this flow is similar to the Dym equation,

$$u_{xxt} = \partial^3 \left( \frac{1}{2\sqrt{u_{xx}}} \right), \tag{13.17}$$

which is also a completely integrable soliton equation [AS06].

## 13.2 The CH equation is isospectral

The isospectral eigenvalue problem associated with eqn (13.1) may be found by using the recursion relation of the bi-Hamiltonian structure, following a standard technique due to Gelfand and Dorfman [GD79]. Let us introduce a spectral parameter $\lambda$ and multiply by $\lambda^n$ the $n$th step of the recursion relation (13.12), then taking the sum yields

$$B_1 \sum_{n=0}^{\infty} \lambda^n \frac{\delta H_n}{\delta m} = \lambda B_2 \sum_{n=0}^{\infty} \lambda^{(n-1)} \frac{\delta H_{n-1}}{\delta m}, \tag{13.18}$$

or, by introducing the squared-eigenfunction $\psi^2$

$$\psi^2(x,t;\lambda) := \sum_{n=-1}^{\infty} \lambda^n \frac{\delta H_n}{\delta m}, \tag{13.19}$$

one finds, formally,

$$B_1 \psi^2(x,t;\lambda) = \lambda B_2 \psi^2(x,t;\lambda). \tag{13.20}$$

This is a third-order eigenvalue problem for the squared-eigenfunction $\psi^2$, which turns out to be equivalent to a second-order ***Sturm–Liouville problem*** for $\psi$.

**Proposition 13.5** *If $\psi$ satisfies*

$$\lambda\left(\frac{1}{4} - \alpha^2 \partial_x^2\right)\psi = \left(\frac{c_0}{4} + \frac{m(x,t)}{2} + \gamma\, \partial_x^2\right)\psi, \tag{13.21}$$

*then $\psi^2$ is a solution of eqn (13.20).*

*Proof* This is straightforward computation. ■

Now, assuming that $\lambda$ will be independent of time, we seek, in analogy with the KdV equation, an evolution equation for $\psi$ of the form,

$$\psi_t = a\psi_x + b\psi, \tag{13.22}$$

where $a$ and $b$ are functions of $u$ and its derivatives. These functions are determined from the requirement that the ***compatibility condition*** $\psi_{xxt} = \psi_{txx}$ between eqns (13.21) and (13.22) implies eqn (13.1). Cross-differentiation shows

$$b = -\frac{1}{2}a_x, \quad \text{and} \quad a = -(\lambda + u). \tag{13.23}$$

Consequently,

$$\psi_t = -(\lambda + u)\psi_x + \frac{1}{2}u_x\psi, \tag{13.24}$$

is the desired evolution equation for the eigenfunction $\psi$.

## Summary of the isospectral property of eqn (13.1)

The Gelfand–Dorfman theory [GD79] determines the isospectral problem for integrable equations via the squared-eigenfunction approach. Its bi-Hamiltonian property implies that the nonlinear shallow-water wave equation (13.1) arises as a compatibility condition for two linear equations. These are the ***isospectral eigenvalue problem***,

$$\lambda\left(\frac{1}{4} - \alpha^2 \partial_x^2\right)\psi = \left(\frac{c_0}{4} + \frac{m(x,t)}{2} + \gamma\, \partial_x^2\right)\psi, \tag{13.25}$$

and the ***evolution equation*** for the eigenfunction $\psi$,

$$\psi_t = -(u + \lambda)\, \psi_x + \frac{1}{2} u_x\, \psi . \tag{13.26}$$

Compatibility of these linear equations ($\psi_{xxt} = \psi_{txx}$) together with isospectrality

$$\mathrm{d}\lambda/\mathrm{d}t = 0,$$

imply eqn (13.1).

**Remark 13.6** *The isospectral eigenvalue problem (13.25) for the nonlinear CH water-wave equation (13.1) restricts to the isospectral problem for KdV (namely, the Schrödinger equation) when $\alpha^2 \to 0$. The evolution equation (13.26) for the isospectral eigenfunctions in the cases of KdV and CH are identical. The isospectral eigenvalue problem and the evolution equation for its eigenfunctions are two linear equations whose compatibility implies a nonlinear equation for the unknowns in the KdV and CH equations. This formulation for the KdV equation led to the famous method of the* ***inverse scattering transform (IST)*** *for the solution of its initial-value problem, reviewed, e.g., in [AS06]. The CH equation also admits the IST solution approach, but for a different isospectral eigenvalue problem that limits to the Schrödinger equation when $\alpha^2 \to 0$. The isospectral eigenvalue problem (13.25) for CH arises in the study of the fundamental oscillations of a non-uniform string under tension.*

## EPDiff is the dispersionless case of CH

In the dispersionless case $c_0 = 0 = \gamma$, the shallow-water equation (13.1) becomes the 1D geodesic equation EPDiff($H^1$)

$$m_t + u m_x + 2 m u_x = 0, \qquad m = u - \alpha^2 u_{xx}. \tag{13.27}$$

The solitary travelling-wave solution of 1D EPDiff (13.27) in this dispersionless case is the ***peakon***,

$$u(x,t) = c\, G(x - ct) = \frac{c}{2\alpha}\, \mathrm{e}^{-|x-ct|/\alpha}.$$

The EPDiff equation (12.1) may also be written as a conservation law for momentum,

$$\partial_t m = -\partial_x \Big( u m + \frac{1}{2} u^2 - \frac{\alpha^2}{2} u_x^2 \Big). \tag{13.28}$$

Its isospectral problem forms the basis for completely integrating the EPDiff equation as a Hamiltonian system and, thus, for finding its soliton solutions. Remarkably, the isospectral problem (13.25) in the dispersionless case $c_0 = 0 = \Gamma$ has a ***purely discrete spectrum*** on the real line and the $N$-soliton solutions for this equation may be expressed as a ***peakon wave train***,

$$u(x,t) = \sum_{i=1}^{N} p_i(t) \mathrm{e}^{-|x-q_i(t)|/\alpha}. \tag{13.29}$$

As before, $p_i(t)$ and $q_i(t)$ satisfy the finite-dimensional geodesic motion equations obtained as Hamilton's canonical equations

$$\dot{q}_i = \frac{\partial H_N}{\partial p_i} \quad \text{and} \quad \dot{p}_i = -\frac{\partial H_N}{\partial q_i}, \tag{13.30}$$

where the Hamiltonian is given by,

$$H_N = \frac{1}{2} \sum_{i,j=1}^{N} p_i \, p_j \, \mathrm{e}^{-|q_i - q_j|/\alpha}. \tag{13.31}$$

Thus, we have proved the following.

**Theorem 13.7** *CH peakons are an integrable subcase of EPDiff pulsons in one dimension for the choice of the $H^1$ norm.*

**Remark 13.8** *The discrete process of peakon creation via the steepening lemma* 12.1.1 *is consistent with the discreteness of the isospectrum for the eigenvalue problem (13.25) in the dispersionless case, when $c_0 = 0 = \gamma$. These discrete eigenvalues correspond in turn to the asymptotic speeds of the peakons. The discreteness of the isospectrum means that only peakons will emerge in the initial-value problem for EPDiff($H^1$) in 1D.*

## Constants of motion for integrable *N*-peakon dynamics

One may verify the integrability of the $N$-peakon dynamics by substituting the $N$-peakon solution (13.29) (which produces the sum of delta functions (12.6) for the momentum $m$) into the isospectral problem (13.25). This substitution reduces (13.25) to an $N \times N$ matrix eigenvalue problem.

In fact, the canonical equations (13.30) for the peakon Hamiltonian (13.31) may be written directly in Lax matrix form,

$$\frac{\mathrm{d}L}{\mathrm{d}t} = [L, A] \quad \Longleftrightarrow \quad L(t) = U(t)L(0)U^{\dagger}(t), \tag{13.32}$$

with $A = \dot{U}U^{\dagger}(t)$ and $UU^{\dagger} = Id$. Explicitly, $L$ and $A$ are $N \times N$ matrices with entries

$$L_{jk} = \sqrt{p_j p_k}\,\phi(q_i - q_j), \quad A_{jk} = -2\sqrt{p_j p_k}\,\phi'(q_i - q_j). \tag{13.33}$$

Here, $\phi'(x)$ denotes derivative with respect to the argument of the function $\phi$, given by $\phi(x) = \mathrm{e}^{-|x|/2\alpha} = 2\alpha G(x/2)$. The Lax matrix $L$ in eqn (13.32) evolves by time-dependent unitary transformations, which leave its spectrum invariant. Isospectrality then implies that the traces $\operatorname{tr} L^n$, $n = 1, 2, \ldots, N$ of the powers of the matrix $L$ (or, equivalently, its $N$ eigenvalues) yield $N$ constants of the motion. These turn out to be functionally independent, non-trivial and in involution under the canonical Poisson bracket. Hence, the canonically Hamiltonian $N$-peakon dynamics (13.30) is completely integrable in the finite-dimensional (Liouville) sense.

**Exercise 13.1**
Verify that the compatibility condition (equality of cross derivatives $\psi_{xxt} = \psi_{txx}$) obtained from the eigenvalue equation (13.25) and the evolution equation (13.26) do indeed yield the CH shallow-water wave equation (13.1) when the eigenvalue $\lambda$ is constant.

**Exercise 13.2**
Show that the peakon Hamiltonian $H_N$ in (13.31) may be expressed as a function of the invariants of the matrix $L$, as

$$H_N = -\operatorname{tr} L^2 + 2(\operatorname{tr} L)^2. \tag{13.34}$$

Show that evenness of $H_N$ implies

1. The $N$ coordinates $q_i$, $i = 1, 2, \ldots, N$ keep their initial ordering.
2. The $N$ conjugate momenta $p_i$, $i = 1, 2, \ldots, N$ keep their initial signs.

This means that no difficulties arise, either due to the non-analyticity of $\phi(x)$, or the sign in the square roots in the Lax matrices $L$ and $A$.

**Exercise 13.3 (Hunter–Saxton equation)**
Retrace the progress of this chapter for the EPDiff equation

$$m_t + um_x + 2mu_x = 0, \quad \text{with} \quad m = -u_{xx}. \tag{13.35}$$

This integrable Hamiltonian partial differential equation arises in the theory of liquid crystals. Its peakon solutions are the compactly supported triangles in Figure 12.2 and Figure 12.4. It may also be regarded as the $\alpha \to \infty$ limit of the CH equation. For more results and discussion of this equation, see [HZ94].

# 14 EPDiff in n dimensions

This chapter discusses the $n$-dimensional generalization of the one-dimensional singular solutions of the EP equation studied in the previous chapter. Much of the one-dimensional structure persists in higher dimensions. For example, the parameters defining the singular solutions of EPDiff in $n$ dimensions still obey canonical Hamiltonian equations. This is understood by identifying the singular solution ansatz as a cotangent-lift momentum map for the left action of the diffeomorphisms on the lower-dimensional support set of the singular solutions.

## 14.1 Singular momentum solutions of the EPDiff equation for geodesic motion in higher dimensions

### 14.1.1 *n*-dimensional EPDiff equation

Eulerian geodesic motion of an ideal continuous fluid in $n$ dimensions is generated as an EP equation via Hamilton's principle, when the Lagrangian is given by the kinetic energy. The kinetic energy defines a norm $\|u\|^2$ for the Eulerian fluid velocity, $\mathbf{u}(\mathbf{x},t) : R^n \times R^1 \to R^n$. As mentioned earlier, the choice of the kinetic energy as a positive functional of fluid velocity $\mathbf{u}$ is a modelling step that depends upon the physics of the problem being studied. Following our earlier procedure, as in eqns (11.11) and (11.14), we shall choose the Lagrangian,

$$\|\mathbf{u}\|^2 = \int \mathbf{u} \cdot Q_{op}\mathbf{u}\, \mathrm{d}^n x = \int \mathbf{u} \cdot \mathbf{m}\, \mathrm{d}^n x, \tag{14.1}$$

so that the positive-definite, symmetric, operator $Q_{op}$ defines the norm $\|\mathbf{u}\|$, under integration by parts for appropriate boundary conditions and the EPDiff equation for Eulerian geodesic motion of a fluid emerges,

$$\frac{\mathrm{d}}{\mathrm{d}t}\frac{\delta \ell}{\delta \mathbf{u}} + \mathrm{ad}^*_{\mathbf{u}}\frac{\delta \ell}{\delta \mathbf{u}} = 0, \quad \text{with} \quad \ell[\mathbf{u}] = \frac{1}{2}\|\mathbf{u}\|^2. \tag{14.2}$$

## Legendre transforming to the Hamiltonian side

The corresponding Legendre transform yields the following invertible relations between momentum and velocity,

$$\mathbf{m} = Q_{op}\mathbf{u} \quad \text{and} \quad \mathbf{u} = G * \mathbf{m}, \tag{14.3}$$

where $G$ is the **Green's function** for the operator $Q_{op}$, assuming appropriate boundary conditions (on $\mathbf{u}$) that allow inversion of the operator $Q_{op}$ to determine $\mathbf{u}$ from $\mathbf{m}$.

The associated **Hamiltonian** is,

$$h[\mathbf{m}] = \langle \mathbf{m}, \mathbf{u} \rangle - \frac{1}{2}\|\mathbf{u}\|^2 = \frac{1}{2}\int \mathbf{m} \cdot G * \mathbf{m} \, \mathrm{d}^n x =: \frac{1}{2}\|\mathbf{m}\|^2, \tag{14.4}$$

which also defines a norm $\|\mathbf{m}\|$ via a convolution kernel $G$ that is symmetric and positive, when the Lagrangian $\ell[\mathbf{u}]$ is a norm. As expected, the norm $\|\mathbf{m}\|$ given by the Hamiltonian $h[\mathbf{m}]$ specifies the velocity $\mathbf{u}$ in terms of its Legendre-dual momentum $\mathbf{m}$ by the variational operation,

$$\mathbf{u} = \frac{\delta h}{\delta \mathbf{m}} = G * \mathbf{m} \equiv \int G(\mathbf{x} - \mathbf{y}) \, \mathbf{m}(\mathbf{y}) \, \mathrm{d}^n y. \tag{14.5}$$

We shall choose the kernel $G(\mathbf{x}-\mathbf{y})$ to be translation-invariant (so Noether's theorem implies that total momentum $\mathbf{M} = \int \mathbf{m} \, \mathrm{d}^n x$ is conserved) and symmetric under spatial reflections (so that $\mathbf{u}$ and $\mathbf{m}$ have the same parity under spatial reflections).

After the Legendre transformation (14.4), the EPDiff equation (11.14) appears in its equivalent **Lie–Poisson Hamiltonian form,**

$$\frac{\partial}{\partial t}\mathbf{m} = \{\mathbf{m}, h\} = -\operatorname{ad}^*_{\delta h/\delta \mathbf{m}}\mathbf{m}. \tag{14.6}$$

Here the operation $\{\cdot, \cdot\}$ denotes the Lie–Poisson bracket dual to the (right) action of vector fields amongst themselves by vector-field commutation. That is,

$$\{f, h\} = -\left\langle \mathbf{m}, \left[\frac{\delta f}{\delta \mathbf{m}}, \frac{\delta h}{\delta \mathbf{m}}\right]\right\rangle. \tag{14.7}$$

For more details and additional background concerning the relation of classical EP theory to Lie–Poisson Hamiltonian equations, see [MR02, HMR98a]. In a moment we will also consider the momentum maps for EPDiff.

### 14.1.2 Pulsons in *n* dimensions

The momentum for the one-dimensional pulson solutions (12.6) on the real line is supported at points via the Dirac delta measures in its solution ansatz,

$$m(x,t) = \sum_{i=1}^{N} p_i(t)\,\delta\big(x - q_i(t)\big), \quad m \in \mathbb{R}. \tag{14.8}$$

We shall develop $n$-dimensional analogues of these one-dimensional pulson solutions for the Euler–Poincaré equation (11.23) by generalizing this solution ansatz to allow measure-valued $n$-dimensional vector solutions $\mathbf{m} \in \mathbb{R}^n$ for which the Euler–Poincaré momentum is supported on codimension-$k$ *subspaces* $\mathbb{R}^{n-k}$ with integer $k \in [1,n]$. For example, one may consider the two-dimensional vector momentum $\mathbf{m} \in \mathbb{R}^2$ in the plane that is supported on one-dimensional curves (momentum fronts). Likewise, in three dimensions, one could consider two-dimensional momentum surfaces (sheets), one-dimensional momentum filaments, etc. The corresponding vector momentum ansatz that we shall use is the following, cf. the pulson solutions (14.8),

$$\mathbf{m}(\mathbf{x},t) = \sum_{i=1}^{N} \int \mathbf{P}_i(s,t)\,\delta\big(\mathbf{x} - \mathbf{Q}_i(s,t)\big)\mathrm{d}s, \quad \mathbf{m} \in \mathbb{R}^n. \tag{14.9}$$

Here, $\mathbf{P}_i, \mathbf{Q}_i \in \mathbb{R}^n$ for $i = 1,2,\ldots,N$. For example, when $n-k=1$, so that $s \in \mathbb{R}$ is one-dimensional, the delta function in solution (14.9) supports an evolving family of vector-valued curves, called ***momentum filaments***. (For simplicity of notation, we suppress the implied subscript $i$ in the arclength $s$ for each $\mathbf{P}_i$ and $\mathbf{Q}_i$.) The Legendre-dual relations (14.3) imply that the velocity corresponding to the momentum filament ansatz (14.9) is,

$$\mathbf{u}(\mathbf{x},t) = G * \mathbf{m} = \sum_{j=1}^{N} \int \mathbf{P}_j(s',t)\,G\big(\mathbf{x} - \mathbf{Q}_j(s',t)\big)\mathrm{d}s'. \tag{14.10}$$

Just as for the 1D case of the pulsons, we shall show that substitution of the $n$-D solution ansatz (14.9) and (14.10) into the EPDiff equation (11.20) produces canonical geodesic Hamiltonian equations for the $n$-dimensional vector parameters $\mathbf{Q}_i(s,t)$ and $\mathbf{P}_i(s,t)$, $i = 1,2,\ldots,N$.

#### Canonical Hamiltonian dynamics of momentum filaments

For definiteness in what follows, we shall consider the example of momentum filaments $\mathbf{m} \in \mathbb{R}^n$ supported on one-dimensional space curves in $\mathbb{R}^n$, so $s \in \mathbb{R}$ is the arclength parameter of one of these curves. This solution ansatz is reminiscent of the Biot–Savart Law for vortex filaments, although the flow is not incompressible. The dynamics of momentum surfaces, for $s \in \mathbb{R}^k$ with $k < n$, follow a similar analysis.

Substituting the momentum filament ansatz (14.9) for $s \in \mathbb{R}$ and its corresponding velocity (14.10) into the Euler–Poincaré equation (11.20), then integrating against a smooth test function $\phi(\mathbf{x})$ implies the following canonical equations (denoting explicit summation on $i, j \in 1, 2, \ldots N$),

$$\frac{\partial}{\partial t}\mathbf{Q}_i(s,t) = \sum_{j=1}^{N} \int \mathbf{P}'_j \, G(\mathbf{Q}_i - \mathbf{Q}'_j) \mathrm{d}s' = \frac{\delta H_N}{\delta \mathbf{P}_i}, \tag{14.11}$$

$$\begin{aligned} \frac{\partial}{\partial t}\mathbf{P}_i(s,t) &= -\sum_{j=1}^{N} \int \left(\mathbf{P}_i \cdot \mathbf{P}'_j\right) \frac{\partial}{\partial \mathbf{Q}_i} G(\mathbf{Q}_i - \mathbf{Q}'_j) \, \mathrm{d}s' \\ &= -\frac{\delta H_N}{\delta \mathbf{Q}_i}, \end{aligned} \tag{14.12}$$

where $\mathbf{P}_i = \mathbf{P}_i(s,t)$, $\mathbf{P}'_j := \mathbf{P}_j(s',t)$ and

$$G(\mathbf{Q}_i - \mathbf{Q}'_j) := G(\mathbf{Q}_i(s,t) - \mathbf{Q}_j(s',t)). \tag{14.13}$$

The dot product $\mathbf{P}_i \cdot \mathbf{P}'_j$ denotes the inner, or scalar, product of the two vectors $\mathbf{P}_i$ and $\mathbf{P}'_j$ in $T^*\mathbb{R}^n$. Thus, the solution ansatz (14.9) yields a closed set of ***integro-partial-differential equations (IPDEs)*** given by (14.11) and (14.12) for the vector parameters $\mathbf{Q}_i(s,t)$ and $\mathbf{P}_i(s,t)$ with $i = 1, 2 \ldots N$. These equations are generated canonically by the Hamiltonian function $H_N : (T^*\mathbb{R}^n)^N \to \mathbb{R}$ given by

$$H_N[\mathbf{P}, \mathbf{Q}] = \frac{1}{2} \iint \sum_{i,j=1}^{N} \left(\mathbf{P}_i \cdot \mathbf{P}'_j\right) G\left(\mathbf{Q}_i - \mathbf{Q}'_j\right) \mathrm{d}s \, \mathrm{d}s' =: \frac{1}{2} \|\mathbf{P}\|^2. \tag{14.14}$$

This Hamiltonian arises by inserting the momentum ansatz (14.9) into the Hamiltonian (14.4) obtained from the Legendre transformation of the Lagrangian corresponding to the kinetic energy norm of the fluid velocity. Thus, the evolutionary IPDE system (14.11) and (14.12) represents canonically Hamiltonian geodesic motion on the space of curves in $R^n$ with respect to the cometric given on these curves in eqn (14.14). The Hamiltonian $H_N = \frac{1}{2}\|\mathbf{P}\|^2$ in eqn (14.14) defines the norm $\|\mathbf{P}\|$ in terms of this cometric that combines convolution using the Green's function $G$ and sum over filaments with the scalar product of momentum vectors in $R^n$.

**Remark 14.1** *Note that the coordinate s is a Lagrangian label moving with the fluid, since*

$$\frac{\partial}{\partial t}\mathbf{Q}_i(s,t) = \mathbf{u}(\mathbf{Q}_i(s,t),t).$$

**Exercise 14.1**
Explain the meaning of the Hamiltonian equation (14.6) with Lie–Poisson bracket (14.7). Discuss the interpretation of $\{\mathbf{m},h\}$ when $\mathbf{m}$ is a vector.
HINT: write $\mathbf{m}(\mathbf{x},t)$ as a spatial integral by inserting a delta function,

$$\mathbf{m}(\mathbf{x},t) = \int_{\mathbb{R}^3} \mathbf{m}(\mathbf{y},t)\,\delta(\mathbf{x}-\mathbf{y})\,\mathrm{d}^3y.$$

Use this representation to show that the Lie–Poisson bracket (14.7) yields dynamics in the form of eqn (14.6),

$$\{\mathbf{m},h\} = -\left\langle \mathrm{ad}^*_{\delta h/\delta \mathbf{m}}\mathbf{m},\ \delta(\mathbf{x}-\mathbf{y})\right\rangle = -\,\mathrm{ad}^*_{\delta h/\delta \mathbf{m}}\mathbf{m}.$$

## 14.2 Singular solution momentum map $\mathbf{J}_{\mathrm{Sing}}$

The momentum filament ansatz (14.9) reduces the solution of the geodesic EP PDE (11.20) in $n+1$ dimensions to the system of eqns (14.11) and (14.12) of $2N$ canonical evolutionary IPDEs. One can summarize the mechanism by which this process occurs, by saying that the map that implements the canonical $(\mathbf{Q},\mathbf{P})$ variables in terms of singular solutions is a (cotangent bundle) momentum map.

Such momentum maps are Poisson maps; so the canonical Hamiltonian nature of the dynamical equations for $(\mathbf{Q},\mathbf{P})$ fits into a general theory that also provides a framework for suggesting other avenues of investigation.

**Theorem 14.2** *The momentum ansatz* (14.9) *for measure-valued solutions of the* EPDiff *equation* (11.20), *defines an equivariant momentum map*

$$\mathbf{J}_{\mathrm{Sing}} : T^*\,\mathrm{Emb}(S,\mathbb{R}^n) \to \mathfrak{X}(\mathbb{R}^n)^*,$$

*called the* ***singular solution momentum map*** *in [HM04].*

We shall explain the notation used in the theorem's statement in the course of its proof. By 'defines' one means that the momentum solution ansatz (14.9) expressing $\mathbf{m}$ (a vector function of spatial position $\mathbf{x}$) in terms of $\mathbf{Q}, \mathbf{P}$ (which are functions of $s$) can be regarded as a map from the space of $(\mathbf{Q}(s), \mathbf{P}(s))$ to the space of $\mathbf{m}$'s. This will turn out to be the Lagrange-to-Euler map for the fluid description of the singular solutions.

Following [HM04], we shall give two proofs of this result from two rather different viewpoints. The first proof below uses the formula for a momentum map for a cotangent lifted action, while the second proof focuses on a Poisson bracket computation. Each proof also explains the context in which one has a momentum map. (See [MR02] for general discussions on momentum maps.)

**First proof.** For simplicity and without loss of generality, let us take $N = 1$ and so suppress the index $a$. That is, we shall take the case of an isolated singular solution. As the proof will show, this is not a real restriction.

*Proof* To set the notation, fix a $k$-dimensional manifold $S$ with a given volume element and whose points are denoted $s \in S$. Let $\mathrm{Emb}(S, \mathbb{R}^n)$ denote the set of smooth embeddings $\mathbf{Q} : S \to \mathbb{R}^n$. (If the EPDiff equations are taken on a manifold $M$, replace $\mathbb{R}^n$ with $M$.) Under appropriate technical conditions, which we shall just treat formally here, $\mathrm{Emb}(S, \mathbb{R}^n)$ is a smooth manifold. (See, for example, [EM70] and [MH94] for a discussion and references.)

The tangent space $T_{\mathbf{Q}}\mathrm{Emb}(S, \mathbb{R}^n)$ to $\mathrm{Emb}(S, \mathbb{R}^n)$ at a point $\mathbf{Q} \in \mathrm{Emb}(S, \mathbb{R}^n)$ is given by the space of ***material velocity fields***, namely the linear space of maps $\mathbf{V} : S \to \mathbb{R}^n$ that are vector fields over the map $\mathbf{Q}$. The dual space to this space will be identified with the space of one-form densities over $\mathbf{Q}$, which we shall regard as maps $\mathbf{P} : S \to (\mathbb{R}^n)^*$. In summary, the cotangent bundle $T^* \mathrm{Emb}(S, \mathbb{R}^n)$ is identified with the space of pairs of maps $(\mathbf{Q}, \mathbf{P})$.

These give us the domain space for the singular solution momentum map. Now we consider the action of the symmetry group. Consider the group $G = \mathrm{Diff}$ of diffeomorphisms of the space $\mathbb{R}^n$ in which the EPDiff equations are operating, concretely in our case this is $\mathbb{R}^n$. Let it act on $\mathbb{R}^n$ by ***composition on the left.*** Namely, for $\eta \in \mathrm{Diff}(\mathbb{R}^n)$, we let

$$\eta \cdot \mathbf{Q} = \eta \circ \mathbf{Q}. \tag{14.15}$$

Now lift this action to the cotangent bundle $T^* \mathrm{Emb}(S, \mathbb{R}^n)$ in the standard way. One may consult, for instance, [MR02] for this cotangent-lift construction. However, it is also given explicitly by the variational construction in eqn (15.4). This lifted action is a symplectic (and hence Poisson) action and has an equivariant momentum map. ***This cotangent-lift momentum map for the left action (14.15) is precisely given by the ansatz (14.9).***

To see this, one only needs to recall and then apply the general formula for the momentum map associated with an action of a general Lie group $G$ on a configuration manifold $Q$ and cotangent lifted to $T^*Q$.

First let us recall the general formula. Namely, the momentum map is the map $\mathbf{J} : T^*Q \to \mathfrak{g}^*$ ($\mathfrak{g}^*$ denotes the dual of the Lie algebra $\mathfrak{g}$ of $G$) defined by

$$\mathbf{J}(\alpha_q) \cdot \xi = \langle \alpha_q, \xi_Q(q) \rangle , \tag{14.16}$$

where $\alpha_q \in T^*_q Q$ and $\xi \in \mathfrak{g}$, where $\xi_Q$ is the infinitesimal generator of the action of $G$ on $Q$ associated to the Lie algebra element $\xi$, and where $\langle \alpha_q, \xi_Q(q) \rangle$ is the natural pairing of an element of $T^*_q Q$ with an element of $T_q Q$.

Now we apply this formula to the special case in which the group $G$ is the diffeomorphism group $\mathrm{Diff}(\mathbb{R}^n)$, the manifold $Q$ is $\mathrm{Emb}(S, \mathbb{R}^n)$ and where the action of the group on $\mathrm{Emb}(S, \mathbb{R}^n)$ is given by eqn (14.15). The Lie algebra of $G = \mathrm{Diff}$ is the space $\mathfrak{g} = \mathfrak{X}$ of vector fields. Hence, its dual is naturally regarded as the space of one-form densities. The momentum map is thus a map $\mathbf{J} : T^* \mathrm{Emb}(S, \mathbb{R}^n) \to \mathfrak{X}^*$.

With $\mathbf{J}$ given by (14.16), we only need to work out this formula. First, we shall work out the infinitesimal generators. Let $X \in \mathfrak{X}$ be a Lie algebra element. By differentiating the action (14.15) with respect to $\eta$ in the direction of $X$ at the identity element we find that the infinitesimal generator is given by

$$X_{\mathrm{Emb}(S,\mathbb{R}^n)}(\mathbf{Q}) = X \circ \mathbf{Q}.$$

Thus, on taking $\alpha_q$ to be the cotangent vector $(\mathbf{Q}, \mathbf{P})$, eqn (14.16) gives

$$\begin{aligned} \langle \mathbf{J}(\mathbf{Q}, \mathbf{P}), X \rangle &= \langle (\mathbf{Q}, \mathbf{P}), X \circ \mathbf{Q} \rangle \\ &= \int_S P_i(s) X^i(\mathbf{Q}(s)) \mathrm{d}^k s. \end{aligned}$$

On the other hand, note that the right-hand side of eqn (14.9) (again with the index $a$ suppressed, and with $t$ suppressed as well), when paired with the Lie algebra element $X$ is

$$\begin{aligned}
&\left\langle \int_S \mathbf{P}(s)\, \delta\left(\mathbf{x}-\mathbf{Q}(s)\right) \mathrm{d}^k s, X \right\rangle \\
&= \int_{\mathbb{R}^n} \int_S \left( P_i(s)\, \delta\left(\mathbf{x}-\mathbf{Q}(s)\right) \mathrm{d}^k s \right) X^i(\mathbf{x}) \mathrm{d}^n x \\
&= \int_S P_i(s) X^i(\mathbf{Q}(s)) \mathrm{d}^k s.
\end{aligned}$$

This shows that the expression given by eqn (14.9) is equal to $\mathbf{J}$ and so the result is proved. ∎

**Second proof.** Momentum maps may be characterized by means of the following relation [MR02], required to hold for all functions $F$ on $T^*\,\mathrm{Emb}(S, \mathbb{R}^n)$. Namely, for all functions $F$ of $\mathbf{Q}$ and $\mathbf{P}$ one requires the Poisson-bracket relation,

$$\{F, \langle \mathbf{J}, \xi \rangle\} = \xi_P[F]. \tag{14.17}$$

For the second proof, we shall take $\mathbf{J}$ to be given by the solution ansatz (14.9) and verify that it satisfies this momentum-map relation.

*Proof* On one hand, let $\xi \in \mathfrak{X}$ so that the left side of eqn (14.17) becomes

$$\begin{aligned}
&\left\{ F, \int_S P_i(s) \xi^i(\mathbf{Q}(s)) \mathrm{d}^k s \right\} \\
&= \int_S \left[ \frac{\delta F}{\delta Q^i} \xi^i(\mathbf{Q}(s)) - P_i(s) \frac{\delta F}{\delta P_j} \frac{\delta}{\delta Q^j} \xi^i(\mathbf{Q}(s)) \right] \mathrm{d}^k s.
\end{aligned}$$

On the other hand, one can compute directly from the definitions that the infinitesimal generator of the action on $T^*\,\mathrm{Emb}(S, \mathbb{R}^n)$ corresponding to the vector field $\xi^i(\mathbf{x}) \frac{\partial}{\partial Q^i}$ (a Lie algebra element), is given by (see [MR02], formula (12.1.14)):

$$\delta \mathbf{Q} = \xi \circ \mathbf{Q}, \quad \delta \mathbf{P} = -\, P_i(s) \frac{\partial}{\partial \mathbf{Q}} \xi^i(\mathbf{Q}(s)),$$

which verifies that eqn (14.17) holds.

An important element left out in this proof so far is that it does not make clear that the momentum map is ***equivariant***, a condition needed for the momentum map to be Poisson. The first proof took care of this property automatically since ***momentum maps for cotangent-lifted actions are always equivariant*** and hence are Poisson.

Thus, to complete the second proof, one should check that the momentum map is equivariant and is thus Poisson. Instead, one may simply check directly that it is a Poisson map from $T^*\,\mathrm{Emb}(S,\mathbb{R}^n)$ to $\mathbf{m} \in \mathfrak{X}^*$ (the dual space of the Lie algebra $\mathfrak{X}$) with its Lie–Poisson bracket.

The following direct computation shows that the singular solution momentum map (14.9) is Poisson. For this, one uses the canonical Poisson brackets for $\{\mathbf{P}\}$, $\{\mathbf{Q}\}$ and applies the chain rule to compute $\{m_i(\mathbf{x}), m_j(\mathbf{y})\}$, with notation $\delta'_k(\mathbf{y}) \equiv \partial\delta(\mathbf{y})/\partial y^k$. The result then follows by a direct substitution of the singular solution (14.9) into the Poisson bracket,

$$
\begin{aligned}
&\{m_i(\mathbf{x}), m_j(\mathbf{y})\} \\
&= \Big\{ \sum_{a=1}^{N} \int \mathrm{d}s\, P_i^a(s,t)\, \delta(\mathbf{x} - \mathbf{Q}^a(s,t)), \\
&\qquad \sum_{b=1}^{N} \int \mathrm{d}s'\, P_j^b(s',t)\, \delta(\mathbf{y} - \mathbf{Q}^b(s',t)) \Big\} \\
&= \sum_{a,b=1}^{N} \iint \mathrm{d}s\,\mathrm{d}s' \Big[ \{P_i^a(s), P_j^b(s')\}\, \delta(\mathbf{x} - \mathbf{Q}^a(s))\, \delta(\mathbf{y} - \mathbf{Q}^b(s')) \\
&\qquad - \{P_i^a(s), Q_k^b(s')\} P_j^b(s')\, \delta(\mathbf{x} - \mathbf{Q}^a(s))\, \delta'_k(\mathbf{y} - \mathbf{Q}^b(s')) \\
&\qquad - \{Q_k^a(s), P_j^b(s')\} P_i^a(s)\, \delta'_k(\mathbf{x} - \mathbf{Q}^a(s))\, \delta(\mathbf{y} - \mathbf{Q}^b(s')) \\
&\qquad + \{Q_k^a(s), Q_\ell^b(s')\} P_i^a(s) P_j^b(s')\, \delta'_k(\mathbf{x} - \mathbf{Q}^a(s))\, \delta'_\ell(\mathbf{y} - \mathbf{Q}^b(s')) \Big].
\end{aligned}
$$

Substituting the canonical Poisson bracket relations

$$
\begin{aligned}
&\{P_i^a(s), P_j^b(s')\} = 0 \\
&\{Q_k^a(s), Q_\ell^b(s')\} = 0, \quad \text{and} \\
&\{Q_k^a(s), P_j^b(s')\} = \delta^{ab}\delta_{kj}\delta(s - s')
\end{aligned}
$$

into the preceding computation yields,

$$
\begin{aligned}
\{m_i(\mathbf{x}), m_j(\mathbf{y})\} &= \sum_{a=1}^{N} \int \mathrm{d}s\, P_j^a(s)\, \delta(\mathbf{x} - \mathbf{Q}^a(s))\, \delta'_i(\mathbf{y} - \mathbf{Q}^a(s)) \\
&\quad - \sum_{a=1}^{N} \int \mathrm{d}s\, P_i^a(s)\, \delta'_j(\mathbf{x} - \mathbf{Q}^a(s))\, \delta(\mathbf{y} - \mathbf{Q}^a(s)) \\
&= -\Big( m_j(\mathbf{x}) \frac{\partial}{\partial x^i} + \frac{\partial}{\partial x^j}\, m_i(\mathbf{x}) \Big) \delta(\mathbf{x} - \mathbf{y}).
\end{aligned}
$$

Thus,

$$\{m_i(\mathbf{x}),\, m_j(\mathbf{y})\} = -\Big(m_j(\mathbf{x})\frac{\partial}{\partial x^i} + \frac{\partial}{\partial x^j}\, m_i(\mathbf{x})\Big)\delta(\mathbf{x}-\mathbf{y}), \tag{14.18}$$

which is readily checked to be the Lie–Poisson bracket on the space of **m**'s, restricted to their singular support. This completes the second proof of the theorem. ■

Each of these two proofs has shown the following.

**Corollary 14.3** *The singular solution momentum map defined by the singular solution ansatz (14.9), namely,*

$$\mathbf{J}_{\mathrm{Sing}} : T^*\,\mathrm{Emb}(S, \mathbb{R}^n) \to \mathfrak{X}(\mathbb{R}^n)^* \tag{14.19}$$

*is a Poisson map from the canonical Poisson structure on the cotangent space* $T^*\,\mathrm{Emb}(S,\mathbb{R}^n)$ *to the Lie–Poisson structure on* $\mathfrak{X}(\mathbb{R}^n)^*$.

This is the fundamental property of the singular solution momentum map. Some of its more sophisticated properties are outlined in [HM04].

## Pulling back the equations

Since the solution ansatz (14.9) has been shown in the preceding Corollary to be a Poisson map, the pull-back of the Hamiltonian from $\mathfrak{X}^*$ to $T^*\,\mathrm{Emb}(S,\mathbb{R}^n)$ gives equations of motion on the latter space that project to the equations on $\mathfrak{X}^*$. The functions $\mathbf{Q}^a(s,t)$ and $\mathbf{P}^a(s,t)$ in eqn (14.9) satisfy canonical Hamiltonian equations. The pull-back of the Hamiltonian $h[\mathbf{m}]$ defined in eqn (14.4) on the dual of the Lie algebra $\mathfrak{g}^*$, to $T^*\,\mathrm{Emb}(S,\mathbb{R}^n)$ is easily seen to be consistent with what we had defined before in eqn (14.14):

$$h[\mathbf{m}] \equiv \frac{1}{2}\langle \mathbf{m},\, G * \mathbf{m}\rangle = \frac{1}{2}\langle\!\langle \mathbf{P},\, G * \mathbf{P}\rangle\!\rangle = H_N[\mathbf{P},\mathbf{Q}]. \tag{14.20}$$

*Since the momentum map* $\mathbf{J}_{\mathrm{Sing}}$ *is Poisson, the functions* $\mathbf{Q}^a(s,t)$ *and* $\mathbf{P}^a(s,t)$ *in eqn (14.9) satisfy canonical Hamiltonian equations.*

**Remark 14.4** *In terms of the pairing*

$$\langle \cdot, \cdot \rangle : \mathfrak{g}^* \times \mathfrak{g} \to \mathbb{R}, \tag{14.21}$$

*between the Lie algebra* $\mathfrak{g}$ *(vector fields in* $\mathbb{R}^n$*) and its dual* $\mathfrak{g}^*$ *(one-form densities in* $\mathbb{R}^n$*), the following relation holds for measure-valued solutions under the momentum map (14.9),*

$$\begin{aligned}
\langle \mathbf{m}, \mathbf{u} \rangle &= \int \mathbf{m} \cdot \mathbf{u} \, d^n\mathbf{x}, \quad L^2 \textit{ pairing for } \mathbf{m} \,\&\, \mathbf{u} \in \mathbb{R}^n, \\
&= \iint \sum_{a,b=1}^{N} \big( \mathbf{P}^a(s,t) \cdot \mathbf{P}^b(s',t) \big) \, G\big( \mathbf{Q}^a(s,t) - \mathbf{Q}^b(s',t) \big) \, ds \, ds' \\
&= \int \sum_{a=1}^{N} \mathbf{P}^a(s,t) \cdot \frac{\partial \mathbf{Q}^a(s,t)}{\partial t} \, ds \\
&\equiv \langle\!\langle \, \mathbf{P}, \dot{\mathbf{Q}} \, \rangle\!\rangle,
\end{aligned} \tag{14.22}$$

*which is the natural pairing between* $(\mathbf{Q}, \mathbf{P}) \in T^* \operatorname{Emb}(S, \mathbb{R}^n)$ *and* $(\mathbf{Q}, \dot{\mathbf{Q}}) \in T \operatorname{Emb}(S, \mathbb{R}^n)$.

**Remark 14.5** *Recall that the coordinate* $s \in \mathbb{R}^k$ *labeling the functions in eqn (14.22) is a 'Lagrangian coordinate' in the sense that it does not evolve in time, but merely labels the solution.*

**Remark 14.6 (Summary)** *In concert with the Poisson nature of the singular solution momentum map, the singular solutions (14.9) in terms of* $\mathbf{Q}$ *and* $\mathbf{P}$ *satisfy Hamiltonian equations and also define an invariant solution set for the EPDiff equations. In fact, this invariant solution set is a special coadjoint orbit for the diffeomorphism group, as we shall discuss in the next section.*

**Exercise 14.2**

Show that the natural pairing relation (14.22) preserves the stationary principle for the Lagrangian $\ell[u]$ under the cotangent lift of Diff($\mathbb{R}^n$). That is, state the conditions under which the stationary principle $\delta S = 0$ for

$$S = \int \ell[u] \, dt$$

will produce equivalent equations of motion for the two expressions for the Lagrangian $\ell[u]$ given by,

$$\ell[u] = \langle \mathbf{m}, \mathbf{u} \rangle - h[\mathbf{m}] \tag{14.23}$$

$$= \langle\!\langle \mathbf{P}, \dot{\mathbf{Q}} \rangle\!\rangle - H_N[\mathbf{P}, \mathbf{Q}], \tag{14.24}$$

upon using equivalence of the geodesic Hamiltonians $h[\mathbf{m}]$ and $H_N[\mathbf{P}, \mathbf{Q}]$ in eqn (14.20) for measure-valued solutions under the momentum map (14.9).

## 14.3 The geometry of the momentum map

In this section we explore the geometry of the singular solution momentum map discussed earlier in a little more detail. The treatment is formal, in the sense that there are a number of technical issues in the infinite-dimensional case that will be left open. We will discuss a few of these issues as we proceed.

**Remark 14.7 (Transitivity)**
*Transitivity of the left action corresponding to $J_{sing}$ holds because, roughly speaking, one can 'move the images of the manifolds $S$ around at will with an arbitrary diffeomorphism of $\mathbb{R}^n$'.*

### 14.3.1 $\mathbf{J}_S$ and the Kelvin circulation theorem

The momentum map $\mathbf{J}_{\rm Sing}$ involves $\mathrm{Diff}(\mathbb{R}^n)$, the left action of the diffeomorphism group on the space of embeddings $\mathrm{Emb}(S, \mathbb{R}^n)$ by smooth maps of the target space $\mathbb{R}^n$, namely,

$$\mathrm{Diff}(\mathbb{R}^n): \ \mathbf{Q} \cdot \eta = \eta \circ \mathbf{Q}, \tag{14.25}$$

where, recall, $\mathbf{Q} : S \to \mathbb{R}^n$. As before, one identifies the cotangent bundle $T^* \mathrm{Emb}(S, \mathbb{R}^n)$ with the space of pairs of maps $(\mathbf{Q}, \mathbf{P})$, with $\mathbf{Q} : S \to \mathbb{R}^n$ and $\mathbf{P} : S \to T^*\mathbb{R}^n$.

#### The momentum map for right action

Another momentum map $\mathbf{J}_S$ exists, associated with the ***right action*** of the diffeomorphism group of $S$ on the embeddings $\mathrm{Emb}(S, \mathbb{R}^n)$ by smooth maps of the ***Lagrangian labels*** $S$ (fluid particle relabeling by $\eta_r : S \to S$). This particle-relabelling action is given by

$$\mathrm{Diff}(S) : \ \mathbf{Q} \cdot \eta_r = \mathbf{Q} \circ \eta_r, \tag{14.26}$$

with parameter $r = 0$ at the identity. The infinitesimal generator of this right action is

$$X_{\mathrm{Emb}(S,\mathbb{R}^n)}(\mathbf{Q}) = \frac{\mathrm{d}}{\mathrm{d}r}\bigg|_{r=0} \mathbf{Q} \circ \eta_r = T\mathbf{Q} \cdot X, \tag{14.27}$$

where $X \in \mathfrak{X}$ is tangent to the curve $\eta_r$ at $r = 0$. Thus, again taking $N = 1$ (so we suppress the index $a$) and also letting $\alpha_q$ in the momentum map formula (14.16) be the cotangent vector $(\mathbf{Q}, \mathbf{P})$, one computes $\mathbf{J}_S$:

$$\begin{aligned}
\big\langle \mathbf{J}_S(\mathbf{Q},\mathbf{P}), X \big\rangle &= \big\langle (\mathbf{Q},\mathbf{P}), T\mathbf{Q} \cdot X \big\rangle \\
&= \int_S P_i(s) \frac{\partial Q^i(s)}{\partial s^m} X^m(s)\, \mathrm{d}^k s \\
&= \int_S X\Big(\mathbf{P}(s) \cdot \mathrm{d}\mathbf{Q}(s)\Big)\, \mathrm{d}^k s \\
&= \bigg( \int_S \mathbf{P}(s) \cdot \mathrm{d}\mathbf{Q}(s) \otimes \mathrm{d}^k s, X(s) \bigg) \\
&= \langle \mathbf{P} \cdot \mathrm{d}\mathbf{Q}, X \rangle.
\end{aligned}$$

Consequently, the momentum map formula (14.16) yields

$$\mathbf{J}_S(\mathbf{Q},\mathbf{P}) = \mathbf{P} \cdot \mathrm{d}\mathbf{Q}, \tag{14.28}$$

with the indicated pairing of the one-form density $\mathbf{P} \cdot \mathrm{d}\mathbf{Q}$ with the vector field $X$.

We have set things up so that the following is true.

**Proposition 14.8** *The momentum map* $\mathbf{J}_S$ *is preserved by the evolution equations (14.11) and (14.12) for* $\mathbf{Q}$ *and* $\mathbf{P}$.

*Proof* It is enough to notice that the Hamiltonian $H_N$ in eqn (14.14) is invariant under the cotangent lift of the action of Diff$(S)$, which amounts to the invariance of the integral over $S$ with respect to reparametrization given by the change of variables formula. (Keep in mind that $\mathbf{P}$ includes a density factor.) ■

### Remark 14.9

- *The result of Proposition 14.8 is similar to the* ***Kelvin–Noether theorem*** *for circulation* $\Gamma$ *of an ideal fluid, which may be written as*

$$\Gamma = \oint_{c(s)} D(s)^{-1}\mathbf{P}(s) \cdot \mathrm{d}\mathbf{Q}(s),$$

*for each Lagrangian circuit* $c(s)$*, where* $D$ *is the mass density and* $\mathbf{P}$ *is again the canonical momentum density. This similarity should come as no surprise, because the Kelvin–Noether theorem for ideal fluids arises from invariance of Hamilton's principle under fluid-parcel relabelling by the same right action of the diffeomorphism group as in (14.26).*

- *Note that, being an equivariant momentum map, the map* $\mathbf{J}_S$*, as with* $\mathbf{J}_{\text{Sing}}$*, is also a Poisson map. That is, substituting the canonical Poisson bracket into relation* (14.28)*; that is, the relation*

$$\mathbf{M}(\mathbf{x}) = \sum_i P_i(\mathbf{x})\nabla Q^i(\mathbf{x})$$

*yields the Lie–Poisson bracket on the space of* $\mathbf{M} \in \mathfrak{X}^*$*. We use the different notations* $\mathbf{m}$ *and* $\mathbf{M}$ *because these quantities are analogous to the body and spatial angular momentum for rigid body mechanics. In fact, the quantity* $\mathbf{m}$ *given by the solution Ansatz* $\mathbf{m} = \mathbf{J}_{\text{Sing}}(\mathbf{Q}, \mathbf{P})$ *gives the singular solutions of the EPDiff equations, while the expression*

$$\mathbf{M}(\mathbf{x}) = \mathbf{J}_S(\mathbf{Q}, \mathbf{P}) = \sum_i P_i(\mathbf{x})\nabla Q^i(\mathbf{x})$$

*is a conserved quantity.*

- *In the language of fluid mechanics, the expression of* $\mathbf{m}$ *in terms of* $(\mathbf{Q}, \mathbf{P})$ *is an example of a* ***Clebsch representation****, which expresses the solution of the EPDiff equations in terms of canonical variables that evolve by standard canonical Hamilton equations. This has been known in the case of fluid mechanics for more than 100 years. For modern discussions of the Clebsch representation for ideal fluids, see, for example, [HK83, MW83, CM87].*

- *One more remark is in order. Namely, the special case in which* $S = M$ *is of course allowed. In this case,* $\mathbf{Q}$ *corresponds to the time evolution map* $\eta_t$ *and* $\mathbf{P}$ *corresponds to its conjugate momentum. The quantity* $\mathbf{m}$ *corresponds to the spatial (dynamic) momentum density (that is, right translation of* $\mathbf{P}$ *to the identity), while* $\mathbf{M}$ *corresponds to the conserved 'body' momentum density (that is, left translation of* $\mathbf{P}$ *to the identity).*

**Exercise 14.3**
To investigate the space-versus-body aspects discussed in the last of these remarks, derive the Euler-Poincaré equation (14.2) as an ***optimal control problem*** obtained by minimising the ***alternative action integral***,

$$S = \int L(u, w, \eta)\, dt = \int l(u) + \frac{1}{2\sigma^2}|w - \mathrm{Ad}_{\eta^{-1}}\, u|^2_{\mathfrak{X}}\, dt, \qquad (14.29)$$

where $w := \eta^{-1}\dot{\eta}$ is a left-invariant vector field under $\mathrm{Diff}(M)$. The second summand imposes a penalty that strengthens as $\sigma^2 \to 0$. This penalty function introduces a Riemannian structure that defines a norm $|\cdot|_{\mathfrak{X}}$ via the $L^2$ inner product $\langle \cdot, \cdot \rangle : \mathfrak{X}^* \times \mathfrak{X} \to \mathbb{R}$. This alternative interpretation of the Euler-Poincaré equation will be investigated further in Chapter 15.

### 14.3.2 Brief summary

$\mathrm{Emb}(S, \mathbb{R}^n)$ admits two group actions. These are: the group $\mathrm{Diff}(S)$ of diffeomorphisms of $S$, which acts by composition on the *right*; and the group $\mathrm{Diff}(\mathbb{R}^n)$, which acts by composition on the *left*. The group $\mathrm{Diff}(\mathbb{R}^n)$ acting from the left produces the singular solution momentum map, $\mathrm{J}_{\mathrm{Sing}}$. The action of $\mathrm{Diff}(S)$ from the right produces the conserved momentum map,

$$\mathrm{J}_S : T^*\,\mathrm{Emb}(S, \mathbb{R}^n) \to \mathfrak{X}(S)^*.$$

The two momentum maps may be assembled into a single figure as follows:

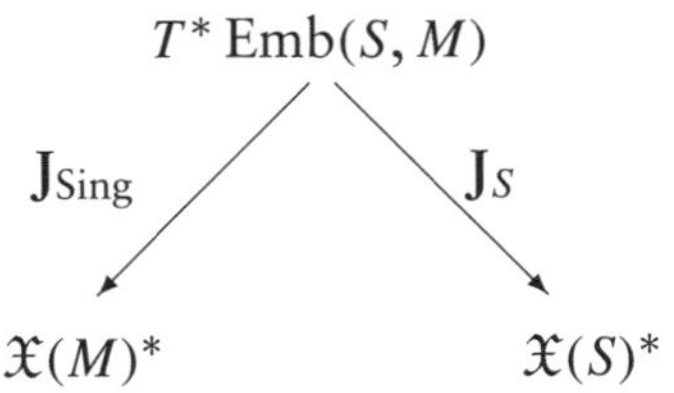


**Remark 14.10 (Images of the momentum maps)**

- *$Im(J_{sing})$ is the set of all 1-form densities on $M$ that are supported on 1-parameter curves.*
- *$(J_S)$ is onto.*
- $J_{sing}(J_S^{-1}\mathbf{M}) = \mathrm{Im}\,(J_{sing})$.
- $J_S(J_{sing}^{-1}\mathbf{m}) = \mathrm{Im}\,(J_S)$.

## 14.4 Numerical simulations of EPDiff in two dimensions

Many open problems and other future applications remain for the EPDiff equation. For example, its analysis requires development of additional methods for PDEs. In particular, while its smooth solutions satisfy a local existence theorem that is analogous to the famous Ebin–Marsden theorem for the Euler fluid equations [EM70], its singular solutions inevitably emerge from smooth initial conditions in its initial-value problem. The implications of this observation are discussed briefly in [HM04], where it is conjectured that these singular solutions may arise from incompleteness of the geodesic flows on the diffeomorphisms. This conjecture emphasizes the opportunities for future analysis of the emergence of measure-valued solutions from smooth initial conditions in nonlinear non-local PDEs. We close this chapter by giving a few examples of the evolutionary behavior of EPDiff singular solutions in simple two-dimensional situations from [HS04].

Figure 14.1 shows the results for the EPDiff equation when a straight peakon segment of finite length and transverse profile $u(x) = \mathrm{e}^{-|x|/\alpha}$ is created initially moving rightward (East). In adjusting to the condition of zero speed at its ends and the finite speed in its interior, the initially straight segment expands outward as it propagates and curves into a peakon "bubble." This adjustment and change of shape requires propagation along the wave crest. (Indeed, the wave crest gets longer.)

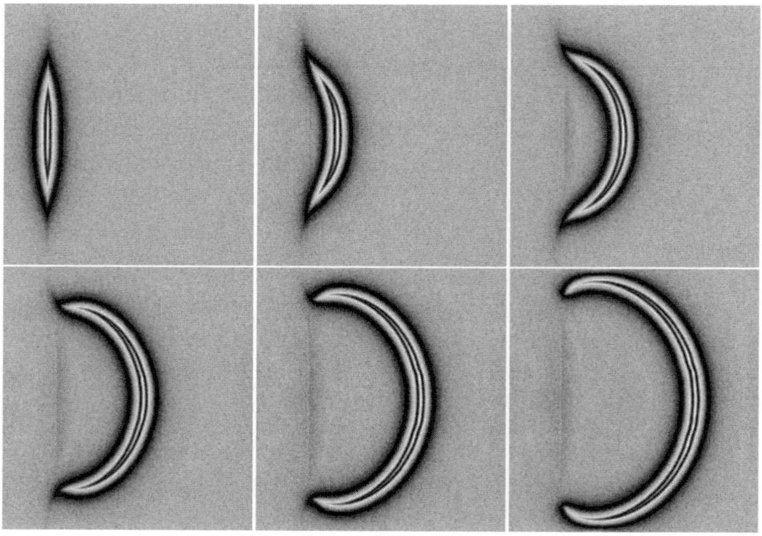

**Figure 14.1** A peakon segment of finite length is initially moving rightward (East). Because its speed vanishes at its ends and it has fully two-dimensional spatial dependence, it expands into a peakon 'bubble' as it propagates. (The colors indicate speed: red is highest, yellow is less, blue low, grey zero.) See Plate 1.

Figure 14.2 shows an initially straight segment whose velocity distribution is exponential in the transverse direction, $u(x) = e^{-|x|/\alpha}$, but the width $\alpha$ is 5 times wider than the lengthscale in the EPDiff equation. This initial velocity distribution evolves under EPDiff to separate into a train of curved peakon 'bubbles,' each of width $\alpha$. This example illustrates the emergent property of the peakon solutions in two dimensions.

Figure 14.3 shows an oblique wave-front collision that produces reconnections for the EPDiff equation in two dimensions. Figure 14.3 shows a single oblique overtaking collision, as a faster expanding peakon wave front overtakes a slower one and reconnects with it at the collision point via flow along the wave crest.

The phenomenon of nonlinear wave reconnection is also observed in Nature. For example, it may be seen in the images taken from the Space Shuttle of trains of internal waves in the South China Sea shown in Figures 14.4 and 14.5. These transbasin oceanic internal waves are some of the most impressive wave fronts seen in Nature. About 200 kilometres in length and separated by about 75 kilometres, they are produced every twelve hours by the tide through the Luzon strait between Taiwan and the Phillipines. They may be observed as they propagate and interact with each other and with geographic features. The characteristic property of these strongly nonlinear

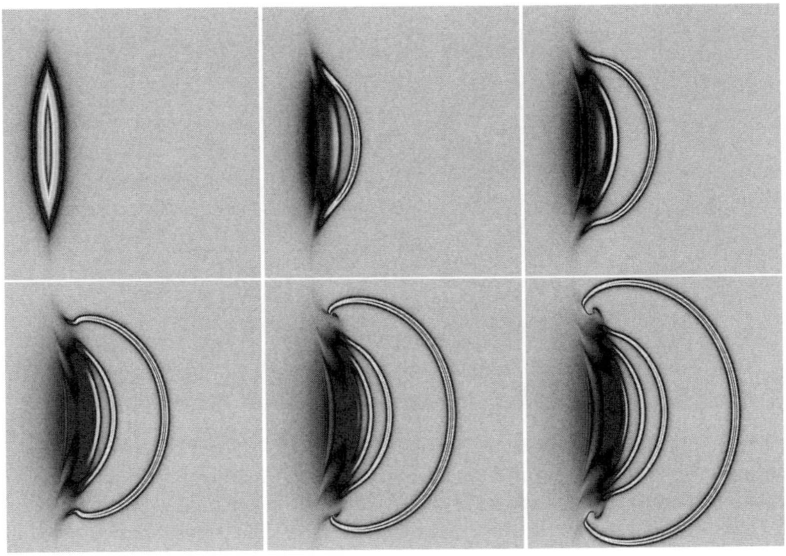

**Figure 14.2** An initially straight segment of velocity distribution whose exponential profile is wider than the width of the peakon solution will break up into a ***train*** of curved peakon 'bubbles'. This example illustrates the emergent property of the peakon solutions in two dimensions. See Plate 2.

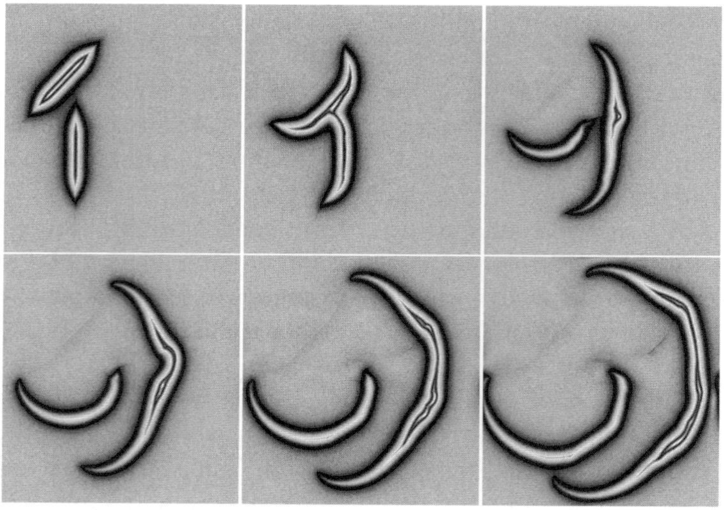

**Figure 14.3** A single collision is shown involving reconnection as the faster peakon segment initially moving Southeast along the diagonal expands, curves and obliquely overtakes the slower peakon segment initially moving rightward (East). This reconnection illustrates one of the collision rules for the strongly two-dimensional EPDiff flow. See Plate 3.

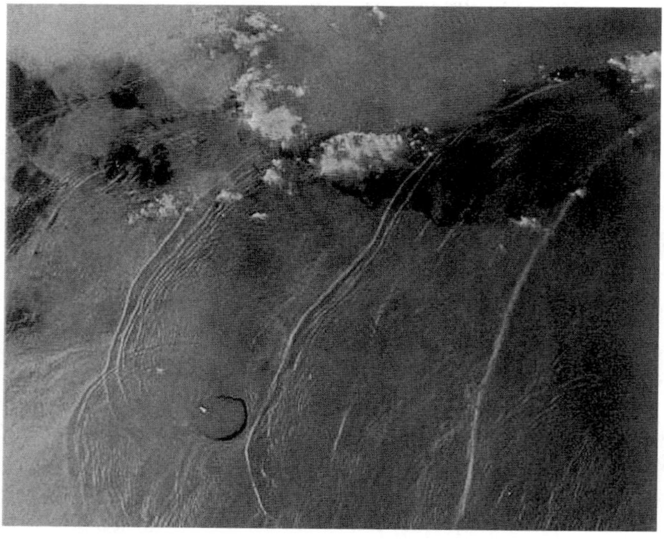

**Figure 14.4** Satellite image using synthetic aperture radar (SAR) of internal wave fronts propagating westward in the South China Sea. A multiwave merger occurs in the region West of the Dong-Sha atoll, which is about 40 km in diameter. An expanded view of this nonlinear wave merger is shown in Figure 14.5. SAR images from A. Liu, private communication.

wavefronts is that they reconnect when two of them collide transversely, as seen in Figures 14.3–14.5.

Figures 14.6–14.9 show additional collision configurations of peakon segments moving in the plane [HS04].

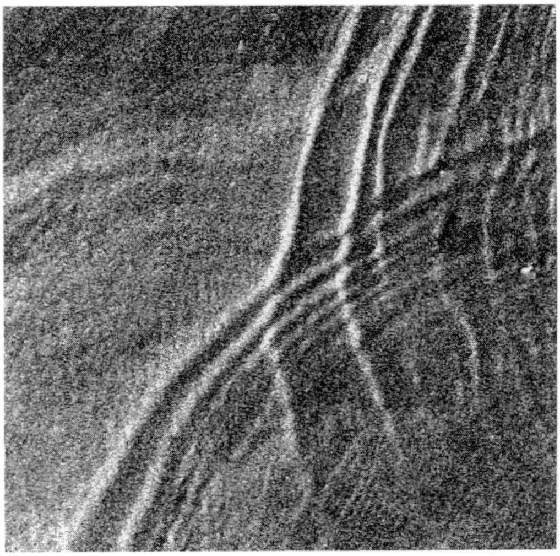

**Figure 14.5** SAR image of nonlinear internal waves West of DongSha in the South China Sea shows merger upon collision due to flow along the wave crests. This sort of merger with flow along the crests of the waves is also seen in the numerical simulations of EPDiff in the plane shown in Figure 14.3.

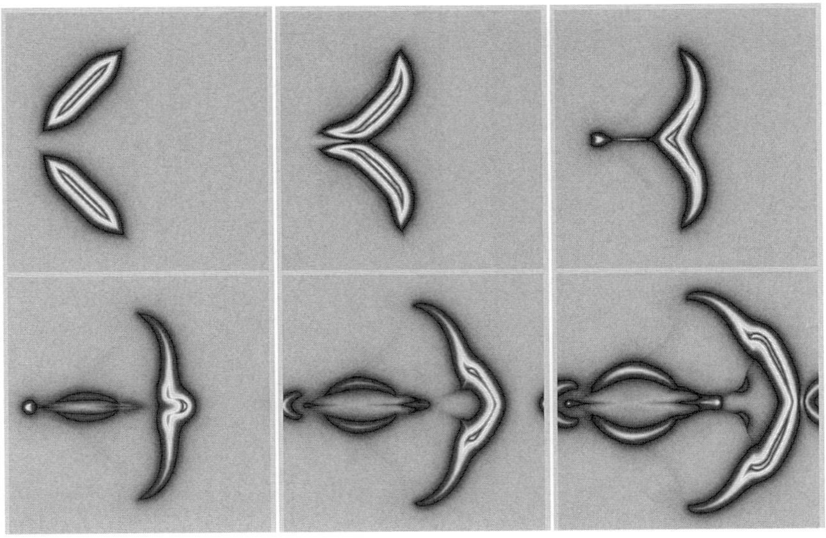

**Figure 14.6** The convergence of two peakon segments moving with reflection symmetry generates considerable acceleration along the midline, which continues to build up after the initial collision. See Plate 4.

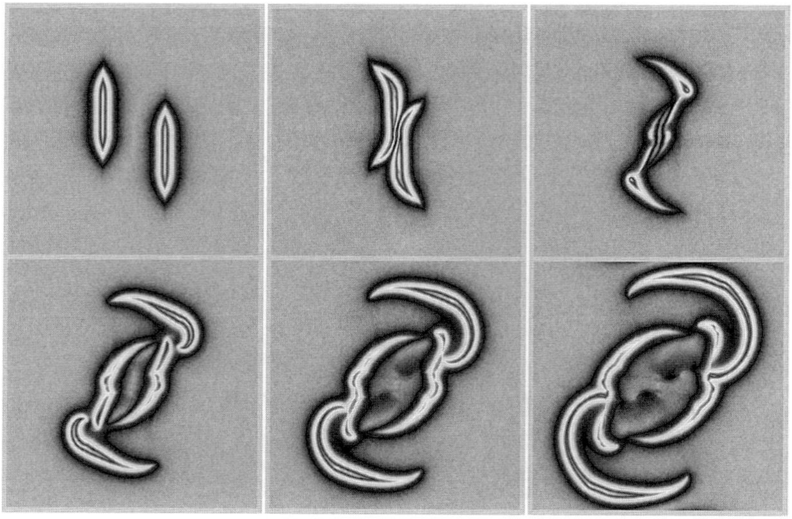

**Figure 14.7** The head-on collision of two offset peakon segments generates considerable complexity. Some of this complexity is due to the process of annihilation and recreation that occurs in the 1D antisymmetric head-on collisions of a peakon with its reflection, the antipeakon, as shown in Figure 12.3. Other aspects of it involve flow along the crests of the peakon segments as they stretch. See Plate 5.

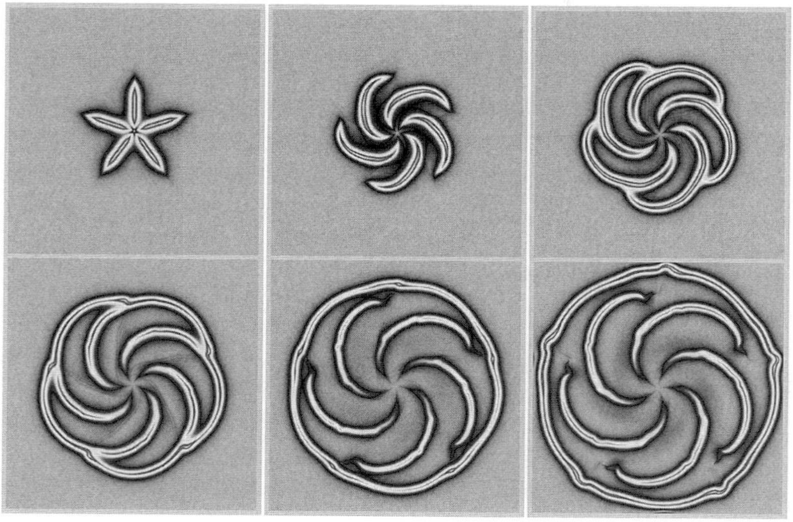

**Figure 14.8** The overtaking collisions of these rotating peakon segments with five-fold symmetry produces many reconnections (mergers), until eventually one peakon ring surrounds five curved peakon segments. If the evolution were allowed to proceed further, reconnections would tend to produce additional concentric peakon rings. See Plate 6.

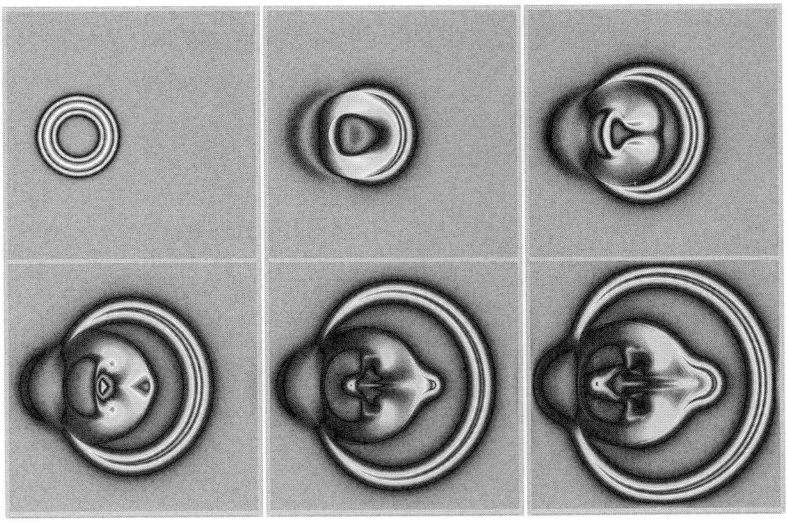

**Figure 14.9** A circular peakon ring initially undergoes uniform rightward translational motion along the $x$ axis. The right outer side of the ring produces diverging peakon curves, which slow as they propagate outward. The left inner side of the ring, however, produces converging peakon segments, which accelerate as they converge. They collide along the midline, then develop into divergent peakon curves still moving rightward that overtake the previous ones and collide with them from behind. These overtaking collisions impart momentum, but they apparently do not produce reconnections. See Plate 7.

## Solutions to selected exercises

**Solution to Exercise 14.3** To avoid confusion with earlier notation $\eta \in \mathrm{Diff}(S)$ for the action of diffeomorphisms on an embedded submanifold $S$, we denote elements of $\mathrm{Diff}(M)$ as $g \in \mathrm{Diff}(M)$. The cross-derivative identities for $\dot{g} = gw$ and $g' = g\xi$ yield the standard formula for variations of the left-invariant velocity,

$$\dot{g}' = g'w + gw' = \dot{g}\xi + g\dot{\xi} \quad \Longrightarrow \quad w' = \dot{\xi} + \mathrm{ad}_w\, \xi, \tag{14.30}$$

where prime ( $'$ ) denotes variational derivative and $w' = \delta w$ is the variation in $w$ inherited from the variation in $g$, $g' = \delta g$ . This formula will be substituted into the variation of the action integral in eqn (14.29) given by

$$0 = \delta S = \delta \int L(u, w, \eta)\, \mathrm{d}t = \int \Big\langle \frac{\partial l}{\partial u}, u' \Big\rangle + \Big\langle p, w' - (\mathrm{Ad}_{g^{-1}}\, u)' \Big\rangle dt \tag{14.31}$$

where the momentum 1-form density, $p$, dual to the vector field $w$, is given by

$$p := \frac{\delta L}{\delta w} = \frac{1}{\sigma^2}\left(w - \mathrm{Ad}_{g^{-1}}\, u\right) \tag{14.32}$$

and the pairing by the $L^2$ inner product $\langle\,\cdot\,,\,\cdot\,\rangle : \mathfrak{X}^* \times \mathfrak{X} \to \mathbb{R}$ is induced by the variational-derivative operation from the Riemannian structure introduced by the penalty term. Formula (14.30) gives the variation $\omega'$ in eqn (14.31) in terms of the vector field $\xi = g^{-1}g' \in \mathfrak{X}$. One calculates the other variation as

$$(\mathrm{Ad}_{g^{-1}}\,\Omega)' = (g^{-1}\Omega g)' = \mathrm{Ad}_{g^{-1}}\,u' + \mathrm{ad}_{(\mathrm{Ad}_{g^{-1}}\,u)}\,\xi .$$

Hence, the variation of the action integral in (14.31) becomes

$$\begin{aligned} 0 = \delta S &= \int \Big\langle \frac{\delta l}{\delta u}\,,\, u' \Big\rangle + \Big\langle p\,,\, \dot{\xi} + \mathrm{ad}_w\,\xi - \mathrm{ad}_{(\mathrm{Ad}_{g^{-1}}\,u)}\,\xi - \mathrm{Ad}_{g^{-1}}\,u')\Big\rangle\, dt \\ &= \int \Big\langle \frac{\delta l}{\delta u} - \mathrm{Ad}^*_{g^{-1}}\,p\,,\, u' \Big\rangle - \Big\langle \dot{p} - \mathrm{ad}^*_{\omega}\,\pi + \mathrm{ad}^*_{(\mathrm{Ad}_{g^{-1}}\,u)}\,p\,,\,\xi\Big\rangle\, dt, \end{aligned}$$

where we assume endpoint terms may be ignored when integrating by parts. Requiring the coefficients of the independent variations to vanish yields the expressions we seek,

$$\begin{aligned} m := \frac{\delta l}{\delta u} &= \mathrm{Ad}^*_{g^{-1}}\,p, \\ \dot{p} - \mathrm{ad}^*_w\,p &= -\,\mathrm{ad}^*_{(\mathrm{Ad}_{g^{-1}}\,u)}\,p. \end{aligned} \tag{14.33}$$

The first of these relates the momenta $p, m \in \mathfrak{X}^*$ dual to the vector fields $w, u \in \mathfrak{X}$ exactly as the spatial and body angular momenta are related for the rigid body.

The variational eqns (14.33) imply, when paired with a fixed vector field $\xi \in \mathfrak{X}$, that

$$\frac{d}{dt}\langle m\,,\,\xi\rangle = \frac{d}{dt}\Big\langle \mathrm{Ad}^*_{g^{-1}}\,p\,,\,\xi\Big\rangle$$

$$
\begin{aligned}
\text{On taking } \tfrac{d}{dt}\,\mathrm{Ad}^*_{g^{-1}} &= \Big\langle \mathrm{Ad}^*_{g^{-1}}\left(\dot{p} - \mathrm{ad}^*_{w}\, p\right),\, \xi \Big\rangle \\
\text{On using } p\text{-eqn (14.33)} &= -\Big\langle \mathrm{Ad}^*_{g^{-1}}\left(\mathrm{ad}^*_{(\mathrm{Ad}_{g^{-1}} u)}\, p\right),\, \xi \Big\rangle \\
\text{On using Ad \& ad definitions} &= -\Big\langle p,\, \mathrm{ad}_{(\mathrm{Ad}_{g^{-1}} u)}\left(\mathrm{Ad}_{g^{-1}}\, \xi\right) \Big\rangle \\
\text{On rearranging} &= -\Big\langle p,\, \mathrm{Ad}_{g^{-1}}\left(\mathrm{ad}_u\, \xi\right) \Big\rangle \\
\text{On taking duals} &= -\Big\langle \mathrm{ad}^*_u\left(\mathrm{Ad}^*_{g^{-1}}\, p\right),\, \xi \Big\rangle \\
\text{On substituting the definition of } m &= -\big\langle \mathrm{ad}^*_u\, m,\, \xi \big\rangle.
\end{aligned}
$$

This recovers EPDiff, the Euler-Poincaré equation,

$$
\frac{d}{dt}\frac{\delta l}{\delta u} = -\,\mathrm{ad}^*_u\,\frac{\delta l}{\delta u}, \quad \text{or} \quad \frac{d}{dt}\left(\mathrm{Ad}^*_g\,\frac{\delta l}{\delta u}\right) = 0.
$$

Thus, using a penalty term in the action integral to impose the action of $\mathrm{Ad}_{g^{-1}}$ on vector fields as a 'soft constraint' when $\sigma^2 > 0$ yields EPDiff dynamics for coadjoint motion on the $L^2$ dual, $\mathfrak{X}^*$, of the right-invariant Lie-algebra, $\mathfrak{X}$.

Equation (14.33) for $p$ may also be written as, cf. eqn (14.32),

$$
\dot{p} - \sigma^2\,\mathrm{ad}^*_p\, p = 0. \tag{14.34}
$$

Since ad* and Lie derivative with respect to a vector field are the same for 1-form densities, this relation for the evolution of the left-invariant momentum density may be interpreted as

$$
\frac{d}{dt}\Big(\mathbf{p}\cdot d\mathbf{x}\otimes dV\Big) = 0 \quad \text{along} \quad \frac{d\mathbf{x}}{dt} = -\sigma^2\mathbf{p}. \tag{14.35}
$$

In Euclidean components, this is

$$
\partial_t p_i = \sigma^2\frac{\partial}{\partial x^j}\left(p_i p^j + \frac{1}{2}\delta_i^j\,|\mathbf{p}|^2\right),
$$

which implies conservation of the integrated left linear momentum,

$$
\frac{d}{dt}\int p_i(\mathbf{x},t)\,d^3x = 0,
$$

for homogeneous boundary conditions. This is the analogue for EPDiff of the conservation of spatial angular momentum for the rigid body.

# 15 Computational anatomy: contour matching using EPDiff

## 15.1 Introduction to computational anatomy (CA)

### Morphology and computational anatomy

Computational anatomy (CA) must measure and analyse a range of variations in shape, or appearance, of highly deformable biological structures. The problem statement for CA was formulated long ago in a famous book by D'Arcy Thompson [Tho92]

> In a very large part of morphology, our essential task lies in the ***comparison of related forms*** rather than in the precise definition of each .... This process of comparison, of recognizing in one form a definite permutation or deformation of another, ... lies within the immediate province of mathematics and finds its solution in ... the Theory of Transformations. ... I learnt of it from Henri Poincaré.
>
> – D'Arcy Thompson, *On Growth and Form* (1917)

D'Arcy Thompson's book [Tho92] examines the idea that the growth and form of all plants and animals can be explained by mathematical principles. His book also acts as a practical guide to understanding how flows of smooth invertible maps may be used to compare shapes. For example, his chapter on transformations contains remarkable diagrams showing how differences in the forms of, say, species of fish can be understood in terms of smooth invertible distortions of the reference coordinate systems onto which they are mapped. A fish is drawn on a square grid, which is then stretched, sheared or shifted so that the deformed image may be identified as that of a related species, as in Figures 15.1 and 15.2.

This chapter explains that EPDiff is the perfect tool to realize D'Arcy Thompson's concept of comparing shapes, upon choosing a norm

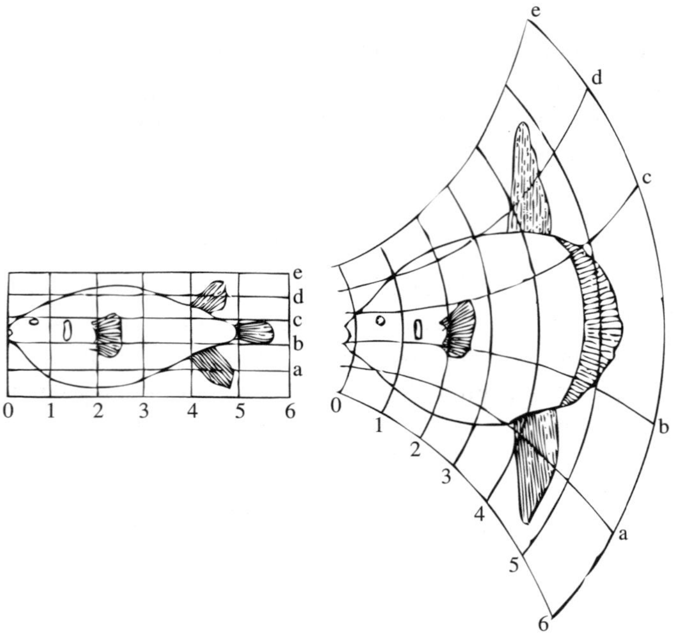


**Figure 15.1** One of D'Arcy Thompson's illustrations of the transformation of two-dimensional shapes from one fish to another, from [Tho92].

that measures the differences between anatomical forms defined by contours.

The flow generated by the EPDiff equation transforms one shape along a curve in the space of smooth invertible maps that takes it optimally into another with respect to the chosen norm. Its application to contours in biomedical imaging, for example, realizes D'Arcy Thompson's concept of quantifying growth and measuring other changes in shape, such as occurs in a beating heart, by providing the transformative mathematical path between the two shapes.

## Computational anatomy (CA)

The pioneering work of Bookstein and Grenander first took up D'Arcy Thompson's challenge by introducing a method called ***template matching*** [Boo91, Gre81]. The past several years have seen an explosion in the use and development of template-matching methods in computer vision and medical imaging that seem to be fulfilling D'Arcy Thompson's expectation. These

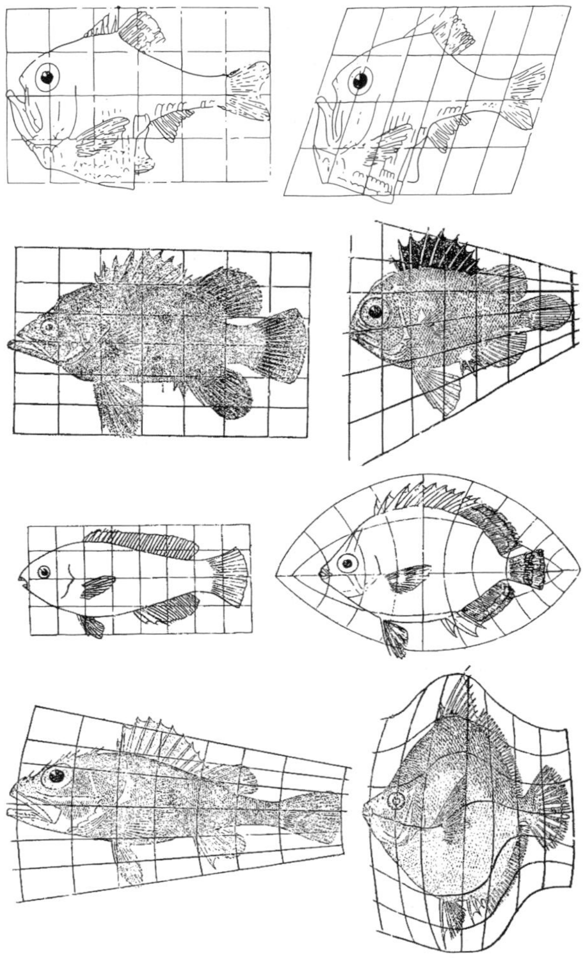

**Figure 15.2** More illustrations by D'Arcy Thompson of the transformation of two-dimensional shapes from one fish to another, from [Tho92].

methods enable the systematic measurement and comparison of anatomical shapes and structures in medical imagery. The mathematical theory of Grenander's deformable template models, when applied to these problems, involves smooth invertible maps (diffeomorphisms). See, e.g., [MTY02] for a review. In particular, the template-matching approach defines classes of Riemannian metrics on the tangent space of the diffeomorphisms and employs their projections onto specific landmark shapes, or image spaces, for the representation of CA data.

The problem for CA then becomes to determine the minimum distance between two images as specified in a certain representation space, $V$, on which the diffeomorphisms act. Metrics are written so that the optimal path in Diff satisfies an evolution equation. This equation turns out to be EPDiff, when $V$ is a closed contour representing the shape of the image. A discussion of EPDiff and the application of its peakons and other singular solutions for matching templates defined by contours of image outlines appears in [HRTY04].

## Objectives

This chapter and the next discuss how the Euler–Poincaré theory may be used to develop new perspectives in CA. In particular, these chapters discuss how CA may be informed by the concept of weak solutions, solitons and momentum maps for geodesic flows [HRTY04, CH93, HS04]. For example, among the geometric structures of interest in CA, the landmark points and image outlines may be identified with the singular solutions of the EPDiff equation. These singular solutions are given by the momentum map $J_{Sing}$ for EPDiff in Chapter 14. This momentum map also yields the canonical Hamiltonian formulation of their dynamics. This evolution, in turn, provides a complete parameterization of the landmarks and image outlines by the ***linear vector space*** comprising their canonical positions and momenta. The singular momentum map $J_{Sing}$ for EPDiff provides an isomorphism between the landmarks and outlines for images and the singular soliton solutions of the EPDiff equation. This isomorphism provides a dynamical paradigm for CA, as well as a basis for anatomical data representation in a linear vector space.

> This chapter introduces the variational formulation of template matching problems in computational anatomy. It makes the connection to the EPDiff evolution equation and discusses the relation of images in CA to the singular momentum map of the EPDiff equation. Then it draws some consequences of EPDiff for the outline matching problem in CA and gives a numerical example. The numerical example is reminiscent of the chapter in D'Arcy Thompson's book where the shapes of fish are related to each other by stretching one shape into another on a square grid.

***Outline of the chapter.*** Section 15.2 describes the template-matching variational problems of computational anatomy and the fundamental role of the EPDiff evolution equation. The singular solutions for the EPDiff equation (15.2) with diffeomorphism group $G$ are discussed in Section 15.3.

They are, in particular, related to the contour-matching problem in CA (or, more generally, in computer vision), examples of which are given in Section 15.4.

## 15.2 Mathematical formulation of template matching for CA

### 15.2.1 Cost

Most problems in CA can be formulated as:

***Find the deformation path (flow) with minimal cost, under the constraint that it carries the template to the target.***

The ***cost*** assigned in template matching for comparing images $\eta_0$ and $\eta_1$ considered as points on a manifold $N$ is assigned as a functional

$$\mathrm{Cost}(t \mapsto g_t) = \int_0^1 \ell(\mathbf{u}_t)\, \mathrm{d}t,$$

which is defined on curves $g_t$ in a Lie group with tangents

$$\frac{\mathrm{d}g_t}{\mathrm{d}t} = \mathbf{u}_t \circ g_t \quad \text{and} \quad \eta_t = g_t \cdot \eta_0. \tag{15.1}$$

In what follows, the function $\mathbf{u}_t \mapsto \ell(\mathbf{u}_t) = \|\mathbf{u}_t\|^2$ will be taken as a squared functional norm on the space of velocity vectors along the flow. The Lie group property specifies the representation space for template matching as a manifold of smooth mappings, which may be differentiated, composed and inverted. The vector space of ***right-invariant*** instantaneous velocities, $\mathbf{u}_t = (\mathrm{d}g_t/\mathrm{d}t) \circ g_t^{-1}$ forms the tangent space at the identity of the considered Lie group, and may be identified as the group's ***Lie algebra***, denoted $\mathfrak{g}$.

### 15.2.2 Mathematical analogy between template matching and fluid dynamics

(I) The frameworks of both CA and fluid dynamics each involve a stationary principle whose action, or cost function, is ***right invariant***. The main difference is that template matching is formulated as an optimal control problem whose cost function is designed for the application, while fluid dynamics is formulated as an initial value problem whose cost function is the fluid's kinetic energy.

(II) The geodesic evolution for both template matching and fluid dynamics is governed by the ***EPDiff equation*** (11.20), rewritten as

$$\Big(\frac{\partial}{\partial t} + \mathbf{u}\cdot\nabla\Big)\mathbf{m} + (\nabla\mathbf{u})^T\cdot\mathbf{m} + \mathbf{m}(\operatorname{div}\mathbf{u}) = 0, \qquad (15.2)$$

in which $(\nabla\mathbf{u})^T\cdot\mathbf{m} = \sum_j m_j \nabla u^j$. Here $\mathbf{u} = K * \mathbf{m}$, where $K*$ denotes convolution with the Green's kernel $K$ for the operator $Q_{op}$, where

$$\mathbf{m} = \frac{\delta \ell}{\delta \mathbf{u}} =: Q_{op}\mathbf{u}.$$

The operator $Q_{op}$ is symmetric and positive-definite for the cost defined by

$$\mathrm{Cost}(t \mapsto g_t) = \int_0^1 \ell(\mathbf{u}_t)\,dt = \frac{1}{2}\int_0^1 \|\mathbf{u}_t\|^2\,dt = \frac{1}{2}\int_0^1 \langle\, \mathbf{u}_t,\, Q_{op}\mathbf{u}_t\,\rangle\,dt,$$

with $L^2$ pairing $\langle\,\cdot,\,\cdot\,\rangle$ whenever $\|\mathbf{u}_t\|^2$ is a norm.

(III) The flows in CA and fluid dynamics both evolve under a left group action on a linear representation space, $\eta_t = g_t \cdot \eta_0$. They differ in the roles of their advected quantities, $a_t = a_0 \circ g_t^{-1}$. The main difference is that image properties are passive and affect the template matching as a constraint in the cost function, while advected quantities may affect fluid flows directly, for example through the pressure, and thereby produce waves.

### 15.2.3 How EPDiff emerges in CA

Choose the cost function for continuously morphing $\eta_0$ into $\eta_1$ along $\eta_t = g_t \cdot \eta_0$ as

$$\mathrm{Cost}(t \mapsto g_t) = \int_0^1 \ell(\mathbf{u}_t)\,dt = \int_0^1 \|\mathbf{u}_t\|^2\,dt,$$

where $u_t$ is the velocity of the fluid deformation at time $t$ and

$$\|\mathbf{u}_t\|^2 = \langle\, \mathbf{u}_t,\, Q_{op}\mathbf{u}_t\,\rangle,$$

and $Q_{op}$ is our positive symmetric linear operator. Then, according to the Euler–Poincaré theory, the momentum governing the process, $\mathbf{m}_t = Q_{op}\mathbf{u}_t$ with Green's kernel $K : \mathbf{u}_t = K * \mathbf{m}_t$ satisfies the EPDiff equation, (15.2). The EPDiff equation arises in both template matching and fluid dynamics, and it informs both fields of endeavour.

## 15.3 Outline matching and momentum measures

***Problem statement for outline matching:***
Given two collections of curves $c_1, \ldots, c_N$ and $C_1, \ldots, C_N$ in $\Omega$, find a time-dependent diffeomorphic process $(t \mapsto g_t)$ of minimal action (or cost) such that $g_0 = \mathrm{id}$ and $g_1(c_i) = C_i$ for $i = 1, \ldots, N$. The matching problem for the image outlines seeks ***singular momentum solutions*** that naturally emerge in the computation of geodesics.

### 15.3.1 Image outlines as singular momentum solutions of EPDiff

For example, in the 2D plane, EPDiff has weak ***singular momentum solutions*** that are expressed as in equation (14.9) [CH93, HM04, HS04]

$$\mathbf{m}(\mathbf{x}, t) = \sum_{a=1}^{N} \int_s \mathbf{P}_a(t, s)\delta\big(\mathbf{x} - \mathbf{Q}_a(t, s)\big)\, ds, \qquad (15.3)$$

where $s$ is a ***Lagrangian coordinate*** defined along a set of $N$ curves in the plane *moving with the flow* by the equations $\mathbf{x} = \mathbf{Q}_a(t, s)$ and supported on the delta functions in the EPDiff solution (15.3). Thus, the evolving singular momentum solutions of EPDiff are supported on delta functions defined along curves $\mathbf{Q}_a(t, s)$ with arclength coordinate $s$ and carrying momentum $\mathbf{P}_a(t, s)$ at each point along the curve as specified in eqn (15.3). These solutions exist in any dimension and they provide a means of performing CA matching for points (landmarks), curves and surfaces, in any combination. For examples, see Figures 14.1–14.3.

### 15.3.2 Leading from geometry to numerics

The basic observation that ties everything together in $n$-dimensions is Theorem 14.2, which we repeat, as follows.

**Theorem 15.1 [HM04])** *The EPDiff singular-solution ansatz (15.3) defines a cotangent-lift momentum map,*

$$T^*\mathrm{Emb}(S, \mathbb{R}^n) \to \mathfrak{g}^* : \ (\mathbf{P}, \mathbf{Q}) \to \mathbf{m},$$

*and such momentum maps are equivariant.*

**Corollary 15.2** *The parameters* $\mathbf{Q}$ *and* $\mathbf{P}$ *in eqn (15.3) evolve by* ***Hamilton's canonical equations.***

*Proof* This is also a property of cotangent-lift momentum maps. ∎

**Remark 15.3** *The parameters* **Q** *and* **P** *in eqn (15.3) provide a* ***complete parameterization*** *of the landmarks and image outlines.*

The proof of this theorem and the mathematics underlying such momentum maps for diffeomorphisms were explained in Chapter 14. For convenience, the main results for template matching are recapped, as follows:

- The embedded manifold $S$ is the support set of $(\mathbf{P}, \mathbf{Q})$.
- The momentum map is for left action of the diffeomorphisms on $S$.
- The whole system is right invariant.
- Consequently, the momentum map for right action is conserved.
- These constructions persist for a certain class of numerical schemes.
- They apply in template matching for every choice of norm.

### 15.3.3 EPDiff dynamics informs optimal control for CA

CA must compare two geometric objects, and thus it is concerned with an ***optimal control problem***. However, the ***initial-value problem*** for EPDiff also has ***important consequences for CA applications***.

- When matching two geometric structures, the ***momentum at time t=0 contains all required information for reconstructing the target from the template***. This is done via ***Hamiltonian geodesic flow***.
- Being canonically conjugate, the momentum has exactly the same dimension as the matched structures, so there is ***no redundancy***.
- Right invariance eliminates the relabelling motions from the optimal solution and also yields a ***conserved momentum map***.
- Besides being one-to-one, the momentum representation is defined on a ***linear space***, being dual to the velocity vectors.
  This means that one may, for example,:
  - study linear instability of CA processes,
  - take averages and
  - apply statistics to the space of image contours.

  Thus, the momentum-map representation (15.3) enables building, sampling and estimating statistical models on a ***linear space***.

### 15.3.4 Summary

Momentum is a key concept in the representation of image data for CA and discussed analogies with fluid dynamics. The ***fundamental idea*** transferring via EPDiff to CA is the idea of ***momentum maps*** corresponding to group actions on image contours that are represented as smoothly embedded subspaces of the ambient space.

## 15.4 Numerical examples of outline matching

Numerical techniques have been developed for applying standard particle-mesh methods from fluid dynamics to the problem of matching outlines. These techniques are based on calculating geodesics in the space of image outlines.

Let $\boldsymbol{Q}_0$ and $\boldsymbol{Q}_1$ be two embeddings of $S^1$ in $\mathbb{R}^2$ that represent two shapes, each a closed planar curve. The contour-matching problem seeks a 1-parameter family of embeddings $\boldsymbol{Q}(t) : S^1 \times [0,1] \to \mathbb{R}^2$ so that $\boldsymbol{Q}(0) = \boldsymbol{Q}_0$ and $\boldsymbol{Q}(1)$ matches $\boldsymbol{Q}_1$ (up to relabelling). The evolution of the contour $\boldsymbol{Q}(t)$ is found by minimizing the constrained norm of its velocity. To find the equation for $\boldsymbol{Q}$ we require extremal values of the action defined by

$$
\begin{aligned}
A &= \int_0^1 L(\boldsymbol{u})\, \mathrm{d}t + \int_0^1 \int_{S^1} \boldsymbol{P}(s,t) \cdot (\dot{\boldsymbol{Q}}(s,t) - \boldsymbol{u}(\boldsymbol{Q}(s,t)))\, \mathrm{d}s\, \mathrm{d}t, \\
L &= \frac{1}{2} \|\boldsymbol{u}(t)\|_{\mathfrak{g}}^2,
\end{aligned}
$$

where the norm $\|\cdot\|_{\mathfrak{g}}$ defines a metric on the tangent space $\mathfrak{g}$ of the diffeomorphisms. That is, we seek timeseries of vector fields $\boldsymbol{u}(t)$ that are minimized in a certain norm subject to the constraint that $\boldsymbol{Q}$ is advected by the flow using the Lagrange multiplier $\boldsymbol{P}$ (the canonical momentum). The extremal solutions are given by

$$
\left.
\begin{aligned}
\frac{\delta L}{\delta \boldsymbol{u}} &= \int_{S^1} \boldsymbol{P}(s,t)\, \delta(\boldsymbol{x} - \boldsymbol{Q}(s,t))\, \mathrm{d}s, \\
\dot{\boldsymbol{P}}(s,t) &= -\left.(\nabla \boldsymbol{u})^T\right|_{(\boldsymbol{Q}(s,t),t)} \cdot \boldsymbol{P}(s,t), \\
\dot{\boldsymbol{Q}}(s,t) &= \boldsymbol{u}(\boldsymbol{Q}(s,t),t),
\end{aligned}
\right\}
\tag{15.4}
$$

subject to $\boldsymbol{Q}(s,0) = \boldsymbol{Q}_0(s)$. As usual, one denotes $(\nabla \mathbf{u})^T \cdot \mathbf{P} = \sum_j P_j \nabla u^j$.

The first equation in the system (15.4) is the momentum map $\mathrm{J}_{\mathrm{Sing}}$ in (14.19) corresponding to the cotangent-lift of the action of vector fields $\boldsymbol{u}$ on embedded curves given by

$$\boldsymbol{Q} \mapsto \boldsymbol{u}(\boldsymbol{Q}).$$

For a suitable test function $\boldsymbol{w}$, the singular momentum solutions $\boldsymbol{m}$ satisfy

$$\frac{\mathrm{d}}{\mathrm{d}t}\langle \boldsymbol{w}, \boldsymbol{m}\rangle - \langle \nabla \boldsymbol{w}, \boldsymbol{u}\boldsymbol{m}\rangle + \langle \boldsymbol{w}, (\nabla \boldsymbol{u})^T \cdot \boldsymbol{m}\rangle = 0, \qquad \boldsymbol{m} = \frac{\delta L}{\delta \boldsymbol{u}},$$

which is the weak form of the EPDiff equation (15.2).

For contour matching, one must seek an initial momentum distribution $\boldsymbol{P}(s,0)$ that takes shape $\boldsymbol{Q}_0(s)$ to shape $\boldsymbol{Q}_1(s)$. For this, a functional $J$ of the advected shape $\boldsymbol{Q}(1,s)$ must be chosen that is minimized when $\boldsymbol{Q}(1,s)$ matches $\boldsymbol{Q}_1(s)$. Following [GTY04], we describe the contours by singular densities:

$$\mu = \int_{S_1} \widehat{\mu}(s)\, \delta(\boldsymbol{x} - \boldsymbol{Q}(1,s))\, \mathrm{d}s\, \mathrm{d}V(\boldsymbol{x}), \tag{15.5}$$

$$\eta = \int_{S_1} \widehat{\eta}(s)\, \delta(\boldsymbol{x} - \boldsymbol{Q}_1(s))\, \mathrm{d}s\, \mathrm{d}V(\boldsymbol{x}), \tag{15.6}$$

and write $J = \|\mu - \eta\|_K^2$, where $\|\cdot\|_K^2$ is a norm for singular densities in a reproducing-kernel Hilbert space with kernel $K$. This approach allows the contours to be matched modulo relabelling.

The remaining matching problem may be solved, for example, by using a gradient algorithm, in which the gradient of the residual error with respect to $\boldsymbol{P}(s,0)$ is calculated using the standard method based on the adjoint equation [Gun03]. Another, more general, approach will be discussed in Chapter 16.

### 15.4.1 Numerical discretization

Various methods, including the Variational particle-mesh (VPM) method [Cot05, CH08] may be used to discretize the equations in (15.4). This method may be sketched, as follows. First, discretize the velocity on an Eulerian grid with $n_g$ points and approximate $\|\boldsymbol{u}\|$ there. Then replace the embedded contour $S^1$ by a finite set of $n_p$ Lagrangian particles $\{\boldsymbol{Q}_\beta\}_{\beta=1}^{n_p}$ (labelled by Greek indices) and interpolate from the grid to the particles by using basis functions

$$u(Q_\beta) = \sum_{k=1}^{n_g} u_k \psi_k(Q_\beta), \quad \text{with} \quad \sum_{k=1}^{n_g} \psi_k(x) = 1, \ \forall\, x,$$

for $n_g$ grid points (labelled by Latin indices). The action for the continuous-time motion on the grid then becomes

$$A = \int_0^1 \frac{1}{2}\|u(t)\|^2 + \sum_\beta P_\beta \cdot \left( \dot{Q}_\beta - \sum_k u_k \psi_k(Q_\beta) \right) \mathrm{d}t,$$

and one can obtain a fully discrete method by discretizing the action in time. For example, a first-order method is obtained by extremizing

$$A = \Delta t \sum_{n=1}^{N} \left( \frac{1}{2}\|u^n\|^2 + \sum_\beta P_\beta^n \cdot \left( \frac{Q_\beta^n - Q_\beta^{n-1}}{\Delta t} - \sum_k u_k^n \psi_k(Q_\beta^{n-1}) \right) \right).$$

This time-stepping method is the (first-order) symplectic Euler-A method for the time-continuous Hamiltonian system for the Lagrangian particles. In general, the method will be symplectic because it arises from a discrete variational principle. See [LR04] for a broad introduction to symplectic numerical methods and their conservation properties. The conservation properties of VPM are discussed in [CH08].

The densities $\mu$ and $\eta$ may be approximated on the grid by using the standard particle-mesh approach (see [FGR02]). This approach produces the representation,

$$\mu_k = \sum_\beta \widehat{\mu}_\beta \psi_k(Q_\beta^N), \qquad \eta_k = \sum_\beta \widehat{\eta}_\beta \psi_k(Q_{1,\beta}),$$

where $Q_{1,\beta}$ are the positions of particles on the target shape. This representation amounts to *pixellating* the singular densities (15.5) and (15.6) on the grid. For a given kernel $K$, one may approximate the norm $J$ by

$$J = \sum_{kl} K(x_k - x_l)(\mu_k - \eta_k)(\mu_l - \eta_l). \tag{15.7}$$

The discrete adjoint equation may then be applied in computing the inversion for the initial conditions for $P_\beta$ that generate the flow. A numerical example calculated by using this method is given in Figure 15.3.

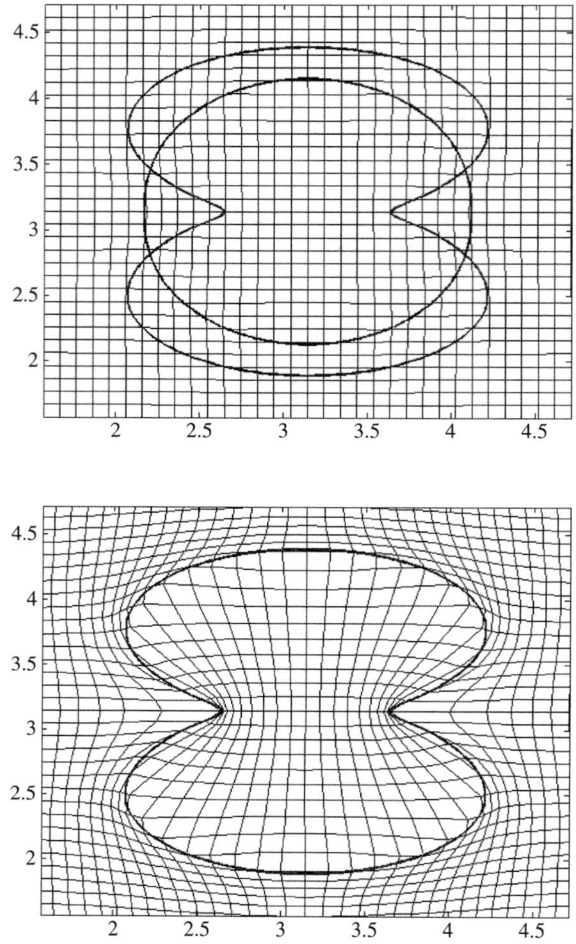


**Figure 15.3** Results from a numerical simulation using the VPM algorithm to calculate the minimal path between a two simple shapes. The initial and final shapes are shown in the top panel. The bottom panel shows the deformation of the initial shape into the final shape, together with a grid that shows how the flow map deforms the space around the shape. The $H^1$ norm for velocity on a $2\pi \times 2\pi$ periodic domain was used on a $128 \times 128$ grid. The fast Fourier transform (FFT) was used in discretizing and the corresponding kernel $K$ was used to calculate $J$ in eqn (15.7). Cubic B-splines were chosen for the basis functions $\psi_k$.

## Outlook

This chapter has sketched an approach for using flows governed by EPDiff to determine the dynamics of contour matching. The EPDiff flows are defined on the smooth invertible maps acting on a spatial domain. They treat dynamical changes of shape and matching of contours as smooth transformations of the domain in precisely the way that D'Arcy Thompson envisioned them. The application of EPDiff to the dynamics of contour matching involves a geodesic curve in the full group of diffeomorphisms with respect to a metric $\|\boldsymbol{u}(t)\|_{\mathfrak{g}}^2$ defined on its tangent space $\mathfrak{g}$. In this scenario, contour matching means connecting two shapes defined by contours in the plane so that the action of the diffeomorphisms on the domain carries one shape to the other smoothly along the geodesic curve. Chapter 16 will discuss an approach to the dynamics of morphology that goes beyond what D'Arcy Thompson envisioned. In this approach, geodesic evolution under the action of the diffeomorphisms carries additional properties, such as colour, intensity or orientation of pixels in the image, as well as shape. Moreover, these additional image properties may evolve under their own optimal dynamics.

**Exercise 15.1**
Verify the equations in the system (15.4).

HINT: Before taking variations in velocity $\boldsymbol{u}$, insert a delta function by writing

$$
\begin{aligned}
&\int_0^1 \int_{S^1} \boldsymbol{P}(s,t) \cdot \boldsymbol{u}(\boldsymbol{Q}(s,t),t)\, \mathrm{d}s\, \mathrm{d}t \\
&= \int_0^1 \int_{S^1} \int_{\mathbb{R}^2} \boldsymbol{P}(s,t) \cdot \boldsymbol{u}(\boldsymbol{x},t)\, \delta\big(\boldsymbol{x} - \boldsymbol{Q}(s,t)\big) \mathrm{d}^2x\, \mathrm{d}s\, \mathrm{d}t.
\end{aligned}
$$

This yields the first equation in (15.4). The others follow by standard manipulations.

**Exercise 15.2**
Find the system of equations defined by minimizing the alternative action,

$$
\begin{aligned}
A &= \int_0^1 L(\boldsymbol{u})\, \mathrm{d}t + \frac{1}{2\sigma^2} \int_0^1 \int_{S^1} \big|\dot{\boldsymbol{Q}}(s,t) - \boldsymbol{u}(\boldsymbol{Q}(s,t),t)\big|^2 \mathrm{d}s\, \mathrm{d}t, \\
L &= \frac{1}{2}\|\boldsymbol{u}(t)\|_{\mathfrak{g}}^2,
\end{aligned}
$$

where $\sigma^2 > 0$ is a constant parameter, the norm $\|\cdot\|_{\mathfrak{g}}$ defines a metric on the tangent space $\mathfrak{g}$ of the diffeomorphisms and $|\cdot|$ without subscript is the Euclidean metric for vectors. Do the equations for the minimizers of this action still admit the momentum-map relation in the first equation of system (15.4)?

HINT: in minimizing this action, the quantity $\dot{\boldsymbol{Q}}(s,t) - \boldsymbol{u}(\boldsymbol{Q}(s,t))$ is minimized by imposing it as a penalty, rather than constraining it to vanish exactly, as in the action for the system (15.4). Before taking variations of this alternative action, it is helpful to rewrite the penalty term equivalently as

$$\int_0^1 \int_{S^1} \left[ \boldsymbol{P}(s,t) \cdot \big(\dot{\boldsymbol{Q}}(s,t) - \boldsymbol{u}(\boldsymbol{Q}(s,t),t)\big) - \frac{\sigma^2}{2} |\boldsymbol{P}|^2(s,t) \right] \mathrm{d}s\, \mathrm{d}t.$$

This form of the penalty term is seen to be equivalent to the original one, upon taking stationary variations with respect to $\boldsymbol{P}$ to find the defining relation

$$\sigma^2 \boldsymbol{P}(s,t) = \dot{\boldsymbol{Q}}(s,t) - \boldsymbol{u}\big(\boldsymbol{Q}(s,t),t\big).$$

# 16 Computational anatomy: Euler–Poincaré image matching

## 16.1 Overview

***Pattern matching*** is an important component of imaging science, and is fundamental in computational anatomy (computerized anatomical analysis of medical images). When comparing images, the purpose is to find an optimal deformation that aligns the images and matches their photometric properties. Diffeomorphic pattern matching methods have been developed to achieve both this objective and the additional goal of defining a (Riemannian) metric structure on spaces of deformable objects [DGM98, Tro98]. This approach has found many applications in medical imaging, where the objects of interest include images, landmarks, measures (supported on point sets) and currents (supported on curves and surfaces). These methods usually address the registration problem by solving a variational problem of the form

$$\text{Minimize}\,\Big(d(\text{id},g)^2 + \text{Error term}(g.n_{temp}, n_{targ})\Big) \qquad (16.1)$$

over all diffeomorphisms $g$, where $n_{temp}$ and $n_{targ}$ are the images being compared (usually referred to as the template and the target), $(g,n) \mapsto g.n$ is the action of diffeomorphisms on the objects and $d$ is a right-invariant Riemannian distance on diffeomorphisms.

> This chapter explains how the pattern-matching problem for images is governed by the Euler–Poincaré equations of geodesic motion for a Lagrangian given by a right-invariant norm on $(T\text{Diff} \times TN)/\text{Diff}$, where $N$ is the manifold of images on which Diff acts.

In problems formulated as in eqn (16.1), the error term breaks the metric aspects inherited from the distance $d$ on the diffeomorphisms because the error term has an inherent template vs. target asymmetry. With the aim of designing a fully metric approach to the template-matching problem, the ***metamorphosis approach*** was formulated in [TY05]. The metamorphosis approach embraces what are called ***morphing*** and ***warping*** in computer graphics while endowing the composition of the two operations with a Riemannian variational structure. The metamorphosis approach provides interesting alternatives to the pattern-matching approach based on eqn (16.1), in the context of a metric framework. This chapter explains the Lagrangian formulation for metamorphosis of images developed in [HTY09] that includes the Riemannian formalism introduced in [TY05].

The discussion will be general enough to include a range of applications. Consider a manifold $N$ that is acted upon by a Lie group $G$. The manifold $N$ contains the ***deformable objects*** and $G$ is the Lie group of deformations, which is taken to be the group of diffeomorphisms. (A few examples of the space $N$ will be discussed later. In particular, $N$ could be another Lie group.)

**Definition 16.1** *A* ***metamorphosis*** *[TY05] is a pair of curves* $(g_t, \eta_t) \in G \times N$ *parameterized by time* $t$*, with* $g_0 = \mathrm{id}$*. Its* ***image*** *is the curve* $n_t \in N$ *defined by the action* $n_t = g_t.\eta_t$*. The quantities* $g_t$ *and* $\eta_t$ *are called the* ***deformation component*** *of the metamorphosis, and its* ***template component****, respectively. When* $\eta_t$ *is constant, the metamorphosis reduces to standard template matching, which is a* ***pure deformation****. In the general case, the image is a composition of deformation and template variation.*

This chapter places the metamorphosis approach into a Lagrangian formulation, and applies the Euler–Poincaré variational framework to derive its evolution equations. Analytical questions about these equations (for example, the existence and uniqueness of their solutions) require additional assumptions on $G$ and the space $N$ of deformed objects that are beyond the scope of the present text. For analytical discussions of the equations in this chapter, see [HTY09].

The next section provides notation and definitions related to the problem of metamorphosis.

## 16.2 Notation and Lagrangian formulation

The letters $\eta$ or $n$ will be used to denote elements of $N$, the former being associated to the template component of a metamorphosis, and the latter to its image under the action of the group.

The variational problem we shall study optimizes over metamorphoses $(g_t, \eta_t)$ by minimizing, for some Lagrangian $L$, the integral

$$\int_0^1 L(g_t, \dot{g}_t, \eta_t, \dot{\eta}_t)\mathrm{d}t, \tag{16.2}$$

with fixed endpoint conditions for the initial and final images $n_0$ and $n_1$ (with $n_t = g_t\eta_t$) and $g_0 = \mathrm{id}_G$ (so only the images are constrained at the endpoints, with the additional normalization $g_0 = \mathrm{id}$).

Let $\mathfrak{g}$ denote the Lie algebra of $G$ and let $(g, U_g, \eta, \xi_\eta) \in TG \times TN$. We will consider Lagrangians defined on $TG \times TN$, that satisfy the following invariance conditions: there exists a function $\ell$ defined on $\mathfrak{g} \times TN$ such that

$$L(g, U_g, \eta, \xi_\eta) = \ell(U_g g^{-1}, g\eta, g\xi_\eta).$$

In other words, $L$ is taken to be invariant under the right action of $G$ on $G \times N$ defined by $(g, \eta)h = (gh, h^{-1}\eta)$.

For a metamorphosis $(g_t, \eta_t)$, the following definitions

$$u_t = \dot{g}_t g_t^{-1}, \qquad n_t = g_t\eta_t, \quad \text{and} \quad \nu_t = g_t\dot{\eta}_t \tag{16.3}$$

lead by right invariance to an expression for the ***reduced Lagrangian***

$$L(g_t, \dot{g}_t, \eta_t, \dot{\eta}_t) = \ell(u_t, n_t, \nu_t). \tag{16.4}$$

## More notation

The Lie derivative with respect to a vector field $X$ will be denoted $\pounds_X$. The Lie algebra of $G$ is identified with the set of right-invariant vector fields $U_g = ug$, $u \in T_eG = \mathfrak{g}$, $g \in G$. If $G$ acts on a set $\tilde{N}$, and $f : \tilde{N} \to \mathbb{R}$, one finds $\pounds_u f(\tilde{n}) = (\mathrm{d}/\mathrm{d}t) f(g_t\tilde{n})$ with $g_0 = \mathrm{id}$ and $\dot{g}_t(0) = u$.

The Lie bracket $[u, v]$ on $\mathfrak{g}$ is defined by

$$\pounds_{[u,v]} = -(\pounds_u\pounds_v - \pounds_v\pounds_u) \tag{16.5}$$

and the associated adjoint operator is $\mathrm{ad}_u\, v = [u, v]$. Letting $I_g(h) = ghg^{-1}$ and $Ad_v g = \pounds_v I_g(\mathrm{id})$ yields $\mathrm{ad}_u\, v = \pounds_u(\mathrm{Ad}_v)(\mathrm{id})$. When $G$ is a group of diffeomorphisms, this defines

$$ad_u v = \mathrm{d}u\, v - \mathrm{d}v\, u, \tag{16.6}$$

as in eqn (11.6).

The pairing between a linear form $l$ and a vector $u$ will be denoted $\langle l,\, u\rangle$. Duality with respect to this pairing will be denoted with an asterisk $(\cdot)^*$.

When the Lie group $G$ acts on a manifold $\tilde{N}$, the associated ***diamond*** operation ($\diamond$) (or dual action) is defined on $T\tilde{N}^* \times \tilde{N}$ and takes values in $\mathfrak{g}^*$, so that

$$\diamond : T\tilde{N}^* \times \tilde{N} \to \mathfrak{g}^*. \tag{16.7}$$

The diamond operation is defined in terms of the pairing $\langle \cdot , \cdot \rangle : \mathfrak{g}^* \times \mathfrak{g} \to \mathbb{R}$ as in eqn (9.32). That is, diamond is defined in this notation by

$$\langle \gamma \diamond \tilde{n}, u \rangle = -\langle \gamma, u\tilde{n} \rangle_{T\tilde{N}^*}, \tag{16.8}$$

where $(\gamma, \tilde{n}) \in T\tilde{N}^* \times \tilde{N}$, $u\tilde{n} = \pounds_u \tilde{\eta}$ and the bracket $\langle \cdot , \cdot \rangle_{T\tilde{N}^*}$ denotes the pairing between $N$ and $TN^*$.

## 16.3 Symmetry-reduced Euler equations

We compute the symmetry-reduced Euler equations as stationarity conditions that extremalize the reduced action, defined in terms of the reduced Lagrangian by,

$$S_{red} := \int_0^1 \ell(u_t, n_t, \nu_t) \mathrm{d}t, \tag{16.9}$$

with respect to variations $\delta u$ and $\omega = \delta n = \delta(g\eta)$ for fixed endpoint conditions $n_0$ and $n_1$. The variation $\delta \nu$ can be obtained from $n = g\eta$ and $\nu = g\dot{\eta}$ yielding

$$\dot{n} = \nu + un \quad \text{and} \quad \dot{\omega} = \delta\nu + u\omega + \delta u n, \tag{16.10}$$

in which Lie algebra action is denoted by ***concatenation from the left***. For example, $un = \pounds_u n$ denotes the Lie derivative of $n$ along the vector field $u$, etc. The computations are performed in a local chart on $TN$ in terms of which partial derivatives are taken.

Taking stationary variations of $S_{red}$ yields

$$\int_0^1 \left( \left\langle \frac{\delta \ell}{\delta u}, \delta u_t \right\rangle + \left\langle \frac{\delta \ell}{\delta n}, \omega_t \right\rangle + \left\langle \frac{\delta \ell}{\delta \nu}, \dot{\omega}_t - u_t\omega_t - \delta u_t\, n_t \right\rangle \right) \mathrm{d}t = 0. \tag{16.11}$$

The $\delta u$-term yields the constant of motion,

$$\frac{\delta \ell}{\delta u} + \frac{\delta \ell}{\delta \nu} \diamond n_t = 0. \tag{16.12}$$

A slight abuse of notation is allowed in writing $\delta\ell/\delta\nu \in T(TN)^*$ as a linear form on $TN$ via $\langle \delta\ell/\delta\nu,\, z\rangle := \langle \delta\ell/\delta\nu,\, (0,z)\rangle$.

After an integration by parts in time, the $\omega$-term in the variation equation (16.11) yields,

$$\frac{\partial}{\partial t}\frac{\delta\ell}{\delta\nu} + u_t \star \frac{\delta\ell}{\delta\nu} - \frac{\delta\ell}{\delta n} = 0, \tag{16.13}$$

with additional notation for the $\star$ operation, defined by

$$\left\langle u \star \frac{\delta\ell}{\delta\nu},\, \omega \right\rangle := \left\langle \frac{\delta\ell}{\delta\nu},\, u\omega \right\rangle. \tag{16.14}$$

The endpoint terms vanish in the integration by parts for the $\omega$-term because $\delta n$ vanishes at the endpoints for $n_0$ and $n_1$ fixed. These manipulations have proven the following.

**Theorem 16.2 (Metamorphosis equations)** *The* ***symmetry-reduced Euler equations*** *associated with extremals of the reduced action $S_{\rm red}$ in eqn (16.9)*

$$\int_0^1 \ell(u_t, n_t, \nu_t)\mathrm{d}t$$

*with fixed endpoint conditions $n_0$ and $n_1$ under variations of the right-invariant velocity ($\delta u$) and image ($\omega = \delta n$) defined in eqn (16.3) consist of the system of* ***metamorphosis equations***

$$\left.\begin{aligned} &\frac{\delta\ell}{\delta u} + \frac{\delta\ell}{\delta\nu} \diamond n_t = 0, \\ &\frac{\partial}{\partial t}\frac{\delta\ell}{\delta\nu} + u_t \star \frac{\delta\ell}{\delta\nu} = \frac{\delta\ell}{\delta n}, \\ &\dot{n}_t = \nu_t + u_t n_t. \end{aligned}\right\} \tag{16.15}$$

**Remark 16.3** *The quantity $\frac{\delta\ell}{\delta u} + \frac{\delta\ell}{\delta\nu} \diamond n$ is the conserved momentum arising from Noether's theorem for right invariance of the Lagrangian. As we shall see, the special form of the endpoint conditions (fixed $n_0$ and $n_1$) ensures that this conserved momentum vanishes identically.*

## 16.4 Euler–Poincaré reduction

A dynamical system equivalent to the metamorphosis equations (16.15) may be obtained by using Euler–Poincaré reduction. In this setting, one takes variations in the group element ($\delta g$) and in the template ($\delta \eta$) instead of the velocity and the image. Set

$$\xi_t = \delta g_t g_t^{-1} \quad \text{and} \quad \varpi_t = g_t \delta \eta_t. \tag{16.16}$$

These definitions lead to expressions for $\delta u$, $\delta n$ and $\delta \nu$. For the velocity one finds the ***constrained variation***,

$$\delta u_t = \dot{\xi}_t + [\xi_t, u_t]. \tag{16.17}$$

This is a standard relation in Euler–Poincaré reduction, as explained in Lemma 7.4. From the definition $n_t = g_t \eta_t$ in (16.3) one has the variational relation

$$\delta n_t = \delta(g_t \eta_t) = \varpi_t + \xi_t n_t. \tag{16.18}$$

From the definition $\nu_t = g_t \dot{\eta}_t$, one finds

$$\delta \nu_t = g_t \delta \dot{\eta}_t + \xi_t \nu_t, \tag{16.19}$$

and from $\varpi_t = g_t \delta \eta_t$ one observes

$$\dot{\varpi}_t = u_t \varpi_t + g_t \dot{\eta}_t, \tag{16.20}$$

which, in turn, yields

$$\delta \nu_t = \dot{\varpi}_t + \xi_t \nu_t - u_t \varpi_t. \tag{16.21}$$

The endpoint conditions for $\xi$ and $\varpi$ are computed, as follows. One starts at $t = 0$ with $g_0 = \text{id}$ and $n_0 = g_0 \eta_0 = \text{cst}$, which implies $\xi_0 = 0$ and $\varpi_0 = 0$. At $t = 1$, the relation $g_1 \eta_1 = \text{cst}$ yields an endpoint condition on the variations at $t = 1$,

$$\delta(g_t \eta_t)\Big|_{t=1} = \xi_1 n_1 + \varpi_1 = 0. \tag{16.22}$$

The variation of the reduced action $S_{\text{red}}$ in eqn (16.9) is now expressed as

$$\int_0^1 \left( \left\langle \frac{\delta \ell}{\delta u}, \dot{\xi}_t - \text{ad}_{u_t} \xi_t \right\rangle + \left\langle \frac{\delta \ell}{\delta n_t}, \varpi_t + \xi_t n_t \right\rangle + \left\langle \frac{\delta \ell}{\delta \nu}, \dot{\varpi}_t + \xi_t \nu_t - u_t \varpi_t \right\rangle \right) \text{d}t = 0.$$

In the integrations by parts in time to eliminate $\dot{\xi}_t$ and $\dot{\varpi}_t$, the endpoint terms sum to

$$\big\langle(\delta\ell/\delta u)_1,\,\xi_1\big\rangle+\big\langle(\delta\ell/\delta \nu)_1,\,\varpi_1\big\rangle.$$

Using the endpoint condition (16.22) on the variations at $t=1$ allows the last term to be rewritten as

$$\big\langle(\delta\ell/\delta\nu)_1,\,\varpi_1\big\rangle=-\big\langle(\delta\ell/\delta\nu)_1,\,\xi_1 n_1\big\rangle=\big\langle(\delta\ell/\delta\nu)_1\diamond n_1,\,\xi_1\big\rangle.$$

One therefore obtains the stationarity relation at time $t=1$,

$$\frac{\delta\ell}{\delta u}(1)+\frac{\delta\ell}{\delta\nu}(1)\diamond n_1=0. \tag{16.23}$$

After another integration by parts, the $\xi$-terms provide the evolution equation for $\delta\ell/\delta u$,

$$\frac{\partial}{\partial t}\frac{\delta\ell}{\delta u}+\mathrm{ad}^*_{u_t}\frac{\delta\ell}{\delta u}+\frac{\delta\ell}{\delta n}\diamond n_t+\frac{\delta\ell}{\delta\nu}\diamond\nu_t=0. \tag{16.24}$$

Likewise, the $\varpi$-terms provide the evolution equation for $\delta\ell/\delta\nu$,

$$\frac{\partial}{\partial t}\frac{\delta\ell}{\delta\nu}+u_t\star\frac{\delta\ell}{\delta\nu}-\frac{\delta\ell}{\delta n}=0. \tag{16.25}$$

These additional manipulations have proven the following.

**Theorem 16.4 (Metamorphosis dynamics)**
*The Euler–Poincaré equations associated with extremals of the reduced action $S_{\rm red}$ in eqn (16.9)*

$$S_{\rm red}=\int_0^1\ell(u_t,n_t,\nu_t)\mathrm{d}t$$

*with fixed endpoint conditions $n_0$ and $n_1$ under variations in the group element $(\delta g)$ and in the template $(\delta\eta)$ consist of the system of equations for*

***Euler–Poincaré metamorphosis dynamics***

$$\left.\begin{aligned} &\frac{\partial}{\partial t}\frac{\delta \ell}{\delta u} + \operatorname{ad}^*_{u_t}\frac{\delta \ell}{\delta u} + \frac{\delta \ell}{\delta n}\diamond n_t + \frac{\delta \ell}{\delta \nu}\diamond \nu_t = 0,\\ &\frac{\partial}{\partial t}\frac{\delta \ell}{\delta \nu} + u_t \star \frac{\delta \ell}{\delta \nu} - \frac{\delta \ell}{\delta n} = 0,\\ &\frac{\delta \ell}{\delta u}(1) + \frac{\delta \ell}{\delta \nu}(1)\diamond n_1 = 0,\\ &\dot{n}_t = \nu_t + u_t n_t. \end{aligned}\right\} \tag{16.26}$$

**Proposition 16.5** *The dynamical system* (16.26) *is equivalent to eqn* (16.15).

*Proof* The equivalence is obvious, since the two systems of equations characterize the same critical points of the reduced action obtained by different independent variations. However, an instructive proof can be given by rewriting the first equation in (16.26) as a ***Kelvin–Noether theorem for images***,

$$\frac{\partial}{\partial t}\Big(\frac{\delta \ell}{\delta u} + \frac{\delta \ell}{\delta \nu}\diamond n\Big) + \operatorname{ad}^*_{u_t}\Big(\frac{\delta \ell}{\delta u} + \frac{\delta \ell}{\delta \nu}\diamond n\Big) = 0. \tag{16.27}$$

Indeed, any solution of (16.26) satisfies,

$$\begin{aligned} &\frac{\partial}{\partial t}\Big(\frac{\delta \ell}{\delta u_t} + \frac{\delta \ell}{\delta \nu}\diamond n_t\Big)\\ &\quad= \frac{\partial}{\partial t}\frac{\delta \ell}{\delta u} + \Big(\frac{\partial}{\partial t}\frac{\delta \ell}{\delta \nu}\Big)\diamond n_t + \frac{\delta \ell}{\delta \nu}\diamond \dot{n}_t\\ &\quad= \frac{\partial}{\partial t}\frac{\delta \ell}{\delta u} + \Big(\frac{\delta \ell}{\delta n} - u_t \star \frac{\delta \ell}{\delta \nu}\Big)\diamond n_t + \frac{\delta \ell}{\delta \nu}\diamond(\nu_t + u_t n_t)\\ &\quad= \frac{\partial}{\partial t}\frac{\delta \ell}{\delta u} + \frac{\delta \ell}{\delta n}\diamond n_t + \frac{\delta \ell}{\delta \nu}\diamond \nu_t - \Big(u_t \star \frac{\delta \ell}{\delta \nu}\Big)\diamond n_t + \frac{\delta \ell}{\delta \nu}\diamond(u_t n_t)\\ &\quad= -\operatorname{ad}^*_{u_t}\frac{\delta \ell}{\delta u} - \operatorname{ad}^*_{u_t}\Big(\frac{\delta \ell}{\delta \nu}\diamond n_t\Big). \end{aligned}$$

In the last equation, we have used the fact that, for any $\alpha \in \mathfrak{g}$,

$$\begin{aligned}\left\langle \frac{\delta \ell}{\delta \nu} \diamond (un) - \left(u \star \frac{\delta \ell}{\delta \nu}\right) \diamond n \,,\, \alpha \right\rangle &= -\left\langle \frac{\delta \ell}{\delta \nu} \,,\, \alpha(un) - u(\alpha n) \right\rangle \\ &= \left\langle \frac{\delta \ell}{\delta \nu} \,,\, [u, \alpha] n \right\rangle \\ &= -\left\langle \frac{\delta \ell}{\delta \nu} \diamond n \,,\, [u, \alpha] \right\rangle \\ &= -\left\langle \mathrm{ad}^*_{u_t} \left( \frac{\delta \ell}{\delta \nu} \diamond n_t \right) , \alpha \right\rangle .\end{aligned}$$

Consequently, the first equation in the system (16.26) is equivalent to eqn (16.27), which in turn may be rewritten equivalently as, cf. Proposition 6.54,

$$\frac{\partial}{\partial t}\left( \mathrm{Ad}^*_{g_t} \left( \frac{\delta \ell}{\delta u} + \frac{\delta \ell}{\delta \nu} \diamond n \right) \right) = 0. \tag{16.28}$$

This equation combined with $(\delta \ell / \delta u)_1 + (\delta \ell / \delta \nu)_1 \diamond n_1 = 0$ implies the first equation in (16.15). ■

**Remark 16.6** *Proposition 16.5 shows that the metamorphosis approach may be regarded as the restriction of Euler–Poincaré dynamics to the zero level set of the conserved momentum that arises from right-invariance of the reduced Lagrangian $\ell$ in eqn (16.4) defined on $\mathfrak{g} \times TN$. In this guise, the metamorphosis approach to shape matching resembles the falling cat problem, in which a cat that falls with zero angular momentum may still turn itself in mid-air to land with its paws pointing downward.* ***Montgomery's falling cat theorem*** *[Mon90, Mon93] relates optimal reorientation of the falling cat to the dynamics of particles in Yang–Mills fields. According to Montgomery's falling cat theorem,*

> *A cat dropped from rest upside down flips itself right side up, even though its angular momentum is zero. It does this by changing its shape. In terms of gauge theory, the shape space of the cat forms the base space of a principal $SO(3)$-bundle, and the statement "angular momentum equals zero" defines a connection on this bundle.*

*That is, as in the falling cat problem, there is an interpretation of the first equation in the system (16.15) as the condition defining the horizontal subspace of the principal G-bundle whose base is the shape space N. We have already seen a similar situation for EPDiff, which may be recognised now by interpreting the penalty term in Exercise 14.3 as a gauge-transformed connection form.*

**Exercise 16.1**
Show that these manipulations prove the claim of Remark 16.3 that the quantity

$$\frac{\delta \ell}{\delta u} + \frac{\delta \ell}{\delta \nu} \diamond n$$

is the conserved momentum arising from ***Noether's theorem*** for right invariance of the reduced Lagrangian in eqn (16.4).

**Exercise 16.2**
Compute the Hamiltonian and Lie–Poisson bracket for the system (16.26) governing metamorphosis dynamics.

## 16.5 Semidirect-product examples

### 16.5.1 Riemannian metric

A primary application of this framework can be based on the definition of a ***Riemannian metric*** on $G \times N$ that is invariant under the right action of $G$: $(g, \eta)h = (gh, h^{-1}\eta)$. The corresponding Lagrangian then takes the form

$$l(u, n, \nu) = \|(u, \nu)\|_n^2. \tag{16.29}$$

The variational problem is now equivalent to the computation of geodesics for the canonical projection of this metric from $G \times N$ onto $N$. This framework was introduced in [MY01]. The evolution equations were derived and studied in [TY05] in the case $l(u, n, \nu) = |u|_{\mathfrak{g}}^2 + |\nu|_n^2$, for a given norm, $|.|_{\mathfrak{g}}$, on $\mathfrak{g}$ and a pre-existing Riemannian structure on $N$. This Riemannian metric on $N$ incorporates the group actions. An example of its application is given below for images. First, though, let us discuss the semidirect-product case $G Ⓢ N$.

### 16.5.2 Semidirect product

Assume that $N$ is a group and that for all $g \in G$, the action of $g$ on $N$ is a group homomorphism: For all $n, \tilde{n} \in N$, $g(n\tilde{n}) = (gn)(g\tilde{n})$ (for example, $N$ can be a vector space and the action of $G$ can be linear). Consider the ***semidirect product*** $G \circledS N$ with

$$(g,n)(\tilde{g},\tilde{n}) = (g\tilde{g}, (g\tilde{n})n), \tag{16.30}$$

and build on $G \circledS N$ a right-invariant metric constrained by its value $\| \ \|_{(\mathrm{id}_G, \mathrm{id}_N)}$ at the identity. Then, optimizing the geodesic energy in $G \circledS N$ between $(\mathrm{id}_G, n_0)$ and $(g_1, n_1)$ with fixed $n_0$ and $n_1$ and free $g_1$ yields a particular case of metamorphosis.

Right invariance for the metric on $G \circledS N$ implies

$$\|(U,\zeta)\|_{(g,n)} = \|(U\tilde{g}, (U\tilde{n})n + (g\tilde{n})\zeta\|_{(g\tilde{g},(g\tilde{n})n)}, \tag{16.31}$$

which yields, upon using $(\tilde{g},\tilde{n}) = (g^{-1}, g^{-1}n^{-1})$ and letting $u = Ug^{-1}$,

$$\begin{aligned}\|(U,\zeta)\|_{(g,n)} &= \|(u, (un^{-1})n + n^{-1}\zeta\|_{(\mathrm{id}_G,\mathrm{id}_N)} \\ &= \|(u, n^{-1}(\zeta - un)\|_{(\mathrm{id}_G,\mathrm{id}_N)},\end{aligned}$$

which may be proved by using the identity

$$0 = u(n^{-1}n) = (un^{-1})n + n^{-1}(un).$$

Consequently, the geodesic energy on $G \circledS N$ for a path of unit length is

$$\int_0^1 \|(u_t, n_t^{-1}(\dot{n}_t - u_t n_t))\|^2_{(\mathrm{id}_G,\mathrm{id}_N)}. \tag{16.32}$$

Optimizing this geodesic energy with fixed $n_0$ and $n_1$ is equivalent to solving the metamorphosis problem with

$$l(u,n,\nu) = \|(u, n^{-1}\nu)\|^2_{(\mathrm{id}_G,\mathrm{id}_N)}. \tag{16.33}$$

This turns out to be a particular case of the previous example. The situation is even simpler when $N$ is a vector space. In this case, $n^{-1}n' = n' - n$ and one computes $(g,n)(\tilde{g},\tilde{n}) = (g\tilde{g}, g\tilde{n} + n)$ so that

$$(\dot{g},\dot{n})(\tilde{g},\tilde{n}) = (\dot{g}\tilde{g}, \dot{g}\tilde{n} + \dot{n}).$$

Consequently,

$$(\dot{g}, \dot{n})(g^{-1}, g^{-1}n^{-1}) = (\dot{g}g^{-1}, \dot{g}g^{-1}n^{-1} + \dot{n}) = (u, -un + \dot{n}) = (u, \nu)$$

and the symmetry-reduced Lagrangian does not depend on $n$. The systems (16.15) and (16.26) take a very simple form when the group operation on $N$ is additive. Namely, they become, respectively,

$$\left.\begin{aligned} &\frac{\delta \ell}{\delta u} + \frac{\delta \ell}{\delta \nu} \diamond n_t = 0, \\ &\frac{\partial}{\partial t}\frac{\delta \ell}{\delta \nu} + u_t \star \frac{\delta \ell}{\delta \nu} = 0, \\ &\dot{n}_t = \nu_t + u_t n_t, \end{aligned}\right\} \tag{16.34}$$

and

$$\left.\begin{aligned} &\frac{\partial}{\partial t}\frac{\delta \ell}{\delta u} + \mathrm{ad}^*_{u_t}\frac{\delta \ell}{\delta u} + \frac{\delta \ell}{\delta \nu} \diamond \nu_t = 0, \\ &\frac{\partial}{\partial t}\frac{\delta \ell}{\delta \nu} + u_t \star \frac{\delta \ell}{\delta \nu} = 0, \\ &\frac{\delta \ell}{\delta u}(1) + \frac{\delta \ell}{\delta \nu}(1) \diamond n_1 = 0, \\ &\dot{n}_t = \nu_t + u_t n_t, \end{aligned}\right\} \tag{16.35}$$

in which concatenation as in $u_t n_t$ denotes Lie algebra action, as before. Thus, when $N$ is a vector space, the evolution of the variable $n \in N$ decouples from the rest of the dynamical system (16.35).

Even when $N$ is not a vector space, metamorphoses that are obtained from the semidirect-product formulation are specific among general metamorphoses, because they satisfy the conservation of momentum property that comes with right invariance of the metric under the Lie group. This conservation equation may be written as, see eqn (16.28),

$$\left(\frac{\delta \ell}{\delta u}, \frac{\delta \ell}{\delta \nu}\right) = \mathrm{Ad}^*_{(g_t, n_t)^{-1}}\left(\frac{\delta \ell}{\delta u}, \frac{\delta \ell}{\delta \nu}\right), \tag{16.36}$$

where the adjoint representation is associated with the semidirect-product Lie group action. This property (that we do not write in the general case) will be illustrated in an example below.

### 16.5.3 Image matching

Consider the case when $N$ is a space of smooth functions from domain $\Omega$ to $\mathbb{R}$, that we will call ***images***, with the action

$$(g,n) \mapsto n \circ g^{-1}. \tag{16.37}$$

A simple case of metamorphoses [MY01] can be obtained with the Lagrangian

$$\ell(u,\nu) = \|u\|_{\mathfrak{g}}^2 + \frac{1}{\sigma^2}\|\nu\|_{L^2}^2. \tag{16.38}$$

If $w \in \mathfrak{g}$ and $n$ is an image, then $wn = -\nabla n^T w$, so that

$$\left\langle \frac{\delta \ell}{\delta \nu} \diamond n\,,\; w \right\rangle = \left\langle \frac{\delta \ell}{\delta \nu}\,,\; \nabla n^T w \right\rangle. \tag{16.39}$$

Thus, since $\delta\ell/\delta\nu = 2\nu/\sigma^2$, the first equation in the system (16.34) is

$$L_{\mathfrak{g}} u_t := \frac{\delta \ell}{\delta u} = -\frac{1}{\sigma^2}\nu_t \nabla n_t, \tag{16.40}$$

where $L_{\mathfrak{g}}$ is the positive symmetric operator associated with the norm $\|u_t\|_{\mathfrak{g}}^2$ by

$$\|u_t\|_{\mathfrak{g}}^2 = \langle L_{\mathfrak{g}} u_t,\, u_t \rangle. \tag{16.41}$$

Now, $u \star (\delta\ell/\delta\nu)$ is defined by

$$\begin{aligned}
\left\langle u \star \left(\frac{\delta \ell}{\delta \nu}\right),\, \omega \right\rangle &= \left\langle \frac{\delta \ell}{\delta \nu}\,,\; u\omega \right\rangle \\
&= -\left\langle \frac{\delta \ell}{\delta \nu}\,,\; \nabla \omega^T u \right\rangle \\
&= -\frac{1}{\sigma^2}\langle \nu,\, \nabla \omega^T u \rangle \\
&= \frac{1}{\sigma^2}\langle \operatorname{div}(\nu u),\, \omega \rangle,
\end{aligned}$$

which yields the second equation in the system (16.34)

$$\dot{\nu}_t + \frac{1}{\sigma^2}\operatorname{div}(\nu_t u_t) = 0. \tag{16.42}$$

We denote $z = \nu/\sigma^2$ and rewrite the three equations in the system (16.34) as

$$\left.\begin{aligned} L_{\mathfrak{g}} u_t &= -z_t \nabla n_t, \\ \dot{z}_t + \operatorname{div}(z_t u_t) &= 0, \\ \dot{n}_t + \nabla n_t^T u_t &= \sigma^2 z_t. \end{aligned}\right\} \tag{16.43}$$

Existence and uniqueness of solutions for this system were proved in [TY05]. From a visual point of view, image metamorphoses are similar to what is usually called ***morphing*** in computer graphics. The evolution of the image over time, $t \mapsto n_t$, is a combination of deformations and image intensity variation. Algorithms and results for the solution of the boundary-value problem (minimize the Lagrangian between two images at the initial and final times) can be found in [MY01]. Two examples of minimizing geodesics between a pair of images are also provided in Figure 16.1.

Image matching can also be seen from the semidirect-product viewpoint, since the action is linear and the Lagrangian takes the form (16.33) with $n^{-1}\nu = \nu$. This implies that the momentum in this case, given by the pair $(L_{\mathfrak{g}} u, z)$, is conserved in a fixed frame and the $n$-equation in (16.43) is absent. Working out the conservation equation $\mathrm{Ad}^*_{(g,n)}(L_{\mathfrak{g}} u, z) = \mathrm{cst}$ in this case yields the equations

$$L_{\mathfrak{g}} u_t + z_t \nabla n_t = \mathrm{cst}$$

and

$$z_t = \det(Dg_t^{-1}) z_0 \circ g_t^{-1}.$$

This last condition is the integrated form of the second equation in the eqn set (16.43), while the first equation of the set (16.43) evaluates the conservation law as $L_{\mathfrak{g}} u_t + z_t \nabla n_t = 0$.

**Exercise 16.3**
Solve Exercise 15.2 again from the viewpoint of metamorphoses by writing the action $A$ using the Lagrangian in Riemannian form given in eqn (16.38) as,

$$A = \frac{1}{2}\int_0^1 \|\boldsymbol{u}(t)\|_{\mathfrak{g}}^2 \,\mathrm{d}t + \frac{1}{2\sigma^2}\int_0^1 \int_{S^1} |\boldsymbol{\nu}(s,t)|^2 \,\mathrm{d}s\,\mathrm{d}t,$$

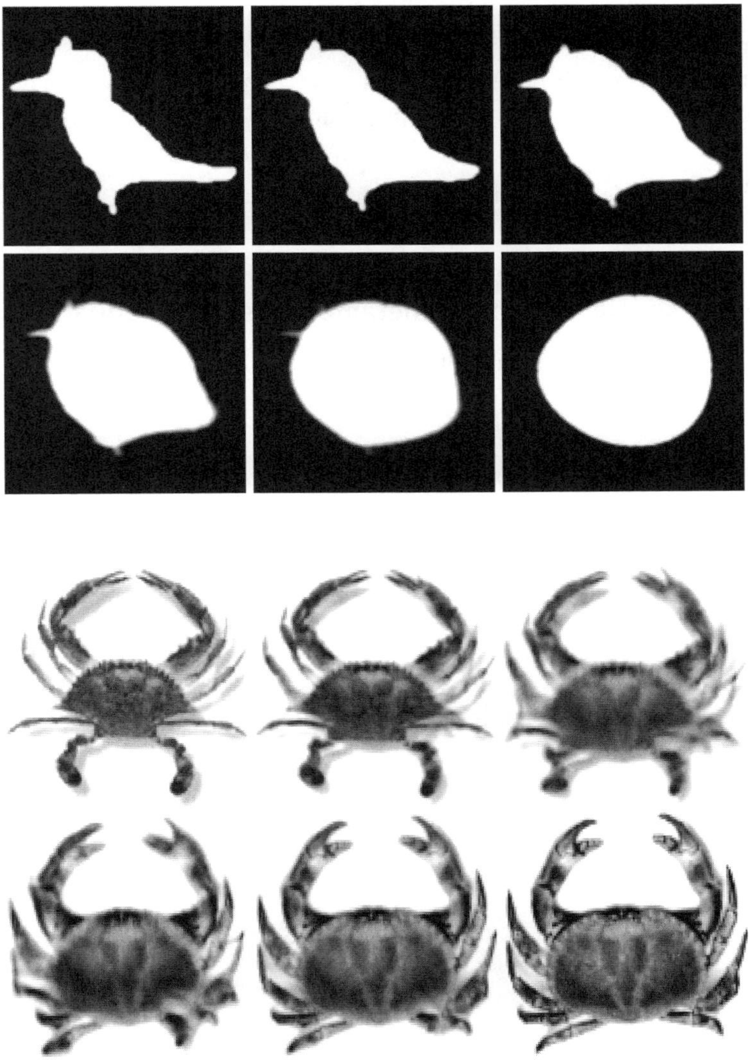

**Figure 16.1** Metamorphoses are shown between two pairs of images [HTY09]. The optimal trajectories for $n_t$ are computed between the first and last images in each case. The remaining images show $n_t$ at intermediate points in time for the two cases.

in which $\boldsymbol{v}(s,t)$ is defined by

$$\boldsymbol{v}(s,t) := \dot{\boldsymbol{Q}}(s,t) - \boldsymbol{u}(\boldsymbol{Q}(s,t),t) \in T\mathbb{R}^2,$$

as was done in Exercise 15.2.

**Exercise 16.4**
Write the metamorphosis equations in the previous exercise when the integral $\int_{S^1} \mathrm{d}s$ over the continuous parameter $s$ is replaced by a sum $\sum_s$ over a finite set of points in $\mathbb{R}^2$ known as the ***landmarks*** $Q_s(t) \in \mathbb{R}^2$ of the image.

### 16.5.4 A special case of 1D metamorphosis: CH2 equations

In 1D, the evolutionary form of the system (16.26) or equivalently (16.43) becomes,

$$\partial_t m + u\partial_x m + 2m\partial_x u = -\rho\partial_x\rho, \qquad \partial_t\rho + \partial_x(\rho u) = 0, \quad (16.44)$$

upon denoting

$$m = L_{\mathfrak{g}}u = (1 - \partial_x^2)u \quad \text{and} \quad \rho = \sigma z.$$

Up to a minus sign in front of $\rho\partial_x\rho$ in the first equation that does not affect its integrability as a Hamiltonian system, this is the ***two-component Camassa–Holm system*** (CH2) studied in [CLZ05, Fal06, Kuz07]. The system (16.44) in our case is equivalent to the compatibility for $\mathrm{d}\lambda/\mathrm{d}t = 0$ of the two linear equations

$$\partial_x^2\psi + \Big(-\frac{1}{4} + m\lambda + \rho^2\lambda^2\Big)\psi = 0, \quad (16.45)$$

$$\partial_t\psi = -\Big(\frac{1}{2\lambda} + u\Big)\partial_x\psi + \frac{1}{2}\psi\partial_x u. \quad (16.46)$$

Because the eigenvalue $\lambda$ in (16.45) is time independent, the evolution of the nonlinear semidirect-product system (16.44) is said to be ***isospectral.*** The second equation (16.46) is the evolution equation for the eigenfunction $\psi$. Thus, the semidirect-product system (16.44) for the metamorphosis of images in 1D is also completely integrable and possesses ***soliton solutions for the CH2 system*** that may be obtained by using the ***inverse scattering transform method.*** An identification of soliton dynamics in image matching of graphical structures, landmarks and image outlines for computational anatomy using the invariant subsystem of system (16.44) with $\rho = 0$ is found in [HRTY04].

### 16.5.5 Modified CH2 equations

The Euler–Poincaré system (16.26) for semidirect-product metamorphosis leads to an interesting modification of the CH2 equations (16.44) when $G = \mathrm{Diff}(\mathbb{R})$ and $N = \mathcal{F}(\mathbb{R})$ (smooth functions). These modified CH2 equations follow from a Lagrangian defined as a norm on $\mathrm{Diff}(\mathbb{R})Ⓢ\mathcal{F}(\mathbb{R})$ in (16.38) given in this notation by

$$\begin{aligned}\ell(u,\rho) &= \frac{1}{2}\|u\|^2_{H^1} + \frac{1}{2}\|\rho\|^2_{H^{-1}} \\ &= \frac{1}{2}\|u\|^2_{L^2} + \frac{\alpha_1^2}{2}\|u_x\|^2_{L^2} + \frac{1}{2}\|(\overline{\rho}-\overline{\rho}_0)\|^2_{L^2} + \frac{\alpha_2^2}{2}\|\overline{\rho}_x\|^2_{L^2},\end{aligned} \tag{16.47}$$

where $\alpha_1$ and $\alpha_2$ are constant length scales, $\overline{\rho}$ is defined in terms of $\rho$ by

$$(1-\alpha_2^2\partial_x^2)\overline{\rho} = \rho + \overline{\rho}_0, \tag{16.48}$$

and $\overline{\rho}_0$ is the constant value of $\overline{\rho}$ as $|x| \to \infty$. Taking stationary variations of the reduced action $S_{\mathrm{red}} = \int_0^1 \ell(u,\rho)\mathrm{d}t$ yields

$$\begin{aligned}0 = \delta S_{\mathrm{red}} &= \int_0^1 \left( \left\langle \frac{\delta \ell}{\delta u}\,,\, \delta u \right\rangle + \left\langle \frac{\delta \ell}{\delta \overline{\rho}}\,,\, \delta\overline{\rho} \right\rangle \right) \mathrm{d}t \\ &= \int_0^1 \left( \left\langle (1-\alpha_1^2\partial_x^2)u\,,\, \delta u \right\rangle + \left\langle \rho\,,\, \delta\overline{\rho} \right\rangle \right) \mathrm{d}t.\end{aligned} \tag{16.49}$$

Hence, the Euler–Poincaré equations (16.24) and (16.25) yield the following modification of the CH2 system (16.44) with $m = (1-\alpha_1^2\partial_x^2)u$

$$\partial_t m + u\partial_x m + 2m\partial_x u = -\rho\partial_x\overline{\rho} \quad \text{with} \quad \partial_t\rho + \partial_x(\rho u) = 0. \tag{16.50}$$

This modification may seem slight, but it has two important effects. First, very likely, it destroys the complete integrability of the 1D CH2 system (16.44), although this has not yet been proven. The apparent loss of integrability may seem unfortunate. However, as a sort of compensation for that loss, direct substitution shows that the modified system gains a property not possessed by the original CH2 system. Namely, the modified system admits a finite-dimensional invariant manifold of ***singular solutions*** in a form that generalizes the peakon solutions of CH to

$$m(t,x) = \sum_{i=1}^{M} P_i(t)\,\delta\big((x - Q_i(t)\big) \quad \text{and} \quad \rho(t,x) = \sum_{i=1}^{M} w_i\,\delta\big((x - Q_i(t)\big), \tag{16.51}$$

in which $w_i = \text{cst}$ for $i = 1, \ldots, M$. Moreover, the functions $P_i(t)$ and $Q_i(t)$ satisfy Hamilton's canonical equations,

$$\frac{dQ_i}{dt} = \frac{\partial H_M}{\partial P_i} \quad \text{and} \quad \frac{dP_i}{dt} = -\frac{\partial H_M}{\partial Q_i}, \tag{16.52}$$

with $M$-particle Hamiltonian,

$$H_M = \frac{1}{2}\sum_{i,j=1}^{M}\Big(P_i P_j\, e^{-|Q_i - Q_j|/\alpha_1} + w_i\, w_j\, e^{-|Q_i - Q_j|/\alpha_2}\Big) + \overline{\rho}_0 \sum_{i=1}^{M} w_i\, e^{-|Q_i - Q_j|/\alpha_2}. \tag{16.53}$$

Just as for the reduced dynamics of the CH equation discussed in Chapter 14, the finite-dimensional invariant manifold of singular solutions (16.51) of the modified CH2 system obeys canonical Hamiltonian equations. As for the case of CH, this canonical reduction occurs because the singular solutions (16.51) represent a cotangent-lift momentum map, this time for the left action of the semidirect-product Diff$\circledS\mathcal{F}$ on points on the real line. Moreover, as in the passage from the CH equation in one dimension to EPDiff in higher dimensions in Chapter 14, the singular solutions and their momentum-map structure for the modified CH2 system also generalize to arbitrary spatial dimensions. These more technical features of the modified CH2 system are explained in [HTY09].

**Exercise 16.5**
As in the falling cat problem [Mon93], there is an interpretation of the first equation in (16.34) as defining the zero-momentum connection form for the horizontal subspace of the quotient space $G\circledS N/G$. Compute the curvature of the zero-momentum connection.

**Exercise 16.6**
Compute the metamorphosis equations (16.15) for the Euclidean metric on the semidirect-product Lie group $SE(3)$.

**Exercise 16.7**
Verify that the conditions $\psi_{xxt} = \psi_{txx}$ and $\mathrm{d}\lambda/\mathrm{d}t = 0$ together imply the CH2 system (16.44).

**Exercise 16.8**
Identify the Lie algebra on whose dual the Lie–Poisson Hamiltonian bracket for the CH2 system in (16.44) is defined in Exercise 16.2.

**Exercise 16.9**
Use Euler–Poincaré theory to derive the higher-dimensional version of the modified CH2 system (16.50).

## Solutions to selected exercises

**Solution to Exercise 16.2** One passes from Euler–Poincaré equations (16.26) on the Lagrangian side to Lie–Poisson Hamiltonian equations via the ***Legendre transformation***, see, e.g., [HMR98a]. In our case, we start with the reduced Lagrangian $\ell(u,v,n)$ in eqn (16.4) and perform a Legendre transformation in the variables $u$ and $v$ only, by writing the Hamiltonian,

$$h(\mu,\sigma,n) = \langle \mu,\, u\rangle + \langle \sigma,\, v\rangle - \ell(u,v,n). \tag{16.54}$$

Variation of the Hamiltonian yields

$$\begin{aligned} \delta h(\mu,\sigma,n) &= \left\langle \frac{\delta h}{\delta \mu},\, \delta\mu \right\rangle + \left\langle \frac{\delta h}{\delta \sigma},\, \delta\sigma \right\rangle + \left\langle \frac{\delta h}{\delta n},\, \delta n \right\rangle \\ &= \langle u,\, \delta\mu\rangle + \langle v,\, \delta\sigma\rangle - \left\langle \frac{\delta \ell}{\delta n},\, \delta n \right\rangle \\ &\quad + \left\langle \mu - \frac{\delta \ell}{\delta u},\, \delta u \right\rangle + \left\langle \sigma - \frac{\delta \ell}{\delta v},\, \delta v \right\rangle. \end{aligned} \tag{16.55}$$

The last two coefficients vanish under the Legendre transformation, so

$$\mu = \frac{\delta \ell}{\delta u}, \qquad \sigma = \frac{\delta \ell}{\delta \nu}, \tag{16.56}$$

which recovers the definitions of the momentum variables $(\mu, \sigma)$ in terms of derivatives of the Lagrangian with respect to the velocities $(u, \nu)$. One then computes the variational derivatives of the Hamiltonian $h$ as

$$\frac{\delta h}{\delta \mu} = u, \qquad \frac{\delta h}{\delta \sigma} = \nu, \qquad \frac{\delta h}{\delta n} = -\frac{\delta \ell}{\delta n}. \tag{16.57}$$

Consequently, the Euler–Poincaré equations (16.26) for metamorphosis in the Eulerian description imply the following equations, for the Legendre-transformed variables, $(\mu, \sigma, n)$, written as a matrix operation, symbolically as

$$\begin{aligned}
\frac{\partial}{\partial t}\begin{bmatrix} \mu \\ \sigma \\ n \end{bmatrix} &= -\begin{bmatrix} \mathrm{ad}^*_{\square}\, \mu & \sigma \diamond \square & -\square \diamond n \\ \square * \sigma & 0 & 1 \\ -\pounds_{\square}\, n & -1 & 0 \end{bmatrix} \begin{bmatrix} \delta h/\delta \mu \\ \delta h/\delta \sigma \\ \delta h/\delta n \end{bmatrix} \\
&=: \mathcal{B} \begin{bmatrix} \delta h/\delta \mu \\ \delta h/\delta \sigma \\ \delta h/\delta n \end{bmatrix},
\end{aligned} \tag{16.58}$$

with boxes $\square$ indicating where the substitutions occur. The Poisson bracket defined by the $L^2$ skew-symmetric Hamiltonian matrix $\mathcal{B}$ is given by

$$\{f, h\} = \int \begin{bmatrix} \delta f/\delta \mu \\ \delta f/\delta \sigma \\ \delta f/\delta n \end{bmatrix}^T \mathcal{B} \begin{bmatrix} \delta h/\delta \mu \\ \delta h/\delta \sigma \\ \delta h/\delta n \end{bmatrix} \mathrm{d}x. \tag{16.59}$$

The pair $(\sigma, n)$ satisfies canonical Poisson-bracket relations. The other parts of the Poisson bracket are linear in the variables $(\mu, \sigma, n)$. This linearity is the signature of a Lie–Poisson bracket. The Lie algebra actions ensure that the Jacobi identity is satisfied. A similar Lie–Poisson bracket was found for complex fluids in [Hol02].

# 17 Continuum equations with advection

This chapter explains how reduction by right-invariance of Hamilton's principle under the diffeomorphisms produces the ***Euler–Poincaré (EP) theorem with advected quantities*** on $(T\mathrm{Diff} \times V)/\mathrm{Diff}$, where $V$ is the vector space of advected quantities whose evolution is generated by right-action of Diff. This EP theorem encompasses most ideal continuum theories.

## 17.1 Kelvin–Stokes theorem for ideal fluids

A fluid flow possesses ***circulation***, if the integral of the tangential component of its velocity $\mathbf{u}$ is non-zero around any smooth closed loop $C_t$ moving with the fluid. A geometrical object such as a closed circulation loop embedded in the fluid flow is an example of a ***Lagrangian quantity***. Such quantities are said to be ***frozen*** into the fluid flow. A theorem of vector calculus due to Kelvin and Stokes links the fluid's circulation with its ***vorticity*** – defined as the curl of its velocity. Namely, the fluid's circulation around the loop $C_t = \partial S$ is given by

$$I(t) = \oint_{C_t=\partial S} \mathbf{u} \cdot d\mathbf{x} = \int_S \mathrm{curl}\mathbf{u} \cdot \hat{\mathbf{n}}\, dS. \tag{17.1}$$

According to this formula, the ***circulation integral*** $I(t)$ around the Lagrangian loop $C_t$ moving with the fluid is equal to the ***vorticity flux*** – the integral of the normal component of the fluid's vorticity – taken over any surface $S$ whose boundary $\partial S$ is the circulation loop, so that each point on the curve $C_t = \partial S$ moves with the fluid velocity. Quantities moving with the fluid velocity are said to be ***advected*** by the flow, or ***frozen*** into the flow.

Thus, circulation loops enclose distributions of vorticity flux, which may be regarded as bundles of vortex lines embedded in the fluid and wrapped around by frozen-in Lagrangian loops. A coherent swirling "blob" of fluid bounded by such material loops containing circulation and, thus, pierced by vortex lines is called an ***eddy***. Eddies stretch themselves into extended shapes as they move by following the flow induced by the circulation corresponding

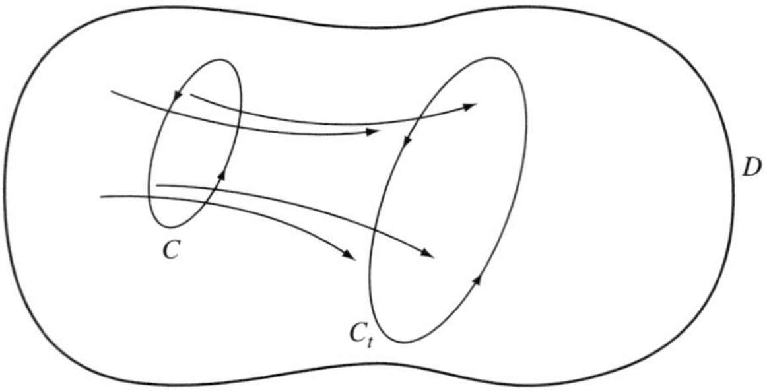


**Figure 17.1** A circulation loop encloses a flux of vorticity as it deforms with the flow from $C$ at an initial time to $C_t$ at time $t$ it. The vorticity flux is represented by arrows that penetrate through the surfaces at the two times whose boundaries are the corresponding circulation loops. According to the Stokes theorem, the normal flux of fluid vorticity through one of the surfaces at any time is equal to the circulation of the fluid velocity around its boundary.

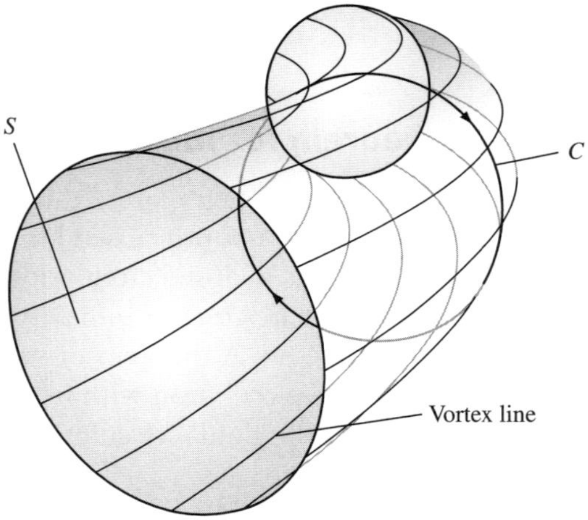


**Figure 17.2** A vortex tube consists of a bundle of vortex lines penetrating through a sequence of surfaces whose boundaries are formed by frozen-in circulation loops.

to the vortex lines that pierce them, as in Figure 17.1. These bundles of vortex lines, wrapped by their Lagrangian circulation loops as sketched in Figure 17.2 are called ***vortex tubes***. The vortex tubes evolve by stretching themselves into finer and finer shapes. These evolving sheets and tubes of vorticity flux wrapped by frozen-in circulation loops comprise the 'sinews' of fluid dynamics.

### 17.1.1 Euler's fluid equations

Euler's equations for the incompressible motion of an ideal flow of a fluid of unit density and velocity $\mathbf{u}$ satisfying the ***divergence-free condition*** $\text{div}\mathbf{u} = 0$

are given by the Newton's-Law expression,

$$\underbrace{\partial_t \mathbf{u} + \mathbf{u} \cdot \nabla \mathbf{u}}_{\text{Acceleration}} = \underbrace{-\nabla p}_{\text{Pressure force}} . \tag{17.2}$$

The requirement that the divergence-free (volume preserving) constraint $\nabla \cdot \mathbf{u} = 0$ be preserved leads to a Poisson equation for pressure $p$, which may be written in several equivalent forms. These are:

$$\begin{aligned} -\Delta p &= \operatorname{div}\big(\mathbf{u} \cdot \nabla \mathbf{u}\big) \\ &= u_{i,j} u_{j,i} \\ &= \operatorname{tr} \mathsf{S}^2 - \frac{1}{2} |\operatorname{curl} \mathbf{u}|^2 , \end{aligned} \tag{17.3}$$

where $\mathsf{S} = \frac{1}{2}(\nabla \mathbf{u} + \nabla \mathbf{u}^T)$ is the ***strain-rate tensor***. Because the velocity $\mathbf{u}$ must be tangent to any fixed boundary, the normal component of the motion equation (17.2) must vanish at such a boundary. This requirement produces a ***Neumann boundary condition*** for the Poisson equation satisfied by the fluid pressure in a domain with fixed boundaries. According to Euler's fluid equations (17.2) the gradient of this fluid pressure then accelerates the fluid at any point in space.

**Theorem 17.1 (Kelvin's circulation theorem)**
*The Euler fluid equations (17.2) preserve the circulation integral $I(t)$ defined in eqn (17.1) as*

$$I(t) = \oint_{C(\mathbf{u})} \mathbf{u} \cdot \mathrm{d}\mathbf{x}, \tag{17.4}$$

*where $C(\mathbf{u})$ is a closed circuit moving with the fluid at velocity* $\mathbf{u}$.

*Proof* The time rate of change of the circulation integral $I(t)$ is computed as,

$$\begin{aligned} \frac{\mathrm{d}}{\mathrm{d}t} \oint_{C(\mathbf{u})} \mathbf{u} \cdot \mathrm{d}\mathbf{x} &= \oint_{C(\mathbf{u})} \left( \frac{\partial \mathbf{u}}{\partial t} + \frac{\partial \mathbf{u}}{\partial x^j} u^j + u_j \frac{\partial u^j}{\partial \mathbf{x}} \right) \cdot \mathrm{d}\mathbf{x} \\ &= - \oint_{C(\mathbf{u})} \nabla \Big( p - \frac{1}{2} |\mathbf{u}|^2 \Big) \cdot \mathrm{d}\mathbf{x} \\ &= - \oint_{C(\mathbf{u})} \mathrm{d} \Big( p - \frac{1}{2} |\mathbf{u}|^2 \Big) = 0 . \end{aligned} \tag{17.5}$$

The second step uses Euler's fluid motion equation (17.2) and the last step in the proof follows, because the integral of an exact differential around a closed loop $C(\mathbf{u})$ vanishes. This proves Kelvin's theorem $dI/dt = 0$ for Euler's fluid equations. ■

**Corollary 17.2 (Stokes theorem for vorticity)**
*The vorticity flux through any circulation loop is constant. That is,*

$$\frac{d}{dt}\int_S \text{curl}\mathbf{u}\cdot\hat{\mathbf{n}}\,dS = 0, \tag{17.6}$$

*for any surface whose boundary is a circulation loop, i.e.* $\partial S = C_t(\mathbf{u})$.

*Proof* The proof of (17.6) follows from Kelvin's circulation Theorem 17.1 by an application of Stokes theorem to the conserved integral in eqn (17.4). ■

## 17.2 Introduction to advected quantities

Material parcels are carried along (advected) in an ideal continuum flow as closed (isolated) thermodynamic subsystems. These parcels do not exchange heat, mass, or other thermodynamic properties with their neighbouring parcels. The advected quantities associated with the material parcels are thus frozen into the continuum flow. However, they may still influence the flow through their contribution to the potential energy. The geometry of these advected quantities varies with the application. For example, mass density, heat, buoyancy and magnetic flux could all be frozen-in quantities in the appropriate physical setting. Each of these advected geometric parameters satisfies an auxiliary equation, whose form depends on how the corresponding frozen-in quantity transforms under the action of diffeomorphisms. Including the frozen-in dynamics of the advected quantities allows the Lagrangian in Hamilton's principle for ideal continuum motion to be extended to include potential energy in the context of geometric mechanics.

As in Section 11.2, let $\text{Diff}(\mathcal{D})$ be the space of diffeomorphisms acting on some domain $\mathcal{D} \in \mathbb{R}^n$.

**Definition 17.3** *The **representation space** $V^*$ of* $\text{Diff}(\mathcal{D})$ *in continuum mechanics is often a subspace of the tensor field densities on* $\mathcal{D}$*, denoted by*

$$V^* = (\mathfrak{T}\otimes dVol)(\mathcal{D}),$$

*and the representation is given by pull-back. It is thus a right representation of* $\text{Diff}(\mathcal{D})$ *on* $V^* = (\mathfrak{T}\otimes dVol)(\mathcal{D})$*. The right action of the Lie algebra of* *smooth vector fields* $\mathfrak{g}(\mathcal{D}) = \mathfrak{X}(\mathcal{D})$ *on* $V^*$ *is denoted as* ***concatenation from the right.*** *That is, we denote*

$$au := \pounds_u a,$$

*which is the Lie derivative of the tensor field density* $a \in V^*$ *along the smooth vector field* $u \in \mathfrak{X}(\mathcal{D})$.

**Definition 17.4** *The* ***Lagrangian of a continuum mechanical system*** *is a function*

$$L : T\,\mathrm{Diff}(\mathcal{D}) \times V^* \to \mathbb{R},$$

*which is right invariant relative to the tangent lift of right translation of* Diff$(\mathcal{D})$ *on itself and pull-back on the tensor field densities. Invariance of the Lagrangian* $L$ *induces a function* $\ell : \mathfrak{g}(\mathcal{D}) \times V^* \to \mathbb{R}$ *given by*

$$\ell(u, a) = L(u \circ g_t, g_t^* a) = L(U, a_0),$$

*where* $u \in \mathfrak{g}(\mathcal{D})$ *and* $a \in V^* \subset \mathfrak{T}(\mathcal{D}) \otimes dVol$, *and where* $g_t^* a$ *denotes the pull-back of the advected quantity* $a$ *by the diffeomorphism of the fluid motion* $g_t \in$ Diff *(also denoted* $g(t)$*) and* $u$ *is the Eulerian velocity. Thus, the formulation of Section 11.2 may be augmented to include potential energy by combining*

$$U = u \circ g_t \quad \textit{and} \quad a_0 = g_t^* a. \tag{17.7}$$

*The evolution of the advected quantity* $a$ *by right action of the vector field of Eulerian velocity* $u$ *satisfies the equation*

$$\dot{a} = -\,\pounds_u\, a = -\,au. \tag{17.8}$$

*That is, the solution of this equation, for the initial condition* $a_0$, *is given by the pull-back,*

$$a(t) = g_{t*} a_0 = a_0 g^{-1}(t), \tag{17.9}$$

*where the lower star denotes the push-forward operation and* $g_t$ *is the flow of* $u = \dot{g} g^{-1}(t)$.

**Exercise 17.1**
Prove by direct substitution that the expression (17.9) solves eqn (17.8).

**Definition 17.5** *Advected Eulerian quantities* *are defined in continuum mechanics to be those variables that are Lie transported by the flow of the Eulerian velocity field. Using this standard terminology, eqn (17.8), or its solution (17.9) states that the tensor field density $a(t)$ (which may include mass density and other Eulerian quantities) is advected.*

**Remark 17.6 (Dual tensors)** *On a general manifold, tensors of a given type have natural duals. For example, symmetric covariant tensors are dual to symmetric contravariant tensor densities, the pairing being given by the integration of the natural contraction of these tensors. Likewise, k-forms are naturally dual to $(n-k)$-forms, the pairing being given by taking the integral of their wedge product. This natural duality emerges from taking the variational derivative of the Lagrangian and using the $L^2$ pairing. For example,*

$$\delta\ell[a] = \left\langle \frac{\delta\ell}{\delta a}, \, \delta a \right\rangle_{V\times V^*} = \int_{\mathcal{D}} \frac{\delta\ell}{\delta a} \cdot \delta a,$$

*where $\langle \cdot, \cdot \rangle_{V\times V^*}$ is the $L^2$ pairing between elements of $V$ and $V^*$, and "dot" in $b \cdot a$ denotes the contraction of $a \in V^*$ and $b \in V$ to make a volume form (density) that may then be integrated over the domain.*

**Definition 17.7** *The* ***diamond operation*** $\diamond$ *between elements $A \in V$ and $a \in V^*$ produces an element of the dual Lie algebra $\mathfrak{g}(\mathcal{D})^*$ (a 1-form density) and is defined as*

$$\left\langle A \diamond a, \, w \right\rangle_{\mathfrak{g}^*\times\mathfrak{g}} = \left\langle A, \, -\pounds_w \, a \right\rangle_{V\times V^*} = -\int_{\mathcal{D}} A \cdot \pounds_w \, a \, , \qquad (17.10)$$

*where $A \cdot \pounds_w a$ denotes the contraction, as described above, of elements $A \in V$ and $a \in V^*$, and $w \in \mathfrak{g}(\mathcal{D})$ is a smooth vector field. (These operations do not depend on a Riemannian structure.)*

## Basic assumptions of the Euler–Poincaré theorem for continua

- There is a *right* representation of a Lie group $G$ on the vector space $V^*$ and $G$ acts in the natural way on the *right* on $TG \times V^*$: $(U_g, a)h = (U_g h, ah)$.
- In particular, if $a_0 \in V^*$, define the Lagrangian $L_{a_0} : TG \to \mathbb{R}$ by

$$L_{a_0}(U_g) = L(U_g, a_0).$$

Then $L_{a_0}$ is right invariant under the lift to $TG$ of the right action of $G_{a_0}$ on $G$, where $G_{a_0}$ is the isotropy group of $a_0$.[1]

- Right $G$-invariance of $L$ permits one to define the Lagrangian on the Lie algebra $\mathfrak{g}$ of the group $G$. Namely, $\ell : \mathfrak{g} \times V^* \to \mathbb{R}$ is defined by,

$$\ell(u, a) = L\big(U_g g^{-1}(t), a_0 g^{-1}(t)\big),$$

where $u = U_g g^{-1}(t)$ and $a = a_0 g^{-1}(t)$. Conversely, this relation defines for any $\ell : \mathfrak{g} \times V^* \to \mathbb{R}$ a right $G_{a_0}$-invariant function $L : TG \times V^* \to \mathbb{R}$.

- For a curve $g(t) \in G$, let $u(t) := \dot{g}(t)g(t)^{-1}$ and define the curve $a(t)$ as the unique solution of the linear differential equation with time-dependent coefficients $\dot{a}(t) = -a(t)u(t)$, where the action of an element of the Lie algebra $u \in \mathfrak{g}$ on an advected quantity $a \in V^*$ is denoted by concatenation from the right. The solution with initial condition $a(0) = a_0 \in V^*$ can be written as $a(t) = a_0 g(t)^{-1}$.

## Notation for reduction of Hamilton's principle for continua

We use the same definitions of geometric fluid quantities as in Section 11.4, but now these are augmented to include advected quantities.

- Let $\mathfrak{g}(\mathcal{D})$ denote the space of smooth vector fields on $\mathcal{D}$. These vector fields are endowed with the ***Lie bracket*** given in components by (summing on repeated indices)

$$[u,\, v]^i = u^j \frac{\partial v^i}{\partial x^j} - v^j \frac{\partial u^i}{\partial x^j}. \tag{17.11}$$

The notation

$$\mathrm{ad}_u\, v := -\,[u,\, v]$$

formally denotes the adjoint action of the *right* Lie algebra of $\mathrm{Diff}(\mathcal{D})$ on itself.

- Identify the Lie algebra of vector fields $\mathfrak{g}$ with its dual $\mathfrak{g}^*$ by using the $L^2$ pairing

$$\langle u,\, v\rangle = \int_{\mathcal{D}} \mathbf{u}\cdot\mathbf{v}\, dV. \tag{17.12}$$

[1] For fluid dynamics, right $G$-invariance of the Lagrangian function $L$ is traditionally called 'particle relabelling symmetry'. Since the initial particle labels $a_0$ may themselves be relabelled, the restriction to an isotropy group $G_{a_0}$ is only a notational convenience in defining $\ell(u, a)$.

- Let $\mathfrak{g}(\mathcal{D})^*$ denote the geometric dual space of $\mathfrak{g}(\mathcal{D})$, that is, $\mathfrak{g}(\mathcal{D})^* := \Lambda^1(\mathcal{D}) \otimes \mathrm{d}Vol$. This is the space of 1-form densities on $\mathcal{D}$. If

$$m = \mathbf{m} \cdot d\mathbf{x} \otimes \mathrm{d}V \in \Lambda^1(\mathcal{D}) \otimes \mathrm{d}V,$$

then, the pairing of $m$ with $u \in \mathfrak{g}(\mathcal{D})$ is given by the $L^2$ pairing,

$$\langle m, u \rangle = \int_{\mathcal{D}} \mathbf{m} \cdot \mathbf{u} \, \mathrm{d}V, \qquad (17.13)$$

where the dot-product $\mathbf{m} \cdot \mathbf{u}$ of the covector $\mathbf{m}$ with the vector $\mathbf{u}$ is also the standard contraction of a 1-form $\mathbf{m} \cdot \mathrm{d}\mathbf{x}$ with a vector field $u = \mathbf{u} \cdot \nabla$.

- For $u \in \mathfrak{g}(\mathcal{D})$ and $m \in \mathfrak{g}(\mathcal{D})^*$, the dual of the adjoint representation is defined by

$$\langle \mathrm{ad}^*_u(m), v \rangle = \langle m, \mathrm{ad}_u \, v \rangle = -\, \langle m, [u, v] \rangle, \qquad (17.14)$$

where $\langle \cdot, \cdot \rangle$ is the $L^2$ pairing. Equivalently, the expression of $\mathrm{ad}^*_u(m)$ is given by

$$\mathrm{ad}^*_u(m) = \pounds_u(m), \qquad (17.15)$$

That is, $\mathrm{ad}^*_u$ coincides with the Lie-derivative $\pounds_u$ for 1-form densities.

- If $u = u^j \partial/\partial x^j$, $m = m_i \mathrm{d}x^i \otimes \mathrm{d}V$, then the preceding formula for $\mathrm{ad}^*_u(m)$ has the ***coordinate expression***

$$\begin{aligned} \left( \mathrm{ad}^*_u \, m \right)_i \mathrm{d}x^i \otimes \mathrm{d}V &= \left( u^j \frac{\partial m_i}{\partial x^j} + m_j \frac{\partial u^j}{\partial x^i} + (\mathrm{div}_{dV} \, \mathbf{u}) m_i \right) \mathrm{d}x^i \otimes \mathrm{d}V \\ &= \left( \frac{\partial}{\partial x^j}(u^j m_i) + m_j \frac{\partial u^j}{\partial x^i} \right) \mathrm{d}x^i \otimes \mathrm{d}V. \end{aligned} \qquad (17.16)$$

The last equality assumes that the divergence is taken relative to the standard measure $\mathrm{d}V = \mathrm{d}^n\mathbf{x}$ in $\mathbb{R}^n$. (On a Riemannian manifold the metric divergence should be used.)

Let the Lagrangian $L_{a_0}(\mathrm{U}) := L(\mathrm{U}, a_0)$ be right invariant under $\mathrm{Diff}_{a_0}(\mathcal{D})$. We can now state the Euler–Poincaré theorem for continua of [HMR98a].

## 17.3 Euler–Poincaré theorem

**Theorem 17.8 (Euler–Poincaré theorem for continua.)**
*Consider a path $g_t$ in $\mathrm{Diff}(\mathcal{D})$ with Lagrangian velocity vector field $U$ and Eulerian velocity vector field $u$. The following are equivalent:*

i *Hamilton's variational principle*

$$\delta \int_{t_1}^{t_2} L\left(U_t(X), a_0(X)\right)\, dt = 0 \tag{17.17}$$

*holds, for variations $\delta g_t$ vanishing at the endpoints in time.*

ii *$g_t$ satisfies the Euler-Lagrange equations for $L_{a_0}$ on $\mathrm{Diff}(\mathcal{D})$.*

iii *The reduced variational principle in Eulerian coordinates*

$$\delta \int_{t_1}^{t_2} \ell(u, a)\, dt = 0 \tag{17.18}$$

*holds on $\mathfrak{g}(\mathcal{D}) \times V^*$, using variations of the form*

$$\delta u = \frac{\partial w}{\partial t} + [u, w] = \frac{\partial w}{\partial t} - \mathrm{ad}_u w, \qquad \delta a = -£_w\, a, \tag{17.19}$$

*where $w_t = \delta g_t \circ g_t^{-1}$ vanishes at the endpoints.*

iv *The Euler–Poincaré equations for continua*

$$\frac{\partial}{\partial t}\frac{\delta \ell}{\delta u} = -\,\mathrm{ad}_u^* \frac{\delta \ell}{\delta u} + \frac{\delta \ell}{\delta a} \diamond a = -£_u \frac{\delta \ell}{\delta u} + \frac{\delta \ell}{\delta a} \diamond a, \tag{17.20}$$

*hold, with auxiliary equations $(\partial_t + £_u)a = 0$ for each advected quantity $a(t)$. The $\diamond$ operation defined in (17.10) needs to be determined on a case by case basis, depending on the nature of the tensor $a(t)$. The variation $m = \delta\ell/\delta u$ is a 1-form density and we have used the relation (17.15) in the last step of eqn (17.20).*

We refer to [HMR98a] for the proof of this theorem in the abstract setting. See also Theorem 7.15 of Chapter 7 for more discussion of Euler–Poincaré with advected parameters. Here, we shall discuss some of its features in the concrete setting of continuum mechanics.

## Discussion of the Euler–Poincaré equations

The following string of equalities shows *directly* that statement **iii** is equivalent to statement **iv**:

$$
\begin{aligned}
0 &= \delta \int_{t_1}^{t_2} l(u,a)\,\mathrm{d}t \\
&= \int_{t_1}^{t_2} \left( \left\langle \frac{\delta l}{\delta u}, \delta u \right\rangle_{\mathfrak{g}^*\times\mathfrak{g}} + \left\langle \frac{\delta l}{\delta a}, \delta a \right\rangle_{V\times V^*} \right) \mathrm{d}t \\
&= \int_{t_1}^{t_2} \left( \left\langle \frac{\delta l}{\delta u}, \frac{\partial w}{\partial t} - \mathrm{ad}_u\, w \right\rangle_{\mathfrak{g}^*\times\mathfrak{g}} - \left\langle \frac{\delta l}{\delta a}, \pounds_w\, a \right\rangle_{V\times V^*} \right) \mathrm{d}t \\
&= \int_{t_1}^{t_2} \left\langle -\frac{\partial}{\partial t}\frac{\delta l}{\delta u} - \mathrm{ad}^*_u \frac{\delta l}{\delta u} + \frac{\delta l}{\delta a} \diamond a,\, w \right\rangle_{\mathfrak{g}^*\times\mathfrak{g}} \mathrm{d}t. \qquad (17.21)
\end{aligned}
$$

The rest of the proof follows essentially the same track as the proof of the pure Euler–Poincaré theorem, modulo slight changes to accomodate the advected quantities, as discussed in Section 11.2 for the left action.

In the absence of dissipation, many Eulerian fluid equations can be written in the EP form in eqn (17.20),

$$
\frac{\partial}{\partial t}\frac{\delta \ell}{\delta u} + \mathrm{ad}^*_u \frac{\delta \ell}{\delta u} = \frac{\delta \ell}{\delta a} \diamond a, \quad \text{with} \quad (\partial_t + \pounds_u)a = 0. \qquad (17.22)
$$

Equation (17.22) is ***Newton's Law***: The Eulerian time derivative of the momentum density $m = \delta\ell/\delta u$ (a 1-form density dual to the velocity $u$) is equal to the force density $(\delta\ell/\delta a) \diamond a$, with the $\diamond$ operation defined in eqn (17.10). Thus, Newton's Law is written in the Eulerian fluid representation as,

$$
\left.\frac{\mathrm{d}}{\mathrm{d}t}\right|_{Lag} m := (\partial_t + \pounds_u)m = \frac{\delta \ell}{\delta a} \diamond a, \qquad (17.23)
$$

$$
\text{with} \quad \left.\frac{\mathrm{d}}{\mathrm{d}t}\right|_{Lag} a := (\partial_t + \pounds_u)a = 0. \qquad (17.24)
$$

In coordinates, a 1-form density takes the form $\mathbf{m} \cdot \mathbf{dx} \otimes dV$ and the EP equation (17.20) is given mnemonically by

$$\frac{d}{dt}\Big|_{Lag}\big(\mathbf{m} \cdot \mathbf{dx} \otimes dV\big) = \underbrace{\frac{d\mathbf{m}}{dt}\Big|_{Lag} \cdot \mathbf{dx} \otimes dV}_{\text{Advection}} + \underbrace{\mathbf{m} \cdot \mathbf{du} \otimes dV}_{\text{Stretching}}$$
$$+ \underbrace{\mathbf{m} \cdot \mathbf{dx} \otimes (\nabla \cdot \mathbf{u}) dV}_{\text{Expansion}} = \frac{\delta \ell}{\delta a} \diamond a,$$

with

$$\frac{d}{dt}\Big|_{Lag} \mathbf{dx} := \big(\partial_t + \pounds_u\big)\mathbf{dx} = \mathbf{du} = \mathbf{u}_{,j} dx^j,$$

upon using commutation of Lie derivative $\pounds_u$ and exterior derivative d. Compare this formula with the definition of $\mathrm{ad}^*_u(\mathbf{m} \cdot \mathbf{dx} \otimes dV)$ in eqns (17.15) and (17.16).

- The left side of the EP equation in eqn (17.23) describes the fluid's dynamics due to its kinetic energy. A fluid's kinetic energy typically defines a norm for the Eulerian fluid velocity, $KE = \frac{1}{2}\|\mathbf{u}\|^2$. The left side of the EP equation is the ***geodesic*** part of its evolution, with respect to this norm. See [AK98] for discussions of this interpretation of ideal incompressible flow and references to the literature. However, in a gravitational field, for example, there will also be dynamics due to potential energy. And this dynamics will by governed by the right side of the EP equation.
- The right side of the EP equation in eqn (17.23) modifies the geodesic motion. Naturally, the right side of the EP equation is also a geometrical quantity. The diamond operation $\diamond$ represents the dual of the Lie algebra action of vectors fields on the tensor $a$. Here, $\delta\ell/\delta a$ is the dual tensor, under the natural pairing (usually, $L^2$ pairing) $\langle \cdot, \cdot \rangle$ that is induced by the variational derivative of the Lagrangian $\ell(u, a)$. The diamond operation $\diamond$ is defined in terms of this pairing in eqn (17.10). For the $L^2$ pairing, this is integration by parts of (minus) the Lie derivative in eqn (17.10).
- The quantity $a$ is typically a tensor (e.g., a density, a scalar, or a differential form) and we shall sum over the various types of tensors $a$ that are involved in the fluid description. The second equation in eqn (17.23) states that each tensor $a$ is carried along by the Eulerian fluid velocity $u$. Thus, $a$ is for fluid 'attribute,' and its Eulerian evolution is given by minus its Lie derivative, $\dot{a} = -\pounds_u a$. That is, $a$ stands for the set of

fluid attributes that each Lagrangian fluid parcel carries around (advects), such as its buoyancy, which is determined by its individual salt, or heat content, in ocean circulation.

- Many examples of how eqn (17.23) arises in the dynamics of continuous media are given in [HMR98a]. The EP form of the Eulerian fluid description in eqn (17.23) is analogous to the classical dynamics of rigid bodies (and tops, under gravity) in body coordinates. Rigid bodies and tops are also governed by Euler–Poincaré equations, as Poincaré showed in a two-page paper with no references, over a century ago [Poi01]. For modern discussions of the EP theory, see, e.g., [MR02], or [HMR98a].

### 17.3.1 Corollary: the Kelvin–Noether theorem

**Corollary 17.9 (Kelvin–Noether circulation theorem)** *Assume $u(x,t)$ satisfies the Euler–Poincaré equations for continua:*

$$\frac{\partial}{\partial t}\left(\frac{\delta \ell}{\delta u}\right) = -\pounds_u\left(\frac{\delta \ell}{\delta u}\right) + \frac{\delta \ell}{\delta a} \diamond a,$$

*and the quantity a satisfies the* ***advection relation***

$$\frac{\partial a}{\partial t} + \pounds_u a = 0. \tag{17.25}$$

*Let $g_t$ be the flow of the Eulerian velocity vector field $u$, that is, $u = (\mathrm{d}g_t/dt) \circ g_t^{-1}$. Define the advected fluid loop $\gamma_t := g_t \circ \gamma_0$ and the circulation map $I(t)$ by*

$$I(t) = \oint_{\gamma_t} \frac{1}{D}\frac{\delta \ell}{\delta u}. \tag{17.26}$$

*In the circulation map $I(t)$ the advected mass density $D_t = D\mathrm{d}V$ satisfies the push forward relation $D_t = g_* D_0$. This implies the advection relation (17.25) with $a = D\mathrm{d}V$, namely, the continuity equation,*

$$(\partial_t + \pounds_u)(D\mathrm{d}V) = (\partial_t D + \operatorname{div} D\mathbf{u})\mathrm{d}V = 0.$$

*Then the map $I(t)$ satisfies the* ***Kelvin–Noether circulation relation,***

$$\frac{\mathrm{d}}{\mathrm{d}t} I(t) = \oint_{\gamma_t} \frac{1}{D}\frac{\delta \ell}{\delta a} \diamond a, \tag{17.27}$$

*with a slight abuse of notation in which $D$ in the expression $D_t = D\mathrm{d}V$ is regarded as a scalar function.*

Both an abstract proof of the Kelvin–Noether circulation theorem and a proof tailored for the case of continuum mechanical systems are given in [HMR98a]. We provide a simplified version of the latter below.

*Proof* A 'bare-hands' proof of the Kelvin–Noether theorem is immediate, upon realizing that the time derivative of the circulation on the loop $\gamma_t$ is the integral around the loop of the Lagrangian time derivative of the integrand, because the loop is following the flow. Consequently,

$$\begin{aligned}\frac{\mathrm{d}}{\mathrm{d}t} I(t) &= \oint_{\gamma_t} \frac{\mathrm{d}}{\mathrm{d}t}\Big|_{Lag} \frac{1}{D}\frac{\delta l}{\delta u} \\ &= \oint_{\gamma_t} \left(\frac{\partial}{\partial t} + £_u\right) \frac{1}{D}\frac{\delta l}{\delta u} = \oint_{\gamma_t} \frac{1}{D}\frac{\delta l}{\delta a} \diamond a,\end{aligned} \tag{17.28}$$

in which the last step uses the advection relation for the mass density $(\partial/\partial t + £_{\mathbf{u}})D = 0$ to write the following alternative form of the EP equation (17.20)

$$\left(\frac{\partial}{\partial t} + £_u\right) \frac{1}{D}\frac{\delta l}{\delta u} = \frac{1}{D}\frac{\delta l}{\delta a} \diamond a. \tag{17.29}$$

The calculation (17.28) proves the Kelvin circulation relation directly. ■

**Remark 17.10** *Two velocity vectors appear in the circulation integrand $I(t)$: These are the fluid velocity vector $\mathbf{u}$ and the specific momentum covector $\mathbf{m}/D$ in the 1-form*

$$\frac{1}{D}\mathbf{m}\cdot \mathrm{d}\mathbf{x} = \frac{1}{D}\frac{\delta l}{\delta \mathbf{u}} \cdot \mathrm{d}\mathbf{x}.$$

*The latter velocity covector is the momentum density in vector notation $\mathbf{m}\cdot \mathrm{d}\mathbf{x} \otimes \mathrm{d}V$ divided by the mass density $D\,\mathrm{d}V$. These two velocities are the basic ingredients for performing modelling and analysis in any ideal fluid problem. These ingredients appear together in the Euler–Poincaré theorem and its corollary, the Kelvin–Noether theorem.*

## Fluid dynamical applications

In some fluid-dynamical applications, the advected Eulerian variables are $a \in \{b, D\,\mathrm{d}^3x\}$, representing the buoyancy $b$ in GFD (or specific entropy, for the compressible case) and volume element (or mass density) $D$, respectively. Equation (17.8) for the advected evolution of the tensor fields $a \in \{b, D\,\mathrm{d}^3x\}$ yields the following Lie-derivative relations and their equivalent partial differential equations,

$$\left(\frac{\partial}{\partial t} + \pounds_{\mathbf{u}}\right) b = 0, \quad \text{or} \quad \frac{\partial b}{\partial t} + \mathbf{u}\cdot\nabla b = 0, \tag{17.30}$$

$$\left(\frac{\partial}{\partial t} + \pounds_{\mathbf{u}}\right) (D\,\mathrm{d}^3x) = 0, \quad \text{or} \quad \frac{\partial D}{\partial t} + \nabla\cdot(D\mathbf{u}) = 0. \tag{17.31}$$

The Lie derivatives $-\pounds_{\mathbf{w}}\, a$ in the diamond operation (17.10) are given by

$$\begin{aligned} -\pounds_{\mathbf{w}}\, b &= -\,\mathbf{w}\cdot\nabla b, \\ -\pounds_{\mathbf{w}}\,(D\,\mathrm{d}^3x) &= -\,\nabla\cdot(D\mathbf{w})\,\mathrm{d}^3x. \end{aligned} \tag{17.32}$$

The corresponding diamond relations defined in (17.10) are given by

$$\left\langle \frac{\delta l}{\delta b} \diamond b,\, w \right\rangle_{\mathfrak{g}^*\times\mathfrak{g}} = \left\langle \frac{\delta l}{\delta b},\, -\,\mathbf{w}\cdot\nabla b \right\rangle_{V\times V^*} = -\int_{\mathcal{D}} \frac{\delta l}{\delta b}\nabla b\cdot\mathbf{w}\,\mathrm{d}^3x\,, \tag{17.33}$$

and

$$\left\langle \frac{\delta l}{\delta D} \diamond D,\, w \right\rangle_{\mathfrak{g}^*\times\mathfrak{g}} = \left\langle \frac{\delta l}{\delta D},\, -\,\nabla\cdot(D\mathbf{w}) \right\rangle_{V\times V^*} = \int_{\mathcal{D}} D\,\nabla\frac{\delta l}{\delta D}\cdot\mathbf{w}\,\mathrm{d}^3x\,. \tag{17.34}$$

Consequently, the Euler–Poincaré equation (17.29) for the Lagrangian,

$$l(u,b,D) : \mathfrak{X}\times\Lambda^0\times\Lambda^3 \mapsto \mathbb{R},$$

whose advected variables satisfy the auxiliary equations (17.30) and (17.31) may be expressed as

$$\left(\frac{\partial}{\partial t} + \pounds_u\right)\frac{1}{D}\frac{\delta l}{\delta u} = -\,\frac{1}{D}\frac{\delta l}{\delta b}\nabla b + \nabla\frac{\delta l}{\delta D}. \tag{17.35}$$

**Exercise 17.2 (Compressible ideal fluids)**

1. Compute the Kelvin–Noether circulation theorem for the Lagrangian,

$$l(u,b,D\,\mathrm{d}^3x) : \mathfrak{X}\times\Lambda^0\times\Lambda^3 \mapsto \mathbb{R},$$

whose advected variables $(b, D\,\mathrm{d}^3x)$ satisfy the auxiliary equations (17.30) and (17.31), respectively.

2. Assume the reduced Lagrangian is given by

$$l(u,b,D) = \int D\Big(\frac{1}{2}|\mathbf{u}|^2 + \mathbf{u}\cdot\mathbf{R}(\mathbf{x}) - e(D,b)\Big)\,\mathrm{d}^3x, \qquad (17.36)$$

with a prescribed function $\mathbf{R}(\mathbf{x})$ and specific internal energy $e(D,b)$ satisfying the ***First Law of Thermodynamics***,

$$\mathrm{d}e = \frac{p}{D^2}\mathrm{d}D + T\,\mathrm{d}b,$$

where $p$ is pressure and $T$ is temperature.
Compute the Legendre transform for this Lagrangian. Does the Legendre transform provide a linear invertible operator on the velocity $\mathbf{u}$?

3. Compute the Hamiltonian arising from this Legendre transform and its variational derivatives.
4. Find the semidirect-product Lie–Poisson bracket for the Hamiltonian formulation of these equations. Assemble the equations into the semidirect-product Hamiltonian form,

$$\frac{\partial}{\partial t}\begin{bmatrix} m_i \\ D \\ b \end{bmatrix} = -\begin{bmatrix} m_j\partial_i + \partial_j m_i & D\partial_i & -b_{,i} \\ \partial_j D & 0 & 0 \\ b_{,j} & 0 & 0 \end{bmatrix}\begin{bmatrix} \delta h/\delta m_j \\ \delta h/\delta D \\ \delta h/\delta b \end{bmatrix}, \qquad (17.37)$$

where $m_i := \delta l/\delta u_i = D(u_i + R_i)$ for the Lagrangian in eqn (17.36).

5. Does the corresponding Lie–Poisson bracket have Casimirs? If so, how are they related to the Kelvin–Noether theorem, symmetries and momentum maps for this system?

**Exercise 17.3**
Assume the reduced Lagrangian is given by

$$l(u,D) = \int \frac{D}{2}\Big(|\mathbf{u}|^2 + \alpha^2|\nabla\mathbf{u}|^2\Big) - p(D-1)\,d^3x, \qquad (17.38)$$

where $|\nabla\mathbf{u}|^2 := u_{i,j}u_{i,j}$. Compute the Euler–Poincaré equation for this reduced Lagrangian. The result is the Lagrangian-averaged Euler-alpha (LAE-alpha) equation, introduced in [HMR98a] as an extension of CH and EPDiff to incompressible flow in higher dimensions.

# 18 Euler–Poincaré theorem for geophysical fluid dynamics

## 18.1 Kelvin circulation theorem for GFD

Geophysical fluid dynamics (GFD) studies the large-scale circulatory motions of the atmosphere and ocean, which are of importance, for example, in modelling climate change.

The history of predicting climate, or rather predicting the general circulation in the atmosphere and ocean, records a series of challenges that were met and overcome in formulating the important questions and finding numerical solutions to the nonlinear partial differential equations (PDE) that govern climate dynamics. One of the first important challenges arose in Lewis Fry Richardson's work on the numerical solution of the partial differential equations that govern weather forecasting [Ric22]. Richardson found that the very smallest scales inherent in the solutions of the problem must be modelled properly (balanced) in a numerical weather forecast, or else numerical instability would lead to explosively growing errors. Ever since Richardson's marvelous calculation (reviewed in [Lyn92]), an important role of applied mathematics in climate modelling has been to find ways of accurately describing the mean behaviour of the fluid equations over length scales that make the problem solvable on computers, without directly computing the smallest scales. This has led to a hierarchy of model equations for approximate solutions that take into account the mean effects of the smallest scales. A comprehensive description of this endeavour for ocean circulation is given in [Val06]. For surveys and discussions of previous and current work on hierarchies of model equations relevant to climate and weather prediction, see [Whi02, WJRS06, Cul07].

What analytical approaches are available in meeting the challenge of modelling the main components of the Earth's climate? One of the main components of the climate comprises the great ocean currents that carry heat from the Sun poleward and thereby determine Earth's climate at high latitudes. How should one proceed to model these currents mathematically? First, let us observe that the depth of the ocean is much less than the widths of its various basins. So we are dealing with fluid flow in a thin domain. Next, let us notice that the large-scale currents in the ocean have persisted over very long times, so their presence must indicate some kind of global balance in the wind-driven ocean circulation. However, this balance is delicate and occasionally adjusts itself in regions of the ocean such as the Southern Ocean, where the currents are seen to suddenly meander and develop complex behaviour at irregular intervals. For example, the Antarctic Circumpolar Current (ACC) flowing eastward meets the westward flowing Agulhas Current off the coast of South Africa. The ACC is strong enough to turn the Agulhas back on itself and make it flow eastward. Where the Agulhas turns back on itself a swirling loop pinches off periodically, about once every two months, releasing an eddy into the South Atlantic Ocean. This eddy or Agulhas ring enters the flow of the Benguela Current and is

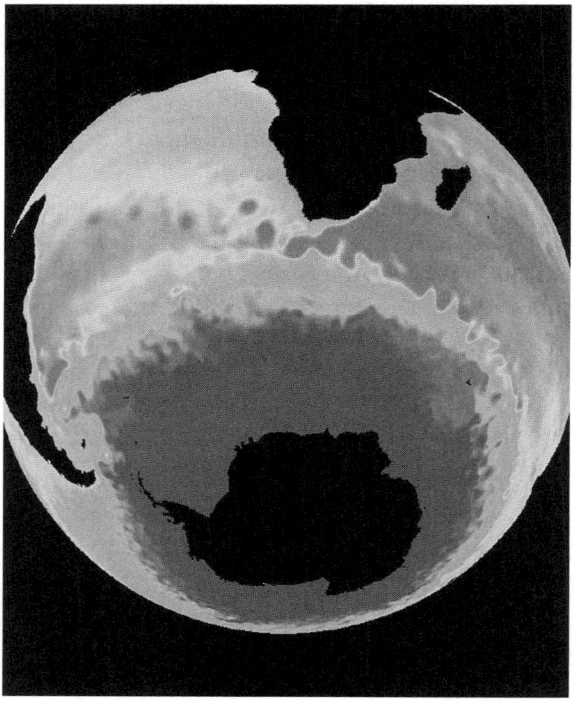

**Figure 18.1** Numerically computed isobars of the sea surface elevation. In the colour scheme of the figure, red is higher sea surface elevation and blue is lower. Geostrophically balanced flow takes place along these isobars. This view of the Southern Ocean shows the collision of the wide eastward Antarctic Circumpolar Current (ACC, blue) with the narrow westward Agulhas Current (red) off the coast of South Africa. The collision with the ACC turns the Agulhas back on itself and makes it flow eastward in an undulating path. Where the Agulhas turns back on itself a swirling loop pinches off and periodically releases eddies into the South Atlantic Ocean. See Plate 8.

translated north-westward all the way across the Atlantic Ocean in about three years.

Ocean currents persist because pressure gradients come into balance in some average sense with the two main forces in global ocean circulation. These are the Coriolis force due to rotation and the buoyancy force due to stratification. Thus, the persistence of patterns in the currents suggests that pressure gradients in the ocean tend to arrange themselves in the horizontal and vertical directions, so as to balance these two main forces. These balances are called:

- ***Geostrophic balance*** between Coriolis force and horizontal pressure gradient;
- ***Hydrostatic balance*** between buoyancy force and vertical pressure gradient.

Following Richardson [Ric22], these are also the balances that meteorologists must introduce in initializing and adjusting their weather forecasts.

To cast these observations of the phenomena into a useful mathematical framework for modelling them, let us begin by identifying the small dimensionless numbers that are associated with these flows. In particular, let us take advantage of having a thin domain that rotates rapidly ($f_0$, once per day). This daily rotation rate is much faster than the rate ($\mathcal{U}_0/L$, once per month) for even relatively small ***mesoscale eddies*** of, say, $L \approx 100\,\text{km}$ to turn around. That is, let us use the available separation of time scales to create the small dimensionless ratio

$$\epsilon = \mathcal{U}_0/(Lf_0) \approx 1/30 \ll 1,$$

called the ***Rossby number***. As we shall see, the Rossby number $\epsilon$ also measures the ratio of the flow's nonlinearity to the Coriolis force.

We are interested in length scales for eddies that are hundreds of kilometres in diameter, but the ocean depth is only about $\mathcal{B}_0 \approx 4\,\text{km}$. So the aspect ratio of our rapidly rotating domain is small. This ***aspect ratio*** provides another small dimensionless number,

$$\sigma = \mathcal{B}_0/L \approx 4/100 \ll 1.$$

The Rossby number, $\epsilon$, and the aspect ratio, $\sigma$, comprise two small ratios of length scales and time scales that may be used as parameters in an asymptotic expansion of the solutions of the exact equations for a rotating, surface-driven, stratified fluid moving under gravity in a neighbourhood of the balanced state. Because the acceleration of gravity $g$ is also acting on these flows, we may introduce the dimensionless ratio, $\mathcal{F} = f_0^2L^2/(g\mathcal{B}_0)$, called the ***rotational Froude*** number. However, oceanic observations show that this parameter is of order unity, $\mathcal{F} = O(1)$, so it is not available as an expansion parameter for ocean circulation dynamics.

At leading order in the asymptotic expansion of the solutions in $\epsilon$, $\sigma \ll 1$, the mathematical model recovers the geostrophic and hydrostatic balances. At the next orders in the expansion, a family of approximate equations emerges. Eventually, the asymptotic expansion and balance considerations at each power of $\epsilon$ and $\sigma$ determine a hierarchy of equations of motion containing the three non-dimensional numbers. This hierarchy of equations is designed to approximate the solutions of the original equations more and more accurately near the geophysically balanced state.

The utility of these equations is that they approximate the solutions of the original Euler fluid equations for the global ocean circulation by replacing the exact equations with a model that is accurate near the balanced state and may also be simulated on a computer over the long periods of time needed to characterize the climate. Such computations are already enormously computationally intensive; but they would be impossible for today's computers if the exact equations were used at the high resolution needed to accurately account for the smallest important scales.

What about dissipation? Motion at such scales is very non-dissipative: large gyres and their smaller eddies in the ocean basins are driven by the winds, but the only significant dissipation mechanism for the currents and eddies in the ocean is their interaction with the bottom. The ocean gyres are thousands of kilometers in diameter. The eddies created by the currents at their edges are about 200–300 km across, so they are also large enough to be in geostrophic balance, although they are much smaller than the gyres. However, these mesoscale eddies still involve a great deal of mass and momentum – so they certainly would not be stopped by molecular viscosity any time soon. In fact, the eddies may persist for years, until they eventually run into a coastline, where they start feeling the drag due to bottom topography. The Reynolds number for an eddy 100 km across turning at 10 cm/s in water of viscosity 0.01 cm$^2$/s is about $10^{10}$. Obviously, the inverse of the Reynolds number is negligible, so viscosity would not enter in these asymptotic balance equations for the large-scale motion. In fact, the only significant dissipation mechanism for this system of gyres, eddies and currents would be their interaction with the bottom, or along the shallows near the lateral boundaries, where turbulent dissipation is generated at much smaller scales. However, the complexities of drag on the ocean flow due to its interaction with the bottom will not be considered here.

A list of leading candidates in the hierarchy of GFD balance equations derived in the literature via this asymptotic procedure was investigated by Allen, Barth and Newberger (ABN) in [ABN90a]. On consulting this 'ABN list' of balance equations reproduced in Figure 18.3, one would want to know the properties of each equation in the hierarchy and the behaviour of its solutions. This question has been asked perennially throughout the GFD modelling community during the past few decades, as the GFD balance models have been developed in concert with the improvement of the computer capabilities for their simulation.

This is where geometric mechanics plays a constructive role in the development of ocean science for climate prediction. Two ***fundamental criteria*** have emerged for selecting 'good' approximate sets of balance equations. Namely, in dealing with the circulation of either the ocean or atmosphere, whatever governing equations one would choose will need two properties:

1. a circulation theorem; and
2. a law of energy conservation.

These two properties are guaranteed in the Euler–Poincaré (EP) framework, both for the exact fluid equations, and for any of their asymptotic approximations in this framework. Thus, the EP framework may be used to develop ocean models that possess these two fundamental properties.

This chapter applies the asymptotic expansion approach to Hamilton's principle in the EP theorem with advected quantities proved in the previous chapter. It then uses this theorem to organize and derive the ***main series of GFD approximations*** for modelling ocean circulation dynamics. The method of applying asymptotics in the Lagrangian for the EP Hamilton's principle is chosen for GFD, rather than the standard approach of applying asymptotic expansions to the fluid equations. This choice is made because ***Hamilton's-principle asymptotics*** in the EP approach will guarantee preservation of two key properties of the GFD balances that are responsible for large-scale ocean and atmosphere circulations. Namely:

(1) the Kelvin circulation theorem, leading to proper potential-vorticity (PV) dynamics; and
(2) the law of energy conservation.

These two important properties are preserved in the EP approach for Hamilton's-principle asymptotics at *every level of approximation*. However, they have often been *lost* when using the standard asymptotic expansions of the fluid equations.

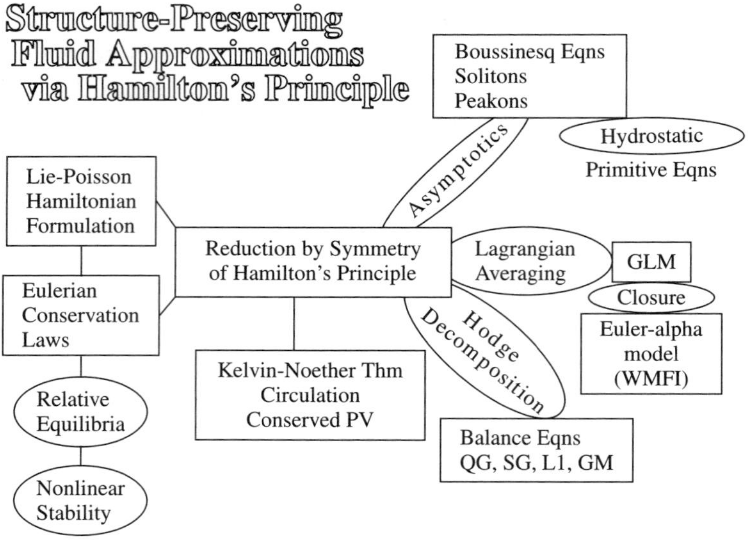


**Figure 18.2** Asymptotics and averaging in Hamilton's principle in the Euler–Poincaré framework produces fluid approximations for GFD that preserve fundamental mathematical structures such as the Kelvin–Noether theorem for circulation leading to conservation of energy and potential vorticity (PV). Legendre transforming the resulting Euler–Poincaré Lagrangian yields the Lie–Poisson Hamiltonian formulation of geophysical fluid dynamics and its Eulerian conservation laws, which may be used to classify steady solutions as relative equilibria and determine sufficient conditions for their nonlinear stability [HMRW85].

The EP approach to mathematical modelling for fluids provides the unified structure that would be required in evaluating any of the balanced models for ocean circulation in the ABN list, and it eliminates those that fail to meet the energy and circulation requirements. Each of the models on the list selected by the EP approach possesses a Kelvin circulation theorem and satisfies an energy law that are consistent with the leading-order geostrophic balance. We shall see that the EP theorem for deriving fluid equations that possess proper circulation and energy laws follows directly from applying the method of asymptotic expansion to the Euler–Poincaré theorem for fluids in the Eulerian description.

In GFD, one is typically dealing with fluids in the Eulerian picture. For example, geostrophic balance is an Eulerian concept. That is, at each point in space the velocity of the fluid may be obtained approximately by assuming that the Coriolis force is balanced by the horizontal pressure gradient. In hydrostatic balance, this horizontal pressure gradient will be proportional to the gradient of the surface elevation. Consequently, one may measure the approximate velocity of the great currents in the sea by measuring the level sets of its surface elevation. These are also ***isobars*** of the hydrostatic

$$\zeta = \nabla^2\psi = \nabla^2\eta - \epsilon 2J(\psi_x, \psi_y), \tag{C3b}$$

$$\nabla^2\chi = -J(\psi, \zeta) - \epsilon(\chi_x\zeta)_x - \epsilon(\chi_y\zeta)_y - \zeta_t. \tag{C3c}$$

QG:

$$(\nabla^2\eta - F\eta)_t = -J(\eta, \zeta_G - \tilde{H}), \tag{C4a}$$

$$\nabla^2\chi = -J(\eta, \zeta_G) - \zeta_{Gt}. \tag{C4b}$$

LBE:

$$(\nabla^2\eta - F\eta)_t = -J(\eta, \zeta_G - \tilde{H}) + \epsilon(\chi_x\tilde{H})_x + \epsilon(\chi_y\tilde{H})_y, \tag{C5a}$$

$$\nabla^2\chi = -J(\eta, \zeta_G) - \zeta_{Gt}. \tag{C5b}$$

LQBE:

$$(\nabla^2\eta - F\eta)_t = -J(\eta, \zeta_G - \tilde{H}) - \epsilon[\chi_x(\zeta_G - \tilde{H})]_x - \epsilon[\chi_y(\zeta_G - \tilde{H})]_y, \tag{C6a}$$

$$\nabla^2\chi = -J(\eta, \zeta_G) - \epsilon(\chi_x\zeta_G)_x - \epsilon(\chi_y\zeta_G)_y - \zeta_{Gt}. \tag{C6b}$$

HBE:

$$(\nabla^2\psi - F\eta)_t = -J(\psi, \zeta - \tilde{H}) + \epsilon(\chi_x\tilde{H})_x + \epsilon(\chi_y\tilde{H})_y, \tag{C7a}$$

$$\zeta = \nabla^2\psi = \nabla^2\eta - \epsilon 2J(\psi_x, \psi_y), \tag{C7b}$$

$$\nabla^2\chi = -J(\psi, \zeta) - \zeta_t. \tag{C7c}$$

BEM:

$$(\nabla^2\psi - F\eta)_t = -J(\psi, \zeta - \tilde{H}) - \epsilon[\chi_x(\zeta - \tilde{H})]_x - \epsilon[\chi_y(\zeta - \tilde{H})]_y, \tag{C8a}$$

$$\zeta = \nabla^2\psi = \nabla^2\eta - \epsilon 2J(\psi_x, \psi_y) - \epsilon^2 J(\zeta, \chi), \tag{C8b}$$

$$\nabla^2\chi = -J(\psi, \zeta) - \epsilon(\chi_x\zeta)_x - \epsilon(\chi_y\zeta)_y - \zeta_t. \tag{C8c}$$

NBE:

$$\psi = \psi_0 + \epsilon^2\psi_1, \tag{C9a}$$

$$(\nabla^2\psi_0 - F\eta)_t = -J(\psi, \zeta_0 - \tilde{H}) - \epsilon[\chi_x(\zeta_0 - \tilde{H})]_x - \epsilon[\chi_y(\zeta_0 - \tilde{H})]_y, \tag{C9b}$$

$$\zeta_0 = \nabla^2\psi_0 = \nabla^2\eta - \epsilon 2J(\psi_{0x}, \psi_{0y}), \tag{C9c}$$

$$\zeta_1 = \nabla^2\psi_1 = -J(\zeta_0, \chi) - \epsilon(\psi_{1x}\zeta_0)_x - \epsilon(\psi_{1y}\zeta_0)_y, \tag{C9d}$$

$$\nabla^2\chi = -J(\psi, \zeta_0) - \epsilon(\chi_x\zeta_0)_x - \epsilon(\chi_y\zeta_0)_y - \zeta_{0t}. \tag{C9e}$$

IM:

$$(\nabla^2\eta - F\eta)_t = -J(\eta, \zeta_G) + J(\psi, \tilde{H}) + \epsilon(\chi_x\tilde{H})_x + \epsilon(\chi_y\tilde{H})_y, \tag{C10a}$$

$$\zeta = \nabla^2\psi = \nabla^2\eta - \epsilon 2J(\eta_x, \eta_y), \tag{C10b}$$

$$\nabla^2\chi = -J(\eta, \zeta_G) - \zeta_{Gt}. \tag{C10c}$$

GV:

$$(\nabla^2\eta - F\eta)_t = -J(\psi, \zeta_G - \tilde{H}) - \epsilon[\chi_x(\zeta_G - \tilde{H})]_x - \epsilon[\chi_y(\zeta_G - \tilde{H})]_y, \tag{C11a}$$

$$\zeta = \nabla^2\psi = \nabla^2\eta + \epsilon\nabla^2K_G - \epsilon(\psi_x\zeta_G)_x - \epsilon(\psi_y\zeta_G)_y - \epsilon^2 J(\zeta_G, \chi), \tag{C11b}$$

$$\nabla^2\chi = -J(\psi, \zeta_G) - \epsilon(\chi_x\zeta_G)_x - \epsilon(\chi_y\zeta_G)_y - \zeta_{Gt}. \tag{C11c}$$

GM:

$$(\zeta_{GM} - F\eta)_t = -J(\psi, \zeta_{GM} - \tilde{H}) - \epsilon[\chi_x(\zeta_{GM} - \tilde{H})]_x - \epsilon[\chi_y(\zeta_{GM} - \tilde{H})]_y, \tag{C12a}$$

$$\zeta = \nabla^2\psi = \nabla^2\eta - \epsilon[J(\eta_x, \psi_y) + J(\psi_x, \eta_y)] + \epsilon^2[J(\eta_x, \chi_x) + J(\eta_y, \chi_y)], \tag{C12b}$$

$$\nabla^2\chi = -J(\psi, \zeta_{GM}) - \epsilon(\chi_x\zeta_{GM})_x - \epsilon(\chi_y\zeta_{GM})_y - \zeta_{GMt}, \tag{C12c}$$

$$\zeta_{GM} = \nabla^2\eta + \epsilon J(\eta_x, \eta_y). \tag{C12d}$$

HP:

$$(\nabla^2\eta - F\eta)_t = -J(\psi, \zeta_G - \tilde{H}) - \epsilon[\chi_x(\zeta_G - \tilde{H})]_x - \epsilon[\chi_y(\zeta_G - \tilde{H})]_y, \tag{C13a}$$

$$\zeta = \nabla^2\psi = \nabla^2\eta + \nabla^2\hat{\mathcal{B}}_{HP} - \epsilon(\psi_x\zeta_G)_x - \epsilon(\psi_y\zeta_G)_y - \epsilon^2 J(\zeta_G, \chi), \tag{C13b}$$

$$\nabla^2\chi = -J(\psi, \zeta_G) - \epsilon(\chi_x\zeta_G)_x - \epsilon(\chi_y\zeta_G)_y - \zeta_{Gt}, \tag{C13c}$$

$$\begin{aligned}\hat{\mathcal{B}}_{HP} &= \mathcal{B}_{HP} - \eta \\ &= \epsilon[\psi_y\eta_y + \psi_x\eta_x - K_G + \epsilon J(\eta, \chi)] + F^{-1}h(\zeta - \zeta_G) \\ &\quad - F^{-1}[(\psi_y - \eta_y)h_{By} + (\psi_x - \eta_x)h_{Bx} + \epsilon J(h_B, \chi)].\end{aligned} \tag{C13d}$$

MSE:

$$(\nabla^2\psi - F\eta)_t = -J(\psi, \zeta - \tilde{H}) - \epsilon[\chi_x(\zeta - \tilde{H})]_x - \epsilon[\chi_y(\zeta - \tilde{H})]_y, \tag{C14a}$$

$$\zeta = \nabla^2\psi = \nabla^2\eta - \epsilon 2J(\psi_x, \psi_y) - \epsilon^2 Z, \tag{C14b}$$

$$Z = 2[J(\chi_x, \psi_x) + J(\chi_y, \psi_y) + \epsilon J(\chi_x, \chi_y)] - J(\psi, \nabla^2\chi) - \epsilon(\chi_x\nabla^2\chi)_x - \epsilon(\chi_y\nabla^2\chi)_y, \tag{C14c}$$

$$\nabla^2\chi = -\epsilon^{-1}D' - J(\psi, \zeta) - \epsilon(\chi_x\zeta)_x - \epsilon(\chi_y\zeta)_y - \zeta_t, \tag{C14d}$$

$$\nabla^2 D' - FD' = \epsilon^2 F[2J(\psi_x, \psi_y) + \epsilon Z]_t. \tag{C14e}$$

SWE:

$$(\nabla^2\psi - F\eta)_t = -J(\psi, \zeta - \tilde{H}) - \epsilon[\chi_x(\zeta - \tilde{H})]_x - \epsilon[\chi_y(\zeta - \tilde{H})]_y, \tag{C15a}$$

**Figure 18.3** The Allen, Barth and Newberger (ABN) list of leading GFD models, from page 1041 of [ABN90a].

pressure. Just as the movement of the atmosphere tends to follow the pressure isobars, the ocean's flow takes place primarily along streamlines defined by the level sets, or isobars of the sea-surface elevation.

Some of the GFD models on the ABN list in Figure 18.3 survive the dual selection criteria of proper energetics and circulation by being EP equations with advected parameters in their own right. These form the ***main sequence*** of GFD model equations and they follow from the Euler–Poincaré form of Hamilton's principle for a fluid Lagrangian that depends parametrically on the advected quantities such as mass, salt and heat, all carried as material properties of the fluid's motion. Thus, the Euler–Poincaré theorem with advected quantities systematically selects and derives the useful GFD fluid models possessing the two main properties of energy balance and the circulation theorem.

## 18.2 Approximate model fluid equations that preserve the Euler–Poincaré structure

The preceding section argues for the applicability of the Euler–Poincaré theorem for ideal continua in deriving approximate fluid models that preserve the Euler–Poincaré structure, and are obtained by making ***asymptotic expansions*** and other approximations in Hamilton's principle for a given set of model equations. As examples, we first discuss the derivation of the quasigeostrophic approximation in geophysical fluid dynamics in two dimensions from an approximation of Hamilton's principle for the motion of a single layer of rotating shallow water. Later, we discuss a sequence of GFD approximations of the same type in three dimensions.

## 18.3 Equations of 2D geophysical fluid motion

### Rotating shallow-water dynamics as Euler–Poincaré equations

We first consider dynamics of rotating shallow water (RSW) in a two-dimensional domain with horizontal coordinates $\mathbf{x} = (x_1, x_2)$. RSW motion is governed by the following non-dimensional equations for horizontal fluid velocity $\mathbf{u} = (u_1, u_2)$ and depth $D$,

$$\epsilon \frac{\mathrm{d}}{\mathrm{d}t}\mathbf{u} + f\hat{\mathbf{z}} \times \mathbf{u} + \nabla\psi = 0, \qquad \frac{\partial D}{\partial t} + \nabla \cdot D\mathbf{u} = 0, \tag{18.1}$$

with notation

$$\frac{d}{dt} \equiv \left(\frac{\partial}{\partial t} + \mathbf{u} \cdot \nabla\right) \quad \text{and} \quad \psi \equiv \left(\frac{D - B}{\epsilon \mathcal{F}}\right). \tag{18.2}$$

These equations include variable Coriolis parameter $f = f(\mathbf{x})$ and bottom topography $B = B(\mathbf{x})$.

The dimensionless scale factors appearing in the non-dimensional RSW equations (18.1) and (18.2) are the Rossby number $\epsilon$ and the rotational Froude number $\mathcal{F}$, given in terms of typical dimensional scales by

$$\epsilon = \frac{\mathcal{U}_0}{f_0 L} \ll 1 \quad \text{and} \quad \mathcal{F} = \frac{f_0^2 L^2}{g B_0} = O(1). \tag{18.3}$$

The dimensional scales $(B_0, L, \mathcal{U}_0, f_0, g)$ denote equilibrium fluid depth, horizontal length scale, horizontal fluid velocity, reference Coriolis parameter, and gravitational acceleration, respectively. Dimensionless quantities in eqns (18.1) are unadorned and are related to their dimensional counterparts (primed), according to

$$\begin{aligned} &\mathbf{u}' = \mathcal{U}_0 \mathbf{u}, \quad \mathbf{x}' = L\mathbf{x}, \quad t' = \left(\frac{L}{\mathcal{U}_0}\right) t, \quad f' = f_0 f, \\ &B' = B_0 B, \quad D' = B_0 D, \quad \text{and} \quad D' - B' = B_0(D - B). \end{aligned} \tag{18.4}$$

Here, dimensional quantities are: $\mathbf{u}'$, the horizontal fluid velocity; $D'$, the fluid depth; $B'$, the equilibrium depth; and $D' - B'$, the free surface elevation.

For barotropic (i.e. $z$-independent) horizontal motions at length scales $L$ in the ocean, say, for which the rotational Froude number $\mathcal{F}$ is order $O(1)$ – as we shall assume – the Rossby number $\epsilon$ is typically quite small ($\epsilon \ll 1$) as indicated in eqn (18.3). Thus, $\epsilon \ll 1$ is a natural parameter for making asymptotic expansions. For example, we shall assume $|\nabla f| = O(\epsilon)$ and $|\nabla B| = O(\epsilon)$, so we may write $f = 1 + \epsilon f_1(\mathbf{x})$ and $B = 1 + \epsilon B_1(\mathbf{x})$. In this scaling, the leading-order implications of equation (18.1) are $\mathbf{u} = \hat{\mathbf{z}} \times \nabla \psi$ and $\nabla \cdot \mathbf{u} = 0$. This is ***geostrophic balance***, which is imposed in the $\epsilon \to 0$ limit of the non-dimensional RSW equations in (18.1).

Substitution into eqn (17.35) with $b$ absent shows that the non-dimensional RSW equations (18.1) arise as Euler–Poincaré equations from Hamilton's principle with action $S_{\mathrm{RSW}}$,

$$S_{\mathrm{RSW}} = \int dt \int dx_1 dx_2 \left[ D\mathbf{u} \cdot \mathbf{R}(\mathbf{x}) - \frac{(D - B)^2}{2\epsilon \mathcal{F}} + \frac{\epsilon}{2} D|\mathbf{u}|^2 \right], \tag{18.5}$$

where $\operatorname{curl}\mathbf{R}(\mathbf{x}) \equiv f(\mathbf{x})\hat{\mathbf{z}}$ yields the prescribed Coriolis parameter. The RSW equations (18.1) themselves can be regarded as being derived from asymptotics in Hamilton's principle for three-dimensional incompressible fluid motion, see [Hol96]. However, this viewpoint is not pursued further here, as we proceed to describe the relation of RSW to the quasigeostrophic approximation of geophysical fluid dynamics.

## Quasigeostrophy

The quasigeostrophic (QG) approximation is a useful model in the analysis of geophysical and astrophysical fluid dynamics, see, e.g., [Ped87]. Physically, QG theory applies when the motion is nearly in geostrophic balance, i.e. when pressure gradients nearly balance the Coriolis force in a rotating frame of reference, as occurs in meso- and large-scale oceanic and atmospheric flows on Earth. Mathematically, the simplest case is for a constant density fluid in a planar domain with Euclidean coordinates $\mathbf{x} = (x_1, x_2)$. QG dynamics for this case is expressed by the following non-dimensional evolution equation for the stream function $\psi$ of the incompressible geostrophic fluid velocity $\mathbf{u} = \hat{\mathbf{z}} \times \nabla\psi$,

$$\frac{\partial(\Delta\psi - \mathcal{F}\psi)}{\partial t} + [\psi, \Delta\psi] + \beta\frac{\partial\psi}{\partial x_1} = 0. \tag{18.6}$$

Here, $\Delta$ is the Laplacian operator in the plane, $\mathcal{F}$ denotes rotational Froude number, $[a,b] \equiv \partial(a,b)/\partial(x_1,x_2)$ is the Jacobi bracket (Jacobian) for functions $a$ and $b$ defined on $\mathbb{R}^2$ and $\beta$ is the gradient of the Coriolis parameter, $f$, taken as $f = 1 + \beta x_2$ in the $\beta$-plane approximation, with constant $\beta$. (Neglecting $\beta$ gives the $f$-plane approximation of QG dynamics.) The QG equation (18.6) may be derived from an asymptotic expansion of the RSW equations (18.1) by truncating at first order in the Rossby number, cf. [Ped87]. Equation (18.6) may be written equivalently in terms of the potential vorticity, $q$, defined as,

$$\frac{\partial q}{\partial t} + \mathbf{u}\cdot\nabla q = 0, \quad \text{where} \quad q \equiv \Delta\psi - \mathcal{F}\psi + f \quad \text{for QG.} \tag{18.7}$$

This form of QG dynamics expresses its basic property: that potential vorticity is conserved on geostrophic fluid parcels.

The QG approximation to the RSW equations thus introduces 'geostrophic particles' that move with geostrophic velocity $\mathbf{u} = \hat{\mathbf{z}} \times \nabla\psi$ and, thus, trace the geostrophic component of the RSW fluid flow. These QG fluid trajectories may be described as functions of Lagrangian mass coordinates $\boldsymbol{\ell} = (\ell_1, \ell_2)$ given by $\mathbf{x}(\boldsymbol{\ell}, t)$ in the domain of flow.

## Hamilton's principle derivation of QG as Euler–Poincaré equations

As in [HZ98], we consider the following action for QG written in the Eulerian velocity representation with the integral operator $(1 - \mathcal{F}\Delta^{-1})$,

$$
\begin{aligned}
S_{\text{red}} &= \int \mathrm{d}t\, l \qquad (18.8)\\
&= \int \mathrm{d}t \int \mathrm{d}x_1 \mathrm{d}x_2 \Big[\frac{\epsilon}{2} D\mathbf{u}\cdot(1 - \mathcal{F}\Delta^{-1})\mathbf{u} + D\mathbf{u}\cdot\mathbf{R}(\mathbf{x}) - \psi(D-1)\Big].
\end{aligned}
$$

This choice can be found as an asymptotic approximation of the RSW action $S_{\text{RSW}}$ in eqn (18.5), in the limit of small wave amplitudes of order $O(\epsilon^2)$ and constant mean depth to the same order, when the wave elevation is determined from the fluid velocity by inverting the geostrophic relation, $\mathbf{u} = \hat{\mathbf{z}} \times \nabla\psi$. The variational derivatives of the reduced Lagrangian $S_{\text{red}}$ at fixed $\mathbf{x}$ and $t$ are

$$
\begin{aligned}
\frac{1}{D}\frac{\delta l}{\delta \mathbf{u}} &= \mathbf{R} + \epsilon\Big[\mathbf{u} - \frac{\mathcal{F}}{2}\Delta^{-1}\mathbf{u} - \frac{\mathcal{F}}{2D}\Delta^{-1}(D\mathbf{u})\Big],\\
\frac{\delta l}{\delta D} &= \frac{\epsilon}{2}\mathbf{u}\cdot(1 - \mathcal{F}\Delta^{-1})\mathbf{u} + \mathbf{u}\cdot\mathbf{R} - \psi, \qquad (18.9)\\
\frac{\delta l}{\delta \psi} &= -\,(D-1),
\end{aligned}
$$

where we have used the symmetry of the Laplacian operator and assumed no contribution arises from the boundary when integrating by parts. For example, we may take the domain to be periodic. Hence, the Euler–Poincaré equation (17.35) for action principles of this type and a vector identity combine to give the Eulerian ***QG motion equation***,

$$
\begin{aligned}
&\epsilon\frac{\partial}{\partial t}(1 - \mathcal{F}\Delta^{-1})\mathbf{u} - \mathbf{u}\times\operatorname{curl}\Big(\epsilon(1 - \mathcal{F}\Delta^{-1})\mathbf{u} + \mathbf{R}\Big)\\
&\quad + \nabla\Big(\psi + \frac{\epsilon}{2}\mathbf{u}\cdot(1 - \mathcal{F}\Delta^{-1})\mathbf{u}\Big) = 0, \qquad (18.10)
\end{aligned}
$$

upon substituting the constraint $D = 1$, imposed by varying $\psi$. The curl of this equation yields

$$
\frac{\partial q}{\partial t} + \mathbf{u}\cdot\nabla q + q\nabla\cdot\mathbf{u} = 0, \qquad (18.11)
$$

where the potential vorticity $q$ is given by

$$q = \epsilon\hat{\mathbf{z}}\cdot \text{curl}\,(1 - \mathcal{F}\Delta^{-1})\mathbf{u} + f = \epsilon(\Delta\psi - \mathcal{F}\psi) + f, \tag{18.12}$$

with

$$f \equiv \hat{\mathbf{z}}\cdot \text{curl}\mathbf{R} = 1 + \beta x_2, \tag{18.13}$$

and $\beta$ is assumed to be of order $O(\epsilon)$. The constraint $D = 1$ implies $\nabla\cdot\mathbf{u} = 0$ (from the kinematic relation $\partial D/\partial t + \nabla\cdot D\mathbf{u} = 0$) and when $\mathbf{u} = \hat{\mathbf{z}}\times\nabla\psi$ is substituted, the equation for $q = \Delta\psi - \mathcal{F}\psi + f$ yields the QG potential vorticity convection equation (18.7). Thus, the QG motion equation follows as the Euler–Poincaré equation for an asymptotic expansion of the action for the RSW equations when the potential energy is modelled by inverting the geostrophic relation.

### 18.3.1 Remarks on 2D fluid models in GFD

The search for tractable models of geophysical fluid dynamics natually leads to considerations of two-dimensional fluid models. Two-dimensional fluid models provide insight for many applications in GFD, because the aspect ratio of the domain ($\sigma$) and the Rossby number ($\epsilon$) of the rotating flow are often small in these applications. Several treatments of two-dimensional GFD models have been developed by using asymptotic expansion methods in Hamilton's principle, see, e.g., [Sal83, Sal85, Sal88]. These treatments tend to focus especially on the rotating shallow-water (RSW) equations, their quasigeostrophic (QG) approximation, and certain intermediate approximations, such as the semigeostrophic (SG) equations [Eli49, Hos75, CP89] and the Salmon [1985] $L_1$ model. Following [AH96], a unified derivation of the RSW, $L_1$, QG and SG equations using Hamilton's principle asymptotics is given in Section 18.6.

Due to their wide applicability in GFD, the properties of the two-dimensional QG equations have been studied extensively. For example, a Lie–Poisson bracket for them was introduced in [Wei83] for studying stability of quasigeostrophic equilibria. The Hamiltonian structure and nonlinear stability of the equilibrium solutions for the QG system and its variants has been thoroughly explored. For references, see [Wei83, HMRW85]. A discussion of the geodesic properties of the QG equations in the framework of Euler–Poincaré theory is given in [HMR98a, HMR98b].

## 18.4 Equations of 3D geophysical fluid motion

### 18.4.1 Euler's fluid equations in dimensional form

The ***motion equation*** for rotating stratified incompressible Euler fluid in three dimensions is,

$$\frac{\partial \mathbf{u}}{\partial t} + \mathbf{u} \cdot \nabla \mathbf{u} - \mathbf{u} \times \operatorname{curl} \mathbf{R} + g\hat{\mathbf{z}} + \frac{1}{\rho_0(1+b)} \nabla p = 0. \tag{18.14}$$

Here, $\operatorname{curl} \mathbf{R} = \mathbf{f}(\mathbf{x})$ is the ***Coriolis parameter***. This equation is augmented by auxiliary equations for ***incompressibility*** and the ***advection of buoyancy***,

$$\operatorname{div} \mathbf{u} = 0 \quad \text{and} \quad \frac{\partial b}{\partial t} + \mathbf{u} \cdot \nabla b = 0. \tag{18.15}$$

Besides spatial position $(x, y, z)$ and time $t$, these equations involve several other quantities with dimensions. These other ***dimensional quantities*** include the fluid's reference density $\rho_0$, velocity $\mathbf{u}$ and pressure $p$, as well as the parameters of rotational frequency $|\mathbf{f}| = f_0$ and gravitational acceleration $g$. (The buoyancy $b$ is defined as a ratio of densities, so it is already non-dimensional.)

### 18.4.2 Rossby, aspect ratio and rotational Froude

To set up the asymptotic expansion that is needed for climate predictions using geostrophic and hydrostatic balance ideas, one first writes the exact Euler equations of incompressible fluid motion (18.14) in ***non-dimensional form*** by identifying the proper ***dimensionless ratios*** in terms of units of $L$ for horizontal distance, $B_0$ for vertical depth, $\mathcal{U}_0$ for horizontal velocity, $B_0\mathcal{U}_0/L$ for vertical velocity, $f_0$ for Coriolis parameter, $\rho_0$ for density and $\rho_0 f_0 L\mathcal{U}_0$ for pressure. As mentioned earlier, these units may be combined into ***three non-dimensional parameters***,

$$\epsilon = \frac{\mathcal{U}_0}{f_0 L}, \quad \sigma = \frac{B_0}{L}, \quad \mathcal{F} = \frac{f_0^2 L^2}{g B_0}, \tag{18.16}$$

corresponding respectively to ***Rossby number***, ***aspect ratio*** and ***(squared) rotational Froude number***.

**Remark 18.1 (Relative sizes)**
*Typically, the Rossby number $\epsilon$ and the aspect ratio $\sigma$ are small,*

$$\epsilon, \sigma \ll 1,$$

*while the rotational Froude number is of order unity,*

$$\mathcal{F} = O(1),$$

*in atmospheric and oceanic dynamics for horizontal length scales at the size $L \approx 100\,km$ of typical eddies and larger, at which the Coriolis force begins to matter.*

### 18.4.3 Non-dimensional notation for velocities and gradients

In this non-dimensional notation, three-dimensional vectors and gradient operators are given the subscript 3, while horizontal vectors and gradient operators are left unadorned. Thus, in three-dimensional Euclidean space we denote,

$$
\begin{aligned}
\mathbf{x}_3 &= (x, y, z), \quad \mathbf{x} = (x, y, 0),\\
\mathbf{u}_3 &= (u, v, w), \quad \mathbf{u} = (u, v, 0),\\
\nabla_3 &= \left(\frac{\partial}{\partial x}, \frac{\partial}{\partial y}, \frac{\partial}{\partial z}\right), \quad \nabla = \left(\frac{\partial}{\partial x}, \frac{\partial}{\partial y}, 0\right),\\
\frac{d}{dt} &= \frac{\partial}{\partial t} + \mathbf{u}_3 \cdot \nabla_3 = \frac{\partial}{\partial t} + \mathbf{u} \cdot \nabla + w\frac{\partial}{\partial z},
\end{aligned}
\tag{18.17}
$$

and $\hat{\mathbf{z}} = (0, 0, 1)^T$ is the unit vector in the vertical $z$ direction. For notational convenience, we also define ***two non-dimensional velocities***,

$$\mathbf{u}_3 = (u, v, w), \quad \text{and} \quad \mathbf{v}_3 = (u, v, \sigma^2 w), \tag{18.18}$$

so that when the aspect ratio tends to zero, $\sigma^2 \to 0$, the second velocity loses its vertical component.

### 18.4.4 Non-dimensional equations

In terms of the two velocities $(\mathbf{u}_3, \mathbf{v}_3)$ and the three non-dimensional parameters in eqn (18.16), the ***non-dimensional Euler fluid equations*** are expressed as,

$$\epsilon\frac{\mathrm{d}\mathbf{v}_3}{\mathrm{d}t} - \mathbf{u}_3 \times \operatorname{curl}\mathbf{R} + \frac{1}{\epsilon\mathcal{F}}\hat{\mathbf{z}} + \frac{1}{1+b}\nabla_3\, p = 0. \tag{18.19}$$

Here the vector is taken $\mathbf{R}(\mathbf{x})$ to be horizontal and independent of the vertical coordinate, so $\operatorname{curl}\mathbf{R} = f(\mathbf{x})\hat{z}$, the frequency of rotation about the vertical direction. As mentioned earlier, the Rossby number $\epsilon$ multiplies both the Eulerian partial time derivative and the nonlinearity $\mathbf{u}_3 \cdot \nabla_3 \mathbf{v}_3$ in the expression for $d/dt$ in equation (18.19).

The non-dimensional motion equation (18.19) is augmented by auxiliary equations for incompressibility and the advection of buoyancy. These are, respectively,

$$\nabla_3 \cdot \mathbf{u}_3 = 0 \quad \text{and} \quad \frac{db}{dt} = \frac{\partial b}{\partial t} + \mathbf{u}_3 \cdot \nabla_3 b = 0. \tag{18.20}$$

## Determining the pressure

The pressure $p$ is found by requiring preservation of the incompressibility constraint. Preservation of incompressibility determines the pressure from a Poisson equation that is obtained by taking the divergence of the motion equation. This Poisson equation determines the pressure upon imposing on it the Neumann boundary conditions that arise from the requirement of tangency of the fluid velocity at any fixed boundaries.

## 18.4.5 Circulation and potential vorticity conservation

**Theorem 18.2 (Potential vorticity conservation)**
*Under evolution by the non-dimensional Euler equations (18.19) with auxiliary conditions (18.15) the* ***Euler potential vorticity*** *defined as,*

$$q_E = \nabla_3 b \cdot \operatorname{curl}(\epsilon \mathbf{v}_3 + \mathbf{R}), \tag{18.21}$$

*is conserved on fluid parcels. That is, the potential vorticity $q_E$ satisfies*

$$\frac{dq_E}{dt} := \frac{\partial q_E}{\partial t} + \mathbf{u}_3 \cdot \nabla q_E = 0. \tag{18.22}$$

*Proof* The curl of the Euler motion equation (18.19) yields

$$\frac{\partial \mathbf{q}_E}{\partial t} - \nabla_3 \times \big(\mathbf{u}_3 \times \mathbf{q}_E\big) + \nabla_3 \frac{1}{1+b} \times \nabla_3 p = 0, \tag{18.23}$$

where $\mathbf{q}_E = \operatorname{curl}\big(\epsilon \mathbf{v}_3 + \mathbf{R}\big)$. A short computation using the auxiliary eqns (18.15) then yields the potential vorticity conservation result in eqn (18.22). ∎

Combined with incompressibility and tangency of the velocity at the boundary, this equation implies the following conservation law.

**Theorem 18.3 (Integral potential vorticity conservation)**
*The non-dimensional Euler equations (18.19) with auxiliary conditions (18.15) conserve the functional,*

$$C_\Phi = \int \Phi(q_E, b)\, \mathrm{d}^3x, \tag{18.24}$$

*integrated over the spatial domain of the flow. That is,* $\mathrm{d}C_\Phi/\mathrm{d}t = 0$ *for any differentiable function* $\Phi$.

*Proof*

$$\begin{aligned}\frac{\mathrm{d}C_\Phi}{\mathrm{d}t} &= \int \Phi_{q_E}\frac{\partial q_E}{\partial t} + \Phi_b\frac{\partial b}{\partial t}\, \mathrm{d}^3x \\ &= -\int \mathbf{u}_3 \cdot \big(\Phi_{q_E}\, \nabla q_E + \Phi_b\, \nabla b\big)\, \mathrm{d}^3x \\ &= -\int \mathbf{u}_3 \cdot \nabla\Phi\, \mathrm{d}^3x = -\int \mathrm{div}\,(\Phi\, \mathbf{u}_3)\, \mathrm{d}^3x \\ &= -\int_{Bdy} \Phi(q_E, b)\, \mathbf{u}_3 \cdot \hat{\mathbf{n}}\, \mathrm{d}S = 0,\end{aligned}$$

which vanishes upon using the condition $\mathbf{u}_3 \cdot \hat{\mathbf{n}}|_{Bdy} = 0$ that the velocity is tangent to the boundary. ■

### 18.4.6 Leading-order balances

The leading-order balances in these equations arise when one sets $\epsilon \to 0$ and $\sigma^2 \to 0$, in which case the second velocity in (18.18) loses its vertical component and the non-dimensional Euler motion equation (18.19) becomes geostrophic in the horizontal direction and hydrostatic in the vertical. Namely,

$$\underbrace{f\hat{\mathbf{z}} \times \mathbf{u} + \nabla p'}_{\text{geostrophic}} = 0 \quad \text{and} \quad \underbrace{b' + \frac{\partial p'}{\partial z}}_{\text{hydrostatic}} = 0. \tag{18.25}$$

The time dependence arises at order $O(\epsilon)$.

Buoyancy in the ocean is typically small ($b \approx 35/1000 \ll 1$); so one may use buoyancy as an ***additional small parameter*** in an asymptotic expansion of the solutions of the Euler equations, as in [Phi69]. For sufficiently small buoyancy, $b = o(\epsilon)$ one expands the pressure gradient term in eqn (18.19) in powers of $\epsilon$. Upon denoting,

$$p' = p + \frac{z}{\epsilon \mathcal{F}} \quad \text{and} \quad b' = \frac{b}{\epsilon \mathcal{F}},$$

and dropping the order $o(\epsilon)$ terms in the expansion of the pressure gradient term in eqn (18.19), one finds the ***non-dimensional Euler–Boussinesq equation*** for fluid motion in three dimensions, namely,

$$\epsilon \frac{\mathrm{d}\mathbf{v}_3}{\mathrm{d}t} - \mathbf{u} \times \operatorname{curl} \mathbf{R} + b'\hat{\mathbf{z}} + \nabla_3 \, p' = 0. \tag{18.26}$$

In horizontal and vertical components, with $\operatorname{curl} \mathbf{R} = f(\mathbf{x})\hat{\mathbf{z}}$, the Euler–Boussinesq motion equation decomposes into,

$$\epsilon \frac{\mathrm{d}\mathbf{u}}{\mathrm{d}t} + \underbrace{f\hat{\mathbf{z}} \times \mathbf{u} + \nabla p'}_{\text{geostrophic}} = 0$$

$$\text{and} \quad \epsilon \, \sigma^2 \frac{\mathrm{d}w}{\mathrm{d}t} + \underbrace{b' + \frac{\partial p'}{\partial z}}_{\text{hydrostatic}} = 0, \tag{18.27}$$

where $\{\mathbf{u}, w, b'\}$ satisfy the auxiliary equations,

$$\frac{\mathrm{d}b'}{\mathrm{d}t} = \frac{\partial b'}{\partial t} + \mathbf{u}_3 \cdot \nabla_3 b' = 0$$

$$\text{and} \quad \nabla_3 \cdot \mathbf{u}_3 = \nabla \cdot \mathbf{u} + \frac{\partial w}{\partial z} = 0. \tag{18.28}$$

Because $\epsilon \ll 1, \sigma^2 \ll 1$ the leading-order balances are still hydrostatic in the vertical, and geostrophic in the horizontal. Equations (18.27) describe the motion of an ideal incompressible stratified fluid in a rotating frame relative to a stable hydrostatic equlibrium in which the density is taken to be constant except in the buoyant force.

If one now neglects the vertical acceleration by setting $\sigma^2 \to 0$ in the Euler–Boussinesq equations (18.27), then the ***Primitive Equations*** (PE) emerge,

$$\epsilon \frac{d\mathbf{u}}{dt} + f\hat{\mathbf{z}} \times \mathbf{u} + \nabla p' = 0 \quad \text{and} \quad b' + \frac{\partial p'}{\partial z} = 0. \tag{18.29}$$

Upon including the auxiliary equations (18.28) one finds that the PE (18.29) may still be solved for the variables $\{\mathbf{u}, w, b', p'\}$ as functions of $(x, y, z, t)$. The PE comprise the basis for the standard model for high-resolution global ocean circulation. For more information about their modern use as the basis of high-resolution numerical simulations of global ocean circulation for climate prediction, see, e.g., http://climate.lanl.gov/ and references there.

In summary, we list the horizontal and vertical decompositions of the Euler equations, the Euler–Boussinesq (EB) equations and the primitive equations (PE), respectively, as

$$\epsilon\,\frac{d\mathbf{u}}{dt} + f\hat{\mathbf{z}}\times\mathbf{u} + \frac{1}{1+b}\nabla\, p = 0 \text{ and } \epsilon\,\sigma^2\,\frac{dw}{dt} + \frac{1}{\epsilon\mathcal{F}} + \frac{1}{1+b}\frac{\partial p}{\partial z} = 0,$$

$$\epsilon\,\frac{d\mathbf{u}}{dt} + f\hat{\mathbf{z}}\times\mathbf{u} + \nabla\, p' = 0 \text{ and } \epsilon\,\sigma^2\,\frac{dw}{dt} + b' + \frac{\partial p'}{\partial z} = 0,$$

$$\epsilon\,\frac{d\mathbf{u}}{dt} + f\hat{\mathbf{z}}\times\mathbf{u} + \nabla\, p' = 0 \text{ and } b' + \frac{\partial p'}{\partial z} = 0.$$

**Remark 18.4** *Various other refinements and approximations of these basic models exist, many of which are topics of ongoing research. Some of these refinements will be discussed after we reformulate the basic models as Euler–Poincaré equations and explain their roles in the main hierarchy of GFD equations. For this, we shall follow [HMR04].*

## 18.5 Variational principle for fluids in three dimensions

In order to apply the Euler–Poincaré theorem in GFD, we begin by computing explicit formulae for the variations $\delta a$ in the cases that the set of tensors $a$ satisfying (17.25) is drawn from a set of scalar fields and densities on $\mathbb{R}^3$. We shall denote this symbolically by writing

$$a \in \{b, D\, d^3x\}. \tag{18.30}$$

We have seen that invariance of the set $a$ in the Lagrangian picture under the dynamics of $\mathbf{u}$ implies in the Eulerian picture that

$$\left(\frac{\partial}{\partial t} + \pounds_{\mathbf{u}}\right) a = 0,$$

where $\pounds_{\mathbf{u}}$ denotes Lie derivative with respect to the velocity vector field $\mathbf{u}$. Hence, for a fluid dynamical action $S = \int dt\; l(\mathbf{u}; b, D)$, the advected variables $\{b, D\,\mathrm{d}^3x\}$ satisfy the following Lie-derivative relations,

$$\left(\frac{\partial}{\partial t} + \pounds_{\mathbf{u}}\right) b = 0, \quad \text{or} \quad \frac{\partial b}{\partial t} = -\,\mathbf{u}\cdot\nabla\, b, \tag{18.31}$$

$$\left(\frac{\partial}{\partial t} + \pounds_{\mathbf{u}}\right) D\, d^3x = 0, \quad \text{or} \quad \frac{\partial D}{\partial t} = -\,\nabla\cdot(D\mathbf{u}). \tag{18.32}$$

In fluid dynamical applications, the advected Eulerian variables $b$ and $D$ represent the buoyancy $b$ (or specific entropy, for the compressible case) and volume element (or mass density) $D$, respectively. According to Theorem 17.8, equation (17.18), the variations of the tensor functions $a$ at fixed $\mathbf{x}$ and $t$ are also given by Lie derivatives, namely $\delta a = -\pounds_{\mathbf{w}}\, a$, or

$$\begin{aligned} \delta b &= -\pounds_{\mathbf{w}}\, b = -\mathbf{w}\cdot\nabla\, b, \\ \delta D\,\mathrm{d}^3x &= -\pounds_{\mathbf{w}}\,(D\,\mathrm{d}^3x) = -\nabla\cdot(D\mathbf{w})\,\mathrm{d}^3x. \end{aligned} \tag{18.33}$$

Hence, Hamilton's principle with this dependence yields

$$\begin{aligned} 0 &= \delta \int \mathrm{d}t\; l(\mathbf{u}; b, D) \\ &= \int \mathrm{d}t \left[\frac{\delta l}{\delta \mathbf{u}}\cdot\delta\mathbf{u} + \frac{\delta l}{\delta b}\,\delta b + \frac{\delta l}{\delta D}\,\delta D\right] \\ &= \int \mathrm{d}t \left[\frac{\delta l}{\delta \mathbf{u}}\cdot\left(\frac{\partial \mathbf{w}}{\partial t} - \mathrm{ad}_{\mathbf{u}}\,\mathbf{w}\right) - \frac{\delta l}{\delta b}\,\mathbf{w}\cdot\nabla\, b - \frac{\delta l}{\delta D}\left(\nabla\cdot(D\mathbf{w})\right)\right] \\ &= \int \mathrm{d}t\; \mathbf{w}\cdot\left[-\frac{\partial}{\partial t}\frac{\delta l}{\delta \mathbf{u}} - \mathrm{ad}^*_{\mathbf{u}}\,\frac{\delta l}{\delta \mathbf{u}} - \frac{\delta l}{\delta b}\,\nabla\, b + D\,\nabla\frac{\delta l}{\delta D}\right] \\ &= -\int \mathrm{d}t\; \mathbf{w}\cdot\left[\left(\frac{\partial}{\partial t} + \pounds_{\mathbf{u}}\right)\frac{\delta l}{\delta \mathbf{u}} + \frac{\delta l}{\delta b}\,\nabla\, b - D\,\nabla\frac{\delta l}{\delta D}\right], \end{aligned} \tag{18.34}$$

where we have consistently dropped boundary terms arising from integrations by parts, by invoking natural boundary conditions. Specifically, we invoke vanishing of $\mathbf{w}$ at the endpoints in time and impose both $\hat{\mathbf{n}}\cdot\mathbf{u} = 0$ and $\hat{\mathbf{n}}\cdot\mathbf{w} = 0$ on the boundary, where $\hat{\mathbf{n}}$ is the boundary's outward unit normal vector.

### 18.5.1 Summary

The Euler–Poincaré equations for continua (17.20) may now be summarized for advected Eulerian variables $a$ in the set (18.30). We adopt the notational convention of the circulation map $\mathcal{K}$ in eqn (17.26) that a 1-form density can be made into a 1-form (no longer a density) by dividing it by the mass density $D\,\mathrm{d}^3x$ and use the Lie-derivative relation for the continuity equation

$$\Big(\frac{\partial}{\partial t} + \pounds_{\mathbf{u}}\Big) D\mathrm{d}^3x = 0.$$

Then, the Euclidean components of the Euler–Poincaré equations for continua in eqn (18.34) are expressed in Kelvin-theorem form (17.29) with a slight abuse of notation as follows.

$$\Big(\frac{\partial}{\partial t} + \pounds_{\mathbf{u}}\Big)\Big(\frac{1}{D}\frac{\delta l}{\delta \mathbf{u}} \cdot d\mathbf{x}\Big) + \frac{1}{D}\frac{\delta l}{\delta b}\nabla b \cdot \mathrm{d}\mathbf{x} \;-\; \nabla\Big(\frac{\delta l}{\delta D}\Big) \cdot \mathrm{d}\mathbf{x} = 0, \tag{18.35}$$

in which the variational derivatives of the Lagrangian $l$ are to be computed according to the usual physical conventions, i.e. as Fréchet derivatives. In vector notation, this equation is

$$\frac{\mathrm{d}}{\mathrm{d}t}\frac{1}{D}\frac{\delta l}{\delta \mathbf{u}} + \frac{1}{D}\frac{\delta l}{\delta u^j}\nabla u^j + \frac{1}{D}\frac{\delta l}{\delta b}\nabla b - \nabla\frac{\delta l}{\delta D} = 0, \tag{18.36}$$

which may also be expressed in three-dimensional curl form as,

$$\frac{\partial}{\partial t}\Big(\frac{1}{D}\frac{\delta l}{\delta \mathbf{u}}\Big) - \mathbf{u} \times \mathrm{curl}\Big(\frac{1}{D}\frac{\delta l}{\delta \mathbf{u}}\Big) + \nabla\Big(\mathbf{u} \cdot \frac{1}{D}\frac{\delta l}{\delta \mathbf{u}} - \frac{\delta l}{\delta D}\Big) + \frac{1}{D}\frac{\delta l}{\delta b}\nabla b = 0. \tag{18.37}$$

In writing the last equation, we have used the ***fundamental vector identity of fluid dynamics***,

$$(\mathbf{b} \cdot \nabla)\mathbf{a} + a_j \nabla b^j = -\,\mathbf{b} \times (\nabla \times \mathbf{a}) + \nabla(\mathbf{b} \cdot \mathbf{a}), \tag{18.38}$$

for any three dimensional vectors **a** and **b** with, in this case, $\mathbf{a} = (D^{-1}\delta l/\delta \mathbf{u})$ and $\mathbf{b} = \mathbf{u}$.

**Exercise 18.1**

Prove eqn (18.38). Hint: the geometric approach would equate the dynamical definition of the Lie derivative of the 1-form $a = \mathbf{a} \cdot \mathrm{d}\mathbf{x}$ by the vector field $b = \mathbf{b} \cdot \nabla$ with Cartan's formula for the Lie derivative, $\pounds_b a = b \,\lrcorner\, \mathrm{d}a + \mathrm{d}(b \,\lrcorner\, a)$.

Formula (18.35) is the Kelvin–Noether form of the equation of motion for ideal continua. Hence, we have the following explicit Kelvin-theorem expression for the rate of change of circulation, cf. equation (17.27),

$$\frac{dI}{dt} = \frac{d}{dt}\oint_{\gamma_t(\mathbf{u})} \frac{1}{D}\frac{\delta l}{\delta \mathbf{u}} \cdot d\mathbf{x} = -\oint_{\gamma_t(\mathbf{u})} \frac{1}{D}\frac{\delta l}{\delta b}\nabla b \cdot d\mathbf{x}\,, \tag{18.39}$$

where the curve $\gamma_t(\mathbf{u})$ moves with the fluid velocity $\mathbf{u}$. Then, by Stokes' theorem, this is equal to

$$\frac{dI}{dt} = \frac{d}{dt}\int_{S_t} \mathrm{curl}\,\frac{1}{D}\frac{\delta l}{\delta \mathbf{u}} \cdot \hat{\mathbf{n}}\, dS = -\int_{S_t} \nabla\left(\frac{1}{D}\frac{\delta l}{\delta b}\right) \times \nabla b \cdot \hat{\mathbf{n}}\, dS\,, \tag{18.40}$$

where the boundary of the surface $S_t$ is the material loop, $\partial S_t = \gamma_t(\mathbf{u})$. Thus, the Euler equations generate circulation of $(D^{-1}\delta l/\delta \mathbf{u})$ whenever the gradients $\nabla b$ and $\nabla(D^{-1}\delta l/\delta b)$ are not collinear. The corresponding ***conservation of potential vorticity*** $q$ on fluid parcels is given by

$$\frac{\partial q}{\partial t} + \mathbf{u}\cdot\nabla q = 0, \quad \text{where} \quad q = \frac{1}{D}\nabla b \cdot \mathrm{curl}\left(\frac{1}{D}\frac{\delta l}{\delta \mathbf{u}}\right). \tag{18.41}$$

**Exercise 18.2**
Prove conservation of potential vorticity in equation (18.41) using the Lie-derivative expression (18.35) and the commutation of Lie derivative and exterior derivative.

**Exercise 18.3**
Prove conservation of potential vorticity in eqn (18.41) using the curl of the vector equation (18.37) and the continuity equation (18.32).

## 18.6 Euler's equations for a rotating stratified ideal incompressible fluid

### The Lagrangian

In the Eulerian velocity representation, we consider Hamilton's principle for fluid motion in a three-dimensional domain with action functional $S = \int dt\, l$ and Lagrangian $l(\mathbf{u}, b, D)$ given by

$$l = \int d^3x\, \rho_0 D(1+b)\Big(\frac{1}{2}|\mathbf{u}|^2 + \mathbf{u}\cdot\mathbf{R}(\mathbf{x}) - gz\Big) - p(D-1), \tag{18.42}$$

where $\rho_{tot} = \rho_0 D(1+b)$ is the total mass density, $\rho_0$ is a dimensional constant and $\mathbf{R}$ is a given function of $\mathbf{x}$. This Lagrangian produces the following variations at fixed $\mathbf{x}$ and $t$

$$\begin{aligned}
\frac{1}{D}\frac{\delta l}{\delta \mathbf{u}} &= \rho_0(1+b)(\mathbf{u}+\mathbf{R}),\\
\frac{\delta l}{\delta b} &= \rho_0 D\Big(\frac{1}{2}|\mathbf{u}|^2 + \mathbf{u}\cdot\mathbf{R} - gz\Big),\\
\frac{\delta l}{\delta D} &= \rho_0(1+b)\Big(\frac{1}{2}|\mathbf{u}|^2 + \mathbf{u}\cdot\mathbf{R} - gz\Big) - p,\\
\frac{\delta l}{\delta p} &= -\,(D-1).
\end{aligned}$$

Hence, from the EP equation (18.35), we find the dimensional form (18.14) of the motion equation for an Euler fluid in three dimensions,

$$\frac{d\mathbf{u}}{dt} - \mathbf{u}\times\operatorname{curl}\mathbf{R} + g\hat{\mathbf{z}} + \frac{1}{\rho_0(1+b)}\nabla p = 0, \tag{18.43}$$

where $\operatorname{curl}\mathbf{R} = f(\mathbf{x})\hat{\mathbf{z}}$ is the Coriolis parameter as before. In writing the equation in this form, we have used the auxiliary equation for the advection of buoyancy,

$$\frac{\partial b}{\partial t} + \mathbf{u}\cdot\nabla b = 0,$$

from eqn (18.31).

## The Kelvin–Noether Theorem

From eqn (18.39), the Kelvin–Noether circulation theorem corresponding to the motion equation (18.43) for an ideal incompressible stratified fluid in three dimensions is,

$$\frac{d}{dt}\oint_{\gamma_t(\mathbf{u})}(\mathbf{u}+\mathbf{R})\cdot d\mathbf{x} = -\oint_{\gamma_t(\mathbf{u})}\frac{1}{\rho_0(1+b)}\nabla p\cdot d\mathbf{x}\,, \tag{18.44}$$

where the curve $\gamma_t(\mathbf{u})$ moves with the fluid velocity $\mathbf{u}$. By Stokes' theorem, the Euler equations generate circulation of $(\mathbf{u}+\mathbf{R})$ around $\gamma_t(\mathbf{u})$ whenever the gradients of bouyancy and pressure are not collinear. Using advection of buoyancy $b$, one finds conservation of potential vorticity $q_{\text{Eul}}$ on fluid parcels, see eqn (18.41),

$$\frac{\partial q_{\text{Eul}}}{\partial t} + \mathbf{u}\cdot\nabla q_{\text{Eul}} = 0, \quad \text{where} \quad q_{\text{Eul}} = \nabla b\cdot\operatorname{curl}(\mathbf{u}+\mathbf{R}). \tag{18.45}$$

The constraint $D = 1$ (volume preservation) is imposed by varying $p$ in Hamilton's principle, according to eqn (18.42). Incompressibility then follows from substituting $D = 1$ into the Lie-derivative relation (18.32) for $D$, which gives $\nabla \cdot \mathbf{u} = 0$. Upon taking the divergence of the motion equation (18.43) and requiring incompressibility to be preserved in time, one finds an elliptic equation for the pressure $p$ with a Neumann boundary condition obtained from the normal component of the motion equation (18.43) evaluated on the boundary.

## 18.6.1 Non-dimensional Euler–Boussinesq equations

### The Lagrangian

The Lagrangian (18.42) for the Euler fluid motion non-dimensionalizes in terms of the Rossby number, aspect ratio and Froude number, as follows:

$$l = \int d^3x \; D(1+b)\Big(\frac{\epsilon}{2}\mathbf{u}_3 \cdot \mathbf{v}_3 + \mathbf{u}\cdot\mathbf{R}(\mathbf{x}) - \frac{z}{\epsilon\mathcal{F}}\Big) - p(D-1). \qquad (18.46)$$

In this notation, the ***non-dimensional Euler fluid equations*** corresponding to the Lagrangian $l$ in eqn (18.46) are recovered from the Euler–Poincaré equation (18.35) with $\mathbf{u} \to \mathbf{u}_3$ and $\nabla \to \nabla_3$. Namely,

$$\epsilon\frac{d\mathbf{v}_3}{dt} - \mathbf{u}\times \operatorname{curl}\mathbf{R} + \frac{1}{\epsilon\mathcal{F}}\hat{\mathbf{z}} + \frac{1}{(1+b)}\nabla_3\, p = 0. \qquad (18.47)$$

Clearly, the leading-order balances in these equations are hydrostatic in the vertical and geostrophic in the horizontal direction.

In this notation, the Kelvin–Noether circulation theorem (18.39) for the Euler fluid becomes

$$\frac{d}{dt}\oint_{\gamma_t(\mathbf{u}_3)} (\epsilon\mathbf{v}_3 + \mathbf{R})\cdot d\mathbf{x}_3 = -\oint_{\gamma_t(\mathbf{u}_3)} \frac{1}{\rho_0(1+b)}\nabla_3 p \cdot d\mathbf{x}_3\,. \qquad (18.48)$$

Likewise, conservation of non-dimensional potential vorticity $q_{\mathrm{Eul}}$ on fluid parcels is given by eqn (18.22) as,

$$\frac{\partial q_E}{\partial t} + \mathbf{u}_3\cdot\nabla_3 q_E = 0, \quad \text{where} \quad q_E = \nabla_3 b \cdot \nabla_3 \times (\epsilon\mathbf{v}_3 + \mathbf{R}). \qquad (18.49)$$

### Hamilton's principle asymptotics

For sufficiently small buoyancy, $b = o(\epsilon)$, we define

$$p' = p + \frac{z}{\epsilon\mathcal{F}} \quad \text{and} \quad b' = \frac{b}{\epsilon\mathcal{F}},$$

and expand the Lagrangian (18.46) in powers of $\epsilon$ as

$$l_{\mathrm{EB}} = \int \mathrm{d}t \int D\Big(\frac{\epsilon}{2}\mathbf{u}_3 \cdot \mathbf{v}_3 + \mathbf{u} \cdot \mathbf{R}(\mathbf{x}) - b'z\Big) \\ - p'(D-1)d^3x + o(\epsilon). \tag{18.50}$$

Upon dropping the order $o(\epsilon)$ term in the Lagrangian $l_{\mathrm{EB}}$ the corresponding Euler–Poincaré equation gives the ***Euler–Boussinesq equation*** for fluid motion in three dimensions, namely,

$$\epsilon \frac{\mathrm{d}\mathbf{v}_3}{\mathrm{d}t} - \mathbf{u} \times \operatorname{curl}\mathbf{R} + b'\hat{\mathbf{z}} + \nabla_3\, p' = 0, \tag{18.51}$$

or, in horizontal and vertical components, with $\operatorname{curl}\mathbf{R} = f(\mathbf{x})\hat{z}$,

$$\epsilon \frac{\mathrm{d}\mathbf{u}}{\mathrm{d}t} + f\hat{z} \times \mathbf{u} + \nabla p' = 0, \tag{18.52}$$

$$\epsilon\, \sigma^2 \frac{\mathrm{d}w}{\mathrm{d}t} + b' + \frac{\partial p'}{\partial z} = 0, \tag{18.53}$$

where

$$\frac{\mathrm{d}b'}{\mathrm{d}t} = 0 \quad \text{and} \quad \nabla_3 \cdot \mathbf{u}_3 = \nabla \cdot \mathbf{u} + \frac{\partial w}{\partial z} = 0.$$

Even for order $O(\epsilon)$ buoyancy, the leading-order balances are still hydrostatic in the vertical, and geostrophic in the horizontal.

### The Kelvin–Noether theorem

The Kelvin–Noether circulation theorem (18.39) for the Euler–Boussinesq equation (18.51) for an ideal incompressible stratified fluid in three dimensions is,

$$\frac{\mathrm{d}}{\mathrm{d}t} \oint_{\gamma_t(\mathbf{u}_3)} (\epsilon \mathbf{v}_3 + \mathbf{R}) \cdot \mathrm{d}\mathbf{x} = - \oint_{\gamma_t(\mathbf{u}_3)} b'\mathrm{d}z\,, \tag{18.54}$$

where the curve $\gamma_t(\mathbf{u}_3)$ moves with the fluid velocity $\mathbf{u}_3$. (The two Kelvin theorems in eqns (18.48) and (18.54) differ in their right-hand sides.) By Stokes' theorem, the Euler–Boussinesq equations generate circulation of $\epsilon\mathbf{v}_3 + \mathbf{R}$ around $\gamma_t(\mathbf{u}_3)$ whenever the gradient of bouyancy is not vertical. Conservation of potential vorticity $q_{\mathrm{EB}}$ on fluid parcels for the Euler–Boussinesq equations is given by

$$\frac{\partial q_{\mathrm{EB}}}{\partial t} + \mathbf{u}_3 \cdot \nabla_3\, q_{\mathrm{EB}} = 0, \tag{18.55}$$

where

$$q_{\mathrm{EB}} = \nabla_3 b' \cdot \nabla_3 \times (\epsilon \mathbf{v}_3 + \mathbf{R}). \tag{18.56}$$

## 18.6.2 Primitive equations

### The Lagrangian

The primitive equations (PE) arise from the Euler–Boussinesq equations, upon imposing the approximation of hydrostatic pressure balance. Setting the aspect ratio parameter $\sigma$ to zero in the Lagrangian $l_{\mathrm{EB}}$ in eqn (18.50), provides the Lagrangian for the non-dimensional primitive equations (PE),

$$\begin{aligned} l_{\mathrm{PE}} = \int \mathrm{d}t \int \mathrm{d}^3x \Big[ \frac{\epsilon}{2} D|\mathbf{u}|^2 + D\mathbf{u} \cdot \mathbf{R}(\mathbf{x}) \\ - Db'z - p'(D-1) \Big]. \end{aligned} \tag{18.57}$$

The Euler–Poincaré equations for $l_{\mathrm{PE}}$ now produce the PE; namely, eqns (18.52) and (18.53) with $\sigma = 0$,

$$\epsilon \frac{\mathrm{d}\mathbf{u}}{\mathrm{d}t} + f\hat{z} \times \mathbf{u} + \nabla p' = 0, \quad b' + \frac{\partial p'}{\partial z} = 0, \tag{18.58}$$

where

$$\frac{\mathrm{d}b'}{\mathrm{d}t} = 0 \quad \text{and} \quad \nabla_3 \cdot \mathbf{u}_3 = \nabla \cdot \mathbf{u} + \frac{\partial w}{\partial z} = 0.$$

Thus, from the viewpoint of Hamilton's principle, imposition of hydrostatic balance corresponds to ignoring the kinetic energy of vertical motion by setting $\sigma = 0$ in the non-dimensional EB Lagrangian (18.50). The hydrostatic pressure in the PE may be interpreted as the weight of water above a given point in the fluid.

### The Kelvin–Noether theorem

The Kelvin–Noether circulation theorem for the primitive equations is obtained from eqn (18.54) for the Euler–Boussinesq equations simply by setting $\sigma = 0$. Namely,

$$\frac{\mathrm{d}}{\mathrm{d}t} \oint_{\gamma_t(\mathbf{u}_3)} (\epsilon \mathbf{u} + \mathbf{R}) \cdot \mathrm{d}\mathbf{x} = - \oint_{\gamma_t(\mathbf{u}_3)} b' \mathrm{d}z \,, \tag{18.59}$$

where the curve $\gamma_t(\mathbf{u}_3)$ moves with the fluid velocity $\mathbf{u}_3$. By Stokes' theorem, the primitive equations generate circulation of $\epsilon\mathbf{u}+\mathbf{R}$ around $\gamma_t(\mathbf{u}_3)$ whenever the gradient of bouyancy is not vertical. The conservation of potential vorticity on fluid parcels for the primitive equations is given by, see eqn (18.41),

$$\frac{\partial q_{\text{PE}}}{\partial t} + \mathbf{u}_3 \cdot \nabla_3\, q_{\text{PE}} = 0, \tag{18.60}$$

where

$$q_{\text{PE}} = \nabla_3 b' \cdot \nabla_3 \times (\epsilon\mathbf{u} + \mathbf{R}). \tag{18.61}$$

**Remark.** In the limit, $\epsilon \to 0$, Hamilton's principle for either $l_{\text{EB}}$, or $l_{\text{PE}}$ gives,

$$f\hat{z} \times \mathbf{u} + b'\hat{z} + \nabla_3\, p' = 0, \tag{18.62}$$

which encodes the leading-order hydrostatic and geostrophic equilibrium balances. These balances form the basis for further approximations for flows are nearly geostrophic and hydrostatic.

### 18.6.3 Hamiltonian balance equations

#### Balanced fluid motions

A fluid motion equation is said to be ***balanced***, if specification of the fluid's stratified buoyancy and divergenceless velocity determines its pressure through the solution of an equation that does not contain partial time derivatives among its highest derivatives. This definition of balance makes pressure a diagnostic variable (as opposed to the dynamic, or prognostic variables such as the horizontal velocity components). The Euler equations (18.47) and the Euler–Boussinesq equations (18.51) for the incompressible motion of a rotating continuously stratified fluid are balanced in this sense, because the pressure in these cases is determined diagnostically from the buoyancy and velocity of the fluid by solving a Neumann problem. However, the hydrostatic approximation of this motion by the primitive equations (PE) is not balanced, because the Poisson equation for the pressure in PE involves the time-derivative of the horizontal velocity divergence, which alters the character of the Euler system from which PE is derived and may lead to rapid time dependence, as discussed in [BHKW90]. Balanced approximations that eliminate this potentially rapid time dependence have been sought and found, usually by using asymptotic expansions of the solutions of the PE in powers of the small Rossby number, $\epsilon \ll 1$, after decomposing the horizontal velocity $\mathbf{u}$ into order $O(1)$ rotational and order $O(\epsilon)$ divergent components, as

$$\mathbf{u} = \hat{z} \times \nabla\psi + \epsilon\nabla\chi, \tag{18.63}$$

where $\psi$ and $\chi$ are the stream function and velocity potential, respectively, for the horizontal motion. (This is just the Helmholtz decomposition with relative weight $\epsilon$.)

Balance equations (BE) are reviewed in [MG80]. Succeeding investigations have concerned the well-posedness and other features of various BE models describing continuously stratified oceanic and atmospheric motions. For example, consistent initial boundary value problems and regimes of validity for BE are determined in Gent and McWilliams [GM83a, GM83b]. In subsequent papers by these authors and their collaborators, balanced models in isentropic coordinates were derived, methods for the numerical solution of BE were developed, and the applications of BE to problems of vortex motion on a $\beta$-plane and wind-driven ocean circulation were discussed. In studies of continuously stratified incompressible fluids, solutions of balance equations that retain terms of order $O(1)$ and order $O(\epsilon)$ in a Rossby number expansion of the PE solutions have been found to compare remarkably well with numerical simulations of the PE; see Allen, Barth, and Newberger [ABN90a, ABN90b] and [BAN90].

## Conservation of energy and potential vorticity

One recurring issue in the early literature was that, when truncated at order $O(\epsilon)$ in the Rossby number expansion, the BE for continuously stratified fluids conserved energy [Lor60], but did not conserve potential vorticity on fluid parcels. A set of BE for continuously stratified fluids that retained additional terms of order $O(\epsilon^2)$ and *did conserve potential vorticity* on fluid parcels was found in [AH96]. This set of equations was derived by using the $\epsilon$-weighted Helmholtz decomposition for $\mathbf{u}$ and expanding Hamilton's principle (HP) for the PE in powers of the Rossby number, $\epsilon \ll 1$. This expansion was truncated at order $O(\epsilon)$, then all terms were retained that result from taking variations. As we have seen, the asymptotic expansion of HP for the Euler–Boussinesq (EB) equations that govern rotating stratified incompressible inviscid fluid flow has two small dimensionless parameters: the aspect ratio of the shallow domain, $\sigma$, and the Rossby number, $\epsilon$. Setting $\sigma$ equal to zero in this expansion yields HP for PE. Setting $\epsilon$ also equal to zero yields HP for equilibrium solutions in both geostrophic and hydrostatic balance. Setting $\sigma = 0$, substituting the $\epsilon$-weighted Helmholtz decomposition for $\mathbf{u}$ in eqn (18.63) and truncating the resulting asymptotic expansion in $\epsilon$ of the HP for the EB equations, yields HP for the set of nearly geostrophic Hamiltonian balance equations (HBE) in [AH96]. The remainder of the chapter will be devoted to explaining this sequence of approximations of Hamilton's principle for applications in GFD.

### The Lagrangian

The Lagrangian for the HBE model is given in [AH96] as,

$$S_{\text{HBE}} = \int \mathrm{d}t \int \mathrm{d}^3x \Big[ D\mathbf{u}\cdot\mathbf{R}(\mathbf{x}) - Dbz - p(D-1) + \epsilon\frac{D}{2}|\mathbf{u}-\epsilon\mathbf{u}_D|^2 \Big], \tag{18.64}$$

where the horizontal fluid velocity is taken in balance equation form as

$$\mathbf{u} = \mathbf{u}_R + \epsilon\mathbf{u}_D = \hat{z}\times\nabla\psi + \epsilon\nabla\chi .$$

In comparison with eqn (18.57) for the PE action, the action $S_{\text{HBE}}$ in eqn (18.64) contains only the non-dimensional contribution to the kinetic energy of the rotational part of the fluid velocity, $\mathbf{u}_R$. The corresponding Euler–Poincaré equations give the dynamics of the HBE model

$$\begin{aligned}
\epsilon\frac{\mathrm{d}}{\mathrm{d}t}\mathbf{u}_R + \epsilon^2 u_{Rj}\nabla u_D^j + f\hat{z}\times\mathbf{u} + \nabla p &= 0,\\
b + \frac{\partial p}{\partial z} + \epsilon^2\mathbf{u}_R\cdot\frac{\partial \mathbf{u}_D}{\partial z} &= 0,\\
\text{with}\quad \frac{\mathrm{d}b}{\mathrm{d}t} = \frac{\partial}{\partial t}b + \mathbf{u}\cdot\nabla b + \epsilon\, w\frac{\partial b}{\partial z} &= 0,\\
\text{and}\quad \nabla\cdot\mathbf{u} + \epsilon\frac{\partial w}{\partial z} &= 0.
\end{aligned} \tag{18.65}$$

Here, the notation is the same as for the PE, except that $w \to \epsilon\, w$ for HBE.

Dropping all terms of order $O(\epsilon^2)$ from the HBE model equations (18.65) recovers the balance equations (BE) introduced in [GM83a, GM83b]. Retaining these order $O(\epsilon^2)$ terms restores the conservation laws due to symmetries of HP at the truncation order $O(\epsilon)$. As explained in [AH96], the resulting HBE model has the same order $O(\epsilon)$ accuracy as the BE, since not *all* of the possible order $O(\epsilon^2)$ terms are retained. Since the HBE model shares the same conservation laws and Euler–Poincaré structure as EB and PE, and differs from them only at order $O(\epsilon^2)$, one may hope for improved accuracy of HBE over that expected for the BE model, which does not share these conservation laws.

### The Kelvin–Noether theorem

The HBE model (18.65) possesses the following Kelvin–Noether circulation theorem,

$$\frac{\mathrm{d}}{\mathrm{d}t}\oint_{\gamma_t(\mathbf{u}_3)}(\mathbf{R}+\epsilon\mathbf{u}_R)\cdot\mathrm{d}\mathbf{x}_3=-\oint_{\gamma_t(\mathbf{u}_3)}b\,\mathrm{d}z, \tag{18.66}$$

for any closed curve $\gamma_t(\mathbf{u}_3)$ moving with the fluid velocity $\mathbf{u}_3$. We compare this result with the Kelvin–Noether circulation theorem for PE in equation (18.59), rewritten as

$$\begin{aligned}\frac{\mathrm{d}}{\mathrm{d}t}\oint_{\gamma_t(\mathbf{u}_3)}\underbrace{(\mathbf{R}+\epsilon\mathbf{u})}_{\text{PE}}\cdot\mathrm{d}\mathbf{x}_3&=\frac{\mathrm{d}}{\mathrm{d}t}\oint_{\gamma_t(\mathbf{u}_3)}\underbrace{(\mathbf{R}+\epsilon\mathbf{u}_R}_{\text{HBE}}+\underbrace{\epsilon^2\mathbf{u}_D)\cdot\mathrm{d}\mathbf{x}_3}_{\text{ZERO}}\\&=-\oint_{\gamma_t(\mathbf{u}_3)}b\,\mathrm{d}z.\end{aligned} \tag{18.67}$$

The circulation of $\mathbf{u}_D$ vanishes

$$\oint\mathbf{u}_D\cdot\mathrm{d}\mathbf{x}_3=\oint\mathrm{d}\chi=0.$$

Therefore, the $\epsilon^2$ term vanishes, and so the HBE circulation integral differs from that of PE only through the differences between the two theories in their solutions for the buoyancy.

The conservation of potential vorticity on fluid parcels for the HBE model is given by, see eqn (18.41),

$$\frac{\partial q_{\text{HBE}}}{\partial t}+\mathbf{u}_3\cdot\nabla_3\,q_{\text{HBE}}=0, \tag{18.68}$$

where

$$q_{\text{HBE}}=\nabla_3\,b\cdot\nabla_3\times(\epsilon\mathbf{u}_R+\mathbf{R}). \tag{18.69}$$

Combining this with advection of $b$ and the tangential boundary coundition on $\mathbf{u}_3$ yields an infinity of conserved quantities,

$$C_\Phi=\int\mathrm{d}^3x\;\Phi(q_{\text{HBE}},b), \tag{18.70}$$

for any function $\Phi$. These are the Casimir functions for the Lie–Poisson Hamiltonian formulation of the HBE given in [Hol96].

## HBE discussion

By their construction as Euler–Poincaré equations from a Lagrangian that possesses the classic fluid symmetries, the HBE conserve integrated energy and conserve potential vorticity on fluid parcels. Their Lie–Poisson Hamiltonian structure endows the HBE with the same type of self-consistency that the PE possess (for the same Hamiltonian reason). After all, the conservation laws in both HBE and PE are not accidental. They correspond to symmetries of the Hamiltonian or Lagrangian for the fluid motion under continuous group transformations, in accordance with Noether's theorem. In particular, energy is conserved because the Hamiltonian in both theories does not depend on time explicitly, and potential vorticity is conserved on fluid parcels because the corresponding Hamiltonian or Lagrangian is right invariant under the infinite set of transformations that relabel the fluid parcels without changing the Eulerian velocity and buoyancy. See, e.g., [Sal85, Sal88] for reviews of these ideas in the GFD context, as well as [HMR98a, HMR98b] and Chapter 17 for the general context for such results.

The vector fields that generate these relabelling transformations turn out to be the *steady flows* of the HBE and PE models. By definition, these steady flows leave invariant the Eulerian velocity and buoyancy as they move the Lagrangian fluid parcels along the flow lines. Hence, as a direct consequence of their shared Hamiltonian structure, the steady flows of both HBE and PE are relative equilibria. That is, steady HBE and PE flows are critical points of a sum of conserved quantities, including the (constrained) Hamiltonian. This shared critical-point property enables one, for example, to use the Lyapunov method to investigate the stability of relative equilibrium solutions of HBE and PE. According to the Lyapunov method, convexity of the constrained Hamiltonian at its critical point (the relative equilibrium) is sufficient to provide a norm that limits the departure of the solution from equilibrium under perturbations. See, e.g., [AHMR86] for applications of this method to the Euler equations for incompressible fluid dynamics and [HMRW85] for a range of other applications in fluid and plasma theories.

Thus, the HBE arise as Euler–Poincaré equations and possess the same Lie–Poisson Hamiltonian structure as EB and PE, and differ in their Hamiltonian and conservation laws by small terms of order $O(\epsilon^2)$. Moreover, the HBE conservation laws are fundamentally of the same nature as those of the EB equations and the PE from which they descend. These conserved quantities – particularly the quadratic conserved quantities – may eventually be useful measures of the deviations of the HBE solutions from EB and PE solutions under time evolution starting from identical initial conditions.

Three-dimensional versions of the QG and SG equations also exist, and recently a continuously stratified $L_1$ model was reviewed in [AHN02] from the present viewpoint of Hamilton's principle asymptotics and the Euler–Poincaré theory. For the suite of idealized, oceanographic, moderate Rossby number, mesoscale flow test problems in [AHN02], this continuously stratified $L_1$ model produces generally accurate approximate solutions. These solutions are not quite as accurate as those from the HBE or BE models, but are substantially more accurate than those from three dimensional SG or QG.

Formulae showing the asymptotic expansion relationships among the Lagrangians for various GFD models are summarized in Tables 18.1 and 18.2.

**Exercise 18.4**
Pick out as many as you can find of the main sequence of GFD models discussed in this chapter that appear in Figure 18.3 from [ABN90a]. The models in the main sequence conserve both energy and circulation.

**Table 18.1** GFD models arising from Hamilton's principle.

$$l_{\text{Euler}} = \int \Big[ D(1+b)\Big( \underbrace{\mathbf{R}(\mathbf{x})\cdot\mathbf{u}}_{\text{Rotation}} + \underbrace{\frac{\epsilon}{2}|\mathbf{u}|^2 + \frac{\epsilon}{2}\sigma^2 w^2}_{\text{Kinetic Energy}}\Big)$$

$$- \underbrace{D(1+b)\Big(\frac{z}{\epsilon\mathcal{F}}\Big)}_{\text{Potential Energy}} - \underbrace{p(D-1)}_{\text{Constraint}} \Big] d^3x$$

- $l_{\text{Euler}} \to l_{\text{EB}}$, for small buoyancy, $b = O(\epsilon)$.
- $l_{\text{EB}} \to l_{\text{PE}}$, for small aspect ratio, $\sigma^2 = O(\epsilon)$.
- $l_{\text{PE}} \to l_{\text{HBE}}$, for horizontal velocity decomposition,

$$\mathbf{u} = \hat{z}\times\nabla\psi + \epsilon\nabla\chi = \mathbf{u}_R + \epsilon\mathbf{u}_D,$$

  and $|\mathbf{u}|^2 \to |\mathbf{u}_R|^2$ in $l_{\text{PE}}$.
- $l_{\text{HBE}} \to l_1$, for horizontal velocity,

$$\mathbf{u} = \mathbf{u}_1 = \hat{z}\times\nabla\tilde{\phi},$$

  where

$$\tilde{\phi}(\mathbf{x}_3, t) = \phi_S(x, y, t) + \int_z^0 dz'\, b,$$

  i.e. $\partial\tilde{\phi}/\partial z = -b$ and dropping terms of order $O(\epsilon^2)$ in $l_{\text{HBE}}$.
- $l_1 \to l_{QG}$, on dropping terms of order $O(\epsilon^2)$ in the Euler–Poincaré equations for $l_1$.

**Table 18.2** Non-dimensional Lagrangians for GFD models.

$$l_{\text{Euler}} = \int \Big[ D(1+b)\Big(\mathbf{R}(\mathbf{x})\cdot\mathbf{u} + \frac{\epsilon}{2}|\mathbf{u}|^2 + \frac{\epsilon}{2}\sigma^2 w^2 - \frac{z}{\epsilon\mathcal{F}}\Big) - p(D-1)\Big] \mathrm{d}^3x$$

$$l_{\text{EB}} = \int \Big[ D\Big(\mathbf{R}\cdot\mathbf{u} + \frac{\epsilon}{2}|\mathbf{u}|^2 + \frac{\epsilon}{2}\sigma^2 w^2 - bz\Big) - p(D-1)\Big] \mathrm{d}^3x$$

$$l_{\text{PE}} = \int \Big[ D\Big(\mathbf{R}\cdot\mathbf{u} + \frac{\epsilon}{2}|\mathbf{u}|^2 - bz\Big) - p(D-1)\Big] \mathrm{d}^3x$$

$$l_{\text{HBE}} = \int \Big[ D\Big(\mathbf{R}\cdot\mathbf{u} + \frac{\epsilon}{2}|\mathbf{u} - \epsilon\mathbf{u}_D|^2 - bz\Big) - p(D-1)\Big] \mathrm{d}^3x$$

$$l_1 = \int \Big[ D\Big((\mathbf{R} + \epsilon\mathbf{u}_1)\cdot\mathbf{u} - \frac{\epsilon}{2}|\mathbf{u}_1|^2 - bz\Big) - p(D-1)\Big] \mathrm{d}^3x$$

$$l_{\text{QG}} = \int_{\mathcal{D}}\int_{z_0}^{z_1} \Big[ D\Big(\mathbf{R}\cdot\mathbf{u} + \frac{\epsilon}{2}\mathbf{u}\cdot(1 - \mathcal{L}(z)\Delta^{-1})\mathbf{u}\Big) - p(D-1)\Big] \mathrm{d}z\,\mathrm{d}^2x,$$

where

$$\mathcal{L}(z) = \Big(\frac{\partial}{\partial z} + B\Big)\frac{1}{\mathcal{S}(z)}\Big(\frac{\partial}{\partial z} - B\Big) - \mathcal{F}$$

and $B = 0$ for standard QG.

## 18.7 Well-posedness, ill-posedness, discretization and regularization

An important issue for any approximate fluid model and particularly for the reduced fluid models of GFD concerns its ***well-posedness***. This is a term introduced by Hadamard [Had02], who proposed that any mathematical model of physical phenomena should have the following properties:

1. A solution exists;
2. The solution is unique;
3. The solution depends continuously on the initial data and boundary conditions, in some reasonable topology.

Examples of well-posed problems include the Dirichlet problem for Laplace's equation and the initial-value problem for the heat equation. These two problems model the familiar physical processes, respectively, of the spatial distribution of electrostatic potential and the diffusion of dissolved chemicals in water. In contrast, the backwards heat equation (that is, deducing a previous distribution of temperature from final data) is not well-posed, because the solution is highly sensitive to changes in the final data. Problems that are not well-posed in the sense of Hadamard are termed ***ill-posed***. An example of an ill-posed fluids problem is an inviscid shear flow with a discontinuous velocity distribution at the interface between two fluid layers with different velocities. (The interface is sometimes called a ***vortex sheet***.) This situation produces the ***Kelvin-Helmholtz instability*** in which the interface rolls up tightly into whorls, and whose unstable growth rate increases with wave number according to Euler's fluid equations; so that the smaller the wavelength of the disturbance, the faster it grows and distorts the interface.

Fluid-dynamics problems must often be represented at discrete points rather than continuously in space and time in order to obtain a numerical solution. Such numerical ***discretizations*** of continuous problems may suffer from numerical instability when solved with finite precision, or with errors in the data. Even if a continuous problem is well-posed, it may still be numerically ***ill-conditioned*** for a given choice of numerical algorithm, meaning that a small error in the initial data can result in much larger errors in the results of numerical simulations.

Although extremely important in applications, the design of optimal techniques for stabilizing numerical algorithms for simulations of fluid equations is beyond the scope of the present text. We only mention that the Euler–Poincaré formulation of fluids may be quite useful in guiding the design of numerical methods that preserve the fundamental mathematical structures of fluid dynamics. This is a fruitful and active field of ongoing study [CH08].

If a continuous problem is well-posed, then it has a chance of being solved on a computer using a stable algorithm for its discrete representation. If it is not well-posed, then it must be reformulated for numerical treatment. Typically this reformulation involves introducing additional assumptions that are designed to guarantee smoothness of the solution. This process is known as ***regularization***.

The famous example of regularization for fluids is Leray's regularization of the Navier–Stokes equations [Ler34], obtained by smoothing the transport velocity relative to the circulation velocity. Various choices among the GFD balance equations may or may not be well-posed, depending on which model is chosen and these, in turn, may or may not be numerically well-conditioned, depending on which algorithm is used. The issue of regularization of models to achieve well-posedness is extremely important for the evaluation of fluid models and their application in problems such as climate prediction. For example, regularization of fluid equations may also be achieved by the introduction of viscous dissipation of energy, as happens physically for the Kelvin–Helmholtz instability. However, regularization by enhanced energy dissipation may produce sluggish fluid response with unrealistically low variability that may under-predict the effects of variations of the driving parameters for the flow. Again, the Euler–Poincaré formulation of fluid dynamics may be quite useful in guiding the design of regularization methods that preserve fluid structure and thus produce solutions whose variability is physically realistic, not over damped because of unphysical added dissipation. The approach of using the Euler–Poincaré formulation for regularization has recently made some advances, for example, in turbulence modelling [FHT01, FHT02]. However, this is another ongoing field of study whose discussion we have decided to place outside the scope of the present text.

# Bibliography


[ABN90a] J. S. Allen, J. A. Barth, and P. A. Newberger. On intermediate models for barotropic continental shelf and slope flow fields. Part I: Formulation and comparison of exact solutions. *J. of Phys. Oceanog.*, 20:1017–1042, 1990.

[ABN90b] J. S. Allen, J. A. Barth, and P. A. Newberger. On intermediate models for barotropic continental shelf and slope flow fields. Part III: Comparison of numerical model solutions in periodic channels. *J. of Phys. Oceanog.*, 20:1949–1973, 1990.

[AH96] J. S. Allen and D. D. Holm. Extended-geostrophic Hamiltonian models for rotating shallow water motion. *Physica D*, 98:229–248, 1996.

[AHMR86] H. D. I. Abarbanel, D. D. Holm, J. E. Marsden, and T. S. Ratiu. Nonlinear stability analysis of stratified ideal fluid equilibria. *Phil. Trans. Roy. Soc. (London) A*, 318:349–409, 1986.

[AHN02] J. S. Allen, D. D. Holm, and P. A. Newberger. Toward an extended-geostrophic Euler–Poincaré model for mesoscale oceanographic flow. In J. Norbury and I. Roulstone, editors, *Large-Scale Atmosphere-Ocean Dynamics 1: Analytical Methods and Numerical Models*, pages 101–125. Cambridge University Press, Cambridge, 2002.

[AK98] V. I. Arnold and B. A. Khesin. *Topological Methods in Hydrodynamics*. Springer, New York, 1998.

[AM78] R. Abraham and J. E. Marsden. *Foundations of Mechanics*. Addison-Wesley, New York, 2nd edition, 1978.

[AMR88] R. Abraham, J. E. Marsden, and T. S. Ratiu. *Manifolds, Tensor Analysis, and Applications*, volume 75 of *Applied Mathematical Sciences*. Springer, New York, 2nd edition, 1988.

[Arn66] V. I. Arnold. Sur la géometrie differentialle des groupes de Lie de dimiension infinie et ses applications à l'hydrodynamique des fluids parfaits. *Ann. Inst. Fourier, Grenoble*, 16:319–361, 1966.

[Arn78] V. I. Arnold. *Mathematical Methods of Classical Mechanics*, volume 60 of *Graduate Texts in Mathematics*. Springer, New York, 1978.

[AS06] M. J. Ablowitz and H. Segur. *Solitons and the Inverse Scattering Transform*. Cambridge University Press, New York, 2006.

[Bak02] A. Baker. *Matrix Groups: An Introduction to Lie Group Theory*. Undergraduate Mathematics Series. Springer, New York, 2002.

[BAN90] J. A. Barth, J. S. Allen, and P. A. Newberger. On intermediate models for barotropic continental shelf and slope flow fields. Part II: Comparison of numerical model solutions in doubly periodic domains. *J. of Phys. Oceanog.*, 20:1044–1076, 1990.

[BG88] M. Berger and B. Gostiaux. *Differential Geometry: Manifolds, Curves and Surfaces*. Springer, New York, 1988.

[BHKW90] G. L. Browning, W. R. Holland, H. O. Kreiss, and S. J. Worley. An accurate hyperbolic system for approximately hydrostatic and incompressible oceanographic flows. *Dyn. Atm. Oceans*, 14:303–332, 1990.

[BKMR96] A. M. Bloch, P. S. Krishnaprasad, J. E. Marsden, and T. S. Ratiu. The Euler-Poincaré equations and double bracket dissipation. *Comm. Math. Phys.*, 175:1–42, 1996.

[BL05] F. Bullo and A. D. Lewis. *Geometric Control of Mechanical Systems*. Springer, New York, 2005.

[Blo03] A. M. Bloch. *Nonholonomic Mechanics and Control*, volume 24 of *Interdisciplinary Applied Mathematics*. Springer-Verlag, New York, 2003. With the collaboration of J. Baillieul, P. Crouch and J. Marsden, With scientific input from P. S. Krishnaprasad, R. M. Murray and D. Zenkov, Systems and Control.

[Boo91] F. L. Bookstein. *Morphometric Tools for Landmark Data; geometry and biology*. Cambridge University Press, Cambridge, 1991.

[CB97] R. H. Cushman and L. M. Bates. *Global Aspects of Integrable Systems*. Birkhäuser, New York, 1997.

[CH93] R. Camassa and D. D. Holm. An integrable shallow water equation with peaked solitons. *Phys. Rev. Lett.*, 71:1661–64, 1993.

[CH08] C. J. Cotter and D. D. Holm. Discrete momentum maps for lattice EPDiff. In R. Temam and J. Tribbia, editors, *Computational Methods for the Atmosphere and the Ocean*, volume Special Volume of *Handbook of Numerical Analysis*, pages 247–278. North-Holland, Amsterdam, 2008.

[Cha87] S. Chandrasekhar. *Ellipsoidal Figures of Equilibrium*. Dover, New York, 1987.

[CLZ05] M. Chen, S. Liu, and Y. Zhang. A two-component generalization of the Camassa-Holm equation and its solutions. *Lett. Math. Phys.*, 75:1–15, 2005.

[CM87] H. Cendra and J. E. Marsden. Lin constraints, Clebsch potentials and variational principles. *Physica D*, 27:63–89, 1987.

[Cot05] C. J. Cotter. A general approach for producing Hamiltonian numerical schemes for fluid equations. *Arxiv preprint math.NA/0501468*, 2005.

[CP89] M. J. P. Cullen and R. J. Purser. Properties of the Lagrangian semigeostrophic equations. *J. Atmos. Sci.*, 46:1477–1497, 1989.

[Cul07] M. J. P. Cullen. Modelling atmospheric flows. *Acta Numerica*, 16:67–154, 2007.

[DGH04] H. Dullin, G. Gottwald, and D. D. Holm. On asymptotically equivalent shallow water wave equations. *Physica D*, 190:1–14, 2004.

[DGM98] P. Dupuis, U. Grenander, and M. Miller. Variational problems on flows of diffeomorphisms for image matching. *Quarterly of Applied Math.*, 56:587–600, 1998.

[DHH03] A. Degasperis, D. D. Holm, and A. N. W. Hone. Integrable and nonintegrable equations with peakons. In M. J. Ablowitz, M. Boiti, F. Pempinelli, and B. Prinari, editors, *Nonlinear Physics: Theory and Experiment (Gallipoli 2002) Vol II*, pages 37–43. World Scientific, Singapore, 2003.

[DK04] J. J. Duistermaat and J. A. C. Kolk. *Lie Groups*. Universitext, Amsterdam, 2004.

[Eli49] A. Eliassen. The quasi-static equations of motion with pressure as independent variable. *Geofysiske Publikasjoner*, 17:327–342, 1949.

[EM70] D. G. Ebin and J. E. Marsden. Groups of diffeomorphisms and the motion of an incompressible fluid. *Ann. of Math.*, 92:102–163, 1970.

[Fal06] G. Falqui. On a Camassa-Holm type equation with two dependent variables. *J. Phys. A: Math. and Theor.*, 39:327–342, 2006.

[FGR02] J. Frank, G. Gottwald, and S. Reich. A Hamiltonian particle-mesh method for the rotating shallow-water equations. In M. Griebel and M. A. Schweitzer, editors, *Meshfree Methods for Partial Differential Equations*, volume 26 of *Lecture Notes in Computational Science and Engineering*, pages 131–142. Springer, 2002.

[FH01] O. B. Fringer and D. D. Holm. Integrable vs nonintegrable geodesic soliton behavior. *Physica D*, 150:237–263, 2001.

[FHT01] C. Foias, D. D. Holm, and E. S. Titi. The Navier–Stokes–alpha model of fluid turbulence. *Physica D*, 152–153:505–519, 2001.

[FHT02] C. Foias, D. D. Holm, and E. S. Titi. The three dimensional viscous Camassa–Holm equations, and their relation to the Navier–Stokes equations and turbulence theory. *J. Dyn. Diff. Eq.*, 14:1–35, 2002.

[Fuc96] B. Fuchssteiner. Some tricks from the symmetry-toolbox for nonlinear equations: Generalizations of the Camassa-Holm equation. *Physica D*, 95:229–243, 1996.

[GD79] I. M. Gelfand and I. Ya. R. Dorfman. Hamiltonian operators and algebraic structures associated with them. *Funct. Anal. Appl.*, 13:248–254, 1979.

[GM83a] P. R. Gent and J. C. McWilliams. Consistent balanced models in bounded and periodic domains. *Dyn. Atmos. Oceans*, 7:67–93, 1983.

[GM83b] P. R. Gent and J. C. McWilliams. Regimes of validity for balanced models. *Dyn. Atmos. Oceans*, 7:167–183, 1983.

[Gol59] H. Goldstein. *Classical Mechanics*. Addison-Wesley Publishing Company, New York, 1959.

[Gre81] U. Grenander. *Lectures in Pattern Theory*, volume 33 of *Applied Mathematical Sciences*. Springer, New York, 1981.

[GS84] V. Guillemin and S. Sternberg. *Symplectic Techniques in Physics*. Cambridge University Press, Cambridge, 1984.

[GTY04] J. Glaunes, A. Trouve, and L. Younes. Diffeomorphic matching of distributions: A new approach for unlabelled point-sets and sub-manifolds matching. In *IEEE Computer Society Conference on Computer Vision and Pattern Recognition*, volume 2, pages 712–718. IEEE, 2004.

[Gun03] M. D. Gunzburger. *Perspectives in Flow Control and Optimization*. Advances in design and control. SIAM, Philadelphia, 2003.

[Had02] J. Hadamard. Sur les problèmes aux drives partielles et leur signification physique. *Princeton University Bulletin*, pages 49–52, 1902.

[HK83] D. D. Holm and B. A. Kupershmidt. Poisson brackets and Clebsch representations for magnetohydrodynamics, multifluid plasmas, and elasticity. *Physica D*, 6:347–363, 1983.

[HM04] D. D. Holm and J. E. Marsden. Momentum maps and measure valued solutions (peakons, filaments, and sheets) of the Euler–Poincaré equations for the diffeomorphism group. In J. E. Marsden and T. S. Ratiu, editors, *The Breadth of Symplectic and Poisson Geometry*, Progr. Math., 232, pages 203–235. Birkhäuser, Boston, 2004.

[HMR98a] D. D. Holm, J. E. Marsden, and T. S. Ratiu. The Euler-Poincaré equations and semidirect products with applications to continuum theories. *Adv. Math.*, 137:1–81, 1998.

[HMR98b] D. D. Holm, J. E. Marsden, and T. S. Ratiu. Euler-Poincaré models of ideal fluids with nonlinear dispersion. *Phys. Rev. Lett.*, 80:4173–4177, 1998.

[HMR04] D. D. Holm, J. E. Marsden, and T. S. Ratiu. The Euler–Poincaré equations in geophysical fluid dynamics. In J. Norbury and I. Roulstone, editors, *Large-Scale*

*Atmosphere-Ocean Dynamics 2: Geometric Methods and Models*, pages 251–299. Cambridge University Press, Cambridge, 2004.

[HMRW85] D. D. Holm, J. E. Marsden, T. S. Ratiu, and A. Weinstein. Nonlinear stability of fluid and plasma equilibria. *Phys. Rep.*, 123:1–116, 1985.

[Hol96] D. D. Holm. Hamiltonian balance equations. *Physica D*, 98:379–414, 1996.

[Hol02] D. D. Holm. Euler-Poincaré dynamics of perfect complex fluids. In P. Holmes P. Newton and A. Weinstein, editors, *Geometry, Mechanics and Dynamics*, pages 113–167. Springer, New York, 2002.

[Hol08] D. D. Holm. *Geometric Mechanics Part 2: Rotating, Translating and Rolling*. Imperial College Press, London, 2008.

[Hos75] B. J. Hoskins. The geostrophic momentum approximation and the semi-geostrophic equations. *J. Atmos. Sci.*, 32:233–242, 1975.

[HRTY04] D. D. Holm, J. T. Ratnanather, A. Trouvé, and L. Younes. Soliton dynamics in computational anatomy. *Neuroimage*, 23:S170–S178, 2004.

[HS74] M. Hirsch and S. Smale. *Differential Equations, Dynamical Systems and Linear Algebra*. Academic Press, New York, 1974.

[HS03] D. D. Holm and M. F. Staley. Wave structures and nonlinear balances in a family of 1+1 evolutionary PDEs. *SIAM J. Appl. Dyn. Syst.*, 2:323–380, 2003.

[HS04] D. D. Holm and M. F. Staley. Interaction dynamics of singular wave fronts. Unpublished, 2004.

[HTY09] D. D. Holm, A. Trouvé, and Y. Younes. The Euler–Poincaré formulation of metamorphosis. *Quart. Appl. Math. and Mech.*, To appear, 2009.

[HZ94] J. K. Hunter and Y. Zheng. On a completely integrable nonlinear hyperbolic variational equation. *Physica D*, 79:361–386, 1994.

[HZ98] D. D. Holm and V. Zeitlin. Hamilton's principle for quasigeostrophic motion. *Phys. Fluids*, 10:800–806, 1998.

[JS98] J. V. José and E. J. Saletan. *Classical Dynamics: A Contemporary Approach*. Cambridge University Press, Cambridge, 1998.

[Kat76] T. Kato. *Perturbation Theory for Linear Operators*. Springer, New York, 1976.

[Kuz07] A. P. Kuz'min. Two-component generalizations of the Camassa-Holm equation. *Math. Notes*, 81:130–134, 2007.

[Lee03] J. Lee. *Introduction to Smooth Manifolds*, volume 218 of *Graduate Texts in Mathematics*. Springer-Verlag, New York, 2003.

[Ler34] J. Leray. Sur le mouvement d'un liquide visqueux emplissant l'espace. *Acta Mathematica*, 63:193–248, 1934.

[LL76] L. D. Landau and E. M. Lifshitz. *Mechanics*. Butterworth-Heinemann, Elsevier, Amsterdam, The Netherlands, 3rd edition, 1976.

[Lor60] E. N. Lorenz. Energy and numerical weather prediction. *Tellus*, 12:364–373, 1960.

[LR04] B. Leimkuhler and S. Reich. *Simulating Hamiltonian Dynamics*, volume 14 of *Cambridge Monographs on Applied and Computational Mathematics*. Cambridge University Press, Cambridge, 2004.

[Lyn92] P. Lynch. Richardson's barotropic forecast: A reappraisal. *Bull. Am. Met. Soc.*, 73:35–47, 1992.

[Mag78] F. Magri. A simple model of the integrable Hamiltonian equation. *J. Math. Phys.*, 19:1156–1162, 1978.

[MG80] J. C. McWilliams and P. R. Gent. Intermediate models of planetary circulations in the atmosphere and ocean. *J. of the Atmos. Sci.*, 37:1657–1678, 1980.

[MH94] J. E. Marsden and T. J. R. Hughes. *Mathematical Foundations of Elasticity*, volume 174 of *London Mathematical Society Lecture Notes*. Prentice Hall. Reprinted by Dover Publications, New York, 1994.

[MN02] A. V. Mikhailov and V. S. Novikov. Perturbative symmetry approach. *J. Phys. A*, 35:4775–4790, 2002.

[Mon90] R. Montgomery. Isoholonomic problems and some applications. *Comm. Math Phys.*, 128:565–592, 1990.

[Mon93] R. Montgomery. Gauge theory of the falling cat. *Fields Inst. Commun.*, 1:193–218, 1993.

[MR02] J. E. Marsden and T. S. Ratiu. *Introduction to Mechanics and Symmetry: A Basic Exposition of Classical Mechanical Systems*. Texts in Applied Mathematics. Springer, New York, 2nd edition, 2002.

[MR03] J. E. Marsden and T. S. Ratiu. Mechanics and Symmetry: Reduction theory. In preparation, 2003.

[MS95] D. McDuff and D. Salamon. *Introduction to Symplectic Topology*. Clarendon Press, Oxford, 1995.

[MTY02] M. I. Miller, A. Trouvé, and L. Younes. On metrics and euler-lagrange equations of computational anatomy. *Ann. Rev. Biomed. Engng.*, 4:375–405, 2002.

[MW83] J. E. Marsden and A. Weinstein. Coadjoint orbits, vortices and Clebsch variables for incompressible fluids. *Physica D*, 7:305–323, 1983.

[MY01] M. I. Miller and L. Younes. Group action, diffeomorphism and matching: a general framework. *Int. J. Comp. Vis.*, 41:61–84, 2001.

[Noe72] E. Noether. Invariante variationsprobleme. *Nachr. v. d. Ges. d. Wiss. zu Göttingen*, 1918:235–257, 1918. See also C. H. Kimberling, *Amer. Math. Monthly*, 79:136–149, 1972.

[Olv00] P. J. Olver. *Applications of Lie Groups to Differential Equations*. Springer, New York, 2000.

[Ped87] J. Pedlosky. *Geophysical Fluid Dynamics*. Springer-Verlag, New York, 1987.

[Phi69] O. M. Phillips. *The Dynamics of the Upper Ocean*. Cambridge Monographs on Mechanics and Applied Mathematics. Cambridge University Press, Cambridge, 1969.

[Poi01] H. Poincaré. Sur une forme nouvelle des équations de la méchanique. *C.R. Acad. Sci.*, 132:369–371, 1901.

[RdSD99] R. M. Roberts and M. E. R. de Sousa-Dias. Symmetries of riemann ellipsoids. *Resenhas- IME. USP*, 4:183–221, 1999.

[Ric22] L. F. Richardson. *Weather Prediction by Numerical Process*. Cambridge Monographs on Mechanics and Applied Mathematics. Cambridge University Press. Reprinted by Dover Publications, N.Y. (1965), Cambridge, 1922.

[Rou60] E. J. Routh. *Treatise of the Dynamics of a System of Rigid Bodies and Advanced Rigid Body Dynamics*. MacMillan and Co., London, 1884; reprinted by Dover, N.Y., 1960.

[Sal83] R. Salmon. Practical use of Hamilton's principle. *J. Fluid Mech.*, 132:431–444, 1983.

[Sal85] R. Salmon. New equations for nearly geostrophic flow. *J. Fluid Mech.*, 153:461–477, 1985.

[Sal88] R. Salmon. Hamiltonian fluid dynamics. *Ann. Rev. Fluid Mech.*, 20:225–256, 1988.

[Shk00] S. Shkoller. Analysis on groups of diffeomorphisms of manifolds with boundary and the averaged motion of a fluid. *J. Diff. Geom.*, 55:145–191, 2000.

[Sin01] S. Singer. *Symmetry in Mechanics: A Gentle, Modern Introduction*. Birkhäuser, New York, 2001.

[Spi65] M. Spivak. *Calculus on manifolds : a modern approach to classical theorems of advanced calculus*. Benjamin/Cummings, New York, 1965.

[Spi79] M. Spivak. *Differential Geometry*. Publish or Perish, Inc., Houston, Texas, 1979.

[Tay96] M. E. Taylor. *Partial Differential Equations I*. Springer-Verlag, New York, 1996.

[Tho92] D'A. W. Thompson. *On Growth and Form*. Dover Reprint, New York, 1992.

[Tro98] A. Trouvé. Diffeomorphism groups and pattern matching in image analysis. *Int. J. Computer Vision*, 28:213–221, 1998.

[TY05] A. Trouvé and L. Younes. Metamorphoses through lie group action. *Found. Comp. Math.*, 5:173–198, 2005.

[Val06] G. K. Vallis. *Atmospheric and Oceanic Fluid Dynamics: Fundamentals and Large-scale Circulation*. Cambridge University Press, Cambridge, 2006.

[War83] F. W. Warner. *Foundation of Differentiable Manifolds and Lie Groups*, volume 94 of *Graduate Texts in Mathematics*. Springer-Verlag, New York, 1983.

[Wei83] A. Weinstein. Hamiltonian structure for drift waves and geostrophic flow. *Phys. Fluids*, 26:388–390, 1983.

[Whi02] A. A. White. A view of the equations of meteorological dynamics and various approximations. In J. Norbury and I. Roulstone, editors, *Large Scale Atmosphere Ocean Dynamics. Volume I: Analytical Methods and Numerical Models*, pages 1–100. Cambridge University Press, Cambridge, 2002.

[WJRS06] A. A. White, Hoskins B. J., I. Roulstone, and A. Staniforth. Consistent approximate models of the global atmosphere: shallow, deep, hydrostatic, quasi-hydrostatic and non-hydrostatic. *Quart. J. Roy. Met. Soc.*, 131:2081–2107, 2006.

# Index